Heat Transfer Solutions

Worked Problems to Supplement
a First Course in Engineering
Heat Transfer

Kirk D. Hagen

Outskirts Press, Inc.
Denver, Colorado

Heat Transfer Solutions
Worked Problems to Supplement a First Course in Engineering Heat Transfer

V2.0

Outskirts Press, Inc.
http://www.outskirtspress.com

ISBN: 978-1-4327-3084-0

PRINTED IN THE UNITED STATES OF AMERICA

Heat, like gravity, penetrates every substance of the universe, its rays
occupy all parts of space. The theory of heat will hereafter form
one of the most important branches of general physics.

Joseph Fourier, *The Analytical Theory of Heat* (1822)

Preface

This book is a collection of worked problems that serves as a supplement to an undergraduate text in engineering heat transfer. When used hand in hand with such a text, the book will help the student solve heat transfer problems that are similar to those encountered in an elementary course on the subject. The book is designed to supplement any basic heat transfer text commonly used in mechanical engineering, chemical engineering or other engineering disciplines in which heat transfer is a part.

For consistency with the most widely used heat transfer texts, the sequence of solved problems follows a traditional path. Chapter 1 contains introductory problems dealing with the fundamentals of conduction, convection and radiation. Chapter 2 presents problems that introduce basic concepts of conduction. Chapters 3 and 4 deal with one-dimensional and two-dimensional steady conduction problems, respectively, while unsteady conduction problems are presented in Chapter 5. Chapter 6 presents some problems that introduce fundamental principles of convection. External and internal convection problems are given in Chapters 7 and 8, respectively, and natural convection problems are presented in Chapter 9. Problems dealing with phase changes are given in Chapter 10. Chapter 11 presents problems dealing with heat exchangers. Finally, Chapter 12 presents problems associated with fundamentals of radiation, while Chapter 13 provides problems related to radiation between surfaces.

A key feature of this book is a well known and consistent problem solving methodology. Most heat transfer texts emphasize a systematic approach for solving heat transfer problems. This book employs the following steps:

Problem Statement
Diagram
Assumptions
Properties
Analysis
Discussion

Final answers are double underlined for easy identification.

Because this book is meant to supplement a heat transfer text, it does not contain tables of thermal properties or other tabular or graphical data. Such data may be found in the appendix of virtually any heat transfer text. Thermal properties used in this book reflect typical values. This book uses SI units exclusively.

Two tables of contents are given, a brief table of contents giving only the chapter title, and a detailed table of contents giving the complete problem title. A list of recommended heat transfer texts is provided immediately following the detailed table of contents.

Brief Table of Contents

Detailed Table of Contents

Recommended Texts

1. Arpaci, V.S., S. Kao and A. Selamet, *Introduction to Heat Transfer*, Prentice Hall, Upper Saddle River, NJ, 1999.

2. Bejan, A., *Heat Transfer*, Wiley, New York, 1993.

3. Cengel, Y.A., *Heat and Mass Transfer--A Practical Approach*, 3rd Edition, McGraw-Hill, New York, 2007.

4. Hagen, K.D., *Heat Transfer with Applications*, Prentice Hall, Upper Saddle River, NJ, 1999.

5. Holman, J.P., *Heat Transfer*, 9th Edition, McGraw-Hill, New York, 2002.

6. Incropera, F.P., DeWitt, D.P., Bergman, T.L. and A.S. Lavine, *Introduction to Heat Transfer*, 5th Edition, Wiley, New York, 2007.

7. Janna, W.S. *Engineering Heat Transfer*, 2nd Edition, CRC Press, Bocan Baton, FL, 2000.

8. Kaviany, M., *Heat Transfer Physics*, Cambridge University Press, Cambridge, 2008.

9. Kreith, F. and M.S. Bohn, *Principles of Heat Transfer*, 6th Edition, Brooks/Cole, Pacific Grove, CA, 2001.

10. Mills, A.F., *Heat Transfer*, 2nd Edition, Prentice Hall, Upper Saddle River, NJ, 1999.

11. Suryanarayana, N.V., *Engineering Heat Transfer*, West, St. Paul, MN, 1995.

12. Thomas, L.C., *Heat Transfer--Professional Version*, Capstone, Knoxville, TN, 2000.

Chapter 1

Fundamental Concepts

PROBLEM 1.1 **One-Dimensional Steady Conduction in a Plane Wall**

The heat transfer through a 8-m^2 plane wall is 4 kW. The temperature of one side of the wall is 250°C, and the wall thickness is 2.0 cm. If the thermal conductivity of the wall is 0.15 W/m·K, what is the temperature of the other side of the wall?

DIAGRAM

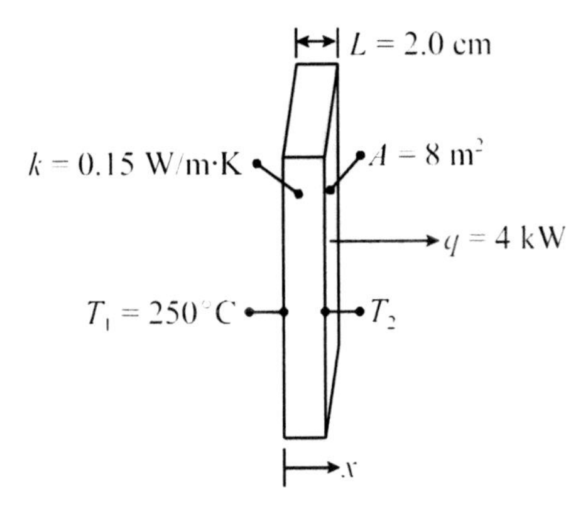

ASSUMPTIONS

1. One-dimensional steady conduction.
2. Constant properties.

PROPERTIES

$k = 0.15$ W/m·K

ANALYSIS

Conduction through the wall is given by Fourier's law,

$$q = -kA\frac{dT}{dx} = kA\frac{(T_1 - T_2)}{L}$$

Solving for T_2,

$$T_2 = T_1 - \frac{qL}{kA}$$

$$T_2 = 250°\text{C} - \frac{(4000 \text{ W})(0.020 \text{ m})}{(0.15 \text{ W/m·K})(8 \text{ m}^2)}$$

$$= \underline{\underline{183°\text{C}}}$$

DISCUSSION

The direction of heat transfer is left to right, which means that $T_1 > T_2$. Using the coordinate system shown in the diagram, the temperature gradient, dT/dx, is negative, which cancels the minus sign in Fourier's law, giving a positive heat transfer. The units for thermal conductivity in the SI system, W/m·K and W/m·°C, are equivalent because a temperature difference is indicated, and $\Delta T(\text{K}) = \Delta T(°\text{C})$.

❐ ❐ ❐

PROBLEM 1.2 Thermal Conductivity Test of a Glass Material

A simple laboratory apparatus is used to determine the thermal conductivity of a specimen of glass. An electrical heater supplies120W to one side of the test specimen while the other side is maintained at a constant temperature by a liquid cooled heat sink. The thickness and surface area of the specimen are 0.30 cm and 60 cm^2, respectively. If a temperature difference of 80°C is measured across the specimen, what is the thermal conductivity of the material?

DIAGRAM

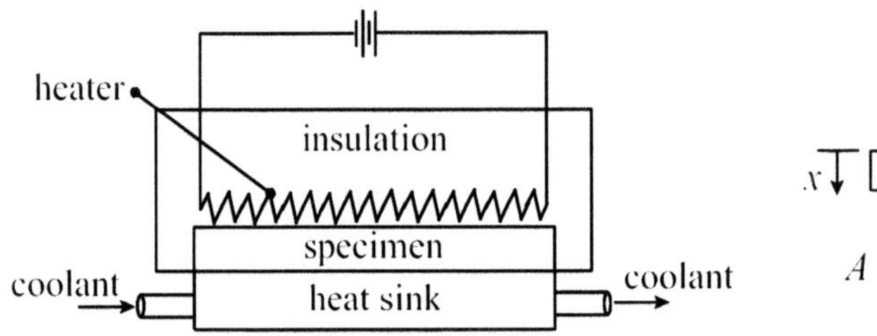

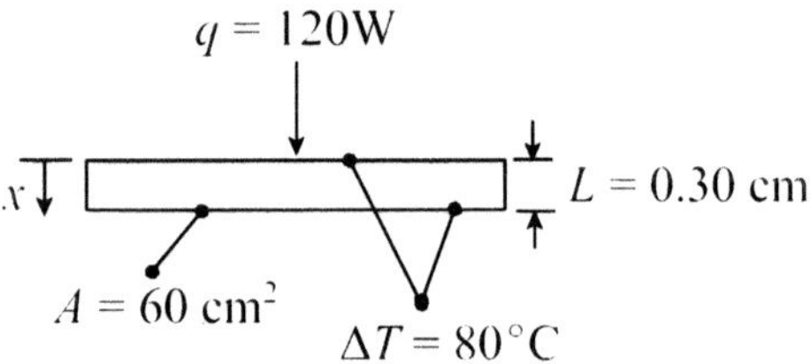

ASSUMPTIONS

1. One-dimensional steady conduction.
2. Constant properties.

PROPERTIES

Thermal conductivity k is to be determined.

ANALYSIS

The electrical power P dissipated by the heater is equivalent to the heat transfer. Conduction through the test specimen is given by Fourier's law,

$$P = q = -kA\frac{dT}{dx} = kA\frac{\Delta T}{L}$$

Solving for k,

$$k = \frac{qL}{A\Delta T}$$

$$k = \frac{(120\text{ W})(0.0030\text{ m})}{(0.0060\text{ m}^2)(80\,^\circ\text{C})}$$

$$= \underline{\underline{0.75\text{ W/m·K}}}$$

DISCUSSION

Using this crude device, one-dimensional conduction is impossible to satisfy in practice because insulating the back side and edges of the specimen will not eliminate heat transfer from these areas but will only reduce it. Hence, not all the power dissipated by the electrical heater will transfer into the specimen, and the heat conducted through the specimen will not flow in one direction only. So, this technique would probably not yield a very accurate result. Actual systems for measuring thermal conductivities of materials employ special measures for practically eliminating heat losses from the back side and edges of the test specimen.

❒ ❒ ❒

PROBLEM 1.3 Minimum Wall Thickness of a Beverage Cooler

A beverage cooler has overall inside dimensions of 40 cm × 40 cm × 25 cm. The six walls of the cooler are constructed of high density polystyrene ($k = 0.13$ W/m·K). If the maximum allowable heat gain by all the beverages inside the cooler is 90 W, what is the minimum wall thickness required for the cooler if the temperatures of the outside and inside surfaces of the walls are 25 °C and 5 °C, respectively?

DIAGRAM

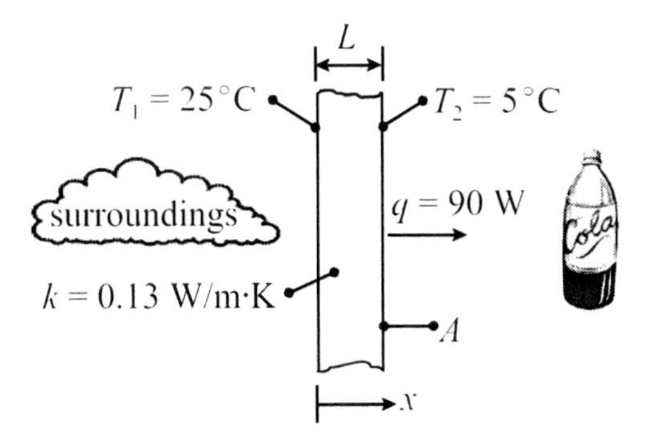

ASSUMPTIONS

1. One-dimensional steady conduction.
2. Conduction through the edges and corners of the cooler walls is neglected.
3. Constant heat gain from the surroundings.
4. Constant properties.

PROPERTIES

polystyrene: $k = 0.13$ W/m·K

ANALYSIS

The cooler consists of four identical sides plus a bottom and a top. The total inside surface area of the cooler is

$$A = 4(40\text{ cm} \times 25\text{ cm}) + 2(40\text{ cm} \times 40\text{ cm})$$

$$= 7200\text{ cm}^2 = 0.72\text{ m}^2$$

Conduction through the cooler walls is given by Fourier's law,

$$q = -kA\frac{dT}{dx} = kA\frac{(T_1 - T_2)}{L}$$

Solving for L,

$$L = kA\frac{(T_1 - T_2)}{q}$$

$$L = (0.13\text{ W/m·K})(0.72\text{ m}^2)\frac{(25 - 5)°\text{C}}{90\text{ W}}$$

$$= 0.0208\text{ m} = \underline{\underline{2.08\text{ cm}}}$$

DISCUSSION

Conduction through the edges and corners was neglected because the heat flow in these regions is not one-dimensional. However, the flat wall surfaces constitute the majority of the cooler's total surface, so neglecting conduction through the edges and corners is a good assumption.

❐ ❐ ❐

PROBLEM 1.4 Temperature Difference Across the Adhesive Under an Electronic Device

A solid state electronic device is bonded to a circuit board using a 0.15-mm thick layer of adhesive. The device dissipates 6 W of power, and measures 2.3 cm × 2.3 cm at its bonding surface. If the thermal conductivity of the adhesive is 0.20 W/m·K, what is the temperature difference across the adhesive layer?

DIAGRAM

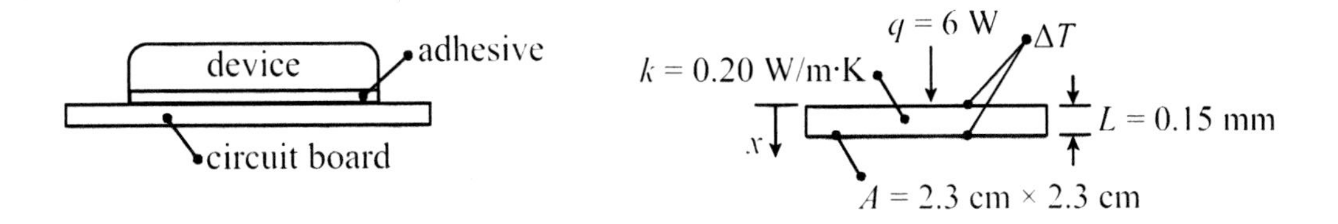

ASSUMPTIONS

1. One-dimensional steady conduction.
2. All power dissipated by the device is transferred to the adhesive.
3. Heat transfer is uniform across the bonding surface of the device.
4. Constant properties.

PROPERTIES

adhesive: $k = 0.20$ W/m·K

ANALYSIS

The bonding surface of the device is

$$A = (2.3 \text{ cm})(2.3 \text{ cm}) = 5.29 \text{ cm}^2 = 5.29 \times 10^{-4} \text{ m}^2$$

Noting that the power, P, dissipated by the device is equivalent to the heat transfer, conduction through the adhesive is given by Fourier's law,

$$P = q = -kA\frac{dT}{dx} = kA\frac{\Delta T}{L}$$

Solving for ΔT,

$$\Delta T = \frac{qL}{kA}$$

$$\Delta T = \frac{(6\ \text{W})(0.15 \times 10^{-3}\ \text{m})}{(0.20\ \text{W/m·K})(5.29 \times 10^{-4}\ \text{m}^2)}$$

$$= \underline{\underline{8.5°\text{C}}}$$

DISCUSSION

From the relation for ΔT, it is evident that the temperature difference across the adhesive layer is directly proportional to q and L and inversely proportional to k and A. In electronics cooling applications of this type, it is desirable to make ΔT as small as possible because this quantity is probably one of several ΔT values that, when added to the temperature of the local environment, yields the surface temperature of the electronic device. Thus, engineers use as little adhesive as possible to minimize the temperature rise across the layer. Since the power dissipation and the size of the device are generally fixed, engineers have control over only the layer thickness and adhesive type. Special adhesives with higher thermal conductivities than standard adhesives are used in electronics cooling applications.

❒ ❒ ❒

PROBLEM 1.5 Heat Losses for Single Pane and Double Pane Windows

Compare the heat loss for a single pane window and a double pane window in typical winter conditions. The single pane window consists of a single sheet of 3.0-mm thick glass, and the double pane window consists of two identical sheets of 3.0-mm thick glass separated by a 2.0-cm air space. In both systems, the temperature of the inside window surface is 18°C, and the temperature of the outside window surface is 3 °C. The thermal conductivity of window glass and air are 0.70 W/m·K and 0.026 W/m·K, respectively. Find the heat loss per unit area for both windows.

DIAGRAM

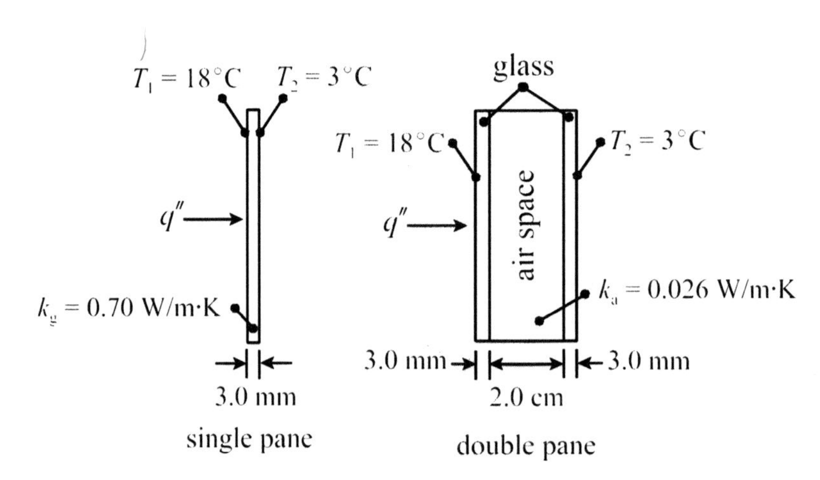

ASSUMPTIONS

1. One-dimensional steady conduction.
2. Stagnant air between the sheets of glass in the double pane window.
3. Constant properties.

PROPERTIES

window glass: $k_g = 0.70$ W/m·K
air: $k_a = 0.026$ W/m·K

ANALYSIS

Single Pane Window

Heat loss per unit area of window surface is the conducted heat trough the single pane of glass. Writing Fourier's law in terms of heat flux,

$$q'' = -k\frac{dT}{dx} = k\frac{(T_1 - T_2)}{L}$$

$$q'' = (0.70 \text{ W/m·K})\frac{(18 - 3)°\text{C}}{0.003 \text{ m}}$$

$$= \underline{\underline{3500 \text{ W/m}^2}}$$

Double Pane Window

The double pane window is a multi-layered system. Because we do not yet know how to analyze conduction through multi-layered systems (see Chapter 3), we apply the surface temperatures across the air space only, neglecting conduction through the two sheets of glass. This approach provides an approximate heat loss that can still be compared to that for the single pane window.

$$q'' = k\frac{(T_1 - T_2)}{L}$$

$$q'' = (0.026 \text{ W/m·K})\frac{(18 - 3)°\text{C}}{0.020 \text{ m}}$$

$$= \underline{\underline{19.5 \text{ W/m}^2}}$$

$q''_{\text{single pane}}/q''_{\text{double pane}} = (3500 \text{ W/m}^2)/(19.5 \text{ W/m}^2) = 179$

DISCUSSION

The insulating effect of the air space in the double pane window is readily apparent. Neglecting conduction through the sheets of glass in the double pane window, the heat loss for the single pane window is nearly two-hundred times that of the double pane window. The drastic reduction in heat loss for the double pane window results primarily from the large difference in thermal conductivities between glass and air. The air space is also thicker than the sheet of glass. If we included the two sheets of glass in the analysis of the double pane window, the heat loss would be even lower because the sheets of glass provide an extra insulating effect.

PROBLEM 1.6 Temperature Difference Across a Concrete Sidewalk

A concrete sidewalk is exposed to a solar radiation of 1050 W/m^2. The thickness of the sidewalk is 10 cm, and the thermal conductivity of concrete is 2.2 W/m·K. Find the temperature difference across the sidewalk.

DIAGRAM

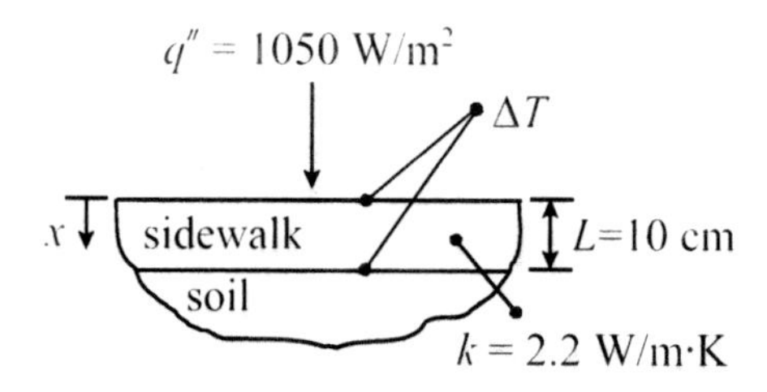

ASSUMPTIONS

1. One-dimensional steady conduction.
2. Steady solar heat flux.
3. All of solar radiation is absorbed by the concrete sidewalk.
4. Emitted radiation and convection from the sidewalk surface is neglected.
5. At a point $x = L$ or deeper, the temperature is constant.
6. Constant properties.

PROPERTIES

concrete: $k = 2.2$ W/m·K

ANALYSIS

Because we are interested in the temperature difference across the sidewalk only, we do not need to analyze heat transfer in the soil. Conduction through the concrete sidewalk is given by Fourier's law, written in terms of heat flux,

$$q'' = -k\frac{dT}{dx} = k\frac{\Delta T}{L}$$

Solving for ΔT,

$$\Delta T = \frac{q''\, L}{k}$$

$$\Delta T = \frac{(1050\ \text{W/m}^2)(0.10\ \text{m})}{2.2\ \text{W/m·K}}$$

$$= \underline{\underline{47.7\,°\text{C}}}$$

DISCUSSION

Assumption 2 is crude because the magnitude of solar radiation changes during the day as the sun moves across the sky. Assumption 3 is somewhat reasonable because concrete typically absorbs over 70 percent of solar radiation incident upon it, and the rest is reflected. According to assumption 4, all of the solar radiation absorbed by the sidewalk is conducted through the concrete and into the soil below. In reality, some of the heat absorbed by the sidewalk would be convected and radiated from the sidewalk surface, making the temperature difference across the sidewalk smaller than the value calculated here. If we did not invoke assumption 5, the temperatures at all values of x would change with time, even with a constant heat flux applied at the sidewalk surface, thereby making steady conduction impossible.

❒ ❒ ❒

PROBLEM 1.7 Heat Transfer Through the Sight Glass of a Boiler

A 15-cm diameter, 2.0-cm thick sight glass installed in the wall of a gas fired boiler allows the operator to observe conditions inside the boiler during operation. During normal boiler operation, the inside surface of the sight glass is 130 °C. Due to natural convection currents in the warm boiler room, air at 35 °C flows vertically across the outside surface of the sight glass, giving a heat transfer coefficient of 25 W/m^2·K. The sight glass is made of soda lime glass with a thermal conductivity of 1.4 W/m·K. Find the temperature of the outside surface of the sight glass and the heat transfer through the sight glass.

DIAGRAM

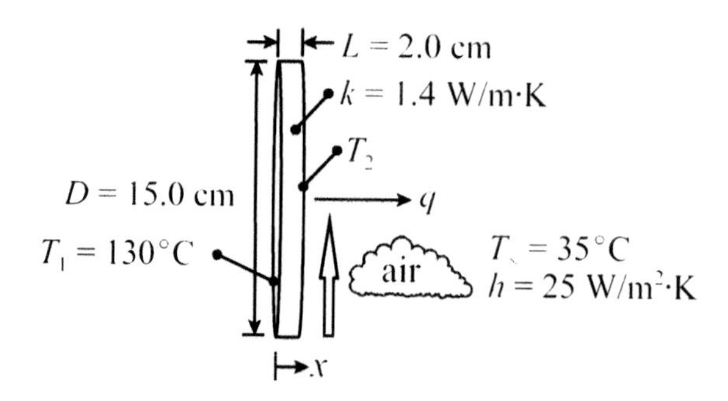

ASSUMPTIONS

1. One-dimensional steady conduction.
2. Radiation from the outside surface of the sight glass is neglected.
3. Constant properties.

PROPERTIES

soda lime glass: $k = 1.4$ W/m·K

ANALYSIS

Conduction through the sight glass is given by Fourier's law,

$$q_{cond} = -kA\frac{dT}{dx} = kA\frac{(T_1 - T_2)}{L}$$

and convection from the outside surface of the sight glass is given by Newton's law of cooling,

$$q_{conv} = hA(T_2 - T_\infty)$$

The heat conducted through the sight glass equals the heat convected from its outside surface.

$$q_{cond} = q_{conv} = q$$

so,

$$kA\frac{(T_1 - T_2)}{L} = hA(T_2 - T_\infty)$$

Dividing out the common surface area and solving for T_2,

$$T_2 = \frac{T_1 + (hL/k)T_\infty}{1 + (hL/k)}$$

$$T_2 = \frac{130°C + [(25\ W/m^2{\cdot}K)(0.02\ m)/(1.4\ W/m{\cdot}K)](35°C)}{1 + (25\ W/m^2{\cdot}K)(0.02\ m)/(1.4\ W/m{\cdot}K)}$$

$$= \underline{\underline{105°C}}$$

Now that T_2 is known, the heat transfer can be found using either Fourier's law or Newton's law of cooling.

$$q = kA\frac{(T_1 - T_2)}{L} = k(\pi D^2/4)\frac{(T_1 - T_2)}{L}$$

$$q = (1.4\ W/m{\cdot}K)\ [\pi\ (0.15\ m)^2/4]\ \frac{(130 - 105)°C}{0.02\ m}$$

$$= \underline{\underline{30.9\ W}}$$

$$q = hA(T_2 - T_\infty) = h(\pi D^2/4)(T_2 - T_\infty)$$

$$q = (25\ W/m^2{\cdot}K)[\pi\ (0.15\ m)^2/4]\ (105 - 35)°C$$

$$= \underline{\underline{30.9\ W}}$$

DISCUSSION

Under steady conditions, the heat conducted through the sight glass equals the heat convected from the outside surface of the sight glass. This can be shown by doing a surface energy balance on the outside surface of the sight glass,

$$\dot{E}_{in} = \dot{E}_{out}$$

$$\dot{E}_{in} = q_{cond} = kA\frac{(T_1 - T_2)}{L}$$

$$\dot{E}_{out} = q_{conv} = hA(T_2 - T_\infty)$$

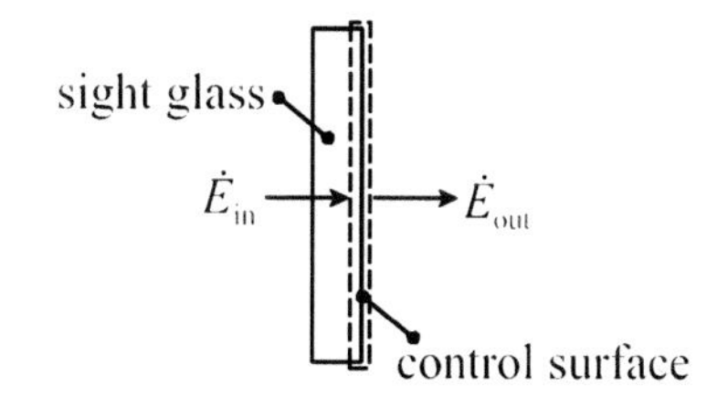

❒ ❒ ❒

PROBLEM 1.8 Surface Temperature of a Cartridge Heater

An electrical cartridge heater is used to heat water to a temperature of 50°C in a manufacturing process. The heater is powered by a 240 V source and draws 15 A of current. The length and diameter of the cartridge heater are 60 cm and 1.5 cm, respectively. If the heater is immersed in the water such that the heat transfer coefficient is 4500 W/m²·K, find the surface temperature of the cartridge heater.

DIAGRAM

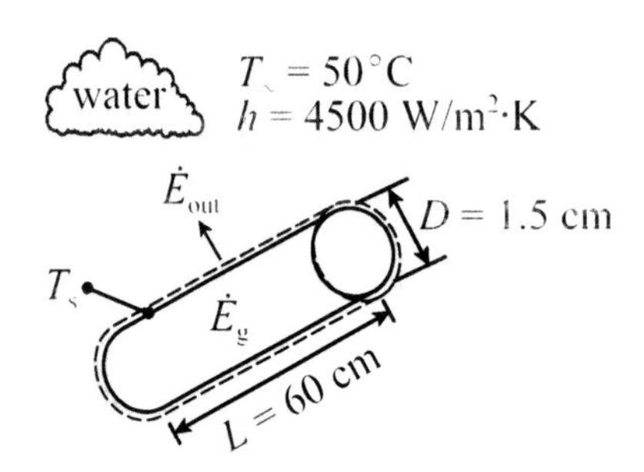

ASSUMPTIONS

1. Steady conditions.
2. Heat transfer from the ends of the heater is neglected.
3. Radiation from the heater is neglected.

PROPERTIES

Not applicable.

ANALYSIS

Under steady conditions, an energy balance on the control volume in the diagram yields

$$\dot{E}_{in} - \dot{E}_{out} = 0$$

The rate of energy generation is the electrical power dissipated by the heater,

$$\dot{E}_g = P = IV$$

$$\dot{E}_g = (15\text{ A})(240\text{ V})$$

$$= 3600\text{ W}$$

The rate of energy transfer out of the control volume is the convective heat transfer from the heater's surface,

$$\dot{E}_{out} = q_{conv} = hA(T_s - T_\infty) = h\pi DL(T_s - T_\infty)$$

Substituting these terms into the energy balance and solving for T_s,

$$T_2 = \frac{\dot{E}_g}{h\pi DL} + T_\infty$$

$$T_2 = \frac{(3600 \text{ W})}{(4500 \text{ W/m}^2\cdot\text{K})\ \pi\ (0.015 \text{ m})(0.60 \text{ m})} + 50°\text{C}$$

$= \underline{\underline{78.3°\text{C}}}$

DISCUSSION

The surface temperature of a cartridge heater reflects the operating temperature of the device and is an important parameter in the design of thermal systems. The casing of a cartridge heater is typically made of copper, stainless steel, or other metal. If the surface temperature of the cartridge heater is too high, the casing will lose its structural integrity or even melt, thereby destroying the heater. We can see that T_s is directly proportional to the dissipated power and inversely proportional to the heat transfer coefficient. If this cartridge heater were to be used to heat air instead of water, the heat transfer coefficient would be much smaller than 4500 W/m^2·K, which would result in an unacceptably high surface temperature.

❒ ❒ ❒

PROBLEM 1.9 Air Cooling a Resistor

A 22-Ω carbon composition resistor is air cooled by a natural convection current near the circuit board on which the resistor is mounted. The convection current produces a heat transfer coefficient of 30 W/m^2·K at the resistor's surface, and the air temperature is 25 °C. The length and diameter of the cylinder-shaped resistor are 2.0 cm and 8.0 mm, respectively. If the electrical current flowing through the resistor is 275 mA, what is the surface temperature of the resistor?

DIAGRAM

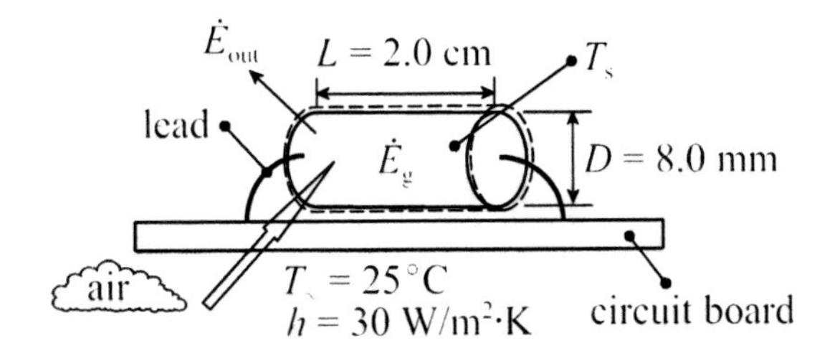

ASSUMPTIONS

1. Steady conditions.
2. Heat transfer from the ends of the resistor is neglected.
3. Conduction in the leads is neglected.
4. Radiation is neglected.

PROPERTIES

Not applicable.

ANALYSIS

Under steady conditions, an energy balance on the control volume in the diagram yields

$$\dot{E}_g - \dot{E}_{out} = 0$$

The rate of energy generation is the electrical power dissipated by the resistor,

$$\dot{E}_g = P = I^2 R$$

$$\dot{E}_g = (0.275 \text{ A})^2 (22 \ \Omega)$$

$$= 1.664 \text{ W}$$

The rate of energy transfer out of the control volume is the convective heat transfer from the resistor's surface,

$$\dot{E}_{out} = q_{conv} = hA(T_s - T_\infty) = h\pi DL(T_s - T_\infty)$$

Solving for T_s,

$$T_s = \frac{\dot{E}_g}{h\pi DL} + T_\infty$$

$$T_s = \frac{(1.664 \text{ W})}{(30 \text{ W/m}^2\cdot\text{K})\ \pi\ (0.008 \text{ m})(0.020 \text{ m})} + 25°\text{C}$$

$$= \underline{\underline{135°\text{C}}}$$

DISCUSSION

Touching this resistor would burn you, so let's reconsider assumptions 3 and 4. A significant amount of heat would be conducted from the resistor through the leads into the circuit board. Also, because natural convection is a rather weak heat transfer mechanism, radiation would be significant. The inclusion of these two effects would drastically reduce the surface temperature of the resistor, making it safe to touch if necessary.

PROBLEM 1.10 Cost of Heat Loss from a Basement Foundation Wall

The basement of an electrically heated home is surrounded by an uninsulated concrete foundation wall with a perimeter and height of 30 m and 2.5 m, respectively. The wall is 20 cm thick, and the concrete has a thermal conductivity of 2.2 W/m·K. On a typical cold day, the soil in contact with the outside surface of the wall has a temperature of 10°C, and the inside surface of the wall has a temperature of 14°C. If the cost of electricity is \$0.08/kWh, find the cost of the heat loss to the homeowner for a period of 24 hours.

DIAGRAM

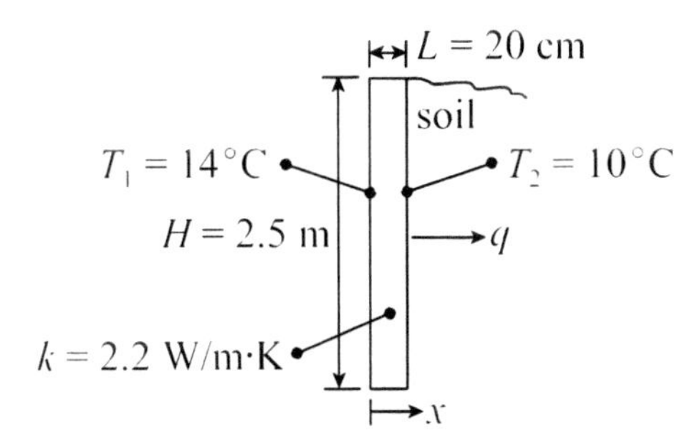

ASSUMPTIONS

1. One-dimensional steady conduction.
2. Constant properties.

PROPERTIES

concrete: $k = 2.2$ W/m·K

ANALYSIS

Conduction through the foundation wall is given by Fourier's law,

$$q = -kA\frac{dT}{dx} = kA\frac{(T_1 - T_2)}{L}$$

$$q = (2.2 \text{ W/m·K})(2.5 \text{ m} \times 30 \text{ m})\frac{(14 - 10)°\text{C}}{0.20 \text{ m}}$$

$$= 3300 \text{ W} = 3.300 \text{ kW}$$

The cost of the heat loss for a period of 24 hours is,

$$\text{cost} = 3.300 \text{ kW} \times 24 \text{ h} \times \frac{\$0.08}{\text{kWh}}$$

$$= \underline{\underline{\$6.43}}$$

DISCUSSION

Heat is also lost through the roof, walls, windows, doors and basement floor. These heat losses are due to conduction through various parts of the building "envelope." Infiltration heat losses are those due to air leakage and can be significant if the home is not sealed properly. Residential basements with uninsulated foundation walls are not usually actively heated because the cost of doing so is too high. The cost of the heat loss for this uninsulated foundation wall for a typical month is $6.34 × 30 = $190.20.

❒ ❒ ❒

PROBLEM 1.11 Surface Temperature of a Walking Person

A typical adult loses about 210 W of heat per square meter of body surface area while engaged in brisk walking. Approximating this person as a 170-cm long, 30-cm diameter cylinder, find the person's surface temperature. The temperature of the surrounding air is 20°C, and the heat transfer coefficient is 18 W/m²·K.

DIAGRAM

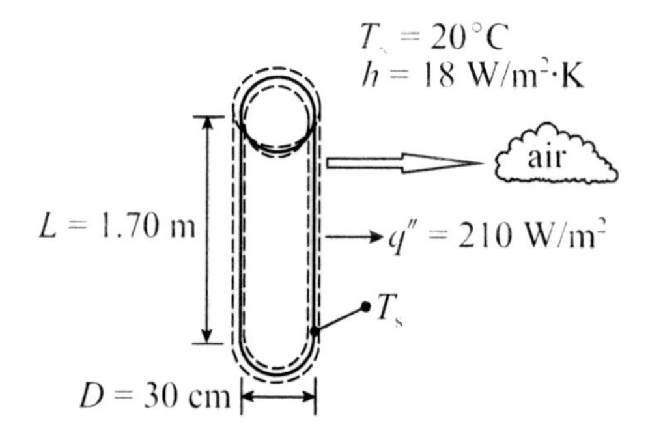

ASSUMPTIONS

1. Steady conditions
2. Radiation and evaporative effects are neglected.
3. Effects of clothing are neglected.

PROPERTIES

Not applicable.

ANALYSIS

An energy balance on the control surface in the diagram yields,

$$\dot{E}_{out} = q''A$$

but the rate of energy transfer out of the control surface equals the convective heat flux,

$$q''A = hA(T_s - T_\infty)$$

Solving for T_s,

$$T_s = \frac{q''}{h} + T_\infty$$

$$T_s = \frac{210\ \text{W/m}^2}{18\ \text{W/m}^2\cdot\text{K}} + 20°\text{C}$$

$$= \underline{\underline{31.7°\text{C}}}$$

DISCUSSION

A body surface temperature of 31.7°C (89.1°F) is reasonable given that a person's deep body temperature is approximately 98.6°F. The dimensions of the body were not required because the heat loss was given as a flux. For a person playing basketball, the heat flux is roughly 350 W/m^2, which yields a surface temperature of

$$T_s = \frac{350\ \text{W/m}^2}{18\ \text{W/m}^2\cdot\text{K}} + 20°\text{C} = 39.4°\text{C}\ (103°\text{F})$$

which is unrealistically high because of assumption 2.

❒ ❒ ❒

PROBLEM 1.12 Radiation Effect on Temperature Measurement of a Gas

A thermocouple is used to measure the temperature of hot air flowing in an uninsulated sheet metal duct located in a cold room. The thermocouple sensing element is a small spherical welded bead that joins two dissimilar wires. Air flows steadily across the bead, giving a heat transfer coefficient at its surface of 150 W/m²·K. The surface temperature of the duct is 40°C, and the thermocouple indicates a temperature of 227°C. If the emissivity of the welded bead is 0.6, what is the true temperature of the air?

DIAGRAM

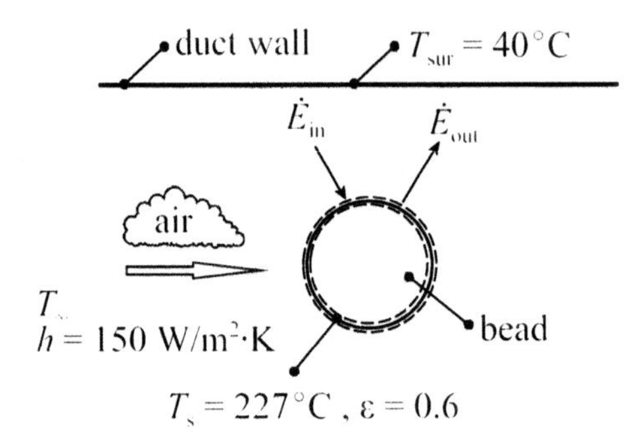

ASSUMPTIONS

1. Steady conditions.
2. The duct surface is much larger than the bead surface.
3. Constant properties.

PROPERTIES

welded bead: $\varepsilon = 0.6$

ANALYSIS

Using the control surface in the diagram, and noting that conditions are steady, an energy balance yields,

$$\dot{E}_{in} - \dot{E}_{out} = 0$$

The rate of energy transfer into the control surface is the convective heat transfer from the air to the bead,

$$\dot{E}_{in} = q_{conv} = hA(T_\infty - T_s)$$

and the rate of energy transfer out of the control surface is the radiative heat transfer from the bead to the cold duct wall,

$$\dot{E}_{out} = q_{rad} = \varepsilon A \sigma (T_s^4 - T_{sur}^4)$$

Substituting these terms into the energy balance and solving for T_∞,

$$T_\infty = \frac{\varepsilon\sigma}{h}(T_s^4 - T_{sur}^4) + T_s$$

$$T_\infty = \frac{(0.6)(5.669 \times 10^{-8} \text{ W/m}^2\cdot\text{K}^4)}{150 \text{ W/m}^2\cdot\text{K}}(500^4 - 313^4)\text{ K}^4 + 500 \text{ K}$$

$= 12.0 \text{ K} + 500 \text{ K}$

$= 512 \text{ K} = \underline{\underline{239°\text{C}}}$

DISCUSSION

This problem illustrates a common radiation effect when measuring the temperature of a hot gas flowing in a cold duct. The first term on the right side of the expression for T_∞ is called the radiation correction term. The thermometer reads a temperature that reflects an overall energy balance on its sensing element. The temperature sensor radiates to the cold duct wall, thereby giving an artificially low reading. In this case, the thermocouple indicates an air temperature of 227°C, but the true air temperature is 239°C. To reduce this radiation effect, radiation shields with a very high reflectivity (low emissivity) are often employed to minimize the radiation correction term. As the correction term goes to zero, the gas temperature approaches the temperature measured by the thermometer.

❐ ❐ ❐

PROBLEM 1.13 Minimum Insulation Thickness for a Kitchen Oven

The walls of a kitchen oven consist of a layer of mineral fiber insulation (k = 0.04 W/m·K) sandwiched by two thin sheets of metal. At the oven's highest setting, the inside surface temperature of the oven wall is 280°C. The maximum temperature of the air in the kitchen is 30°C, and the heat transfer coefficient on the outside oven wall surface is 15 W/m²·K. Find the minimum insulation thickness required to limit the outside surface temperature of the oven wall to 45°C.

DIAGRAM

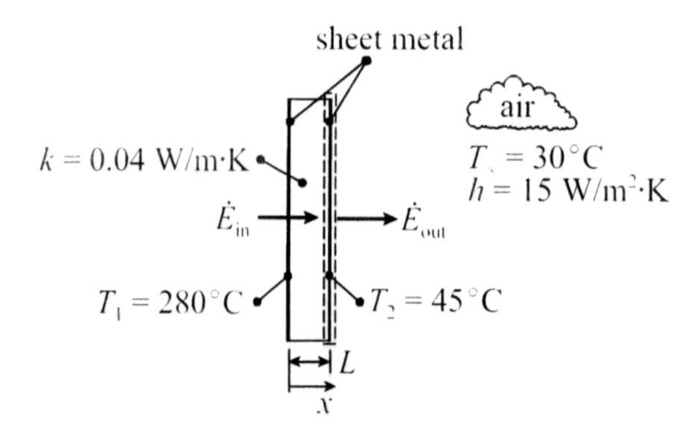

ASSUMPTIONS

1. Steady conditions.
2. Conduction in the sheet metal layers is neglected.
3. Radiation is neglected.
4. Constant properties.

PROPERTIES

mineral fiber insulation: $k = 0.04$ W/m·K

ANALYSIS

Using the control surface in the diagram, and noting that conditions are steady, an energy balance yields,

$$\dot{E}_{in} - \dot{E}_{out} = 0$$

Based on assumption 2, we apply the surface temperatures across the insulation only. The rate of energy transfer into the control surface is the conduction through the insulation, given by Fourier's law,

$$\dot{E}_{in} = q_{cond} = -kA\frac{dT}{dx} = kA\frac{(T_1 - T_2)}{L}$$

The rate of energy transfer out of the control surface is the convection from the outside surface of the wall, given by Newton's law of cooling,

$$\dot{E}_{out} = q_{cond} = hA(T_1 - T_2)$$

Substituting these terms into the energy balance and solving for L,

$$L = \frac{k(T_1 - T_2)}{h(T_2 - T_\infty)}$$

$$L = \frac{(0.04 \text{ W/m·K})(280 - 45)°\text{C}}{(15 \text{ W/m}^2\text{·K})(45 - 30)°\text{C}}$$

$$= 0.0418 \text{ m} = \underline{\underline{4.18 \text{ cm}}}$$

DISCUSSION

Heat transfer is an important consideration in the design of kitchen ovens and other household appliances. Of particular concern in the design of kitchen ovens is safety. The insulation thickness calculated here is the minimum value needed such that the outside surface of the oven wall does not exceed a safe touch temperature. If assumptions 2 and 3 were relaxed, the insulation thickness would be somewhat greater than the value calculated here.

❒ ❒ ❒

PROBLEM 1.14 Surface Temperature of a Convectively Cooled Plate

One side of a 3.5-mm thick plate of stainless steel (k = 15 W/m·K) is heated by an electrical heater that produces a heat flux of 2500 W/m^2. The other side of the plate is exposed to a low velocity flow of 25°C air where the heat transfer coefficient is 40 W/m^2·K. Find the temperature of the surface of the plate exposed to the air.

DIAGRAM

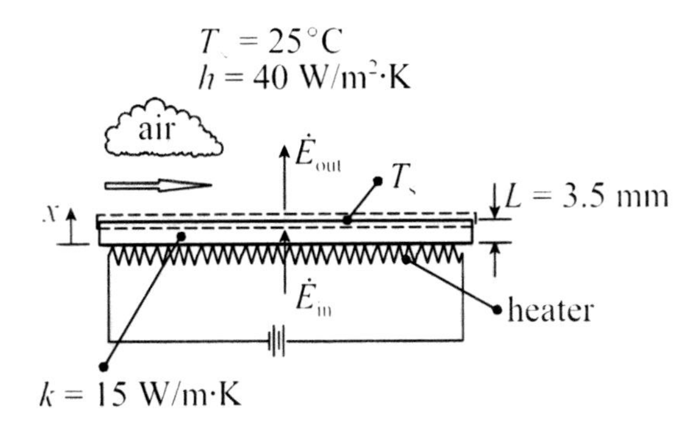

ASSUMPTIONS

1. Steady conditions.
2. All heat dissipated by the heater transfers into the plate.
3. Radiation is neglected.
4. Constant properties.

PROPERTIES

stainless steel: $k = 15$ W/m·K

ANALYSIS

Using the control surface in the diagram, and noting that conditions are steady, an energy balance yields,

$$\dot{E}_{in} - \dot{E}_{out} = 0$$

The rate of energy transfer into the control surface is the conduction through the plate. Since all the heat dissipated by the heater transfers into the plate, the heat conducted through the plate equals the heat dissipated by the electrical heater,

$$\dot{E}_{in} = q_{cond} = q'' A$$

where $q'' = 2500$ W/m^2. The rate of energy transfer out of the control surface is the convective heat transfer to the surrounding air.

$$\dot{E}_{out} = q_{conv} = hA(T_s - T_\infty)$$

Substituting these terms into the energy balance and solving for T_s,

$$T_s = \frac{q''}{h} + T_\infty$$

$$T_s = \frac{2500 \text{ W/m}^2}{40 \text{ W/m}^2\cdot\text{K}} + 25°\text{C}$$

$$= \underline{\underline{87.5°\text{C}}}$$

DISCUSSION

Note that neither the plate's thickness nor its thermal conductivity were needed to solve this problem, which means that the surface temperature of the plate is independent of these parameters. Using Fourier's law, we can find the temperature of the other surface of the plate, which we will call T_1,

$$q'' = k\frac{(T_1 - T_2)}{L}$$

Solving for T_1,

$$T_1 = \frac{q''\,L}{k} + T_s$$

$$T_1 = \frac{(2500\ \text{W/m}^2)(0.0035\ \text{m})}{15\ \text{W/m·K}} + 87.5°\text{C}$$

$$= 88.1°\text{C}$$

This temperature is reasonable because we would not expect a large temperature difference across a thin metal plate. The temperature difference is only 0.6°C.

❒ ❒ ❒

PROBLEM 1.15 Convection from a Current-Carrying Wire

A long nichrome wire with a diameter of 0.81 mm carries 3 A of electrical current. The wire is surrounded by 20°C air, and the heat transfer coefficient at the surface of the wire is 50 W/m^2·K. If the electrical resistance of the wire is 2.16 Ω/m, find the surface temperature of the wire.

DIAGRAM

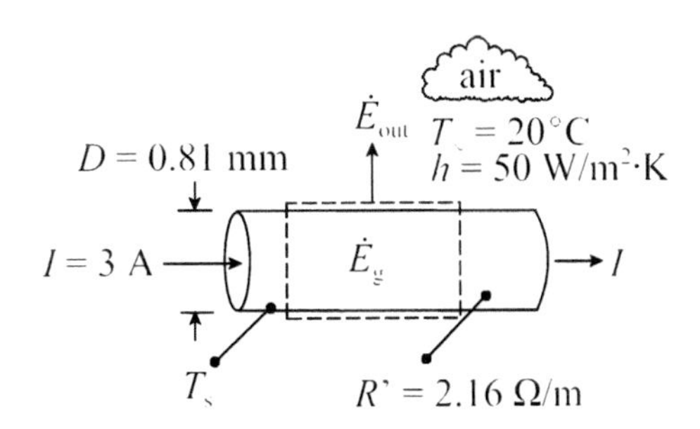

ASSUMPTIONS

1. Steady conditions.
2. Radiation is neglected.
3. Constant properties.

PROPERTIES

nichrome wire resistance: $R' = 2.16\ \Omega/\text{m}$

ANALYSIS

Under steady conditions, an energy balance on the control volume in the diagram yields

$$\dot{E}_g - \dot{E}_{out} = 0$$

The rate of energy generation is the electrical power dissipation,

$$\dot{E}_g = I^2 R' L$$

where L is an arbitrary length of wire. The rate of energy transfer out of the control volume is the convection from the wire,

$$\dot{E}_{out} = q_{conv} = h\pi DL(T_s - T_\infty)$$

Substituting these terms into the energy balance and solving for T_s,

$$T_s = \frac{I^2 R'}{h\pi D} + T_\infty$$

$$T_s = \frac{(3\ \text{A})^2\,(2.16\ \Omega/\text{m})}{(50\ \text{W/m}^2\cdot\text{K})\ \pi\ (0.81 \times 10^{-3}\ \text{m})} + 20°\text{C}$$

$$= \underline{\underline{173°\text{C}}}$$

DISCUSSION

This answer is the surface temperature of the wire, but the wire diameter is so small that there would be a negligible temperature difference across the radius of the wire. Note that the length of the wire was not required in the analysis because the length term divided out. The convection per unit length of wire is

$$q'_{conv} = q_{conv} / L = h\pi D(T_s - T_\infty)$$

$$q'_{conv} = (50\ \text{W/m}^2\cdot\text{K})\ \pi\ (0.81 \times 10^{-3}\ \text{m})(173 - 20)°\text{C}$$

$$= 19.5\ \text{W/m}$$

❒ ❒ ❒

PROBLEM 1.16 Conduction and Convection Losses from an Electronic Device

Air is forced across a cylindrical-shaped electronic device. The air temperature is 25°C, and the associated heat transfer coefficient is 60 W/m^2·K. The device dissipates 3.7 W and is mounted on a circuit board by three drawn copper (k = 287 W/m·K) leads whose length and diameter are 6.0 mm and 0.18 mm, respectively. A heat sink, connected to the circuit board, maintains the circuit board temperature at 30°C. Find the surface temperature of the device.

DIAGRAM

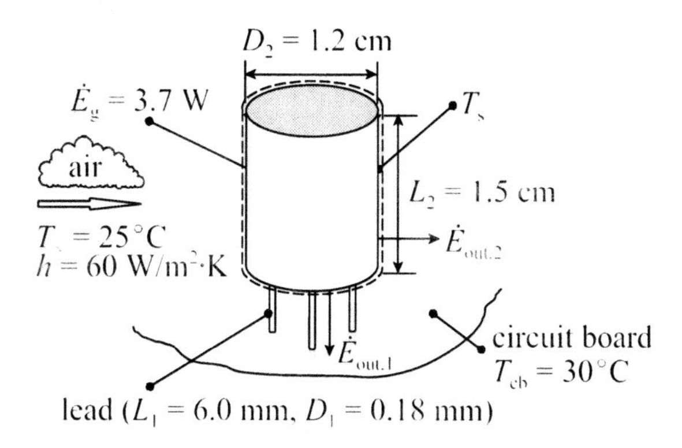

ASSUMPTIONS

1. Steady conditions.
2. The surface of the device is isothermal.
3. Radiation from the device is neglected.
4. Convection and radiation from the leads are neglected.
5. Constant properties.

PROPERTIES

drawn copper: k = 287 W/m·K

ANALYSIS

Under steady conditions, an energy balance on the control volume in the diagram yields

$$\dot{E}_g - \dot{E}_{out,1} - \dot{E}_{out,2} = 0$$

The rate of energy generation is the device's power dissipation. The rate of energy transfer out of the control volume consists of two parts, the conduction in the three leads and convection from the surface of the device,

$$\dot{E}_{out,1} = q_{cond} = 3kA_1 \frac{(T_s - T_{cb})}{L_1}$$

$$\dot{E}_{out,2} = q_{conv} = hA_2 (T_s - T_\infty)$$

where A_1 is the cross sectional area of one lead, and A_2 is the total surface area of the device. Substituting these terms into the energy balance,

$$\dot{E}_g - 3kA_1 \frac{(T_s - T_{cb})}{L_1} - hA_2 (T_s - T_\infty) = 0$$

Solving for T_s,

$$T_s = \frac{\dot{E}_g + hA_2 T_\infty + 3kA_1 T_{cb} / L_1}{hA_2 + 3kA_1 / L_1}$$

Neglecting the small surface area occupied by the leads in the calculation of A_2,

$$A_1 = \pi D_1^2/4$$

$$= \pi (0.81 \times 10^{-3}\ \text{m})^2/4 = 5.153 \times 10^{-7}\ \text{m}^2$$

$$A_2 = \pi D_2 L_2 + 2 \pi D_2^2/4$$

$$= \pi (0.012\ \text{m})(0.015\ \text{m}) + 2 \pi (0.012\ \text{m})^2/4 = 7.917 \times 10^{-4}\ \text{m}^2$$

The surface temperature is

$$T_s = \frac{(3.7) + (60)(7.917 \times 10^{-4})(25) + 3(287)(5.153 \times 10^{-7})(30)/0.006}{(60)(7.917 \times 10^{-4}) + 3(287)(5.153 \times 10^{-7})/0.006}$$

$$= \underline{\underline{58.5°\text{C}}}$$

DISCUSSION

It is instructive to calculate and compare the two heat losses from the device. The conduction loss is

$$q_{cond} = 3kA_1 \frac{(T_s - T_{cb})}{L_1}$$

$$q_{cond} = 3(287 \text{ W/m·K})(5.153 \times 10^{-7} \text{ m}^2)\frac{(58.5 - 30)°\text{C}}{0.006 \text{ m}}$$

$$= 2.11 \text{ W}$$

and the convection loss is

$$q_{conv} = hA_2(T_s - T_\infty)$$

$$q_{conv} = (60 \text{ W/m}^2\text{·K})(7.917 \times 10^{-4} \text{ m}^2)(58.5 - 25)°\text{C}$$

$$= 1.59 \text{ W}$$

Thus, most of the heat is lost by conduction through the leads. As a check of the calculations, we see that the two heat losses add to 3.7 W, the power dissipation of the device,

$$q = q_{cond} + q_{conv}$$

$$q = 2.11 \text{ W} + 1.59 \text{ W} = 3.70 \text{ W}$$

❒ ❒ ❒

PROBLEM 1.17 Cooling Rates of a Heat Treated Steel Part

A heat treated steel part has a uniform temperature of 400°C as it is brought out of a heat treating oven into a large room where the air temperature is 25°C and the walls have a temperature of 15°C. The steel part has rectangular dimensions of 15 cm × 6 cm × 1 cm and an emissivity of 0.8. If the density and specific heat of the steel are 7850 kg/m^3 and 430 J/kg·K, respectively, find the cooling rates of the part as it passes through temperatures of 300°C, 200°C, 100°C and 50°C. For all part temperatures, assume a heat transfer coefficient of 20 W/m^2·K.

DIAGRAM

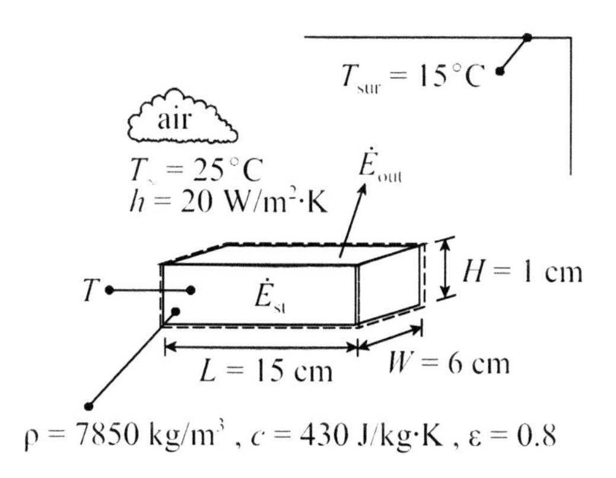

ASSUMPTIONS

1. At any given time, the part is isothermal.
2. Heat transfer coefficient is constant.
3. Constant properties.

PROPERTIES

steel: $\rho = 7850$ kg/m^3 , $c = 430$ J/kg·K , $\varepsilon = 0.8$

ANALYSIS

An energy balance on the control volume in the diagram yields,

$$-\dot{E}_{out} = \dot{E}_{st}$$

The rate of energy transfer out of the control volume is the convection and radiation from the part, and the rate of energy storage within the part is given by,

$$\dot{E}_{out} = hA(T - T_\infty) + \varepsilon A\sigma(T^4 - T_{sur}^4)$$

$$\dot{E}_{st} = \rho c V \frac{dT}{dt}$$

Substituting these terms into the energy balance and solving for the cooling rate dT/dt,

$$\frac{dT}{dt} = \frac{-hA(T - T_\infty) - \varepsilon A\sigma(T^4 - T_{sur}^4)}{\rho c V}$$

The surface area of the part is

$$A = 2(L\,W + L\,H + W\,H)$$

$$= 2[(0.15 \text{ m})(0.06 \text{ m}) + (0.15 \text{ m})(0.01 \text{ m}) + (0.06 \text{ m})(0.01 \text{ m})]$$

$$= 0.0222 \text{ m}^2$$

and the volume of the part is

$$V = L\,W\,H$$

$$= (0.15 \text{ m})(0.06 \text{ m})(0.01 \text{ m})$$

$$= 9.00 \times 10^{-5} \text{ m}^3$$

The cooling rates for the various part temperatures are

300°C:

$$\frac{dT}{dt} = \frac{-(20)(0.0222)(300 - 25) - (0.8)(0.0222)(5.669 \times 10^{-8})(573^4 - 288^4)}{(7850)(430)(9.00 \times 10^{-5})}$$

$$= \underline{\underline{-0.736°C/s}}$$

200°C:

$$\frac{dT}{dt} = \frac{-(20)(0.0222)(200 - 25) - (0.8)(0.0222)(5.669 \times 10^{-8})(473^4 - 288^4)}{(7850)(430)(9.00 \times 10^{-5})}$$

$$= \underline{\underline{-0.399°C/s}}$$

100°C:

$$\frac{dT}{dt} = \frac{-(20)(0.0222)(100 - 25) - (0.8)(0.0222)(5.669 \times 10^{-8})(373^4 - 288^4)}{(7850)(430)(9.00 \times 10^{-5})}$$

$$= \underline{\underline{-0.151°C/s}}$$

50°C:

$$\frac{dT}{dt} = \frac{-(20)(0.0222)(50 - 25) - (0.8)(0.0222)(5.669 \times 10^{-8})(323^4 - 288^4)}{(7850)(430)(9.00 \times 10^{-5})}$$

$$= \underline{\underline{-0.050°C/s}}$$

DISCUSSION

The negative sign on the answers indicates that the temperature of the part is decreasing with time, i.e., the part is cooling. Note that the magnitude of the cooling rate decreases with part temperature because the temperature difference between the part and its surroundings is decreasing. The temperature at which the cooling rate is zero is found by solving the following equation for T,

$$hA(T - T_\infty) + \varepsilon A\sigma(T^4 - T_{sur}^4) = 0$$

Solving this equation numerically, $T = 23.2°C$, which, as expected, is between T_∞ and T_{sur}.

❒ ❒ ❒

PROBLEM 1.18 Surface Temperature of a Deep Space Probe

The onboard power system of a small deep space probe steadily generates 3 kW. The probe's components are enclosed by a 90-cm diameter spherical aluminum shell. To minimize the surface temperature of the shell, and therefore the components within it, the shell's outside surface is covered by a very thin layer of a special paint with an emissivity of 0.95. What is the outside surface temperature of the aluminum shell? What is the heat flux?

DIAGRAM

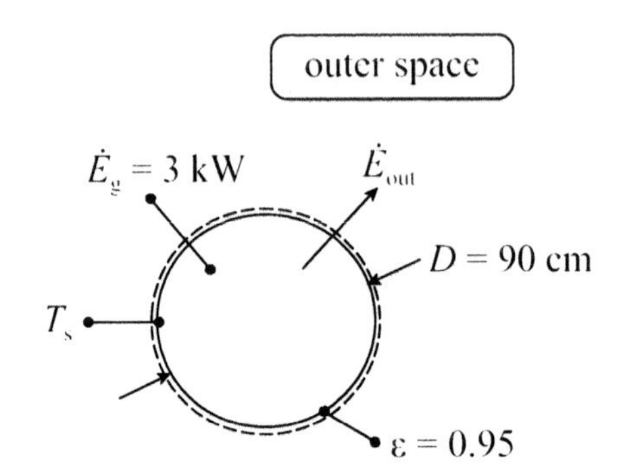

ASSUMPTIONS

1. Steady conditions.
2. Surface of the probe is isothermal.
3. Outer space is treated as a very large enclosure at 0 K.
4. Probe operates far from any source of solar radiation.
5. Constant properties.

PROPERTIES

paint: $\varepsilon = 0.95$

ANALYSIS

An energy balance on the control volume in the diagram yields,

$$\dot{E}_g - \dot{E}_{out} = 0$$

The rate of energy generation is the power generated by the probe's power system. The rate of energy transfer from the probe is the radiation to outer space, which is treated as a very large enclosure,

$$\dot{E}_{out} = \varepsilon A \sigma (T_s^4 - T_{sur}^4)$$

Noting that $T_{sur} = 0$ K, substituting the above terms into the energy balance, and solving for T_s,

$$T_s = \left(\frac{\dot{E}_g}{\varepsilon A \sigma} \right)^{1/4} = \left(\frac{\dot{E}_g}{\varepsilon \pi D^2 \sigma} \right)^{1/4}$$

$$T_s = \left(\frac{3000 \text{ W}}{(0.95)\, \pi\, (0.90 \text{ m})^2\, (5.669 \times 10^{-8} \text{ W/m}^2 \cdot \text{K}^4)} \right)^{1/4}$$

$$= 385 \text{ K} = \underline{\underline{112°\text{C}}}$$

The heat flux is the rate of energy generation divided by the surface area of the spherical shell,

$$q'' = \dot{E}_g / A = \dot{E}_g / (\pi D^2)$$

$$= (3000 \text{ W})/[\pi\, (0.90 \text{ m})^2]$$

$$= \underline{\underline{1179 \text{ W/m}^2}}$$

DISCUSSION

It is instructive to determine the sensitivity of the shell's surface temperature to the emissivity of the paint on its surface. The following values are obtained:

$\varepsilon = 0.05$: $T_s = 803 \text{ K} = 530°\text{C}$

$\varepsilon = 0.5$: $T_s = 452 \text{ K} = 179°\text{C}$

$\varepsilon = 0.7$: $T_s = 415 \text{ K} = 142°\text{C}$

$\varepsilon = 1.0$: $T_s = 380 \text{ K} = 107°\text{C}$

The shell's surface temperature is a strong function of its emissivity, and a high emissivity is desirable for minimizing the shell's temperature. If the paint behaves as a black body ($\varepsilon = 1.0$), the lowest surface temperature possible, assuming the probe's power dissipation and surface area remain unchanged, is 107°C.

❐ ❐ ❐

PROBLEM 1.19 Chilled Feeling of a Person Indoors During Winter

The air in a house is maintained at 21 °C by a heating system during the winter and a cooling system during the summer. Due to different outdoor air temperatures during these seasons, the inside surface temperature of the walls is 16 °C during the winter and 28 °C during the summer. Using this information, show how a person can feel chilled during the winter even though the indoor air temperature is constant all year around. Assume that the surface temperature and emissivity of the person are 29 °C and 0.85, respectively, and that the heat transfer coefficient at the surface of the person is 10 W/m²·K.

DIAGRAM

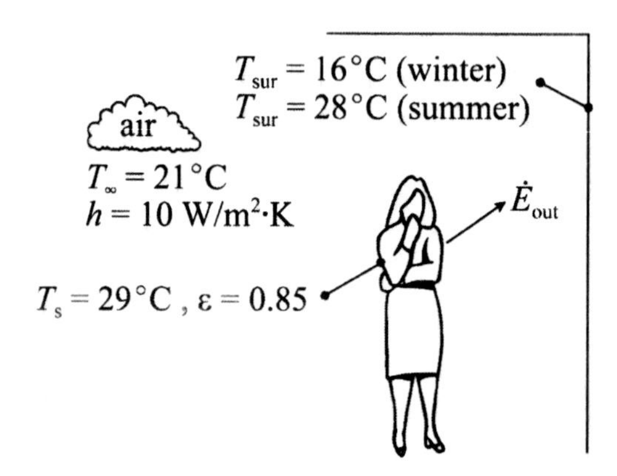

ASSUMPTIONS

1. Steady conditions.
2. Surface area of the person is much less than the surface area of the walls.
3. Conduction from the person is neglected.
4. Constant properties.

PROPERTIES

body surface: ε = 0.85

ANALYSIS

The rate of energy transfer from the person consists of convection and radiation. The convective heat flux, which is constant all year around, is

$$q''_{conv} = h(T_s - T_\infty)$$

$$q'' = (10 \text{ W/m}^2\cdot\text{K})(29 - 21)°\text{C}$$

$$= 80 \text{ W/m}^2$$

The radiative heat flux for the winter and summer are

winter:
$$q''_{rad} = \varepsilon\sigma(T_s^4 - T_{sur}^4)$$

$$q''_{rad} = (0.85)(5.669 \times 10^{-8}\ \mathrm{W/m^2{\cdot}K^4})(302^4 - 289^4)\mathrm{K}^4$$

$$= \underline{\underline{64.7\ \mathrm{W/m^2}}}$$

summer:
$$q''_{rad} = \varepsilon\sigma(T_s^4 - T_{sur}^4)$$

$$q''_{rad} = (0.85)(5.669 \times 10^{-8}\ \mathrm{W/m^2{\cdot}K^4})(302^4 - 301^4)\mathrm{K}^4$$

$$= \underline{\underline{5.28\ \mathrm{W/m^2}}}$$

DISCUSSION

The winter radiative heat flux is over ten times greater than the summer radiative heat flux. The chilled feeling is attributable to the effect of colder walls on radiation. This chilled effect during the winter can be particularly pronounced if the person is located near a window because the inside surface temperature of windows is even lower than that of walls.

❒ ❒ ❒

PROBLEM 1.20 Surface Temperature of an Incandescent Light Bulb

The surface of an illuminated 60-W incandescent light bulb is very hot to the touch. Show this by approximating the light bulb as a 6.5-cm diameter sphere with an emissivity of 0.9. The light bulb operates in a large room where the walls, ceiling and floor have a temperature of 18°C. If the temperature of the air in the room is 22°C and the heat transfer coefficient at the light bulb's surface is 16 W/m²·K, what is the surface temperature of the light bulb?

DIAGRAM

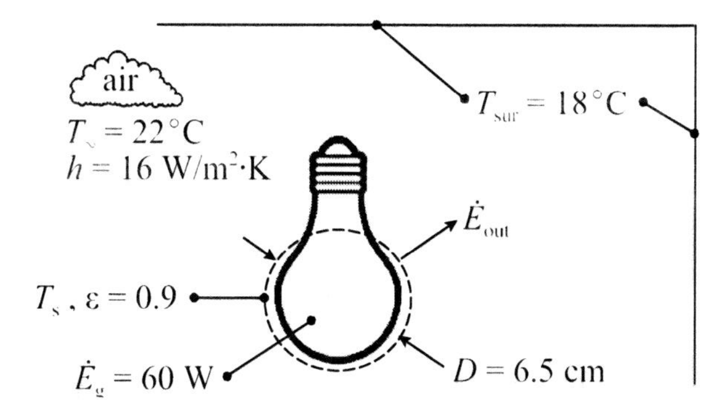

ASSUMPTIONS

1. Steady conditions.
2. Light bulb surface area is much less than the surface area of the surroundings.
3. Conduction into the light bulb's base is neglected.
4. Constant properties.

PROPERTIES

light bulb surface: $\varepsilon = 0.9$

ANALYSIS

An energy balance on the control volume in the diagram yields

$$\dot{E}_{in} - \dot{E}_{out} = 0$$

The rate of energy generation is the power dissipated by the light bulb. The rate of energy transfer from the light bulb is the sum of the convective and radiative heat transfer,

$$\dot{E}_{out} = q_{conv} + q_{rad} = hA(T_s - T_\infty) + \varepsilon A\sigma(T_s^4 - T_{sur}^4)$$

Substituting these terms into the energy balance,

$$\dot{E}_g - hA(T_s - T_\infty) - \varepsilon A\sigma(T_s^4 - T_{sur}^4) = 0$$

Substituting values into this equation,

$$60 - (16)\,\pi\,(0.065)^2(T_s - 295) - (0.9)\,\pi\,(0.065)^2(5.669 \times 10^{-8})(T_s^4 - 291^4) = 0$$

Because T_s is raised to the fourth power in the radiation term, the surface temperature cannot be found using simple algebraic manipulations. A numerical method must be employed. The most efficient way to solve for T_s is to use an equation solver software package or a root finding routine on a scientific calculator. Solving for T_s numerically,

$$T_s = 459 \text{ K} = \underline{\underline{186°\text{C}}}$$

DISCUSSION

Touching this light bulb could result in a severe burn. If conduction into the light bulb's base was taken into account, the bulb's surface temperature would be much lower than the value calculated here.

PROBLEM 1.21 Cost of Heat Loss for an Incinerator

The walls of a gas-fired incinerator consist of fire clay brick (k = 1.0 W/m·K). Combustion of waste materials within the incinerator results in a temperature of 400°C on the inside surface of the walls. Outside the incinerator the air temperature is 30°C, and the heat transfer coefficient and emissivity associated with the wall outside surface are 25 W/m^2·K and 0.75, respectively. If the walls are 16 cm thick, find the temperature of the outside surface of the incinerator wall and the heat loss per unit area of wall surface. If the cost of natural gas is \$0.35/m^3 and the total wall surface area is 250 m^2, what is the cost of the incinerator's heat loss for a 24-hour period? Use 37 MJ/m^3 as the heat of combustion of natural gas.

DIAGRAM

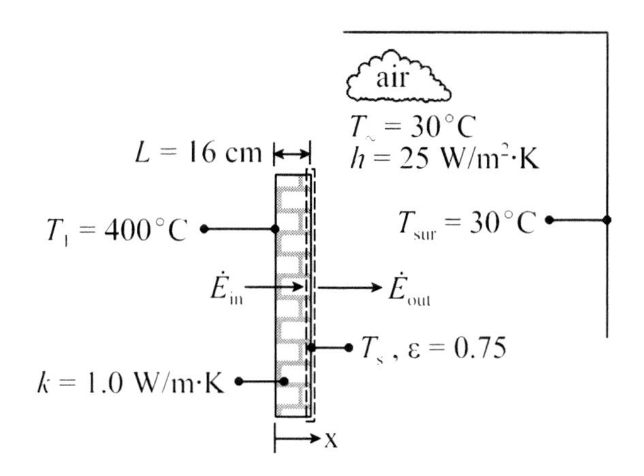

ASSUMPTIONS

1. Steady conditions.
2. One-dimensional conduction in the wall.
3. Surroundings for radiation have the same temperature as the air.
4. Constant properties.

PROPERTIES

fire clay brick: k = 1.0 W/m·K , ε = 0.75

ANALYSIS

An energy balance on the control surface in the diagram yields

$$\dot{E}_{in} - \dot{E}_{out} = 0$$

The rate of energy transfer into the control surface is the conduction through the wall,

$$\dot{E}_{in} = q_{cond} = -kA\frac{dT}{dx} = kA\frac{(T_1 - T_s)}{L}$$

and the rate of energy transfer out of the control surface is the convective and radiative heat transfer,

$$\dot{E}_{out} = q_{conv} = hA(T_s - T_\infty) + \varepsilon A\sigma(T_s^4 - T_{sur}^4)$$

Substituting these terms into the energy balance,

$$k\frac{(T_1 - T_s)}{L} - h(T_s - T_\infty) - \varepsilon\sigma(T_s^4 - T_{sur}^4) = 0$$

Substituting values into this equation,

$$(1.0)\frac{(673 - T_s)}{0.16} - (25)(T_s - 303) - (0.75)(5.669 \times 10^{-8})(T_s^4 - 303^4) = 0$$

Because T_s is raised to the fourth power in the radiation term, the surface temperature cannot be found using simple algebraic manipulations. A numerical method must be employed. The most efficient way to solve for T_s is to use an equation solver software package or a root finding routine on a scientific calculator. Solving for T_s numerically,

$$T_s = 315.1 \text{ K} = \underline{\underline{42.0°\text{C}}}$$

Now that T_s is known, the heat flux can be calculated using Fourier's law,

$$q'' = k(T_1 - T_s)$$

$$q'' = (1.0 \text{ W/m}\cdot\text{K})(400 - 42.0)°\text{C}$$

$$= \underline{\underline{358 \text{ W/m}^2}}$$

or,

$$q'' = h(T_s - T_\infty) + \varepsilon\sigma(T_s^4 - T_{sur}^4)$$

$$q'' = (25 \text{ W/m}^2\cdot\text{K})(42.0 - 30)°\text{C} + (0.75)(5.669 \times 10^{-8} \text{ W/m}^2\cdot\text{K}^4)(315.1^4 - 303.2^4)\text{K}^4$$

$$= 360 \text{ W/m}^2$$

where the small difference in the answers is due to numerical roundoff. The cost of the heat loss for a 24-hour period is

$$\text{cost} = \frac{(358 \text{ J/m}^2\cdot\text{s})(250 \text{ m}^2)(3600 \text{ s/h})(24 \text{ h})(\$0.35/\text{m}^3)}{(37 \times 10^6 \text{ J/m}^3)}$$

$$= \underline{\underline{\$73.15}}$$

DISCUSSION

As required by the energy balance, the calculations yield the same heat flux within the wall and at the wall surface. Insulating the walls of the incinerator would significantly reduce this cost. In addition to the cost of heat losses through the walls of the incinerator, there are also heat losses associated with the hot combustion products that flow out of the stack.

❒ ❒ ❒

PROBLEM 1.22 Amount of Ice Melted in a Storage Tank

A thin-walled spherical storage tank with a diameter of 2 m contains water ice at 0 °C. The tank wall is constructed of stainless steel with an emissivity of 0.25. The tank is located within a large warehouse where the air temperature is 30 °C and the surrounding walls, ceiling and floor have an average temperature of 22 °C. If the heat transfer coefficient at the surface of the tank is 40 W/m^2·K, find the heat transfer to the ice and the amount of ice at 0 °C that melts during a 24-hour period. For the latent heat of fusion, use h_{if} = 334 kJ/kg.

DIAGRAM

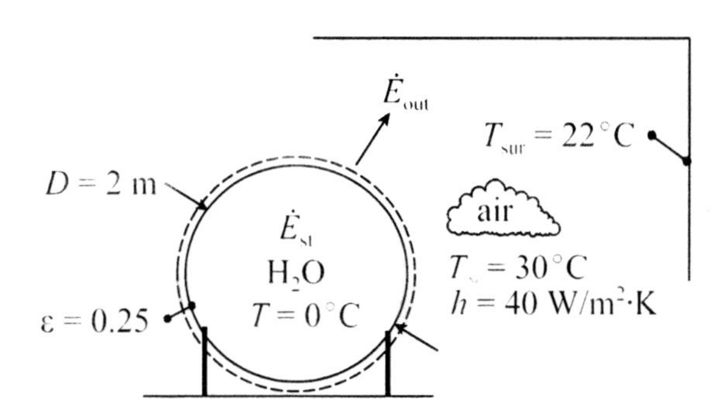

ASSUMPTIONS

1. Steady conditions.
2. Conduction in the tank wall is neglected.
3. Constant properties.

PROPERTIES

stainless steel: ε = 0.25

ANALYSIS

An energy balance on the control volume in the diagram yields,

$$-\dot{E}_{out} = \dot{E}_{st}$$

The rate of energy transfer out of the control volume is

$$\dot{E}_{out} = q_{conv} + q_{rad} = hA(T_s - T_\infty) + \varepsilon A\sigma(T_s^4 - T_{sur}^4)$$

$$= h\pi D^2(T_s - T_\infty) + \varepsilon\pi D^2\sigma(T_s^4 - T_{sur}^4)$$

$$\dot{E}_{out} = (40)\,\pi\,(2)^2(273 - 303) + (0.25)\,\pi\,(2)^2(5.669 \times 10^{-8})(273^4 - 295^4)$$

$$= \underline{\underline{-15{,}439 \text{ W}}}$$

where the minus sign signifies that heat is transferred *to* the ice and not *from* the ice. Thus,

$$\dot{E}_{st} = -(-15{,}439 \text{ W})$$

$$= 15{,}439 \text{ W}$$

The amount of ice that melts during a 24-hour period is

$$m = \frac{\dot{E}_{st}\,t}{h_{if}}$$

$$m = \frac{(15{,}439 \text{ W})(24 \text{ h} \times 3600 \text{ s/h})}{334 \times 10^3 \text{ J/kg}}$$

$$= \underline{\underline{3994 \text{ kg}}}$$

DISCUSSION

The amount of ice melted during a given period of time is directly proportional to the heat transfer to the ice, so if the environmental temperatures were higher, a greater amount of ice would melt. Furthermore, if the storage tank was located outdoors, solar radiation effects would have to be considered.

PROBLEM 1.23 Temperature of a Thin Metal Plate in Outer Space

At a distance of one astronomical unit (AU) from the sun, the distance at which the earth's orbit is located, the radiative heat flux from the sun is approximately 1350 W/m^2. Consider a thin copper plate at this distance, one side of which faces directly into the sun and the other side of which faces deep space. The side facing the sun is highly polished with a solar absorptivity of 0.16 and an emissivity of 0.03, whereas the side facing deep space is oxidized with an emissivity of 0.80. Find the temperature of the plate.

DIAGRAM

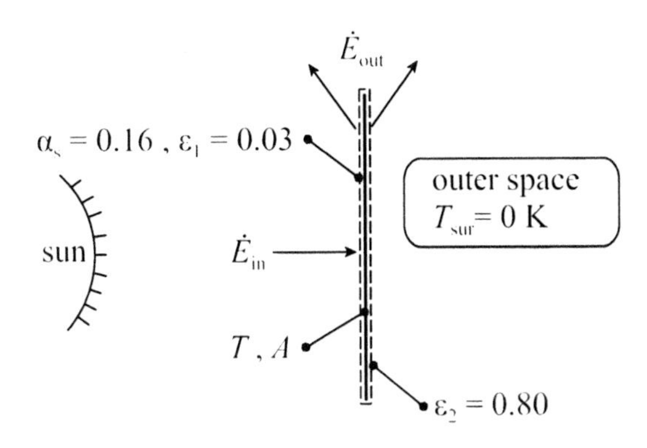

ASSUMPTIONS

1. Steady conditions.
2. Conduction in the plate is neglected.
3. Plate is isothermal.
4. Outer space is treated as a large enclosure at 0 K.
5. Constant properties.

PROPERTIES

copper, polished: $\alpha_s = 0.16$, $\varepsilon_1 = 0.03$
copper, oxidized: $\varepsilon_2 = 0.80$

ANALYSIS

An energy balance on the control volume in the diagram yields,

$$\dot{E}_{in} - \dot{E}_{out} = 0$$

Note that the control volume has a negligible width because the plate is thin. The rate of energy transfer into the control volume is the solar radiation absorbed by the plate,

$$\dot{E}_{in} = \alpha_s q'' A$$

where $q'' = 1350\ \text{W/m}^2$. The rate of energy transfer out of the control volume is the radiation exchange between the plate and outer space,

$$\dot{E}_{out} = \varepsilon_1 A\sigma(T^4 - T_{sur}^4) + \varepsilon_2 A\sigma(T^4 - T_{sur}^4)$$

Noting that outer space is treated as a large enclosure at 0 K, substituting the energy transfer terms into the energy balance, and solving for T,

$$T = \left[\frac{\alpha_s q''}{\sigma(\varepsilon_1 + \varepsilon_2)}\right]^{1/4}$$

$$T = \left[\frac{(0.16)(1350\ \text{W/m}^2)}{(5.669 \times 10^{-8}\ \text{W/m}^2\cdot\text{K}^4)(0.03 + 0.80)}\right]^{1/4}$$

$$= 260.3\ \text{K} = \underline{\underline{-12.9°\text{C}}}$$

DISCUSSION

As demanded by the physics of the problem, the temperature of the plate is directly proportional to the solar heat flux and absorptivity, and inversely proportional to the emissivities. It is instructive to examine a special case in which the plate behaves as a blackbody. If the plate behaves as a blackbody, which is a perfect absorber and emitter of radiation, the surface properties have a value of 1. The plate temperature for this case is

$$T = \left[\frac{(1.0)(1350\ \text{W/m}^2)}{(5.669 \times 10^{-8}\ \text{W/m}^2\cdot\text{K}^4)(1.0 + 1.0)}\right]^{1/4}$$

$$= 330.3\ \text{K} = 57.2°\text{C}$$

❐ ❐ ❐

PROBLEM 1.24 Maximum Power Dissipation of Electronic Devices

Identical electronic devices measuring 18.0 mm on a side are mounted to a vertically oriented circuit board installed in a large enclosure whose temperature is 25 °C. The maximum allowable surface temperature of the devices is 85 °C, and they have an emissivity of 0.70. The devices transfer heat by radiation and natural convection, where the heat transfer coefficient, which depends on surface temperature, is approximated as $h = C\,(T_s - T_\infty)^{1/4}$, where $C = 4.0\ \text{W/m}^2\cdot\text{K}^{5/4}$. If the surrounding air temperature equals that of the enclosure, find the maximum power dissipation of the devices.

DIAGRAM

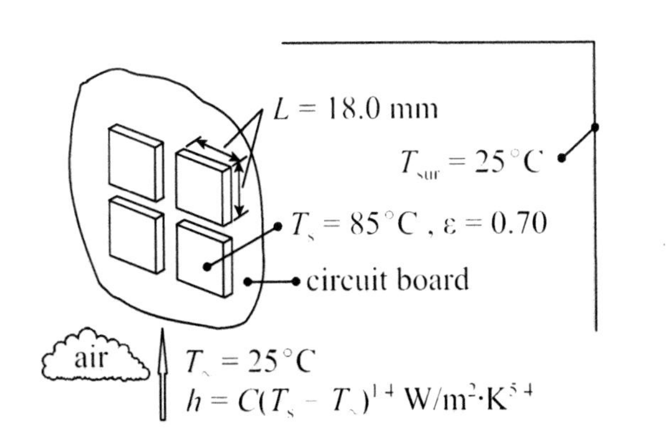

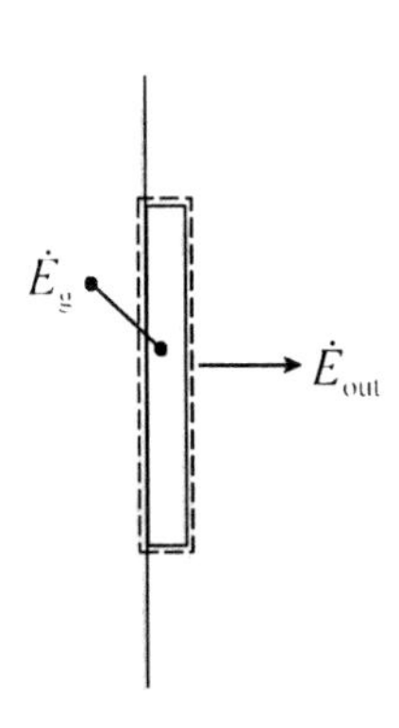

ASSUMPTIONS

1. Steady conditions.
2. Conduction into the circuit board is neglected.
3. Surface area of the devices is much smaller than that of the enclosure.
4. Convection and radiation from the sides of the devices are neglected.
5. Constant properties.

PROPERTIES

device surface: $\varepsilon = 0.70$

ANALYSIS

Applying an energy balance on the control volume of one device yields,

$$\dot{E}_g - \dot{E}_{out} = 0$$

The rate of energy transfer out of the control volume is the convection and radiation from the device,

$$\dot{E}_{out} = q_{conv} + q_{rad} = hA(T_s - T_\infty) + \varepsilon A\sigma(T_s^4 - T_{sur}^4)$$

$$= CA(T_s - T_\infty)^{5/4} + \varepsilon A\sigma(T_s^4 - T_{sur}^4)$$

Substituting these terms into the energy balance and solving for the power dissipation,

$$\dot{E}_g = (4.0)(0.018)^2(85 - 25)^{5/4} + (0.70)(0.018)^2(5.669 \times 10^{-8})(358^4 - 298^4)$$

$$= 0.216 \text{ W} + 0.110 \text{ W}$$

$$= \underline{\underline{0.236 \text{ W}}}$$

DISCUSSION

Most electronic devices of this size dissipate more power than that calculated here. The heat transfer coefficient is

$$h = C(T_s - T_\infty)^{5/4}$$

$$= (4.0 \text{ W/m}^2\cdot\text{K}^{5/4})(85\,°\text{C} - 25\,°\text{C})^{1/4}$$

$$= 11.1 \text{ W/m}^2\cdot\text{K}$$

Natural convection and radiation are weak heat transfer mechanisms for this system, and are insufficient for transferring the dissipated power while maintaining the device's temperature below 85 °C. Even if a fan is used to force air over the devices, producing a heat transfer coefficient of 300 W/m²·K, the maximum power dissipation is

$$\dot{E}_{out} = hA(T_s - T_\infty) + q_{rad}$$

$$\dot{E}_{\text{out}} = (300 \text{ W/m}^2\cdot\text{K})(0.018 \text{ m})^2(85 - 25)°\text{C} + 0.110 \text{ W}$$

$$= 5.83 \text{ W} + 0.110 \text{ W}$$

$$= 5.94 \text{ W}$$

which is still lower than many high-power microprocessors. If the device is to survive thermally, conduction into the circuit board and/or a heat sink that extends the surface area of the device would have to be considered.

❐ ❐ ❐

PROBLEM 1.25 Cooling Time of a Can of Cooked Carrots

During the commercial processing of vegetables, racks of canned diced carrots are cooked in large pressure cookers at 110 °C. The cans have a length and diameter of 11 cm and 7.6 cm, respectively, and are filled with small carrot pieces and water. After cooking, the cans of carrots are allowed to cool in air at 25 °C where natural convection currents produce a heat transfer coefficient of 16 W/m²·K. From the energy balance on a can, write the differential equation that gives the change in temperature of the carrots with respect to time. Solve the differential equation, and find the time required for the carrots to reach 40 °C, a safe handling temperature.

DIAGRAM

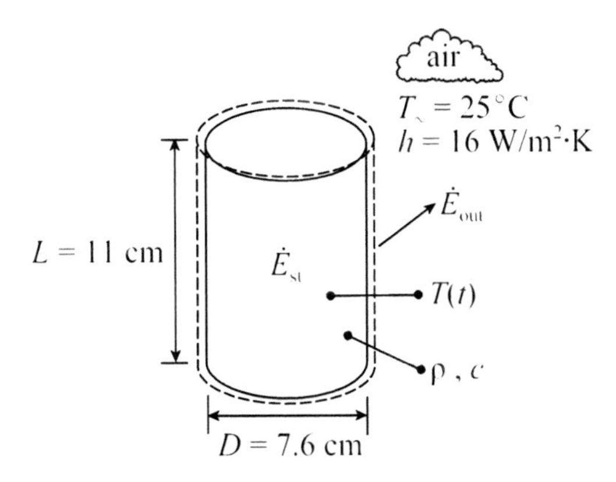

ASSUMPTIONS

1. Contents of the can are isothermal.
2. Thermal capacity of the can wall is neglected.
3. Radiation from the can is neglected.
4. Contents of the can have constant thermal properties of liquid water at 300 K.

PROPERTIES

liquid water at 300 K: $\rho = 997$ kg/m^3 , $c = 4179$ J/kg·K

ANALYSIS

An energy balance on the control volume in the diagram yields,

$$-\dot{E}_{out} = \dot{E}_{st}$$

The rate of energy transfer out of the control volume is the convection from the can,

$$\dot{E}_{out} = hA(T - T_\infty)$$

and the rate at which energy is stored is

$$\dot{E}_{st} = \rho c V \frac{dT}{dt}$$

Substituting these terms into the energy balance,

$$-hA(T - T_\infty) = \rho c V \frac{dT}{dt}$$

It is mathematically convenient to introduce the temperature difference,

$$\theta = T - T_\infty$$

Substituting the temperature difference into the differential equation and rearranging,

$$\frac{d\theta}{dt} = -\left(\frac{hA}{\rho c V}\right)\theta$$

The initial condition for this differential equation is

$$t = 0: \quad \theta = T_i - T_\infty = \theta_i$$

Separating variables and integrating the differential equation,

$$\frac{\rho c V}{hA}\int_{\theta_i}^{\theta}\frac{d\theta}{\theta} = -\int_0^t dt$$

Evaluating the integrals,

$$t = -\left(\frac{\rho c V}{hA}\right)\ln\left(\frac{\theta}{\theta_i}\right)$$

The total surface area of the can is

$$A = \pi DL + 2\pi D^2/4$$

$$= \pi\,(0.076 \text{ m})(0.11 \text{ m}) + 2\,\pi\,(0.076 \text{ m})^2/4$$

$$= 0.0353 \text{ m}^2$$

and the volume of the can is

$$V = (\pi D^2/4)L$$

$$= \pi\,[(0.076 \text{ m})^2/4]\,(0.11 \text{ m})$$

$$= 4.990 \times 10^{-4} \text{ m}^3$$

The time required for the carrots to reach 40°C is

$$t = -\frac{(997 \text{ kg/m}^3)(4179 \text{ J/kg·K})(4.990 \times 10^{-4} \text{ m}^3)}{(16 \text{ W/m}^2\text{·K})(0.0353 \text{ m}^2)}\ln\left(\frac{40°\text{C} - 25°\text{C}}{110°\text{C} - 25°\text{C}}\right)$$

$= 6385 \text{ s} = \underline{\underline{1.77 \text{ h}}}$

DISCUSSION

If conduction and radiation effects were included, the cooling time would be significantly lower. The cooling time could also be reduced by forcing air over the can using a blower. Assumption 1 is suspect because there are probably temperature gradients in the contents of the can at any given time.

❒ ❒ ❒

PROBLEM 1.26 Asphalt Parking Lot Temperature on a Hot Sunny Day

On a hot sunny day, an asphalt parking lot is exposed to a solar radiative heat flux of 1000 W/m^2, and the air temperature is 40°C. Natural convection currents near the surface of the asphalt yield a heat transfer coefficient of 15 W/m^2·K. Neglecting conduction in the asphalt and underlying soil, find the surface temperature of the asphalt. For the solar absorptivity and emissivity of asphalt, use 0.9. The effective sky temperature for radiation is 30°C.

DIAGRAM

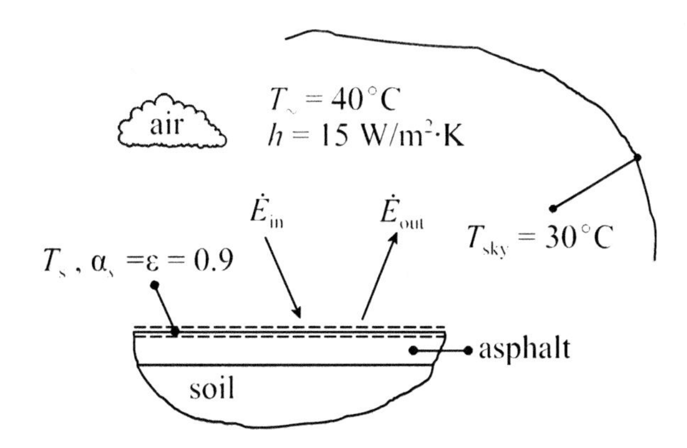

ASSUMPTIONS

1. Steady conditions.
2. Conduction in the asphalt and soil is neglected.
3. Constant properties.

PROPERTIES

asphalt: $\alpha_s = \varepsilon = 0.90$.

ANALYSIS

An energy balance on the control surface in the diagram yields,

$$\dot{E}_{in} - \dot{E}_{out} = 0$$

The rate of energy transfer into the control surface is the solar radiation absorbed by the asphalt,

$$\dot{E}_{in} = \alpha_s q'' A$$

where $q'' = 1000\ \mathrm{W/m^2}$. The rate of energy transfer out of the control surface is the convection and radiation from the asphalt,

$$\dot{E}_{out} = hA(T_s - T_\infty) + \varepsilon A\sigma(T_s^4 - T_{sky}^4)$$

Substituting these terms into the energy balance, and noting that surface area divides out,

$$(0.9)(1000) = (15)(T_s - 313) + (0.9)(5.669 \times 10^{-8})(T_s^4 - 303^4)$$

Numerically solving for T_s,

$$T_s = 350.4\ \mathrm{K} = \underline{\underline{77.3\,^\circ\mathrm{C}}}$$

DISCUSSION

People could not reasonably walk or work on a parking lot with a surface temperature this high. Neither assumption is physically reasonable from a practical point of view. Solar radiation changes throughout the day, so conditions are not steady, and much of the absorbed solar radiation would be conducted into the asphalt and soil below, giving a lower surface temperature than that calculated here. An analysis that includes unsteady effects and conduction is required to give a physically reasonable result.

❐ ❐ ❐

PROBLEM 1.27 Radiative Cooling of a Lead Sphere

A 1.6-cm diameter solid sphere of lead has a uniform initial temperature of 280°C when it is placed in a large vacuum chamber whose walls are maintained at 30°C. The density and specific heat of lead are 11,340 kg/m³ and 129 J/kg·K, respectively. If the emissivity of lead is 0.85, find the time required for the sphere to cool to 40°C.

DIAGRAM

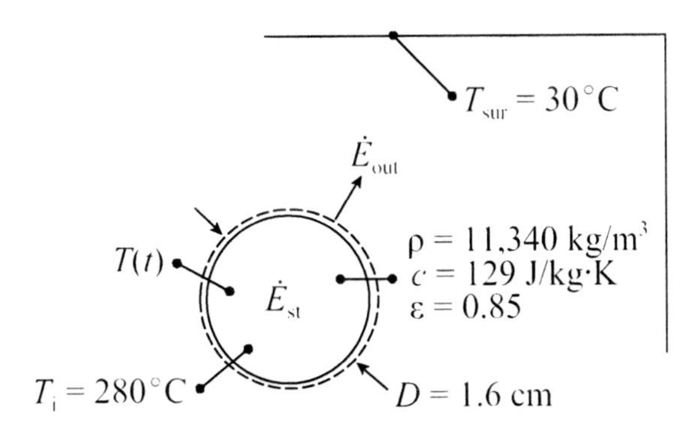

ASSUMPTIONS

1. At any given time, the sphere is isothermal.
2. Radiation is the only heat transfer mechanism present.
3. Constant properties at 300 K.

PROPERTIES

lead: $\rho = 11{,}340$ kg/m^3 , $c = 129$ J/kg·K , $\varepsilon = 0.85$

ANALYSIS

An energy balance on the control volume in the diagram yields,

$$-\dot{E}_{out} = \dot{E}_{st}$$

The rate of energy transfer from the control volume is the radiation between the sphere and the chamber walls,

$$\dot{E}_{out} = \varepsilon A \sigma (T^4 - T_{sur}^4)$$

The rate at which energy is stored is

$$\dot{E}_{st} = \rho c V \frac{dT}{dt}$$

Substituting these energy terms into the energy balance,

$$-\varepsilon A \sigma (T^4 - T_{sur}^4) = \rho c V \frac{dT}{dt}$$

The initial condition is

$$t = 0: \quad T = T_i$$

Separating variables and integrating,

$$\int_{T_i}^{T} \frac{dT}{T^4 - T_{sur}^4} = -C \int_0^t dt$$

where,

$$C = \frac{\varepsilon A \sigma}{\rho c V} = \frac{\varepsilon \pi D^2 \sigma}{\rho c \pi D^3 / 6} = \frac{6 \varepsilon \sigma}{\rho c D}$$

$$C = \frac{6(0.85)(5.669 \times 10^{-8}\ \text{W/m}^2\cdot\text{K}^4)}{(11{,}340\ \text{kg/m}^3)(129\ \text{J/kg}\cdot\text{K})(0.016\ \text{m})}$$

$$= 1.235 \times 10^{-11}\ \text{s}^{-1}\cdot\text{K}^{-3}$$

Carrying out the integrations,

$$\frac{1}{4T_{sur}^3} \ln\left(\frac{T - T_{sur}}{T + T_{sur}}\right) - \frac{1}{2T_{sur}^3} \tan^{-1}\left(\frac{T}{T_{sur}}\right)$$

$$-\left[\frac{1}{4T_{sur}^3} \ln\left(\frac{T_i - T_{sur}}{T_i + T_{sur}}\right) - \frac{1}{2T_{sur}^3} \tan^{-1}\left(\frac{T_i}{T_{sur}}\right)\right] = -Ct$$

Dividing both sides of this equation by $-C$, substituting values of the temperatures, and solving for t,

$$t = 1713\ \text{s} = \underline{\underline{28.5\ \text{min}}}$$

DISCUSSION

If the sphere was in the presence of a cold fluid where convection could occur, the cooling time would be significantly lower than the value calculated here.

Chapter 2

Principles of Conduction

PROBLEM 2.1 Three-Dimensional Temperature Field and Heat Flux Vector

At a given instant of time, the scalar temperature field in a solid is

$$T(x, y, z) = 2x^2 + y^3 + 4z^2 - xy + 3yz \ \ ^\circ\mathrm{C}$$

If the thermal conductivity of the solid is k = 10 W/m·K, find the heat flux vector $\boldsymbol{q}''$ at the coordinates (2,4,3) m. What is the temperature of the solid at this point?

DIAGRAM

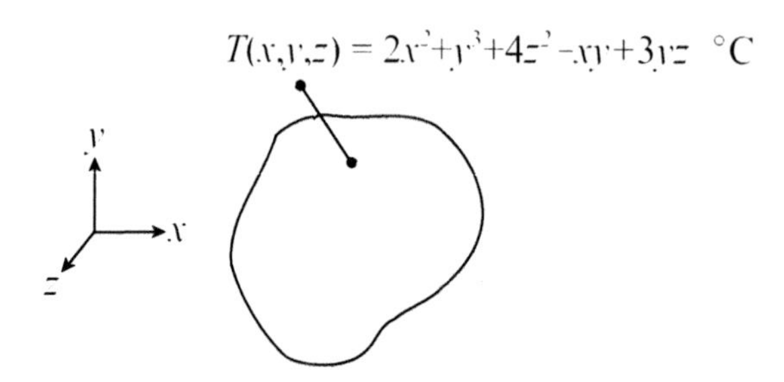

ASSUMPTIONS

1. Conduction is three-dimensional and unsteady.

PROPERTIES

thermal conductivity: k = 10 W/m·K

ANALYSIS

Heat flux is a vector quantity which can be written in terms of a general form of Fourier's law,

$$\boldsymbol{q}'' = -k\nabla T = -k\left(\boldsymbol{i}\frac{\partial T}{\partial x} + \boldsymbol{j}\frac{\partial T}{\partial y} + \boldsymbol{k}\frac{\partial T}{\partial z}\right)$$

where ∇ is the del operator, and $\boldsymbol{i}, \boldsymbol{j}$ and $\boldsymbol{k}$ are the Cartesian unit vectors. The partial derivatives are

$$\frac{\partial T}{\partial x} = 4x - y$$

$$\frac{\partial T}{\partial y} = 3y^2 - x + 3z$$

$$\frac{\partial T}{\partial z} = 8z + 3y$$

The heat flux vector is

$$\boldsymbol{q}'' = -k\left[\boldsymbol{i}(4x - y) + \boldsymbol{j}(3y^2 - x + 3z) + \boldsymbol{k}(8z + 3y)\right]$$

Substituting the coordinates and thermal conductivity,

$$\boldsymbol{q}''(2,4,3) = -(10 \text{ W/m·K})(4\,\boldsymbol{i} + 55\,\boldsymbol{j} + 36\,\boldsymbol{k}) \text{ K/m}$$

$$= \underline{\underline{-40\,\boldsymbol{i} - 550\,\boldsymbol{j} - 360\,\boldsymbol{k} \text{ W/m}^2}}$$

The temperature of the solid at the given coordinates is

$$T(x,y,z) = 2x^2 + y^3 + 4z^2 - xy + 3yz$$

$$= 2(2)^2 + (4)^3 + 4(3)^2 - (2)(4) + 3(4)(3)$$

$$= \underline{\underline{136°\text{C}}}$$

DISCUSSION

The scalar components of the heat flux vector are

$$q_x'' = -40 \text{ W/m}^2$$

$$q_y'' = -550 \text{ W/m}^2$$

$$q_z'' = -360 \text{ W/m}^2$$

The temperature calculated here is the value for a given instant of time at the specified spatial coordinates. Because the thermal conductivity is not a function of coordinate direction, the solid is said to be *isotropic*.

❒ ❒ ❒

PROBLEM 2.2 Three-Dimensional Steady Conduction with no Energy Generation

Consider a solid material in which the temperature field is

$$T(x,y,z) = x^3 - 3xy^2 + z \text{ °C}$$

Show that this temperature field satisfies a problem where the conduction is three-dimensional and steady in a material with constant thermal conductivity and no energy generation.

DIAGRAM

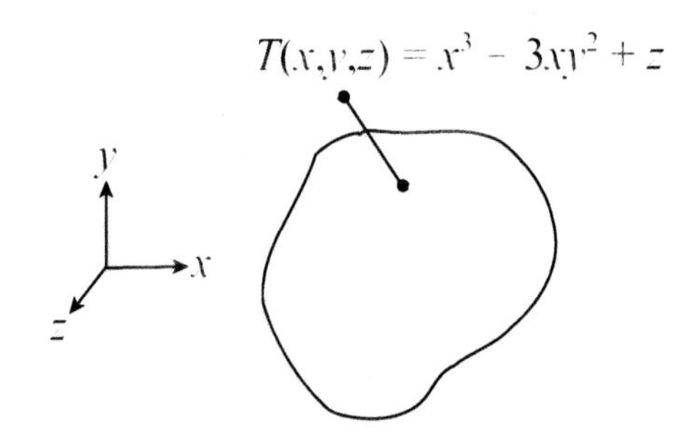

ASSUMPTIONS

1. Three-dimensional steady conduction.
2. Constant thermal conductivity.
3. No energy generation.

PROPERTIES

thermal conductivity: k

ANALYSIS

The heat equation in Cartesian coordinates is

$$\frac{\partial}{\partial x}\left(k\frac{\partial T}{\partial x}\right)+\frac{\partial}{\partial y}\left(k\frac{\partial T}{\partial y}\right)+\frac{\partial}{\partial z}\left(k\frac{\partial T}{\partial z}\right)+\dot{q}=\rho c_{\mathrm{p}}\frac{\partial T}{\partial t}$$

For constant thermal conductivity and no internal generation, the heat equation reduces to

$$\frac{\partial^2 T}{\partial x^2}+\frac{\partial^2 T}{\partial y}+\frac{\partial^2 T}{\partial z^2}=0$$

Substituting the temperature field into the heat equation and performing the derivatives,

$$6x-6x+0=0$$

The temperature field satisfies the heat equation and thus the physical problem.

DISCUSSION

A nonzero term on the left side of the heat equation would obviously violate the equality, indicating that the conduction is either unsteady and/or that internal generation is present.

❒ ❒ ❒

PROBLEM 2.3 **One-Dimensional Steady Conduction in a Plane Wall with Variable Area and Temperature-Dependent Thermal Conductivity**

Consider a plane wall of thickness L whose area for conduction is a function of the coordinate x. The surfaces of the wall are maintained at temperatures T_1 and T_2, where $T_1 > T_2$. Assuming one-dimensional steady conduction in the wall, use Fourier's law to derive a general expression for the heat transfer if thermal conductivity is a function of temperature.

DIAGRAM

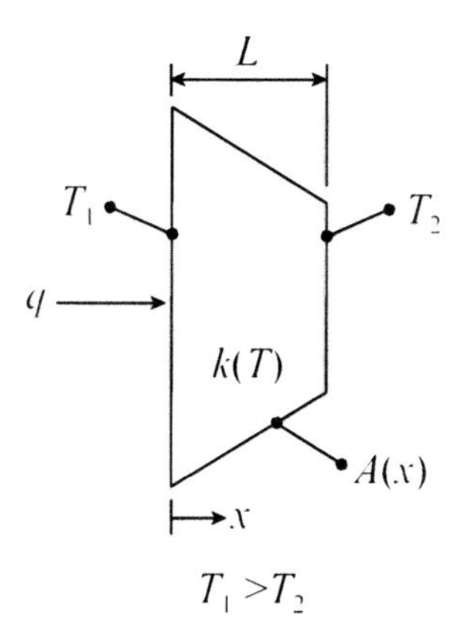

ASSUMPTIONS

1. One-dimensional steady conduction.

PROPERTIES

thermal conductivity: $k(T)$

ANALYSIS

Fourier's law is

$$q = -k(T)A(x)\frac{dT}{dx}$$

The heat transfer q is constant, so the variables can be separated as,

$$\frac{q\,dx}{A(x)} = -k(T)dT$$

Integrating the term on the left from 0 to L and the term on the right from T_1 to T_2,

$$q\int_0^L \frac{dx}{A(x)} = -\int_{T_1}^{T_2} k(T)dT$$

Thus, the heat transfer is

$$q = \frac{-\int_{T_1}^{T_2} k(T)dT}{\int_0^L \frac{dx}{A(x)}}$$

DISCUSSION

Given the functions $A(x)$ and $k(T)$ and numerical values of all quantities, the integrals could be evaluated and the heat transfer calculated. Note that if $A(x)$ and $k(T)$ are constant, the expression for q reduces to the familiar form of Fourier's law,

$$q = kA\frac{(T_1 - T_2)}{L}$$

❐ ❐ ❐

PROBLEM 2.4 **One-Dimensional Steady Conduction in a Plane Wall with Variable Area**

Consider a plane wall of thickness L = 1m whose area for conduction varies as $A(x) = A_o(1+ax)$ m^2, where A_o = 1 m^2 and a = -0.3 m^{-1}. The wall surfaces are maintained at temperatures T_1 = 100°C and T_2 = 0°C. Assuming one-dimensional steady conduction in the wall, use Fourier's law to find the heat transfer if the thermal conductivity is 10 W/m·K. Also, find the heat flux at both surfaces of the wall.

DIAGRAM

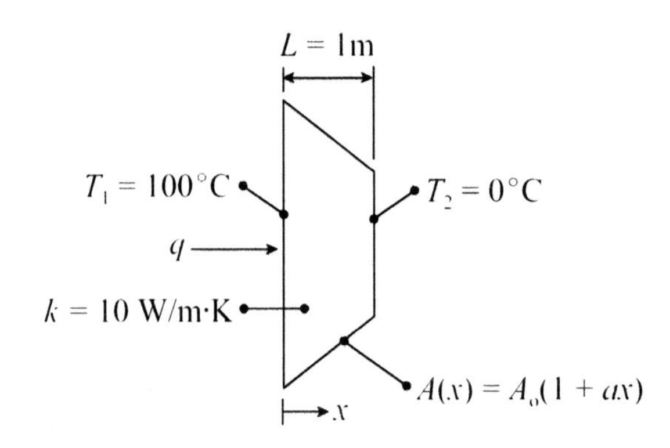

ASSUMPTIONS

1. One-dimensional steady conduction.

PROPERTIES

thermal conductivity: k

ANALYSIS

Fourier's law for this problem is

$$q = -kA(x)\frac{dT}{dx}$$

Recognizing that q is constant and separating variables,

$$\frac{-q dx}{kA_o(1+ax)} = dT$$

Integrating,

$$\frac{-q}{kA_o}\int_0^L \frac{dx}{1+ax} = \int_{T_1}^{T_2} dT$$

Evaluating the integrals and simplifying, the heat transfer is

$$q = \frac{kA_o a(T_1 - T_2)}{\ln(1+aL)}$$

Substituting values for the quantities into this relation,

$$q = \frac{(10\ \text{W/m·K})(1\ \text{m}^2)(-0.3\ \text{m}^{-1})(100 - 0)°\text{C}}{\ln[1 + (-0.3\ \text{m}^{-1})(1\ \text{m})]}$$

$$= \underline{\underline{841\ \text{W}}}$$

The heat flux at the 100°C surface is

$$q'' = \frac{q}{A_o}$$

$$q'' = \frac{841\ \text{W}}{1\ \text{m}^2}$$

$$= \underline{\underline{841\ \text{W/m}^2}}$$

The heat flux at the 0°C surface is

$$q'' = \frac{q}{A_o(1+aL)}$$

$$q'' = \frac{841 \text{ W}}{(1 \text{ m}^2)[1 + (-0.3 \text{ m}^{-1})(1 \text{ m})]}$$

$$= \underline{\underline{1202 \text{ W/m}^2}}$$

DISCUSSION

The heat transfer q is constant at all locations in the wall, but due to the decrease in area A with x, the heat flux q'' increases with x. The wall area varies from $A = 1$ m^2 at $x = 0$ to $A = 0.7$ m^2 at $x = L$, which means that the conduction is two-dimensional. Hence, the assumption is somewhat questionable.

❒ ❒ ❒

PROBLEM 2.5 **One-Dimensional Steady Conduction in a Plane Wall with Temperature-Dependent Thermal Conductivity and Variable Area**

Consider a plane wall of thickness $L = 8.0$ cm whose area for conduction varies according to the function $A(x) = A_o(1+ax)^2$ m^2, where $A_o = 1$ m^2 and $a = 5$ m^{-1}. The wall surfaces are maintained at temperatures $T_1 = 50$°C and $T_2 = 10$°C. Assuming one-dimensional steady conduction in the wall, use Fourier's law to find the heat transfer if the thermal conductivity varies with temperature as $k(T) = k_o(1+\beta T)$ W/m·K, where $k_o = 3$ W/m·K and $\beta = 0.02$ °C^{-1}.

DIAGRAM

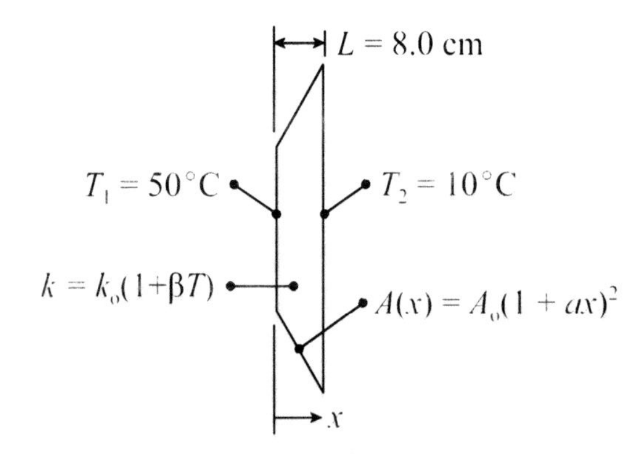

ASSUMPTIONS

1. One-dimensional steady conduction.

PROPERTIES

thermal conductivity: $k = k_o(1 + \beta T)$ W/m·K

ANALYSIS

Fourier's law is

$$q = -k(T)A(x)\frac{dT}{dx}$$

Substituting the expressions for thermal conductivity and area into Fourier's law, and separating variables,

$$\frac{q}{A_o}\frac{dx}{(1+ax)^2} = -k_o(1+\beta T)dT$$

Integrating,

$$\frac{q}{A_o}\int_0^L \frac{dx}{(1+ax)^2} = -\int_{T_1}^{T_2} k_o(1+\beta T)dT$$

Carrying out the integrations and simplifying, the heat transfer is

$$q = \frac{A_o k_o (aL+1)}{L}\left[T_1 - T_2 + \frac{\beta}{2}(T_1^2 - T_2^2)\right]$$

$$q = \frac{(1\text{ m}^2)(3\text{ W/m·K})[(5.0\text{ m}^{-1})(0.08\text{ m}) + 1]}{0.08\text{ m}}[50°\text{C} - 10°\text{C} + \frac{0.02}{2}(50^2 - 10^2)°\text{C}^2$$

$$= \underline{\underline{3360\text{ W}}}$$

DISCUSSION

The heat transfer q is constant at any location in the wall, but the heat flux q'' is not constant because the area changes with x. What is the heat flux at $x = 0$ and $x = L$?

❐ ❐ ❐

PROBLEM 2.6 Thermal Conductivities of an Anisotropic Material

An anisotropic material is a material in which the mechanical or thermal properties are dependent on coordinate direction. Consider an anisotropic solid in which the heat flux components in the x, y and z directions at the point $(-1,1,-4)$ m are 30 W/m^2, 10 W/m^2 and -25 W/m^2, respectively. If the scalar temperature field is

$$T(x, y, z) = 2x - 3xy + yz \ ^\circ\text{C}$$

what are the values of the corresponding thermal conductivities at these coordinates?

DIAGRAM

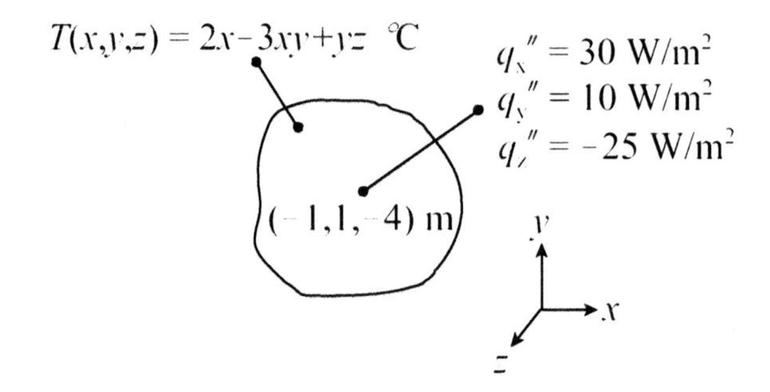

ASSUMPTIONS

1. Material is anisotropic.

PROPERTIES

k_x , k_y and k_z are to be found.

ANALYSIS

From Fourier's law, the thermal conductivities associated with conduction in each coordinate direction are defined as

$$k_x \equiv -\frac{q_x''}{\left(\dfrac{\partial T}{\partial x}\right)} \qquad k_y \equiv -\frac{q_y''}{\left(\dfrac{\partial T}{\partial y}\right)} \qquad k_z \equiv -\frac{q_z''}{\left(\dfrac{\partial T}{\partial z}\right)}$$

Evaluating the partial derivatives,

$$\frac{\partial T(-1,1,-4)}{\partial x} = 2 - 3y = 2 - 3(1) = -1\,^\circ\mathrm{C/m}$$

$$\frac{\partial T(-1,1,-4)}{\partial y} = -3x + z = -3(-1) - 4 = -1\,^\circ\mathrm{C/m}$$

$$\frac{\partial T(-1,1,-4)}{\partial z} = y = 1\,^\circ\mathrm{C/m}$$

The thermal conductivities are

$$k_x = \frac{-30\ \mathrm{W/m^2}}{-1\ ^\circ\mathrm{C/m}}$$

$$= \underline{\underline{30\ \mathrm{W/m{\cdot}K}}}$$

$$k_y = \frac{-10\ \mathrm{W/m^2}}{-1\ ^\circ\mathrm{C/m}}$$

$$= \underline{\underline{10\ \mathrm{W/m{\cdot}K}}}$$

$$k_z = \frac{-(-25\ \mathrm{W/m^2})}{1\ ^\circ\mathrm{C/m}}$$

$$= \underline{\underline{25\ \mathrm{W/m{\cdot}K}}}$$

DISCUSSION

The thermal conductivities could have also been expressed in units of W/m·°C because a temperature difference is indicated. Common anisotropic materials are fiber-wound composites used in aerospace and automotive applications.

❐ ❐ ❐

PROBLEM 2.7 **Steady Temperature Distribution in a Plane Wall with Specified Surface Temperatures**

Consider a plane wall of thickness L and constant thermal conductivity k. The left surface of the wall is maintained at temperature T_1, while the right surface is maintained at a lower temperature T_2. On a graph with T-x coordinates, sketch the temperature distribution in the wall assuming one-dimensional steady conduction. Explain the shape of the distribution.

DIAGRAM

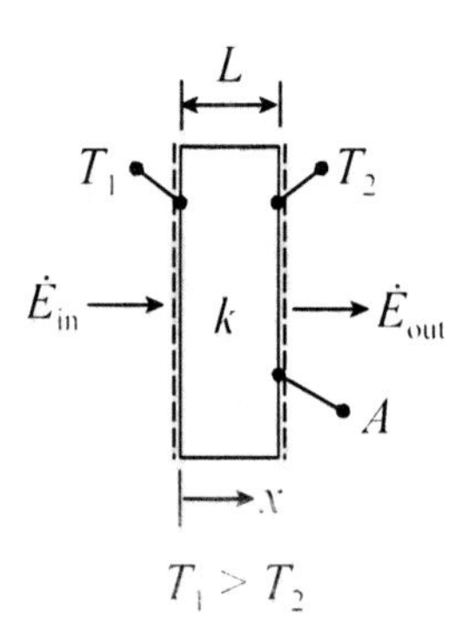

ASSUMPTIONS

1. One-dimensional steady conduction.
2. Constant properties.

PROPERTIES

thermal conductivity: k

ANALYSIS

An energy balance on the control surfaces in the diagram yields,

$$\dot{E}_{in} - \dot{E}_{out} = 0$$

The rates of energy transfer are therefore equal, and can be expressed using Fourier's law,

$$\dot{E}_{in} = \dot{E}_{out} = q_{cond} = -kA\frac{dT}{dx} = kA\frac{(T_1 - T_2)}{L}$$

Under steady conditions, the heat transfer q_{cond} is constant and positive, as are the quantities k, A and L. Hence, the temperature gradient dT/dx must be *negative*. Furthermore, the functional relationship between T and x is *linear*. Therefore, the temperature distribution is a straight line with a negative slope.

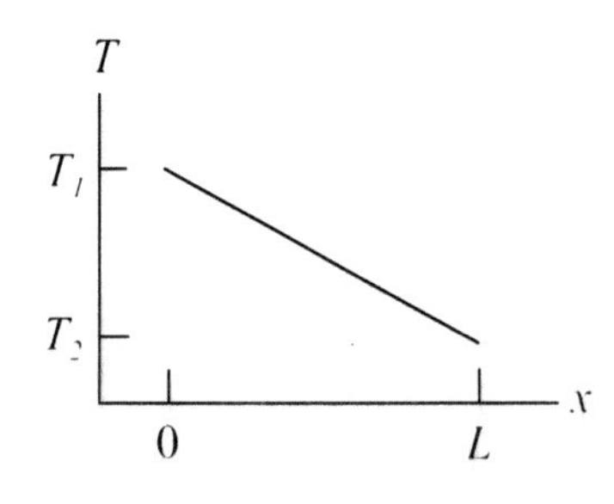

DISCUSSION

The direction of heat transfer is from left to right, which is positive according to the coordinate system chosen. It is instructive to remember that, according to the second law of thermodynamics, heat flows from a region of low temperature to a region of high temperature. Stated another way, heat flows *down* the temperature gradient, i.e., heat flows "down hill." Thus, the temperature decreases in the direction of heat flow.

PROBLEM 2.8 Steady Temperature Distribution in a Cylindrical Wall with Specified Surface Temperatures

Consider a cylindrical wall of inside radius r_1, outside radius r_2 and constant thermal conductivity k. The inside surface of the wall is maintained at temperature T_1, while the outside surface is maintained at a lower temperature T_2. On a graph with T-r coordinates, sketch the temperature distribution in the wall assuming one-dimensional steady conduction. Explain the shape of the distribution.

DIAGRAM

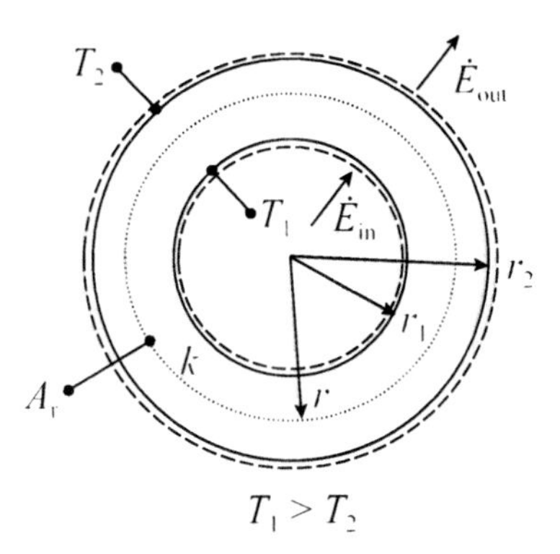

ASSUMPTIONS

1. One-dimensional steady conduction.
2. Constant properties.

PROPERTIES

thermal conductivity: k

ANALYSIS

An energy balance on the control surfaces in the diagram yields,

$$\dot{E}_{in} - \dot{E}_{out} = 0$$

The rates of energy transfer are therefore equal, and can be expressed using Fourier's law,

$$\dot{E}_{in} = \dot{E}_{out} = q_{cond} = -kA_r \frac{dT}{dr} = -k(2\pi rL)\frac{dT}{dr}$$

Under steady conditions, the heat transfer q_{cond} is constant and positive, as are the quantities k and L, thermal conductivity and length of the cylinder. The surface area A_r is not constant but increases with r, so

$$r\frac{dT}{dr} = C$$

where C is a constant. Thus, the temperature gradient is

$$\frac{dT}{dr} = \frac{C}{r}$$

The slope of the temperature distribution is inversely proportional to r, which means that as r increases, the slope dT/dr decreases.

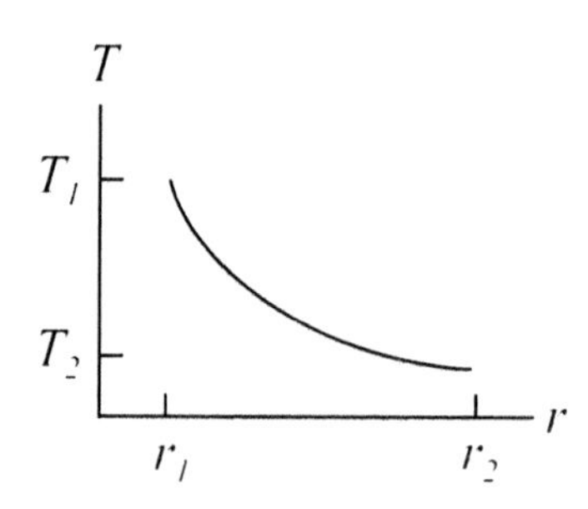

DISCUSSION

While the heat transfer q_{cond} is constant and independent of r, the heat flux q'' is not. Heat flux is

$$q'' = \frac{q_{cond}}{A_r}$$

The area A_r increases with r, so q'' decreases with r.

❐ ❐ ❐

PROBLEM 2.9 **Thermal Response Comparison of Two Metal Alloys**

Consider two metal alloys with the following thermal properties:

$\rho_1 = 7850\ \text{kg/m}^3$
$c_{p,1} = 430\ \text{J/kg·K}$
$k_1 = 65\ \text{W/m·K}$

$\rho_2 = 2800\ \text{kg/m}^3$
$c_{p,2} = 875\ \text{J/kg·K}$
$k_2 = 170\ \text{W/m·K}$

Two geometrically identical samples of these alloys are suddenly exposed to a 200°C gas and allowed to achieve thermal equilibrium. If the samples have the same initial thermal conditions, which alloy will attain a temperature of 200°C first? Justify the answer by analysis.

DIAGRAM

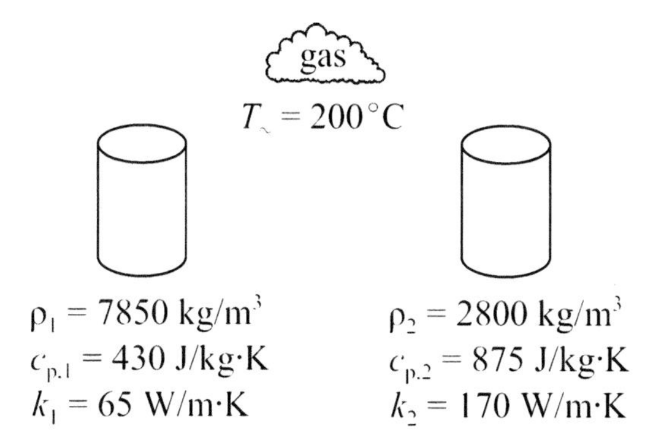

ASSUMPTIONS

1. Both samples are exposed to the same thermal environment at the same time.
2. Constant properties.

PROPERTIES

alloy 1: $\rho_1 = 7850\ \text{kg/m}^3$, $c_{p,1} = 430\ \text{J/kg·K}$, $k_1 = 65\ \text{W/m·K}$

alloy 2: $\rho_2 = 2800\ \text{kg/m}^3$, $c_{p,2} = 875\ \text{J/kg·K}$, $k_2 = 170\ \text{W/m·K}$

ANALYSIS

Thermal diffusivity α is the thermal property that reflects the ability of a material to conduct thermal energy relative to its ability to store thermal energy. Materials of large α respond rapidly to changes in the thermal environment, whereas materials of small α respond slowly to changes in the thermal environment. The thermal diffusivity of the first alloy is

$$\alpha_1 = \frac{k_1}{\rho_1 c_{p,1}}$$

$$\alpha_1 = \frac{(65 \text{ W/m·K})}{(7850 \text{ kg/m}^3)(430 \text{ J/kg·K})}$$

$$= \underline{\underline{1.926 \times 10^{-5} \text{ m}^2\text{/s}}}$$

and the thermal diffusivity of the second alloy is

$$\alpha_2 = \frac{k_2}{\rho_2 c_{p,2}}$$

$$\alpha_2 = \frac{(170 \text{ W/m·K})}{(2800 \text{ kg/m}^3)(875 \text{ J/kg·K})}$$

$$= \underline{\underline{6.939 \times 10^{-5} \text{ m}^2\text{/s}}}$$

Because the second alloy has the larger thermal diffusivity of the two alloys, it will reach 200°C first.

DISCUSSION

The first alloy has thermal properties similar to those of carbon steel, and the second alloy has thermal properties similar to those of aluminum. Thus, aluminum responds more quickly to a change in thermal environment than steel. Using the thermal properties in the appendix of a heat transfer text, the student is encouraged to calculate the thermal diffusivities for a variety of materials to obtain a general sense of how the thermal responses of these materials compare to one another.

❐ ❐ ❐

PROBLEM 2.10 **Heat Equation for One-Dimensional Steady Conduction in a Plane Wall with Specified Surface Temperatures**

Solve the heat equation for the temperature distribution in a plane wall with one-dimensional steady conduction. The temperatures of the left and right surfaces of the wall are T_1 and T_2, respectively, where $T_1 > T_2$, and the thickness of the wall is L.

DIAGRAM

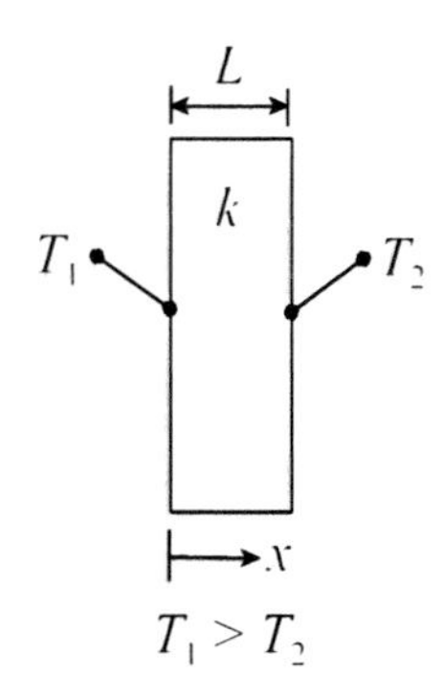

ASSUMPTIONS

1. One-dimensional steady conduction.
2. No energy generation.
3. Constant thermal conductivity.
4. Constant surface temperatures.

PROPERTIES

thermal conductivity: k

ANALYSIS

The heat equation, for constant thermal conductivity, in Cartesian coordinates is

$$\frac{\partial^2 T}{\partial x^2} + \frac{\partial^2 T}{\partial y^2} + \frac{\partial^2 T}{\partial z^2} + \frac{\dot{q}}{k} = \frac{1}{\alpha}\frac{\partial T}{\partial t}$$

For one-dimensional steady conduction with no energy generation, the heat equation reduces to

$$\frac{d^2 T}{dx^2} = 0$$

The boundary conditions are

$$T(0) = T_1 \quad , \quad T(L) = T_2$$

Integrating the differential equation twice, the general solution is

$$T(x) = C_1 x + C_2$$

where C_1 and C_2 are constants of integration. Using the boundary conditions to evaluate these constants,

$$C_1 = \frac{T_2 - T_1}{L} \quad , \quad C_2 = T_1$$

Substituting these results into the general solution, we obtain the solution for the temperature at any location x in the wall.

$$T(x) = \left(\frac{T_2 - T_1}{L}\right)x + T_1$$

The solution may be written in dimensionless form as

$$\frac{T(x) - T_1}{T_2 - T_1} = \frac{x}{L}$$

DISCUSSION

This solution is the equation for a straight line with a slope of $-(T_1 - T_2)/L$. Thus, the temperature distribution in the wall is linear. Note that the temperature distribution is independent of thermal conductivity. We can readily see that this solution is correct by substituting the boundary conditions into it. If we differentiate the expression for $T(x)$ and substitute the result into Fourier's law, we obtain

$$q = -kA\frac{dT}{dx} = kA\frac{(T_1 - T_2)}{L}$$

which is the familiar heat transfer relation for one-dimensional steady conduction in a plane wall.

❒ ❒ ❒

PROBLEM 2.11 Heat Equation for One-Dimensional Steady Conduction in a Plane Wall with Constant Temperature on One Side and Insulation on the Other

Consider a plane wall of thickness L and thermal conductivity k. One side of the wall has a constant surface temperature T_s, and the back side is perfectly insulated. By solving the heat equation, show that the temperature at all locations in the wall is T_s.

DIAGRAM

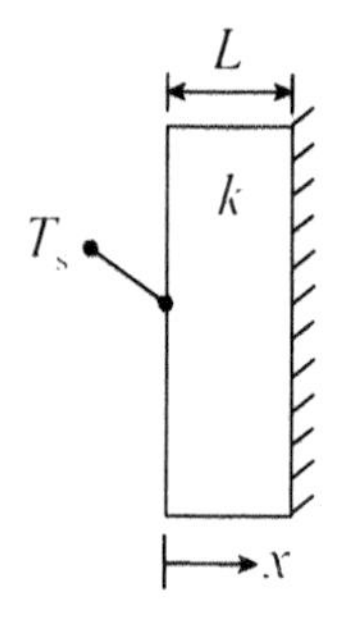

ASSUMPTIONS

1. One-dimensional steady conduction.
2. No energy generation.
3. Constant properties.

PROPERTIES

thermal conductivity: k

ANALYSIS

The heat equation for one-dimensional steady conduction in a plane wall with no internal generation is

$$\frac{d^2T}{dx^2} = 0$$

The boundary conditions are

$$T(0) = T_s \quad , \quad -k\frac{dT(L)}{dx} = 0$$

Integrating the differential equation twice, the general solution is

$$T(x) = C_1 x + C_2$$

where C_1 and C_2 are constants of integration. Using the boundary conditions to evaluate these constants,

$$C_1 = 0 \quad , \quad C_2 = T_s$$

Substituting these results into the general solution, the final solution is

$$T(x) = T_s$$

which proves that the temperature is T_s at all locations in the wall.

DISCUSSION

From simple physical reasoning we could have surmised this result. If the back side of the wall is insulated and the surface is maintained at T_s, the steady temperature of the wall at all values of x must be T_s. Note that the solution is independent of wall thickness L and thermal conductivity k.

❐ ❐ ❐

PROBLEM 2.12 **Heat Equation for One-Dimensional Steady Conduction in a Plane Wall with Constant Heat Flux on One Side and Insulation on the Other**

Consider a plane wall of thickness *L* and thermal conductivity *k*. One side is exposed to a constant heat flux q'', while the back side is perfectly insulated. By attempting to solve the heat equation, show that this problem is physically impossible.

DIAGRAM

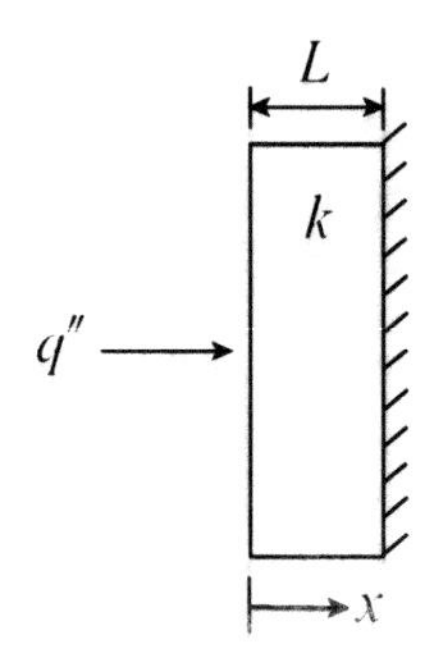

ASSUMPTIONS

1. One-Dimensional steady conduction.
2. No energy generation.
3. Heat flux is uniform.
4. No convective or radiative heat loss from the front side of the wall.
5. Constant properties.

PROPERTIES

thermal conductivity: k

ANALYSIS

The heat equation for one-dimensional steady conduction in a plane wall with no internal generation is

$$\frac{d^2T}{dx^2} = 0$$

The boundary conditions are

$$-k\frac{dT(0)}{dx} = q'' \quad , \quad -k\frac{dT(L)}{dx} = 0$$

Integrating the differential equation twice, the general solution is

$$T(x) = C_1 x + C_2$$

where C_1 and C_2 are constants of integration. Using the first boundary condition, we find the first integration constant to be

$$C_1 = \frac{q''}{k}$$

and using the second boundary condition, we find

$$C_1 = 0$$

which is a mathematical contradiction. Moreover, the second integration constant C_2 cannot be evaluated because it vanishes when the derivatives are taken.

DISCUSSION

The nonphysical character of this problem is not only shown in the mathematics, it is revealed in the diagram. Under steady conditions, a plane wall with a constant heat flux imposed on one side with perfect insulation on the other side cannot have a finite temperature. If we were to solve the unsteady version of this problem, we would find that after an infinite period of time following the application of the heat flux, the temperatures of the wall become infinitely large because the heat flux continually supplies thermal energy to the wall, but, because the back side is perfectly insulated, thermal energy is not allowed to leave.

❐ ❐ ❐

PROBLEM 2.13 **Heat Equation for One-Dimensional Steady Conduction in a Plane Wall with Constant Heat Flux on One Side and Constant Temperature on the Other**

Consider a plane wall of thickness L and thermal conductivity k. One side of the wall is exposed to a steady heat flux q'', while the back side is maintained at temperature T_s. By solving the heat equation, find the temperature distribution in the wall.

DIAGRAM

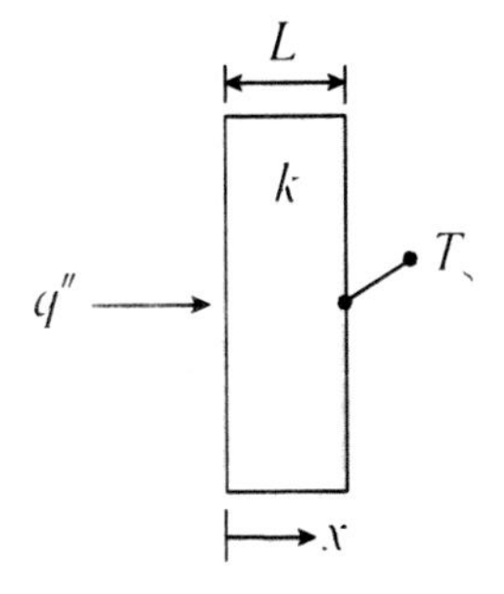

ASSUMPTIONS

1. One-Dimensional steady conduction.
2. No energy generation.
3. Heat flux is uniform.
4. No convective or radiative heat loss from the front side of the wall.
5. Constant properties.

PROPERTIES

thermal conductivity: k

ANALYSIS

The heat equation for one-dimensional steady conduction in a plane wall with no internal generation is

$$\frac{d^2T}{dx^2} = 0$$

The boundary conditions are

$$-k\frac{dT(0)}{dx} = q'' \quad , \quad T(L) = T_s$$

Integrating the differential equation twice, the general solution is

$$T(x) = C_1 x + C_2$$

where C_1 and C_2 are constants of integration. Using the boundary conditions to evaluate these constants,

$$C_1 = -\frac{q''}{k} \quad , \quad C_2 = \frac{q''}{k}L + T_s$$

Substituting these results into the general solution, the temperature distribution, expressed in dimensionless form, is

$$\frac{T(x) - T_s}{\left(\dfrac{q''L}{k}\right)} = 1 - \left(\frac{x}{L}\right)$$

DISCUSSION

Setting $x = 0$, we see that the temperature of the front side of the wall is

$$T(0) = \frac{q'' L}{k} + T_o$$

Setting $x = L$, it is clear that the temperature of the back side of the wall is

$$T(L) = T_s$$

in agreement with the second boundary condition.

❐ ❐ ❐

PROBLEM 2.14 Heat Equation for One-Dimensional Steady Conduction in a Plane Wall with Variable Thermal Conductivity

Solve the heat equation for the temperature distribution in a plane wall with one-dimensional steady conduction and variable thermal conductivity. The surfaces of the wall are maintained at temperatures 200°C and 50°C, and the wall thickness is 10 cm. The thermal conductivity varies as $k(x) = k_o\,(1 + ax)$ W/m·K, where $k_o = 12$ W/m·K and $a = -5\ \text{m}^{-1}$, and x is measured from the 200°C surface. On a graph with T-x coordinates, plot the temperature distribution and compare it with the linear temperature distribution for constant thermal conductivity.

DIAGRAM

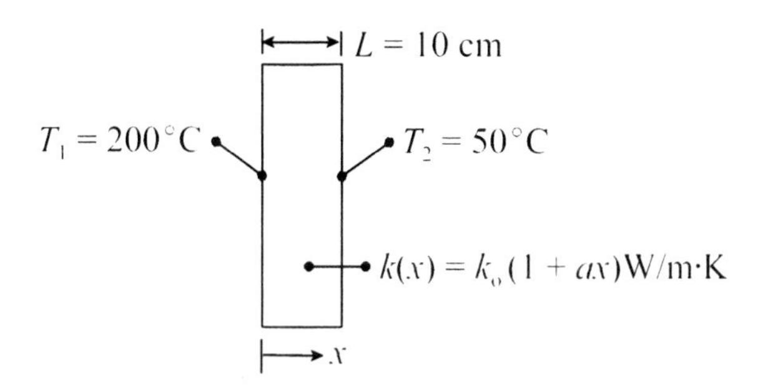

ASSUMPTIONS

1. One-dimensional steady conduction.
2. No energy generation.

PROPERTIES

thermal conductivity: $k(x) = k_o\,(1 + ax)$ W/m·K

ANALYSIS

The heat equation for one-dimensional steady conduction in a plane wall with no energy generation but with variable thermal conductivity is

$$\frac{d}{dx}\left(k(x)\frac{dT}{dx}\right) = 0$$

The boundary conditions are

$$T(0) = T_1 \quad , \quad T(L) = T_2$$

Integrating the differential equation once,

$$k(x)\frac{dT}{dx} = C_1$$

Integrating again, and recognizing that $k(x) = k_o\,(1 + ax)$,

$$\int dT = \int \frac{C_1}{k_o(1+ax)}dx$$

The general solution is

$$T(x) = \frac{C_1}{k_o a}\ln(1+ax) + C_2$$

Substituting the boundary conditions into the solution and evaluating the integration constants,

$$C_1 = \frac{(T_2 - T_1)k_o a}{\ln(1+aL)}$$

$$C_2 = T_1$$

Substituting these expressions into the solution and simplifying, the final solution is written in dimensionless form as

$$\frac{T(x) - T_1}{T_2 - T_1} = \frac{\ln(1+ax)}{\ln(1+aL)}$$

Temperature distributions for variable and constant thermal conductivities are shown in the graph below.

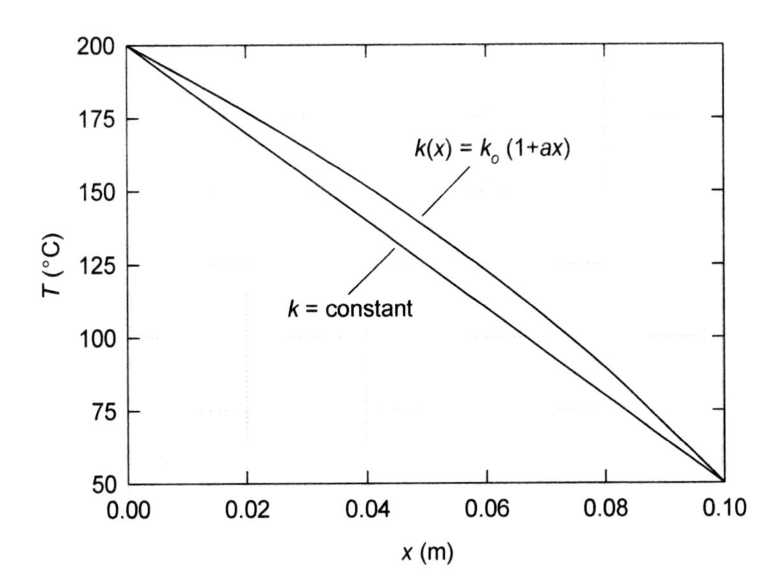

DISCUSSION

As shown in the graph, the temperature distribution for constant thermal conductivity is linear, whereas the temperature distribution for linear thermal conductivity is logarithmic. Since $a < 0$, thermal conductivity decreases with x, varying from 12 W/m·K at $x = 0$ to 6 W/m·K at $x = L$, which explains why the temperatures are higher for variable thermal conductivity than for constant thermal conductivity. However, if $a > 0$, thermal conductivity would increase with x, resulting in lower temperatures than those for a constant thermal conductivity.

❐ ❐ ❐

PROBLEM 2.15 Heat Equation for One-Dimensional Steady Conduction in a Plane Wall with Temperature-Dependent Thermal Conductivity

Solve the heat equation for the temperature distribution in a plane wall with one-dimensional steady conduction and temperature-dependent thermal conductivity. The surfaces of the wall are maintained at temperatures 200°C and 50°C, and the wall thickness is 10 cm. The thermal conductivity varies as $k(T) = k_o (1 + \beta T)$ W/m·K, where $\beta = 0.005$ °C^{-1}, and x is measured from the 200°C surface. On a graph with T-x coordinates, plot the temperature distribution and compare it with the linear temperature distribution for constant thermal conductivity.

DIAGRAM

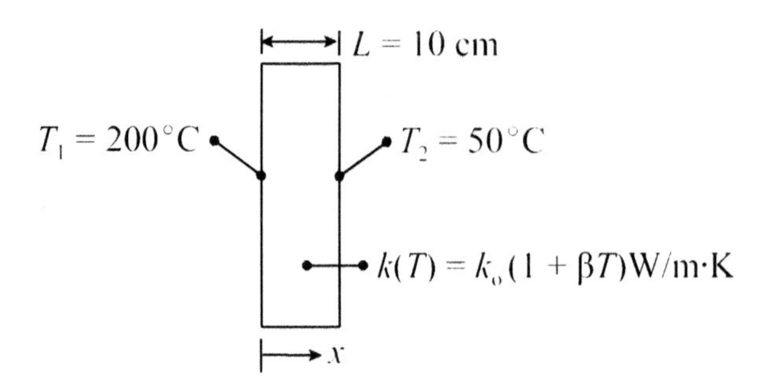

ASSUMPTIONS

1. One-dimensional steady conduction.
2. No energy generation.

PROPERTIES

thermal conductivity: $k(T) = k_o\,(1 + \beta T)$ W/m·K

ANALYSIS

The heat equation for one-dimensional steady conduction in a plane wall with no internal generation but with temperature-dependent thermal conductivity is

$$\frac{d}{dx}\left(k(T)\frac{dT}{dx}\right) = 0$$

The boundary conditions are

$$T(0) = T_1 \quad , \quad T(L) = T_2$$

Integrating the differential equation once,

$$\int d\left(k(T)\frac{dT}{dx}\right) = 0$$

$$k(T)\frac{dT}{dx} = C_1$$

Integrating again,

$$\int k_o(1+\beta T)dT = \int C_1 dx$$

$$k_o\left(T+\frac{\beta}{2}T^2\right) = C_1 x + C_2$$

Using the boundary conditions to evaluate the integration constants,

$$C_1 = \frac{k_o}{L}\left[\left(T_2+\frac{\beta}{2}T_2^2\right)-\left(T_1+\frac{\beta}{2}T_1^2\right)\right]$$

$$C_2 = k_o\left(T_1+\frac{\beta T_1^2}{2}\right)$$

Substituting these relations and simplifying, the solution is

$$T(x) = \frac{-1 \pm \left[1-2\beta\phi(x)\right]^{1/2}}{\beta}$$

where,

$$\phi(x) = \left[\left(T_1+\frac{\beta T_1^2}{2}\right)-\left(T_2+\frac{\beta T_2^2}{2}\right)\right]\left(\frac{x}{L}\right)-\left(T_1+\frac{\beta T_1^2}{2}\right)$$

In order to satisfy the boundary conditions, the root with a positive sign in front of the radical must be chosen. Thus, the final solution is

$$T(x) = \frac{-1 + \left[1-2\beta\phi(x)\right]^{1/2}}{\beta}$$

Temperature distributions for temperature-dependent and constant thermal conductivities are shown in the graph below.

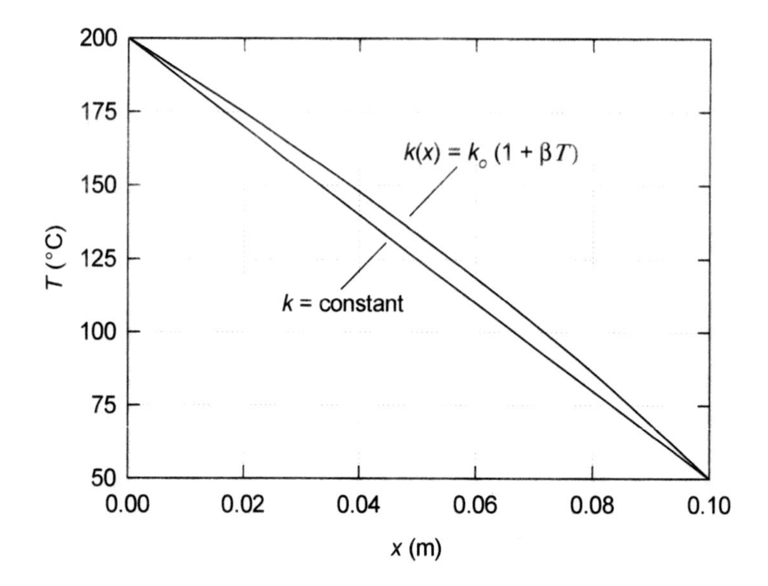

DISCUSSION

Note that the solution is independent of the constant k_o. If $\beta < 0$, the temperatures would be lower than those for a constant thermal conductivity.

❒ ❒ ❒

PROBLEM 2.16 Equivalency of Cylinder Solution to Plane Wall Solution for Very Large Radii

Consider a long hollow cylinder of inside radius r_1 and outside radius r_2. The inside and outside surfaces are maintained at temperatures T_1 and T_2, respectively. By solving the heat equation, find the one-dimensional steady temperature distribution in the cylinder. Show that this solution is equivalent to the solution for a plane wall as the cylinder radii approach infinity.

DIAGRAM

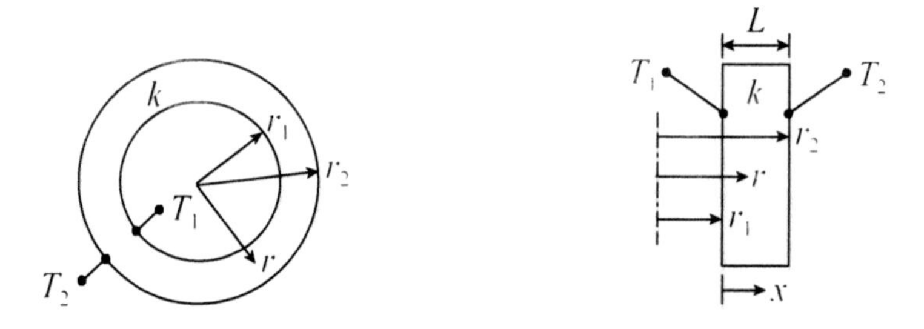

ASSUMPTIONS

1. One-dimensional steady conduction.
2. No energy generation.
3. Constant properties.

PROPERTIES

thermal conductivity: k

ANALYSIS

The heat equation in cylindrical coordinates for one-dimensional steady conduction in the radial direction is

$$\frac{1}{r}\frac{d}{dr}\left(kr\frac{dT}{dr}\right) = 0$$

The boundary conditions are

$$T(r_1) = T_1 \quad , \quad T(r_2) = T_2$$

Integrating the differential equation once,

$$\frac{dT}{dr} = \frac{C_1}{r}$$

Integrating again, the general solution is

$$T(x) = C_1 \ln r + C_2$$

where C_1 and C_2 are integration constants. Substituting the boundary conditions in to the general solution, and evaluating the integration constants,

$$C_1 = \frac{T_2 - T_1}{\ln(r_2/r_1)}$$

$$C_2 = T_1 - \frac{T_2 - T_1}{\ln(r_2/r_1)}\ln r_1$$

Substituting these constants into the general solution and simplifying, the final solution is

$$\frac{T(x) - T_1}{T_2 - T_1} = \frac{\ln(r/r_1)}{\ln(r_2/r_1)}$$

Using the second diagram, we note that

$$L = r_2 - r_1$$

$$r = x + r_1$$

Substituting these coordinate relations into the solution and simplifying,

$$\frac{T(x) - T_1}{T_2 - T_1} = \frac{\ln(1 + x/r_1)}{\ln(1 + L/r_1)}$$

We see that as $r_1 \to \infty$, the right side of this equation becomes ln(1)/ln(1) = 0/0, which is indeterminate. Applying L'Hôpital's rule by differentiating numerator and denominator with respect to r_1, and letting $r_1 \to \infty$,

$$\frac{T(x) - T_1}{T_2 - T_1} \to \frac{\left(\dfrac{1}{1 + x/r_1}\right)\left(-\dfrac{x}{r_1^2}\right)}{\left(\dfrac{1}{1 + L/r_1}\right)\left(-\dfrac{L}{r_1^2}\right)} \to \frac{(1)}{(1)}\frac{x}{L}$$

Thus, as $r_1 \to \infty$, the solution becomes

$$\frac{T(x) - T_1}{T_2 - T_1} = \frac{x}{L}$$

which is the solution for a plane wall (see Problem 2.10).

DISCUSSION

The same result can be obtained by expressing the solution in terms of r_2 instead of r_1. A related problem is to show that the solution for one-dimensional conduction in a spherical wall approaches the plane wall solution as the sphere radii approach infinity.

❒ ❒ ❒

PROBLEM 2.17 Heat Equation for One-Dimensional Steady Conduction in a Plane Wall with Constant Temperature on One Side and Convection on the Other

Using the heat equation, derive a relation for the temperature distribution in a plane wall of thickness L and constant thermal conductivity k. The conduction is one-dimensional, and the wall has a constant surface temperature T_s on one side and experiences convection on the other side where the fluid temperature is T_∞ and the heat transfer coefficient is h. Let $T_s > T_\infty$.

DIAGRAM

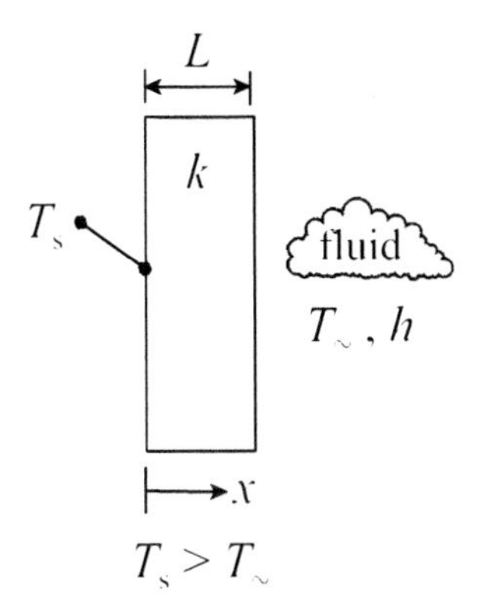

ASSUMPTIONS

1. One-dimensional steady conduction.
2. No energy generation.
3. Constant properties.

PROPERTIES

thermal conductivity: k

ANALYSIS

The heat equation for one-dimensional steady conduction in a plane wall is

$$\frac{d^2T}{dx^2} = 0$$

Integrating the differential equation twice, the general solution is

$$T(x) = C_1 x + C_2$$

where C_1 and C_2 are integration constants. The boundary conditions are

$$T(0) = T_s \quad , \quad -k\frac{dT(L)}{dx} = h\big(T(L) - T_\infty\big)$$

Using the boundary conditions to find the integration constants,

$$C_1 = \frac{h(T_\infty - T_s)}{k + hL}$$

$$C_2 = T_s$$

Substituting these results into the general solution and simplifying, the final solution is

$$\frac{T(x)-T_s}{T_\infty - T_s} = \frac{Bi}{1+Bi}\left(\frac{x}{L}\right)$$

where the dimensionless Biot number Bi is defined as

$$Bi \equiv \frac{hL}{k}$$

DISCUSSION

It is instructive to examine the behavior of the solution for very large values of the heat transfer coefficient h. As $h \to \infty$, $Bi \to \infty$, and

$$\frac{Bi}{1+Bi} \to 1$$

Thus, for very large values of h,

$$\frac{T(x)-T_s}{T_\infty - T_s} \to \frac{x}{L}$$

and at $x = L$ we have

$$T(L) \to T_\infty$$

In any heat transfer problem in which the heat transfer coefficient is very large, such as when phase changes occur, the surface of a solid in contact with a fluid is approximately the temperature of the fluid.

❒ ❒ ❒

PROBLEM 2.18 Heat Equation for One-Dimensional Steady Conduction in a Plane Wall with Constant Heat Flux on One Side and Convection on the Other

Using the heat equation, derive a relation for the temperature distribution in a plane wall of thickness L and constant thermal conductivity k. The conduction is one-dimensional, and the wall has a constant heat flux q'' on one side and experiences convection on the other side where the fluid temperature is T_∞ and the heat transfer coefficient is h.

DIAGRAM

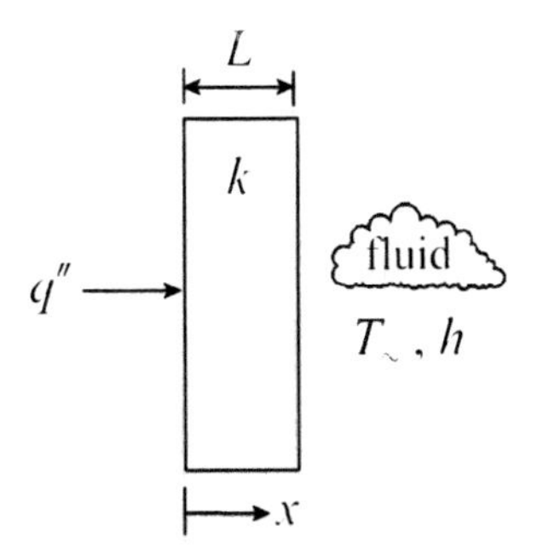

ASSUMPTIONS

1. One-dimensional steady conduction.
2. No energy generation.
3. Constant properties.

PROPERTIES

thermal conductivity: k

ANALYSIS

The heat equation for one-dimensional steady conduction in a plane wall is

$$\frac{d^2T}{dx^2} = 0$$

The boundary conditions are

$$-k\frac{dT(0)}{dx} = q'' \quad , \quad -k\frac{dT(L)}{dx} = h\left(T(L) - T_\infty\right)$$

Integrating the differential equation twice, the general solution is

$$T(x) = C_1 x + C_2$$

where C_1 and C_2 are integration constants. Using the boundary conditions to find the integration constants,

$$C_1 = -\frac{q''}{k}$$

$$C_2 = q''\left(\frac{1}{h} + \frac{L}{k}\right) + T_\infty$$

Substituting these results into the general solution and simplifying, the final solution is

$$\frac{T(x) - T_\infty}{\left(\frac{q'' L}{k}\right)} = \left(\frac{1 + Bi}{Bi}\right) - \frac{x}{L}$$

where the Biot number Bi is defined as

$$Bi \equiv \frac{hL}{k}$$

DISCUSSION

As in Problem 2.18, as $h \to \infty$, $Bi \to \infty$, so $T(L) \to T_\infty$. The Biot number is a dimensionless quantity that often appears in the analysis of conduction problems when convection is present.

❒ ❒ ❒

PROBLEM 2.19 Heat Equation for One-Dimensional Steady Conduction in a Plane Wall with Convection on Both Sides

Using the heat equation, derive a relation for the temperature distribution in a plane wall of thickness L and constant thermal conductivity k. The conduction is one-dimensional, and both sides of the wall are exposed to different convective conditions in which the fluid temperatures and heat transfer coefficients are $T_{\infty,1}$ and h_1, and $T_{\infty,2}$ and h_2. Let $T_{\infty,1} > T_{\infty,2}$.

DIAGRAM

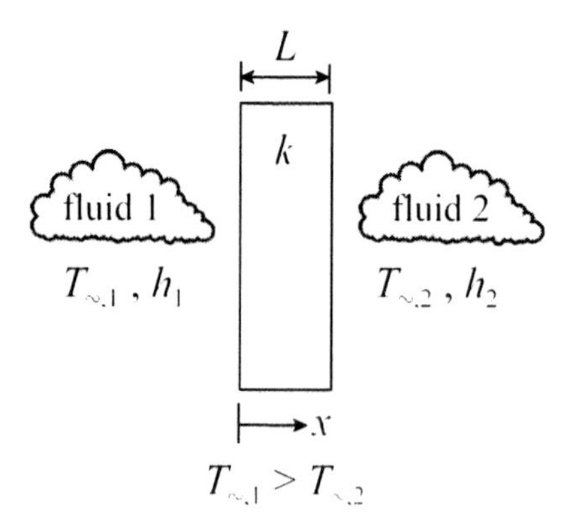

ASSUMPTIONS

1. One-dimensional steady conduction.
2. No energy generation.
3. Constant properties.

PROPERTIES

thermal conductivity: k

ANALYSIS

The heat equation for one-dimensional steady conduction in a plane wall is

$$\frac{d^2T}{dx^2} = 0$$

The boundary conditions are

$$-k\frac{dT(0)}{dx} = h_1\left(T_{\infty,1} - T(0)\right) \quad , \quad -k\frac{dT(L)}{dx} = h_2\left(T(L) - T_{\infty,2}\right)$$

Integrating the differential equation twice, the general solution is

$$T(x) = C_1 x + C_2$$

where C_1 and C_2 are integration constants. Using the boundary conditions to find the integration constants,

$$C_1 = \frac{1}{L}\left(\frac{T_{\infty,2} - T_{\infty,1}}{1 + 1/Bi_1 + 1/Bi_2}\right)$$

$$C_2 = \frac{1}{Bi_1}\left(\frac{T_{\infty,2} - T_{\infty,1}}{1 + 1/Bi_1 + 1/Bi_2}\right) + T_{\infty,1}$$

where the Biot numbers Bi_1 and Bi_2 are defined as

$$Bi_2 \equiv \frac{h_1 L}{k} \quad , \quad Bi_1 \equiv \frac{h_2 L}{k}$$

Substituting these results into the general solution and simplifying, the final solution is

$$\frac{T(x) - T_{\infty,1}}{\left(\dfrac{T_{\infty,2} - T_{\infty,1}}{1 + 1/Bi_1 + 1/Bi_2}\right)} = \frac{x}{L} + \frac{1}{Bi_1}$$

DISCUSSION

The temperature distribution in the wall is linear. The sketch below shows the temperature distribution in the wall and in the thermal boundary layers on each side of the wall. A thermal boundary layer is a narrow region near a solid surface where the fluid temperature varies from the value at the solid surface to the free stream fluid temperature. Precisely how the temperature varies in the boundary layer and the boundary layer thickness depends on the flow conditions and other factors.

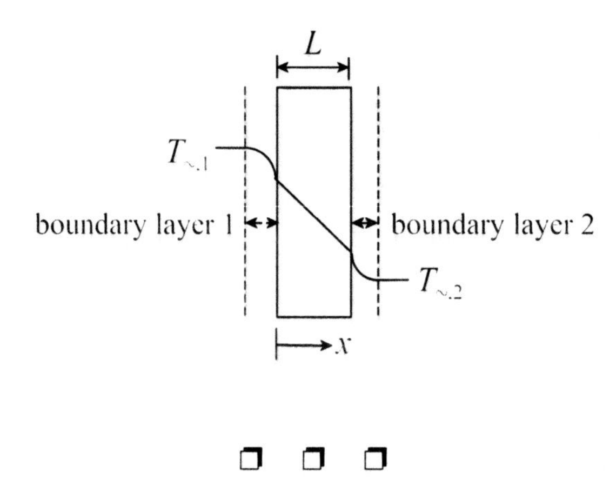

❒ ❒ ❒

PROBLEM 2.20 **Heat Equation for One-Dimensional Steady Conduction in a Plane Wall with Specified Surface Temperatures and Energy Generation**

Using the heat equation, derive a relation for the temperature distribution in a plane wall of thickness $2L$ and constant thermal conductivity k. The conduction is one-dimensional and steady, and the sides of the wall have temperatures T_1 and T_2. Energy is generated within the wall at a constant rate $\dot{q}$.

DIAGRAM

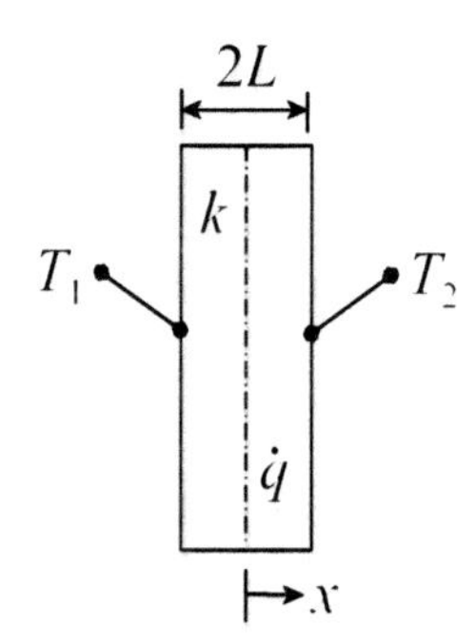

ASSUMPTIONS

1. One-dimensional steady conduction.
2. Constant energy generation.
3. Constant properties.

PROPERTIES

thermal conductivity: k

ANALYSIS

It is mathematically convenient to locate the origin at the midplane of the wall. The heat equation for one-dimensional steady conduction in a plane wall with constant thermal conductivity and internal generation is

$$\frac{d^2T}{dx^2} + \frac{\dot{q}}{k} = 0$$

The boundary conditions are

$$T(-L) = T_1 \quad , \quad T(L) = T_2$$

Integrating the differential equation twice, the general solution is

$$T(x) = -\frac{\dot{q}}{2k}x^2 + C_1 x + C_2$$

where C_1 and C_2 are integration constants. Using the boundary conditions to find the integration constants,

$$C_1 = \frac{T_2 - T_1}{2L}$$

$$C_2 = \frac{\dot{q}L^2}{2k} + \frac{T_1 + T_2}{2}$$

Substituting these results into the general solution and simplifying, the final solution is

$$T(x) = \frac{\dot{q}L^2}{2k}\left[1 - \left(\frac{x}{L}\right)^2\right] + \left(\frac{T_2 - T_1}{2}\right)\frac{x}{L} + \frac{T_1 + T_2}{2}$$

DISCUSSION

A special case of this solution is when $T_1 = T_2 = T_s$. For this case, the solution is

$$\frac{T(x) - T_s}{\left(\dfrac{\dot{q}L^2}{2k}\right)} = 1 - \left(\frac{x}{L}\right)^2$$

We can readily see that the maximum temperature occurs at $x = 0$, the midplane of the wall. At this location the temperature is

$$T(0) = T_o = \frac{\dot{q}L^2}{2k} + T_s$$

and the solution can be written as

$$\frac{T(x) - T_s}{T_o - T_s} = 1 - \left(\frac{x}{L}\right)^2$$

or,

$$\frac{T(x) - T_o}{T_s - T_o} = \left(\frac{x}{L}\right)^2$$

The temperature distribution for this special case is parabolic and symmetric about the midplane. Furthermore, the slope of the temperature distribution at the midplane is zero, so $dT/dx = 0$, which means that the midplane may be treated as an adiabatic surface. Thus, the two situations illustrated below are thermally equivalent.

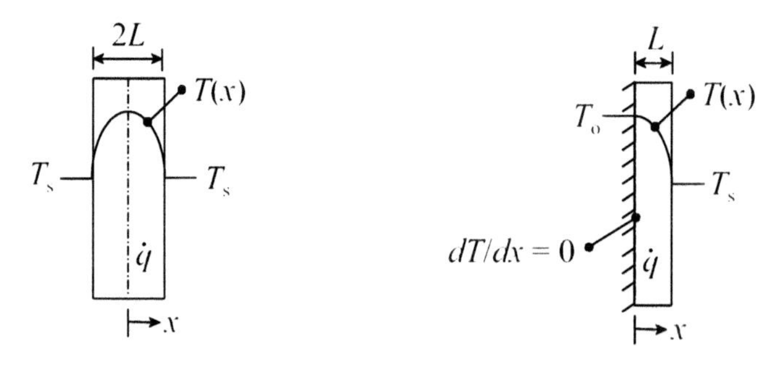

❒ ❒ ❒

PROBLEM 2.21 **Heat Equation for One-Dimensional Steady Conduction in a Spherical Wall with Specified Surface Temperatures**

Solve the heat equation for the temperature distribution in a spherical wall with one-dimensional steady conduction and no energy generation. The inside and outside radii of the wall are r_1 and r_2, respectively, and the corresponding surface temperatures are T_1 and T_2, where $T_1 > T_2$.

DIAGRAM

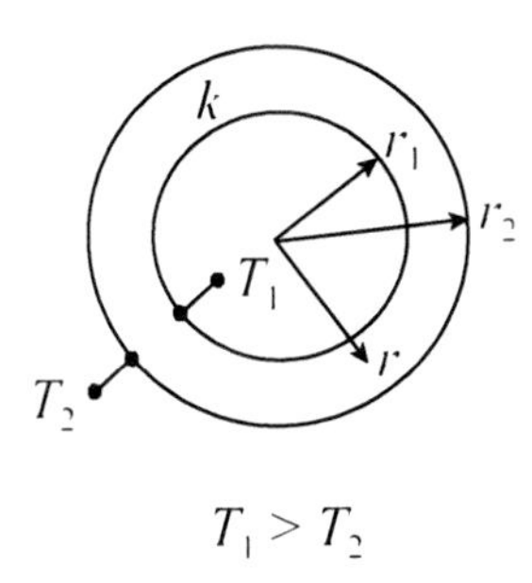

$T_1 > T_2$

ASSUMPTIONS

1. One-dimensional steady conduction.
2. No energy generation.
3. Constant thermal conductivity.
4. Constant surface temperatures.

PROPERTIES

thermal conductivity: k

ANALYSIS

If thermal conductivity is constant, the heat equation in spherical coordinates is

$$\frac{1}{r^2}\frac{\partial}{\partial r}\left(r^2\frac{\partial T}{\partial r}\right)+\frac{1}{r^2\sin^2\theta}\frac{\partial}{\partial\phi}\left(\frac{\partial T}{\partial\phi}\right)+\frac{1}{r^2\sin\theta}\frac{\partial}{\partial\theta}\left(\sin\theta\frac{\partial T}{\partial\theta}\right)+\frac{\dot{q}}{k}=\frac{1}{\alpha}\frac{\partial T}{\partial t}$$

For one-dimensional steady conduction in the radial direction with no energy generation, the heat equation reduces to

$$\frac{1}{r^2}\frac{d}{dr}\left(kr^2\frac{dT}{dr}\right)=0$$

The boundary conditions are

$$T(r_1) = T_1 \quad , \quad T(r_2) = T_2$$

Integrating the differential equation twice, the general solution is

$$T(r) = \frac{C_1}{r} + C_2$$

where C_1 and C_2 are integration constants. Using the boundary conditions to evaluate the integration constants,

$$C_1 = \frac{T_2 - T_1}{1/r_2 - 1/r_1}$$

$$C_2 = T_1 - \frac{T_2 - T_1}{\left(1/r_2 - 1/r_1\right)} \frac{1}{r_1}$$

Substituting these results into the general solution and simplifying, the final solution is

$$\frac{T(r) - T_1}{T_2 - T_1} = \frac{1 - r_1/r}{1 - r_1/r_2}$$

DISCUSSION

We can readily see that $T(r_1) = T_1$ and $T(r_2) = T_2$, as required by the boundary conditions. Spherical walls are found in thermal-fluid systems such as pressure vessels and storage tanks.

❒ ❒ ❒

PROBLEM 2.22 Heat Equation for One-Dimensional Steady Conduction in a Cylindrical Wall with Heat Flux and Convective Surface Conditions

Solve the heat equation for the temperature distribution in a cylindrical wall with one-dimensional steady conduction in the radial direction and no energy generation. The inside and outside radii of the wall are r_1 and r_2, respectively, and the thermal conductivity is k. The inside surface is exposed to a uniform heat flux q'', and the outside surface is exposed to a fluid at temperature T_∞ where the heat transfer coefficient is h. From this solution, find the temperatures of the inside and outside surfaces of the wall. Use the following values:

$r_1 = 12$ cm $\qquad r_2 = 14$ cm
$q'' = 1.8$ kW/m^2 $\qquad k = 0.5$ W/m·K
$T_\infty = 10$°C $\qquad h = 30$ W/m^2·K

DIAGRAM

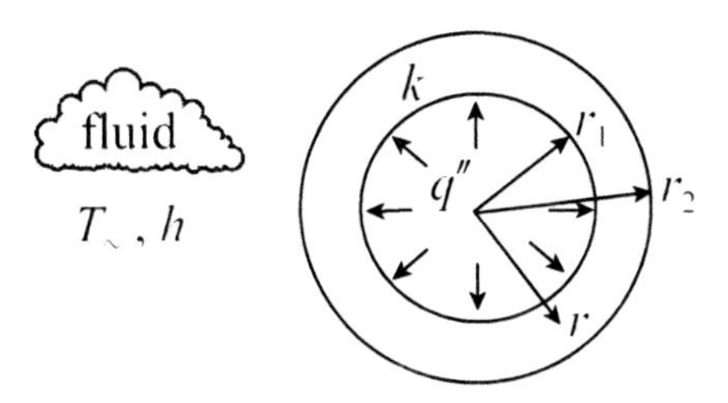

ASSUMPTIONS

1. One-dimensional steady conduction.
2. No energy generation.
3. Constant thermal conductivity.

PROPERTIES

thermal conductivity: $k = 0.5$ W/m·K

ANALYSIS

If thermal conductivity is constant, the heat equation in cylindrical coordinates is

$$\frac{1}{r}\frac{\partial}{\partial r}\left(r\frac{\partial T}{\partial r}\right)+\frac{1}{r^2}\frac{\partial}{\partial \phi}\left(\frac{\partial T}{\partial \phi}\right)+\frac{\partial}{\partial z}\left(\frac{\partial T}{\partial z}\right)+\frac{\dot{q}}{k}=\frac{1}{\alpha}\frac{\partial T}{\partial t}$$

For one-dimensional steady conduction in the radial direction with no energy generation, the heat equation reduces to

$$\frac{1}{r}\frac{d}{dr}\left(r\frac{dT}{dr}\right)=0$$

The boundary conditions are

$$-k\frac{dT(r_1)}{dx}=q" \quad , \quad -k\frac{dT(r_2)}{dx}=h\left[T(r_2)-T_\infty\right]$$

Integrating the differential equation twice, the general solution is

$$T(r)=C_1\ln r+C_2$$

where C_1 and C_2 are integration constants. Using the boundary conditions to evaluate the integration constants,

$$C_1 = -\frac{q''}{k} r_1$$

$$C_2 = q'' r_1 \left(\frac{1}{h r_2} + \frac{1}{k} \ln r_2 \right) + T_\infty$$

Substituting these results into the general solution and simplifying, the final solution is

$$\frac{T(r) - T_\infty}{\left(\frac{q'' r_1}{k} \right)} = \ln(r_2/r) + \frac{1}{Bi}$$

where the Biot number Bi is defined as

$$Bi \equiv \frac{h r_2}{k}$$

The temperature of the inside surface is

$$T(r_1) = \left(\frac{q'' r_1}{k} \right) \left(\ln(r_2/r_1) + \frac{1}{Bi} \right) + T_\infty$$

$$T(r_1) = \frac{(1.8 \times 10^3 \text{ W/m}^2)(0.12 \text{ m})}{0.5 \text{ W/m·K}} \left(\ln(0.14 \text{ m}/0.12 \text{ m}) + \frac{(0.5 \text{ W/m·K})}{(30 \text{ W/m}^2\text{·K})(0.14 \text{ m})} \right) + 10°\text{C}$$

$$= \underline{\underline{128°\text{C}}}$$

The temperature of the outside surface is

$$T(r_2) = \left(\frac{q'' r_1}{k} \right) \frac{1}{Bi} + T_\infty = \frac{q'' r_1}{h r_2} + T_\infty$$

$$T(r_2) = \frac{(1.8 \times 10^3 \text{ W/m}^2)(0.12 \text{ m})}{(30 \text{ W/m}^2\text{·K})(0.14 \text{ m})} + 10°\text{C}$$

$$= \underline{\underline{61.4°\text{C}}}$$

DISCUSSION

The temperature distribution in the wall is logarithmic. Our calculated temperatures provide a way to check the physics of the problem in the respect that the inside surface where the heat flux is applied is hotter than the outside surface in contact with the fluid. Cylindrical walls with an applied heat flux on the inside surface and convection on the outside surface correspond to a number of thermal systems such as electrical wire insulation and nuclear fuel rod casings.

❒ ❒ ❒

PROBLEM 2.23 Heat Equation for One-Dimensional Steady Conduction in a Spherical Wall with Convection on Both Surfaces

Solve the heat equation for the temperature distribution in a spherical wall with one-dimensional steady conduction and no energy generation. The inside and outside radii of the wall are r_1 and r_2, respectively. The inside surface is exposed to a fluid at temperature $T_{\infty,1}$ where the heat transfer coefficient is h_1, and the outside surface is exposed to a fluid at temperature $T_{\infty,2}$ where the heat transfer coefficient is h_2. The thermal conductivity of the wall is k. Let $T_{\infty,1} > T_{\infty,2}$.

DIAGRAM

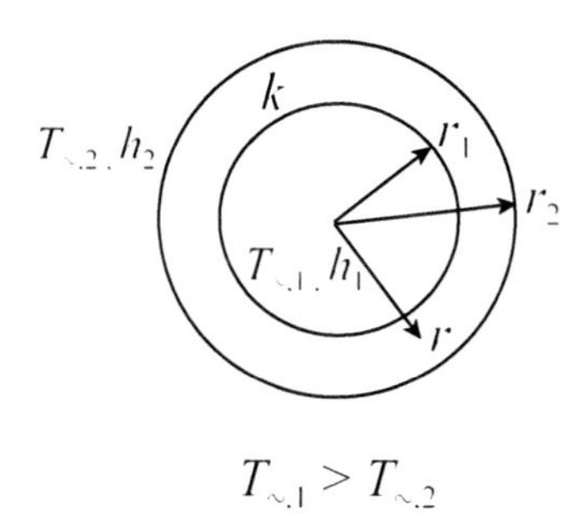

$T_{\infty,1} > T_{\infty,2}$

ASSUMPTIONS

1. One-dimensional steady conduction.
2. No energy generation.
3. Constant thermal conductivity.

PROPERTIES

thermal conductivity: k

ANALYSIS

The heat equation for one-dimensional steady conduction in the radial direction in spherical coordinates for constant thermal conductivity and no energy generation is

$$\frac{1}{r^2}\frac{d}{dr}\left(r^2\frac{dT}{dr}\right) = 0$$

The boundary conditions are

$$-k\frac{dT(r_1)}{dr}=h\left[T_{\infty,1}-T(r_1)\right] \quad , \quad -k\frac{dT(r_2)}{dr}=h\left[T(r_2)-T_{\infty,2}\right]$$

Integrating the differential equation twice, the general solution is

$$T(r)=-\frac{C_1}{r}+C_2$$

where C_1 and C_2 are integration constants. Using the boundary conditions to evaluate the integration constants,

$$C_1=\frac{T_{\infty,1}-T_{\infty,2}}{\frac{1}{r_2}\left(1-\frac{1}{Bi_2}\right)-\frac{1}{r_1}\left(1+\frac{1}{Bi_1}\right)}$$

$$C_2=\frac{(T_{\infty,1}-T_{\infty,2})}{\frac{1}{r_2}\left(1-\frac{1}{Bi_2}\right)-\frac{1}{r_1}\left(1+\frac{1}{Bi_1}\right)}\left(1+\frac{1}{Bi_1}\right)\frac{1}{r_1}$$

where the Biot numbers Bi_1 and Bi_2 are defined as

$$Bi_1\equiv\frac{hr_1}{k} \quad , \quad Bi_2\equiv\frac{hr_2}{k}$$

Substituting these relations into the general solution and simplifying, the final solution is

$$\frac{T(r)-T_{\infty,1}}{T_{\infty,2}-T_{\infty,1}}=\frac{\frac{1}{r}-\frac{1}{r_1}\left(1+\frac{1}{Bi_1}\right)}{\frac{1}{r_2}\left(1-\frac{1}{Bi_2}\right)-\frac{1}{r_1}\left(1+\frac{1}{Bi_1}\right)}$$

DISCUSSION

We can readily see that as $Bi_1\to\infty$ and $Bi_2\to\infty$, $T(r_1)\to T_{\infty,1}$ and $T(r_2)\to T_{\infty,2}$.

PROBLEM 2.24 **Heat Equation for One-Dimensional Steady Conduction in a Two-Layer Plane Wall with Specified Surface Temperatures**

A two-layer plane wall has constant surface temperatures $T_{s,1}$ and $T_{s,2}$, thicknesses L_1 and L_2, and thermal conductivities k_1 and k_2. By solving the heat equation, derive a relationship for the temperature distribution in each layer. Assume one-dimensional steady conduction in the wall. Sketch the temperature distribution in the wall if $k_1 > k_2$.

DIAGRAM

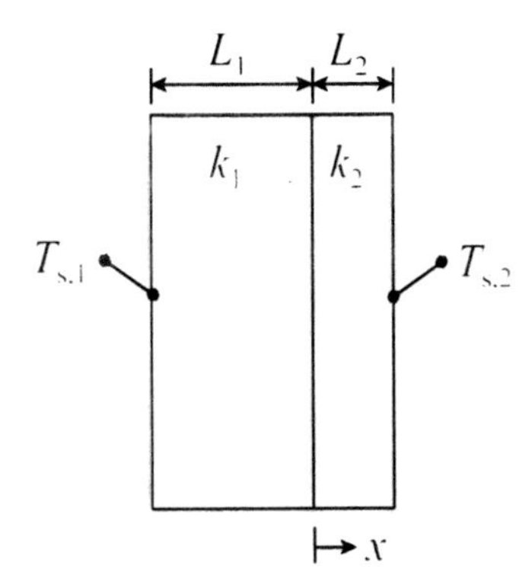

ASSUMPTIONS

1. One-dimensional steady conduction.
2. No energy generation.
3. Layers are in intimate thermal contact.
4. Constant thermal conductivities.

PROPERTIES

thermal conductivities: k_1 and k_2

ANALYSIS

Because the wall consists of two layers, an analysis of each layer is needed, but each analysis must be mathematically "connected" via the boundary conditions. It is mathematically convenient to place the origin at the interface of the two layers. The heat equations for one-dimensional steady conduction in layers 1 and 2, respectively, are

$$\frac{d^2T_1}{dx^2} = 0 \quad , \quad \frac{d^2T_2}{dx^2} = 0$$

The boundary conditions are

$$T_1(0) = T_2(0) \quad , \quad T_1(-L_1) = T_{s,1} \quad , \quad T_2(L_2) = T_{s,2}$$

$$-k_1 \frac{dT_1(0)}{dx} = -k_2 \frac{dT_2(0)}{dx}$$

The first boundary condition follows from assumption 3, the second and third boundary conditions represent the outside surface temperatures, and the fourth boundary condition follows from the continuity of heat conduction across the interface. Integrating each differential equation twice, the general solutions are

$$T_1(x) = C_1 x + C_2$$

$$T_2(x) = C_3 x + C_4$$

where C_1, C_2, C_3 and C_4 are integration constants. Substituting the boundary conditions into the general solutions, the integration constants are

$$C_1 = \frac{T_{s,2} - T_{s,1}}{\left(k_1/k_2\right)L_2 + L_1}$$

$$C_2 = C_4 = T_{s,1} + \frac{T_{s,2} - T_{s,1}}{\left(\dfrac{k_1 L_2}{k_2 L_2}\right) + 1}$$

$$C_3 = \frac{T_{s,2} - T_{s,1}}{\left(k_2/k_1\right)L_1 + L_2}$$

Substituting these results into the general solutions and simplifying, the final solutions are

$$\frac{T_1(x) - T_{s,1}}{T_{s,2} - T_{s,1}} = \frac{1}{\left(\dfrac{k_1 L_2}{k_2 L_1}\right) + 1}\left(\frac{x}{L_1} + 1\right)$$

$$\frac{T_2(x) - T_{s,1}}{T_{s,2} - T_{s,1}} = \frac{1}{\left(\dfrac{k_1 L_2}{k_2 L_1}\right) + 1}\left[\left(\frac{k_1 L_2}{k_2 L_1}\right)\frac{x}{L_2} + 1\right]$$

The temperature distribution in each layer is linear. If $k_1 > k_2$, the slope of the temperature distribution in layer 1 is steeper than that in layer 2, so the temperature distribution in the wall resembles the one shown below.

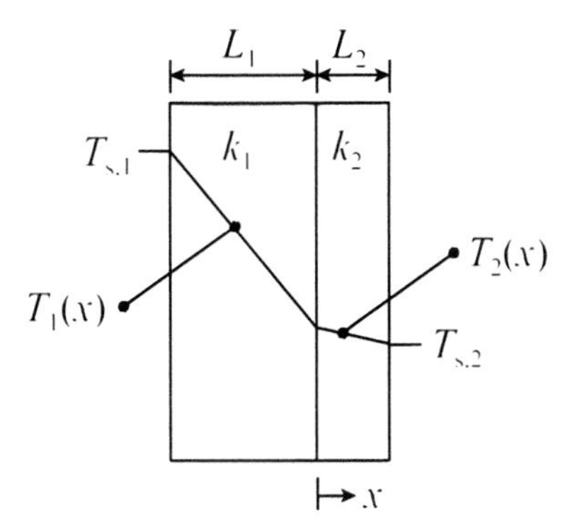

DISCUSSION

As a check on the solution, we can readily see that $T_1(-L_1) = T_{s,1}$ and that $T_2(L_2) = T_{s,2}$. We can also readily see that $T_1(0) = T_2(0)$. The approach used here could be extended to plane walls with three or more layers.

❒ ❒ ❒

PROBLEM 2.25 Energy-Generating Cylinder Buried in the Ground

A long solid cylinder is buried deep in the ground where the soil temperature is 10°C. Energy is generated uniformly within the cylinder at a rate of 3 MW/m^3. The radius and thermal conductivity of the cylinder are 6 cm and 2 W/m·K, respectively. By solving the heat equation, derive an expression for the temperature distribution in the cylinder. Using this solution, find the maximum temperature of the cylinder.

DIAGRAM

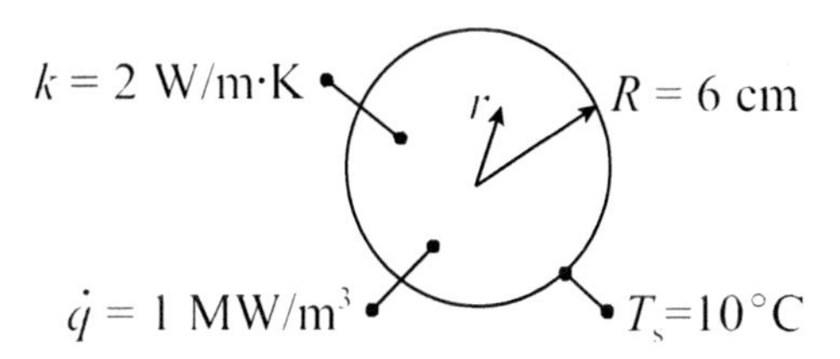

ASSUMPTIONS

1. One-dimensional steady conduction.
2. Surface temperature of the cylinder equals the soil temperature.
3. Uniform energy generation.
4. Constant thermal conductivity.

PROPERTIES

thermal conductivity: $k = 2$ W/m·K

ANALYSIS

The heat equation for one-dimensional steady conduction in the radial direction in cylindrical coordinates with energy generation and constant thermal conductivity is

$$\frac{1}{r}\frac{d}{dr}\left(r\frac{dT}{dr}\right)+\frac{\dot{q}}{k}=0$$

The boundary conditions are

$$T(R)=T_s \quad , \quad \frac{dT(0)}{dr}=0$$

Integrating the differential equation once,

$$\frac{dT}{dr}=-\frac{\dot{q}}{2k}r+\frac{C_1}{r}$$

Integrating again yields the general solution,

$$T(r)=-\frac{\dot{q}}{4k}r^2+C_1\ln r+C_2$$

where C_1 and C_2 are integration constants. At the center of cylinder, $r = 0$, the second term on the right side of the equation for the first derivative becomes undefined, so $C_1 = 0$. The second integration constant is

$$C_2=T_s+\frac{\dot{q}R^2}{4k}$$

Substituting these constants into the general solution and simplifying, the final solution for the temperature distribution is

$$\frac{T(r)-T_s}{\left(\frac{\dot{q}R^2}{4k}\right)} = 1-\left(\frac{r}{R}\right)^2$$

The maximum temperature occurs at the center of the cylinder at $r = 0$. Hence,

$$T_{\max} = T_s + \frac{\dot{q}R^2}{4k}$$

Substituting values into this expression,

$$T_{\max} = 10°\text{C} + \frac{(1 \times 10^6 \text{ W/m}^3)(0.06 \text{ m})^2}{4(2 \text{ W/m·K})}$$

$$= \underline{\underline{460°\text{C}}}$$

DISCUSSION

The maximum temperature of the cylinder is directly proportional to the surface temperature and energy generation. The temperature distribution in the cylinder is sketched below.

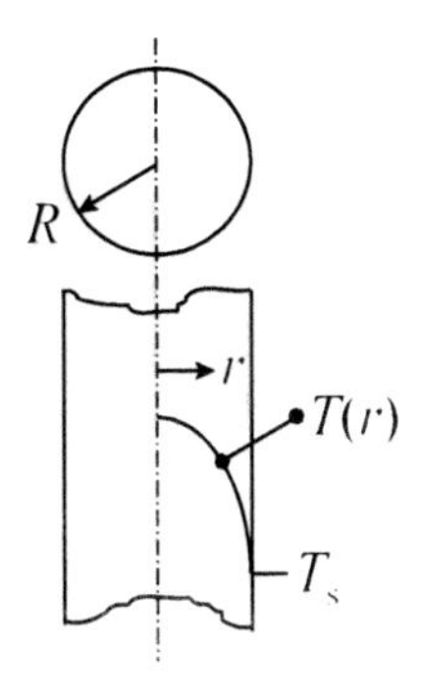

Chapter 3

One-Dimensional Steady Conduction

PROBLEM 3.1 One-Dimensional Steady Conduction in a Composite Plane Wall

A composite plane wall consists of three dissimilar layers. One surface of the wall is exposed to a fluid at 250°C where the heat transfer coefficient is 75 W/m^2·K, and the other surface is exposed to a 40°C fluid where the heat transfer coefficient is 30 W/m^2·K. Find the heat transfer through the wall, the surface temperatures, and the temperatures at the layer interfaces. For the area of the wall, let A = 1 m^2. The values of all known quantities are:

$L_A = 2.0$ cm	$k_A = 0.5$ W/m·K
$L_B = 5.0$ cm	$k_B = 8$ W/m·K
$L_C = 3.5$ cm	$k_C = 40$ W/m·K
$T_{\infty,1} = 250°C$	$h_1 = 75$W/m^2·K
$T_{\infty,2} = 40°C$	$h_2 = 30$ W/m^2·K

DIAGRAM

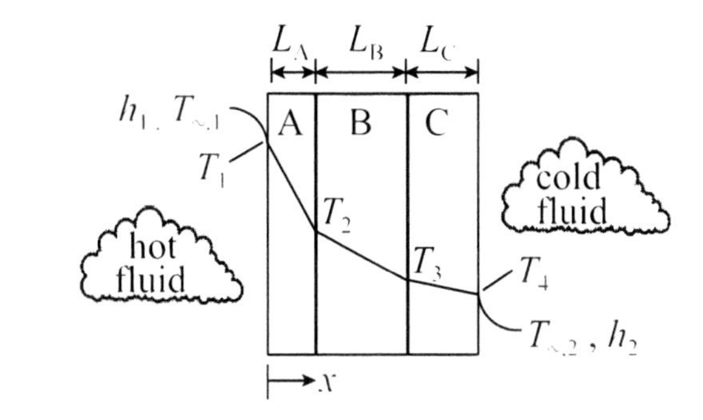

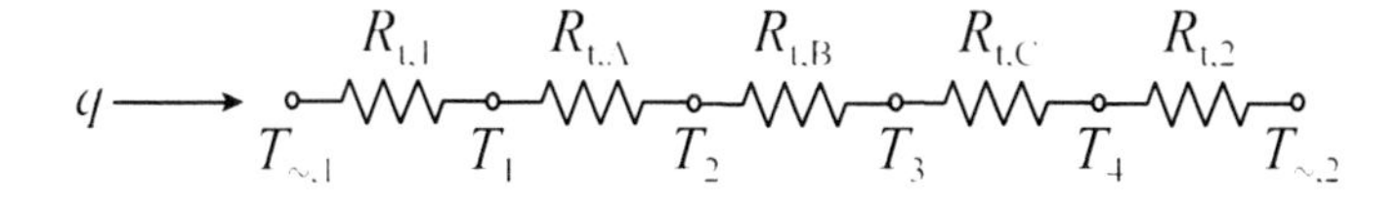

ASSUMPTIONS

1. One-dimensional steady conduction.
2. Wall layers are in intimate thermal contact.
3. Radiation is neglected.
4. Constant properties.

PROPERTIES

material A: $k_A = 0.5$ W/m·K
material B: $k_B = 8$ W/m·K
material C: $k_C = 40$ W/m·K

ANALYSIS

First, we find the thermal resistances. The thermal resistance of a plane wall layer is given by the relation

$$R_t = \frac{L}{kA}$$

Thus, we have

$$R_{t,\mathrm{A}} = \frac{(0.02\ \mathrm{m})}{(0.5\ \mathrm{W/m{\cdot}K})(1\ \mathrm{m}^2)}$$

$$= 0.040\ \mathrm{K/W}$$

$$R_{t,\mathrm{B}} = \frac{(0.05\ \mathrm{m})}{(8\ \mathrm{W/m{\cdot}K})(1\ \mathrm{m}^2)}$$

$$= 6.250 \times 10^{-3}\ \mathrm{K/W}$$

$$R_{t,\mathrm{C}} = \frac{(0.035\ \mathrm{m})}{(40\ \mathrm{W/m{\cdot}K})(1\ \mathrm{m}^2)}$$

$$= 8.750 \times 10^{-4}\ \mathrm{K/W}$$

Thermal resistance associated with convection is given by the relation

$$R_t = \frac{1}{hA}$$

Thus, we have

$$R_{t,1} = \frac{1}{(75\ \mathrm{W/m^2{\cdot}K})(1\ \mathrm{m}^2)}$$

$$= 0.01333\ \mathrm{K/W}$$

$$R_{t,2} = \frac{1}{(30\ \mathrm{W/m^2{\cdot}K})(1\ \mathrm{m}^2)}$$

$$= 0.0333\ \mathrm{K/W}$$

Adding the individual resistances, the total thermal resistance is

$$R_{\mathrm{tot}} = R_{t,1} + R_{t,A} + R_{t,B} + R_{t,C} + R_{t,2}$$

$R_{tot} = (0.01333 + 0.040 + 6.250 \times 10^{-3} + 8.750 \times 10^{-4} + 0.0333)$ K/W

$= 0.0938$ K/W

Using the electrical analogy for heat conduction, the heat transfer through the wall is given by the relation

$$q = \frac{T_{\infty,1} - T_{\infty,2}}{R_{tot}}$$

The heat transfer is

$$q = \frac{(250 - 40)\text{K}}{0.0938 \text{ K/W}}$$

$= \underline{\underline{2239 \text{ W}}}$

Because the heat transfer is the same in the wall layers and the boundary layers, we can find the surface temperatures and the interface temperatures by using the electrical analogy for each layer.

$$q = \frac{T_{\infty,1} - T_1}{R_{t,1}} = \frac{T_1 - T_2}{R_{t,A}} = \frac{T_2 - T_3}{R_{t,B}} = \frac{T_3 - T_4}{R_{t,C}} = \frac{T_4 - T_{\infty,2}}{R_{t,2}}$$

Solving for T_1,

$$T_1 = T_{\infty,1} - qR_{t,1}$$

$T_1 = 250°\text{C} - (2239 \text{ W})(0.01333 \text{ K/W})$

$= \underline{\underline{220.2°\text{C}}}$

Solving for T_2,

$$T_2 = T_1 - qR_{t,A}$$

$T_2 = 220.2°\text{C} - (2239 \text{ W})(0.040 \text{ K/W})$

$= \underline{\underline{130.6°\text{C}}}$

Solving for T_3,

$$T_3 = T_2 - qR_{t,B}$$

$T_3 = 130.6°\text{C} - (2239 \text{ W})(6.250 \times 10^{-3} \text{ K/W})$

$= \underline{\underline{116.6°\text{C}}}$

Solving for T_4,

$$T_4 = T_3 - qR_{t,C}$$

$$T_4 = 116.6°\text{C} - (2239 \text{ W})(8.750 \times 10^{-4} \text{ K/W})$$

$$= \underline{\underline{114.6°\text{C}}}$$

DISCUSSION

The second surface temperature T_4 could have been calculated using the relation

$$T_4 = qR_{t,2} + T_{\infty,2}$$

$$T_4 = (2239 \text{ W})(0.0333 \text{ K/W}) + 40°\text{C}$$

$$= \underline{\underline{114.6°\text{C}}}$$

which is in agreement with the previous result. A check of the results shows continuously decreasing temperatures from one surface of the wall to the other, which must be the case for steady heat conduction through the wall. The largest temperature difference is across layer A because this layer has the largeset thermal resistance of any layer in the system, and the smallest temperature difference is across layer C because this layer has the smallest thermal resistance of any layer in the system. It is important to note that the temperature unit in thermal resistance denotes a temperature difference, so it does not matter whether °C or K is used.

❒ ❒ ❒

PROBLEM 3.2 One-Dimensional Steady Conduction in a Composite Cylindrical Wall

A composite cylindrical wall consists of three dissimilar layers. The inside surface of the cylinder is exposed to a fluid at 300°C where the heat transfer coefficient is 180 W/m²·K, and the outside surface is exposed to a 20°C fluid where the heat transfer coefficient is 25 W/m²·K. Find the heat transfer through the wall, the inside and outside surface temperatures, and the temperatures at the layer interfaces. For the length of the cylinder, let L = 1m. The values of all known quantities are:

$r_1 = 8$ cm	$k_A = 16$ W/m·K
$r_2 = 10$ cm	$k_B = 1.5$ W/m·K
$r_3 = 15$ cm	$k_C = 0.3$ W/m·K
$r_4 = 30$ cm	
$T_{\infty,1} = 300°$C	$h_1 = 180$ W/m²·K
$T_{\infty,2} = 20°$C	$h_2 = 25$ W/m²·K

DIAGRAM

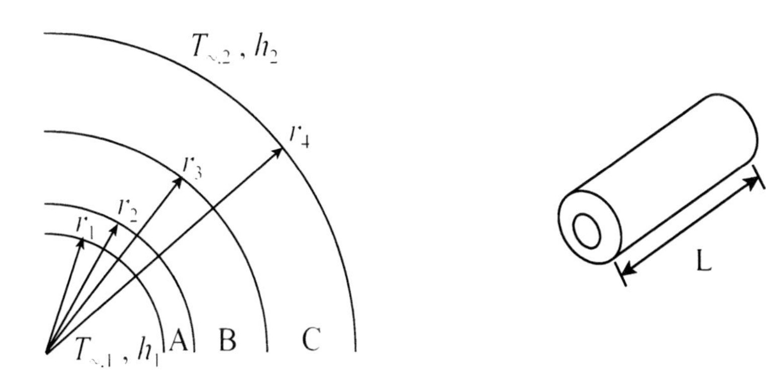

$R_{t,1}$ $R_{t,A}$ $R_{t,B}$ $R_{t,C}$ $R_{t,2}$

$q \longrightarrow$ $T_{\infty,1}$ T_1 T_2 T_3 T_4 $T_{\infty,2}$

ASSUMPTIONS

1. One-dimensional steady conduction.
2. Wall layers are in intimate thermal contact.
3. Radiation is neglected.
4. Constant properties.

PROPERTIES

material A: $k_A = 16$ W/m·K
material B: $k_B = 1.5$ W/m·K
material C: $k_C = 0.3$ W/m·K

ANALYSIS

First, we find the thermal resistances. The thermal resistance of a cylindrical wall layer is given by the relation

$$R_t = \frac{\ln(r_o/r_i)}{2\pi Lk}$$

where r_o and r_i denote the outside and inside radius, respectively, of a layer. Thus, we have

$$R_{t,A} = \frac{\ln(0.10\text{ m}/0.08\text{ m})}{2\pi(1\text{ m})(16\text{ W/m·K})}$$

$= 0.0124$ K/W

$$R_{t,B} = \frac{\ln(0.15 \text{ m}/0.10 \text{ m})}{2\pi(1 \text{ m})(1.5 \text{ W/m·K})}$$

$= 0.0430$ K/W

$$R_{t,C} = \frac{\ln(0.30 \text{ m}/0.15 \text{ m})}{2\pi(1 \text{ m})(0.3 \text{ W/m·K})}$$

$= 0.3677$ K/W

Thermal resistance associated with convection is given by the relation

$$R_t = \frac{1}{2\pi rLh}$$

Thus, we have

$$R_{t,1} = \frac{1}{2\pi(0.08 \text{ m})(1 \text{ m})(180 \text{ W/m}^2\text{·K})}$$

$= 0.0111$ K/W

$$R_{t,2} = \frac{1}{2\pi(0.30 \text{ m})(1 \text{ m})(25 \text{ W/m}^2\text{·K})}$$

$= 0.0212$ K/W

Adding the individual resistances, the total thermal resistance is

$$R_{\text{tot}} = R_{t,1} + R_{t,A} + R_{t,B} + R_{t,C} + R_{t,2}$$

$R_{\text{tot}} = (0.0111 + 0.0124 + 0.0430 + 0.3677 + 0.0212)$ K/W

$= 0.4554$ K/W

Using the electrical analogy for heat conduction, the heat transfer through the wall is given by the relation

$$q = \frac{T_{\infty,1} - T_{\infty,2}}{R_{tot}}$$

The heat transfer is

$$q = \frac{(300 - 20)\text{K}}{0.4554 \text{ K/W}}$$

$= \underline{\underline{614.8 \text{ W}}}$

Because the heat transfer is the same in the wall layers and the boundary layers, we can find the surface temperatures and the interface temperatures by using the electrical analogy for each layer.

$$q = \frac{T_{\infty,1} - T_1}{R_{t,1}} = \frac{T_1 - T_2}{R_{t,A}} = \frac{T_2 - T_3}{R_{t,B}} = \frac{T_3 - T_4}{R_{t,C}} = \frac{T_4 - T_{\infty,2}}{R_{t,2}}$$

Solving for T_1,

$$T_1 = T_{\infty,1} - qR_{t,1}$$

$T_1 = 300°\text{C} - (614.8 \text{ W})(0.0111 \text{ K/W})$

$= \underline{\underline{293.2°\text{C}}}$

Solving for T_2,

$$T_2 = T_1 - qR_{t,A}$$

$T_2 = 293.2°\text{C} - (614.8 \text{ W})(0.0124 \text{ K/W})$

$= \underline{\underline{285.6°\text{C}}}$

Solving for T_3,

$$T_3 = T_2 - qR_{t,B}$$

$T_3 = 285.6°\text{C} - (614.8 \text{ W})(0.0430 \text{ K/W})$

$= \underline{\underline{259.2°\text{C}}}$

Solving for T_4,

$$T_4 = T_3 - qR_{t,C}$$

$T_4 = 259.2°\text{C} - (614.8 \text{ W})(0.3677 \text{ K/W})$

$= \underline{\underline{33.1°\text{C}}}$

DISCUSSION

The second surface temperature T_4 could have been calculated using the relation

$$T_4 = qR_{t,2} + T_{\infty,2}$$

$T_4 = (614.8 \text{ W})(0.0212 \text{ K/W}) + 20°\text{C}$

$= \underline{\underline{33.0°\text{C}}}$

which, within roundoff errors, agrees with the previous result. A check of the results shows continuously decreasing temperatures from the inside surface to the outside surface, which must be the case for steady heat conduction through the cylinder wall. The largest temperature difference is across layer C because this layer has the largest thermal resistance of any layer in the system.

❐ ❐ ❐

PROBLEM 3.3 One-Dimensional Steady Conduction in a Composite Spherical Wall

A composite spherical wall consists of three dissimilar layers. The inside surface of the sphere is exposed to a fluid at 100°C where the heat transfer coefficient is 550 W/m²·K, and the outside surface is exposed to a 0°C fluid where the heat transfer coefficient is 60 W/m²·K. Find the heat transfer through the wall, the inside and outside surface temperatures, and the temperatures at the layer interfaces. The values of all known quantities are:

$r_1 = 20$ cm	$k_A = 2.5$ W/m·K
$r_2 = 30$ cm	$k_B = 0.30$ W/m·K
$r_3 = 55$ cm	$k_C = 0.75$ W/m·K
$r_4 = 90$ cm	
$T_{\infty,1} = 100$°C	$h_1 = 550$ W/m²·K
$T_{\infty,2} = 0$°C	$h_2 = 60$ W/m²·K

DIAGRAM

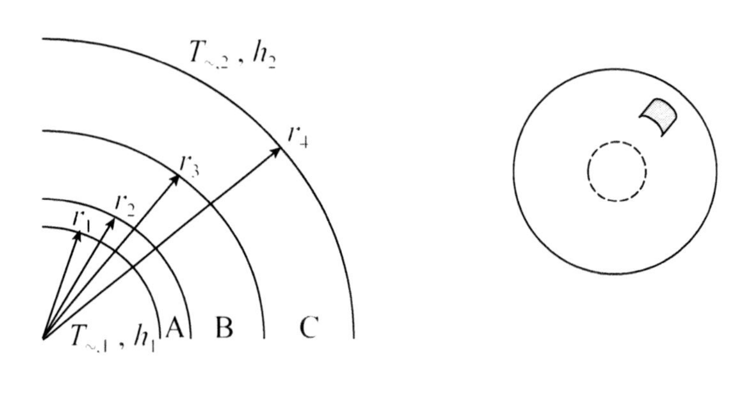

ASSUMPTIONS

1. One-dimensional steady conduction.
2. Wall layers are in intimate thermal contact.
3. Radiation is neglected.
4. Constant properties.

PROPERTIES

material A: $k_A = 2.5$ W/m·K
material B: $k_B = 0.30$ W/m·K
material C: $k_C = 0.75$ W/m·K

ANALYSIS

First, we find the thermal resistances. The thermal resistance of a spherical wall layer is given by the relation

$$R_t = \frac{1}{4\pi k}\left(\frac{1}{r_i} - \frac{1}{r_o}\right)$$

where r_i and r_o denote the inside and outside radius, respectively, of a layer. Thus, we have

$$R_{t,A} = \frac{1}{4\pi\,(2.5\ \text{W/m·K})}\,(1/0.20\ \text{m} - 1/0.30\ \text{m})$$

$$= 0.0531\ \text{K/W}$$

$$R_{t,B} = \frac{1}{4\pi\,(0.30\ \text{W/m·K})}\,(1/0.30\ \text{m} - 1/0.55\ \text{m})$$

$$= 0.4019\ \text{K/W}$$

$$R_{t,C} = \frac{1}{4\pi\,(0.75\ \text{W/m·K})}\,(1/0.55\ \text{m} - 1/0.90\ \text{m})$$

$$= 0.0750\ \text{K/W}$$

Thermal resistance associated with convection is given by the relation

$$R_t = \frac{1}{4\pi r^2 h}$$

Thus, we have

$$R_{t,1} = \frac{1}{4\pi\,(0.20\text{ m})^2\,(550\text{ W/m}^2\cdot\text{K})}$$

$$= 3.617 \times 10^{-3}\text{ K/W}$$

$$R_{t,2} = \frac{1}{4\pi\,(0.90\text{ m})^2\,(60\text{ W/m}^2\cdot\text{K})}$$

$$= 1.637 \times 10^{-3}\text{ K/W}$$

Adding the individual resistances, the total thermal resistance is

$$R_{\text{tot}} = R_{t,1} + R_{t,A} + R_{t,B} + R_{t,C} + R_{t,2}$$

$$R_{\text{tot}} = (3.617 \times 10^{-3} + 0.0531 + 0.4019 + 0.0750 + 1.637 \times 10^{-3})\text{ K/W}$$

$$= 0.5353\text{ K/W}$$

Using the electrical analogy for heat conduction, the heat transfer through the wall is given by the relation

$$q = \frac{T_{\infty,1} - T_{\infty,2}}{R_{tot}}$$

The heat transfer is

$$q = \frac{(100 - 0)\text{K}}{0.5353\text{ K/W}}$$

$$= \underline{\underline{186.8\text{ W}}}$$

Because the heat transfer is the same in the wall layers and the boundary layers, we can find the surface temperatures and the interface temperatures by using the electrical analogy for each layer.

$$q = \frac{T_{\infty,1} - T_1}{R_{t,1}} = \frac{T_1 - T_2}{R_{t,A}} = \frac{T_2 - T_3}{R_{t,B}} = \frac{T_3 - T_4}{R_{t,C}} = \frac{T_4 - T_{\infty,2}}{R_{t,2}}$$

Solving for T_1,

$$T_1 = T_{\infty,1} - qR_{t,1}$$

$$T_1 = 100°\text{C} - (186.8\text{ W})(3.617 \times 10^{-3}\text{ K/W})$$

$$= \underline{\underline{99.3°\text{C}}}$$

Solving for T_2,

$$T_2 = T_1 - qR_{t,A}$$

$T_2 = 99.3°\text{C} - (186.8 \text{ W})(0.0531 \text{ K/W})$

$= \underline{\underline{89.4°\text{C}}}$

Solving for T_3,

$$T_3 = T_2 - qR_{t,B}$$

$T_3 = 89.4°\text{C} - (186.8 \text{ W})(0.4019 \text{ K/W})$

$= \underline{\underline{14.3°\text{C}}}$

Solving for T_4,

$$T_4 = T_3 - qR_{t,C}$$

$T_4 = 14.3°\text{C} - (186.8 \text{ W})(0.0750 \text{ K/W})$

$= \underline{\underline{0.29°\text{C}}}$

DISCUSSION

The largest temperature difference is across layer B because this layer has the largest thermal resistance of any layer in the system.

❐ ❐ ❐

PROBLEM 3.4 Cost of Heat Loss Through a Residential Frame Wall

A typical frame wall used in residential construction consists of regularly-spaced fir studs with plaster board on the inside and sheathing and hardboard siding on the outside, as shown in the diagram below. To reduce heat loss, spaces between the studs are occupied with fiberglass batts. For the month of January, the average outdoor air temperature is 5°C. The total wall area of an electrically-heated frame house is 100 m^2, and the homeowner pays $0.085/ kWh for electricity. If the homeowner keeps his thermostat set at 22°C, what is the heat transfer through the walls of his house? How much does the homeowner pay during the month of January for the heat loss through the walls? The values of all known quantities are:

$L_A = 1.27$ cm , $L_B = L_C = 8.9$ cm , $L_D = 1.42$ cm , $L_E = 1.0$ cm
$W_A = W_D = W_E = 40.6$ cm , $W_B = 3.8$ cm
$k_A = 0.17$ W/m·K , $k_B = 0.12$ W/m·K , $k_C = 0.04$ W/m·K , $k_D = 0.05$ W/m·K
$k_E = 0.09$ W/m·K
$T_{\infty,1} = 22°\text{C}$, $h_1 = 11$ W/m^2·K , $T_{\infty,2} = 5°\text{C}$, $h_2 = 18$ W/m^2·K

DIAGRAM

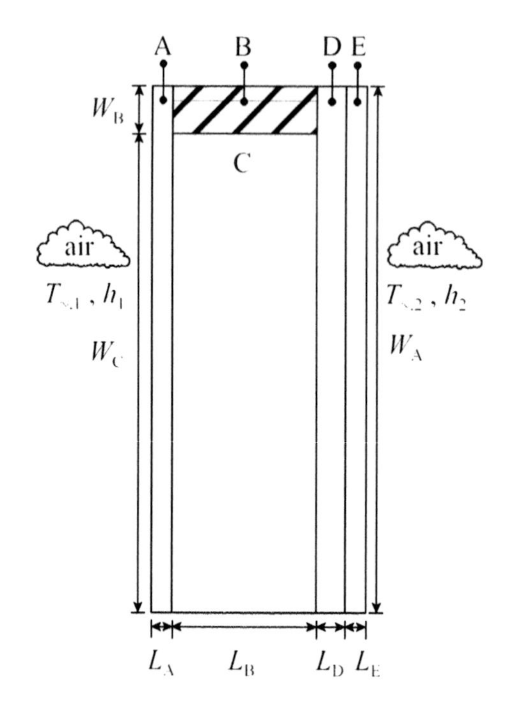

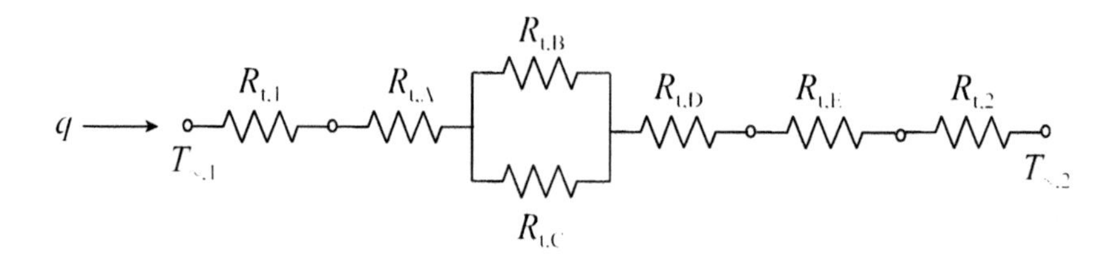

ASSUMPTIONS

1. One-dimensional steady conduction.
2. Wall members are in intimate thermal contact.
3. Radiation is neglected.
4. Constant properties.

PROPERTIES

plaster board (A): $k_A = 0.17$ W/m·K
fir stud (B): $k_B = 0.12$ W/m·K
fiberglass batt (C): $k_C = 0.04$ W/m·K
sheathing (D): $k_D = 0.05$ W/m·K
hardboard siding (E): $k_E = 0.09$ W/m·K

ANALYSIS

First, we find the thermal resistances. The thermal resistance of a plane wall layer is given by the relation

$$R_t = \frac{L}{kA}$$

where the area is given by $A = W\,H$, the product of width and height. For the height of the wall layers, we arbitrarily let $H = 1$ m. Thus, we have

$$R_{t,\mathrm{A}} = \frac{(0.0127\ \mathrm{m})}{(0.17\ \mathrm{W/m{\cdot}K})(0.406\ \mathrm{m})(1\ \mathrm{m})}$$

$$= 0.1840\ \mathrm{K/W}$$

$$R_{t,\mathrm{B}} = \frac{(0.089\ \mathrm{m})}{(0.12\ \mathrm{W/m{\cdot}K})(0.038\ \mathrm{m})(1\ \mathrm{m})}$$

$$= 19.52\ \mathrm{K/W}$$

$$R_{t,\mathrm{C}} = \frac{(0.089\ \mathrm{m})}{(0.04\ \mathrm{W/m{\cdot}K})(0.368\ \mathrm{m})(1\ \mathrm{m})}$$

$$= 6.046\ \mathrm{K/W}$$

$$R_{t,\mathrm{D}} = \frac{(0.0142\ \mathrm{m})}{(0.05\ \mathrm{W/m{\cdot}K})(0.406\ \mathrm{m})(1\ \mathrm{m})}$$

$$= 0.6995\ \mathrm{K/W}$$

$$R_{t,\mathrm{E}} = \frac{(0.01\ \mathrm{m})}{(0.09\ \mathrm{W/m{\cdot}K})(0.406\ \mathrm{m})(1\ \mathrm{m})}$$

$$= 0.2737\ \mathrm{K/W}$$

Thermal resistance associated with convection is given by the relation

$$R_t = \frac{1}{hA}$$

Thus, we have

$$R_{t,1} = \frac{1}{(11\ \mathrm{W/m^2{\cdot}K})(0.406\ \mathrm{m})(1\ \mathrm{m})}$$

$$= 0.2239\ \mathrm{K/W}$$

$$R_{t,2} = \frac{1}{(18\ \text{W/m}^2\cdot\text{K})(0.406\ \text{m})(1\ \text{m})}$$

$$= 0.1368\ \text{K/W}$$

As shown in the diagram, all the thermal resistors, except those associated with the fir stud (B) and fiberglass batt (C), which are connected in parallel, are connected in series. Thus, adding the individual resistances, the total thermal resistance of the wall section considered is

$$R_{\text{tot}} = R_{t,1} + R_{t,A} + \frac{1}{\dfrac{1}{R_{t,B}} + \dfrac{1}{R_{t,C}}} + R_{t,D} + R_{t,E} + R_{t,2}$$

$$R_{\text{tot}} = (0.2239 + 0.1840 + 4.6162 + 0.6995 + 0.2737 + 0.1368)\ \text{K/W}$$

$$= 6.134\ \text{K/W}$$

Using the electrical analogy for heat conduction, the heat transfer through the wall section is given by the relation

$$q = \frac{T_{\infty,1} - T_{\infty,2}}{R_{\text{tot}}}$$

The heat transfer is

$$q = \frac{(22 - 5)\text{K}}{6.134\ \text{K/W}}$$

$$= 2.771\ \text{W}$$

This result is the heat loss for a section of the wall measuring 0.406 m × 1 m = 0.406 m^2. The heat loss for all the walls of the frame house is

$$q_{\text{tot}} = \frac{(100\ \text{m}^2)}{0.406\ \text{m}^2}(2.771\ \text{W})$$

$$= \underline{\underline{682.5\ \text{W}}}$$

There are 31 days in the month of January, so the homeowner pays the following amount for the heat loss through the walls of his house:

$$\text{cost} = 682.5\ \text{W} \times \frac{1\ \text{kW}}{1000\ \text{W}} \times \frac{24\ \text{h}}{\text{day}} \times \frac{\$0.085}{\text{kWh}} \times 31\ \text{day}$$

$$= \underline{\underline{\$43.16}}$$

DISCUSSION

Heat is also lost by conduction through the roof, windows, doors, basement foundation walls and basement floors. Another significant source of heat loss for a house is infiltration, thermal energy carried by air leaking to the outdoors through cracks and other openings in the building envelope.

❐ ❐ ❐

PROBLEM 3.5 Temperatures of an Insulated Copper Rod with Energy Generation

A long cylindrical rod of pure copper is clad with a 1.5-cm thick layer of rigid phenolic insulation ($k = 0.03$ W/m·K). The phenolic insulation was installed by cooling the copper cylinder, inserting it into a phenolic tube, and letting the cylinder expand against the inside of the tube, producing a thermal contact conductance of 400 W/m^2·K. The passage of an electrical current in the rod generates 175 kW/m^3 of energy within the rod. Surrounding the phenolic insulation is room air at 30°C, and the heat transfer coefficient is 20 W/m^2·K. If the radius of the copper rod is 2.5 cm, find the temperature of the outside surface of the rod and the inside and outside surfaces of the phenolic insulation.

DIAGRAM

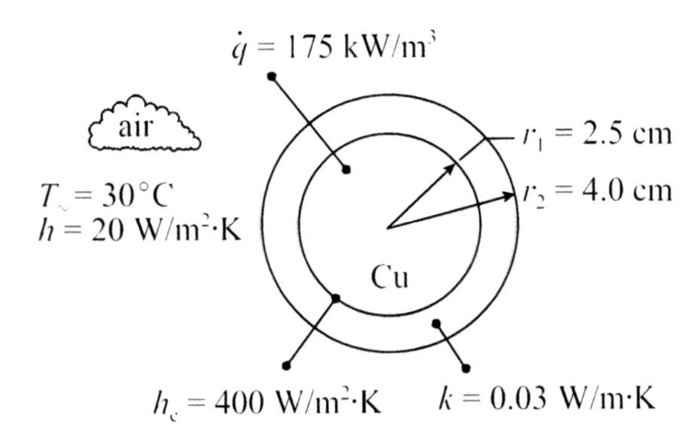

$q \rightarrow$ T_1 — $R_{t,c}$ — T_2 — $R_{t,1}$ — T_3 — $R_{t,2}$ — T_∞

ASSUMPTIONS

1. One-dimensional steady conduction.
2. Electrical energy is generated uniformly within the rod.
3. Radiation is neglected.
4. Constant properties.

PROPERTIES

phenolic insulation: $k = 0.03$ W/m·K

ANALYSIS

The thermal resistance of the phenolic insulation is given by the relation

$$R_{t,1} = \frac{\ln(r_2/r_1)}{2\pi Lk}$$

A unit length, $L = 1$ m, will be used in subsequent calculations. The thermal resistance of the phenolic insulation is

$$R_{t,1} = \frac{\ln(0.040 \text{ m}/0.025 \text{ m})}{2\pi\,(0.03 \text{ W/m·K})(1 \text{ m})}$$

$$= 2.493 \text{ K/W}$$

Thermal resistance associated with the contact of the rod and insulation is given by the relation

$$R_{t,c} = \frac{1}{2\pi r_1 L h_c}$$

$$R_{t,c} = \frac{1}{2\pi\,(0.025 \text{ m})(1 \text{ m})(400 \text{ W/m}^2\text{·K})}$$

$$= 0.0159 \text{ K/W}$$

Thermal resistance associated with convection is given by the relation

$$R_{t,2} = \frac{1}{2\pi r_2 L h}$$

$$R_{t,2} = \frac{1}{2\pi\,(0.04 \text{ m})(1 \text{ m})(20 \text{ W/m}^2\text{·K})}$$

$$= 0.1989 \text{ K/W}$$

Adding the individual resistances, the total thermal resistance is

$$R_{tot} = R_{t,C} + R_{t,1} + R_{t,2}$$

$$R_{tot} = (0.0159 + 2.493 + 0.1989) \text{ K/W}$$

$$= 2.7078 \text{ K/W}$$

The heat transfer is the energy generation multiplied by the volume of the cylinder.

$$q = \dot{q}V = \dot{q}\pi r_1^2 L$$

$q = (175 \times 10^3\ \text{W/m}^3)\ \pi\ (0.025\ \text{m})^2\ (1\ \text{m})$

$= 343.6\ \text{W}$

Using the electrical analogy for heat conduction, the heat transfer is given by the relation

$$q = \frac{T_1 - T_\infty}{R_{\text{tot}}}$$

Solving for T_1, the temperature of the outside surface of the copper rod,

$$T_1 = qR_{\text{tot}} + T_\infty$$

$T_1 = (343.6\ \text{W})(2.7078\ \text{K/W}) + 30°\text{C}$

$= \underline{\underline{960.4°\text{C}}}$

Because the heat transfer is the same in all layers, we can find the other temperatures by using the electrical analogy for each layer.

$$q = \frac{T_1 - T_2}{R_{t,C}} = \frac{T_2 - T_3}{R_{t,1}} = \frac{T_3 - T_\infty}{R_{t,2}}$$

Solving for T_2, the temperature of the inside surface of the phenolic insulation,

$$T_2 = T_1 - qR_{t,C}$$

$T_2 = 960.4°\text{C} - (343.6\ \text{W})(0.0159\ \text{K/W})$

$= \underline{\underline{954.9°\text{C}}}$

Solving for T_3, the temperature of the outside surface of the phenolic insulation,

$$T_3 = T_2 - qR_{t,1}$$

$T_3 = 954.9°\text{C} - (343.6\ \text{W})(2.493\ \text{K/W})$

$= \underline{\underline{98.3°\text{C}}}$

DISCUSSION

The temperature difference across the phenolic insulation is about 857°C, the largest temperature difference for any layer in the system. The melting point of pure copper is 1083°C, only 123°C higher than the surface temperature of the rod. Hence, the structural integrity of the rod will be compromised to some extent, which may be significant if the insulated rod is under a load.

❒ ❒ ❒

PROBLEM 3.6 Effect of Thermal Contact Resistance on Maximum Power Dissipation of an Integrated Circuit

An integrated circuit (IC) device with a foot print of 5 cm × 5 cm is mounted to a 2.0-mm thick glass-epoxy circuit board (k_b = 1 W/m·K). The back side of the circuit board is maintained at 10°C by a liquid-cooled heat sink, and the thermal contact resistance of the dry interface between the device and circuit board is 3×10^{-4} m^2·K/W. If the maximum allowable case temperature of the device is 85°C, how much power can the device dissipate? If the dry interface is replaced by a 0.02-mm layer of thermally-conductive grease (k_g = 0.2 W/m·K), how much power can the device dissipate?

DIAGRAM

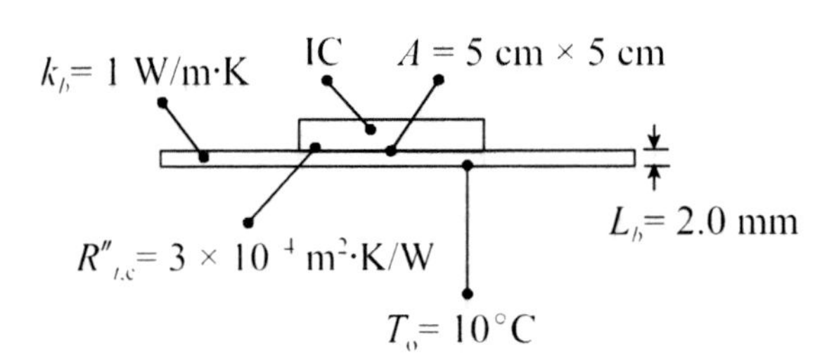

$R_{t,1}$ $R_{t,2}$

q → T_c T_1 T_o

ASSUMPTIONS

1. One-dimensional steady conduction.
2. Radiation is neglected.
3. Constant properties.

PROPERTIES

glass-epoxy board: k_b = 1 W/m·K
thermal grease: k_g = 0.2 W/m·K

ANALYSIS

The thermal contact resistance of the interface is related to its thermal resistance by the relation

$$R''_{t,c} = AR_{t,1}$$

The thermal resistances of the interface and circuit board, respectively, are

$$R_{t,1} = \frac{R''_{t,c}}{A}$$

$$R_{t,1} = \frac{3 \times 10^{-4}\ \text{m}^2\cdot\text{K/W}}{(0.05\ \text{m})^2}$$

$$= 0.120\ \text{K/W}$$

$$R_{t,2} = \frac{L_b}{k_b A}$$

$$R_{t,2} = \frac{(0.002\ \text{m})}{(1\ \text{W/m}\cdot\text{K})(0.05\ \text{m})^2}$$

$$= 0.800\ \text{K/W}$$

The total resistance is

$$R_{\text{tot}} = R_{t,1} + R_{t,2}$$

$$R_{\text{tot}} = (0.120 + 0.800)\ \text{K/W}$$

$$= 0.920\ \text{K/W}$$

The maximum power that the device can dissipate is

$$q = \frac{T_{c,\max} - T_o}{R_{\text{tot}}}$$

$$q = \frac{(85 - 10)^\circ\text{C}}{0.920\ \text{K/W}}$$

$$= \underline{\underline{81.5\ \text{W}}}$$

The thermal resistance of a 0.02-mm layer of thermal grease is

$$R_{t,1} = \frac{(2 \times 10^{-5}\ \text{K/W})}{(0.2\ \text{W/m·K})(0.05\ \text{m})^2}$$

$$= 0.040\ \text{K/W}$$

and the total resistance is

$$R_{tot} = (0.040 + 0.800)\ \text{K/W}$$

$$= 0.840\ \text{K/W}$$

so the power dissipation is

$$q = \frac{(85 - 10)°\text{C}}{0.840\ \text{K/W}}$$

$$= \underline{\underline{89.3\ \text{W}}}$$

DISCUSSION

We see that by using a thin layer of thermally conductive grease, the allowable power dissipation of the integrated circuit device increases by 7.8 W, about ten percent. The advantage of using a thermally conductive grease in electronics cooling applications is that the thermal conductivity of the grease is greater than atmospheric air in a dry gap. This is achieved at the manufacturer by filling the grease with small particles of alumina or some other thermally conductive material. The other advantage of grease is that it flows into the tiny voids in the mating surfaces. In applications where grease is not appropriate, thin foils of a soft metal such as aluminum, copper, lead or indium may be used.

❒ ❒ ❒

PROBLEM 3.7 Heat Transfer in Two Plane Walls Separated by a Thin Heater

Two plane walls are separated by a thin electrical heater that dissipates 2 kW. The thickness and thermal conductivity of the first wall are 10 cm and 3 W/m·K, and the thickness and thermal conductivity of the second wall are 6 cm and 0.08 W/m·K. The outside surface of the first wall is exposed to a 50°C-fluid where the heat transfer coefficient is 200 W/m^2·K, and the outside surface of the second wall is exposed to a 20°C-fluid where the heat transfer coefficient is 50 W/m^2·K. Find the temperature of the heater and the heat transfer in each wall.

DIAGRAM

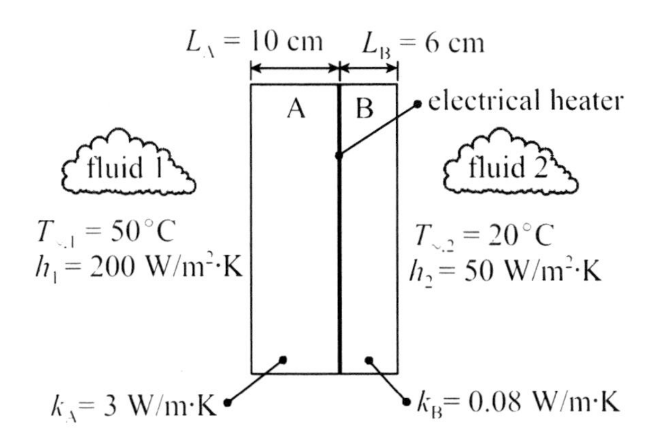

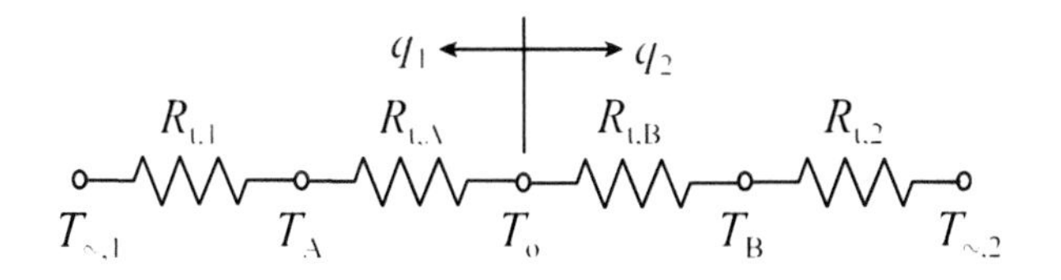

ASSUMPTIONS

1. One-dimensional steady conduction.
2. Radiation is neglected.
3. Heater is in intimate thermal contact with each wall.
4. Heater is isothermal.
5. Constant properties.

PROPERTIES

material A: $k_A = 3$ W/m·K
material B: $k_B = 0.08$ W/m·K

ANALYSIS

The thermal resistance of a plane wall layer is given by the relation

$$R_t = \frac{L}{kA}$$

In all calculations, we let $A = 1\text{m}^2$. Thus, we have

$$R_{t,\mathrm{A}} = \frac{(0.10\ \mathrm{m})}{(3\ \mathrm{W/m\cdot K})(1\ \mathrm{m^2})}$$

$$= 0.0333\ \mathrm{K/W}$$

$$R_{t,\mathrm{B}} = \frac{(0.06\ \mathrm{m})}{(0.08\ \mathrm{W/m\cdot K})(1\ \mathrm{m^2})}$$

$$= 0.750\ \mathrm{K/W}$$

Thermal resistance associated with convection is given by the relation

$$R_t = \frac{1}{hA}$$

Thus, we have

$$R_{t,1} = \frac{1}{(200\ \mathrm{W/m^2\cdot K})(1\ \mathrm{m^2})}$$

$$= 5 \times 10^{-3}\ \mathrm{K/W}$$

$$R_{t,2} = \frac{1}{(50\ \mathrm{W/m^2\cdot K})(1\ \mathrm{m^2})}$$

$$= 0.020\ \mathrm{K/W}$$

Heat dissipated by the electrical heater is split into two fractions of the total dissipation. One fraction passes through layer A and boundary layer 1, whereas the other fraction passes through layer B and boundary layer 2. These two fractions are represented as q_1 and q_2, respectively, and, when added, equal the total dissipation of 2 kW.

$$q = q_1 + q_2$$

The total thermal resistance corresponding to q_1 is

$$R_{\mathrm{tot},1} = R_{t,1} + R_{t,\mathrm{A}}$$

$$= (5 \times 10^{-3} + 0.0333)\ \mathrm{K/W}$$

$$= 0.0383\ \mathrm{K/W}$$

and the total thermal resistance corresponding to q_2 is

$$R_{\mathrm{tot},2} = R_{t,2} + R_{t,\mathrm{B}}$$

$= (0.020 + 0.750)$ K/W

$= 0.770$ K/W

Using the electrical analogy for one-dimensional conduction and the relation above for the total heat transfer, we have

$$q = \frac{T_o - T_{\infty,1}}{R_{\text{tot},1}} + \frac{T_o - T_{\infty,2}}{R_{\text{tot},2}}$$

Solving for T_o, the temperature of the heater,

$$T_o = \frac{q + \dfrac{T_{\infty,1}}{R_{\text{tot},1}} + \dfrac{T_{\infty,2}}{R_{\text{tot},2}}}{\dfrac{1}{R_{\text{tot},1}} + \dfrac{1}{R_{\text{tot},2}}}$$

$$T_o = \frac{2000\text{ W} + (50°\text{C}/0.0383\text{ K/W}) + (20°\text{C}/0.770\text{ K/W})}{1/(0.0383\text{ K/W}) + 1/(0.770\text{ K/W})}$$

$= \underline{\underline{121.55°\text{C}}}$

Note that either °C or K units could be used in the above relation for $T_{\infty,1}$ and $T_{\infty,2}$. The two heat transfer fractions are

$$q_1 = \frac{(121.55 - 50)°\text{C}}{0.0383\text{ K/W}}$$

$= \underline{\underline{1868\text{ W}}}$

$$q_2 = \frac{(121.55 - 20)°\text{C}}{0.770\text{ K/W}}$$

$= \underline{\underline{132\text{ W}}}$

DISCUSSION

The second heat transfer fraction could also be obtained by subtraction.

$q_2 = q - q_1$

$= 2000\text{ W} - 1868\text{ W} = 132\text{ W}$

❐ ❐ ❐

PROBLEM 3.8 **Heat Transfer in a Composite Plane Wall with Energy Generation in One Layer**

A composite wall consists of three layers of different thicknesses and thermal conductivities, as shown in the diagram. The outer surface of the first layer is insulated, and energy is generated in this layer at a rate of 850 kW/m^3. The outer surface of the third layer is in contact with 25°C-water where the heat transfer coefficient is 700 $W/m^2\cdot K$. If the surface of the composite wall measures 10 cm × 10 cm, find the temperatures of the layer interfaces and the outer surface of the third layer. Sketch the temperature distribution in the system.

DIAGRAM

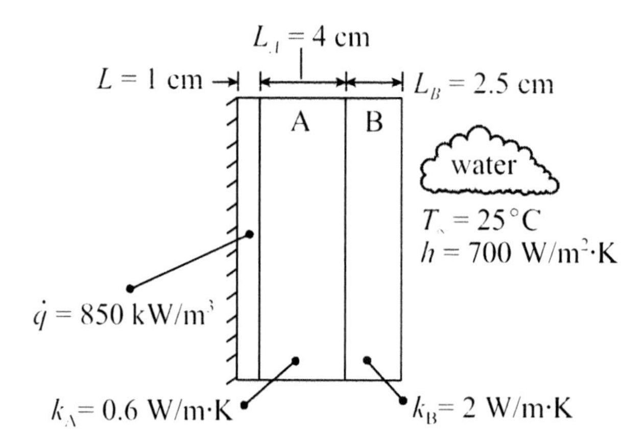

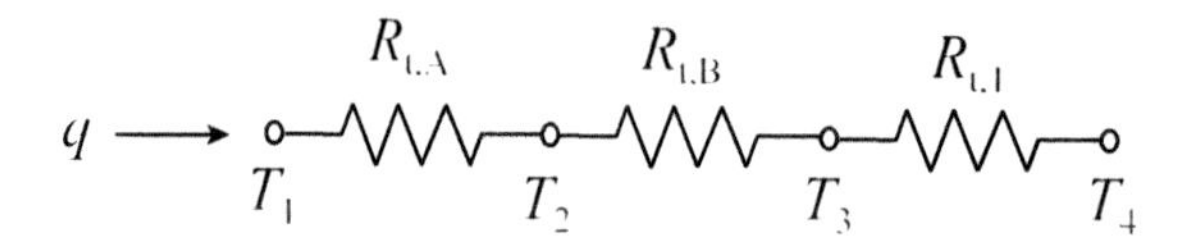

ASSUMPTIONS

1. One-dimensional steady conduction.
2. Radiation is neglected.
3. Wall layers are in intimate thermal contact.
4. Outer surface of the first layer is perfectly insulated.
5. Energy is generated uniformly in the first layer.
6. Constant properties.

PROPERTIES

material A: $k_A = 0.6$ W/m·K
material B: $k_B = 2$ W/m·K

ANALYSIS

Because the outer surface of the first layer is insulated, all the energy generated in this layer will pass through the other layers. The value of this heat transfer is the energy generation times the volume of the layer.

$$q = \dot{q}\, A\, L$$

$$= (850 \times 10^3 \text{ W/m}^3)(0.01 \text{ m}^2)(0.01 \text{ m})$$

$$= 85 \text{ W}$$

The thermal resistance of a plane layer is given by the relation

$$R_t = \frac{L}{kA}$$

Thus, we have

$$R_{t,\text{A}} = \frac{(0.04 \text{ m})}{(0.6 \text{ W/m·K})(0.01 \text{ m}^2)}$$

$$= 6.667 \text{ K/W}$$

$$R_{t,\text{B}} = \frac{(0.025 \text{ m})}{(2 \text{ W/m·K})(0.01 \text{ m}^2)}$$

$$= 1.250 \text{ K/W}$$

Thermal resistance associated with convection is given by the relation

$$R_t = \frac{1}{hA}$$

Thus, we have

$$R_{t,1} = \frac{1}{(700 \text{ W/m}^2\text{·K})(0.01 \text{ m}^2)}$$

$$= 0.1429 \text{ K/W}$$

The total thermal resistance is

$$R_{\text{tot}} = R_{t,A} + R_{t,B} + R_{t,1}$$

$$R_{\text{tot}} = (6.667 + 1.250 + 0.1429) \text{ K/W}$$

$$= 8.060 \text{ K/W}$$

Using the electrical analogy,

$$q = \frac{T_1 - T_\infty}{R_{\text{tot}}} = \frac{T_1 - T_2}{R_{t,A}} = \frac{T_2 - T_3}{R_{t,B}} = \frac{T_3 - T_\infty}{R_{t,1}}$$

Solving for T_1,

$$T_1 = q\, R_{\text{tot}} + T_\infty$$

$$= (85 \text{ W})(8.060 \text{ K/W}) + 25°\text{C}$$

$$= \underline{\underline{710°\text{C}}}$$

Solving for T_2,

$$T_2 = T_1 - q\, R_{t,\text{A}}$$

$$= 710°\text{C} - (85 \text{ W})(6.667 \text{ K/W})$$

$$= \underline{\underline{143°\text{C}}}$$

Solving for T_3,

$$T_3 = T_2 - q\, R_{t,\text{B}}$$

$$= 143°\text{C} - (85 \text{ W})(1.250 \text{ K/W})$$

$$= \underline{\underline{36.8°\text{C}}}$$

The temperature distribution in the system is sketched below. The slope of the temperature at the outer surface of the first layer is zero, according to Fourier's law, because this surface is insulated. In the solid layers, the temperature distribution is linear. In the boundary layer, the temperature asymptotically approaches the free stream value T_∞.

DISCUSSION

In order to find the temperature of the outer surface of the layer that generates energy, it is necessary to solve the heat conduction equation. This approach is illustrated in Problem 3.9.

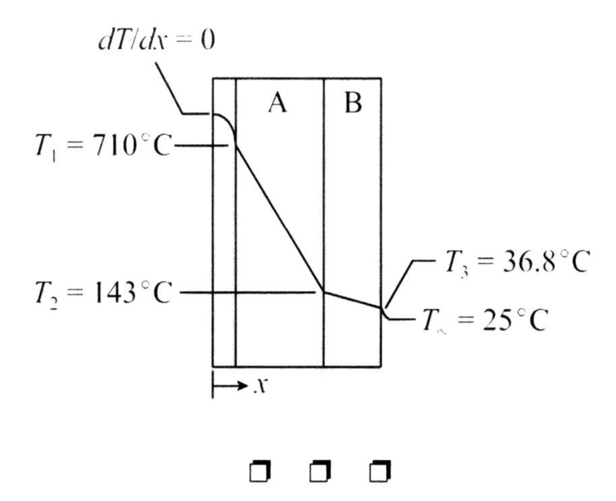

❐ ❐ ❐

PROBLEM 3.9 **Heat Equation for One-Dimensional Steady Conduction in a Two-Layer Plane Wall with Energy Generation in One Layer**

A plane wall consisting of two dissimilar layers is insulated on one side and exposed to a fluid on the other. Energy is generated uniformly in the layer that is insulated on one side. By solving the heat equation for each layer, find the one-dimensional steady temperature distribution in the wall. Sketch the temperature distribution.

DIAGRAM

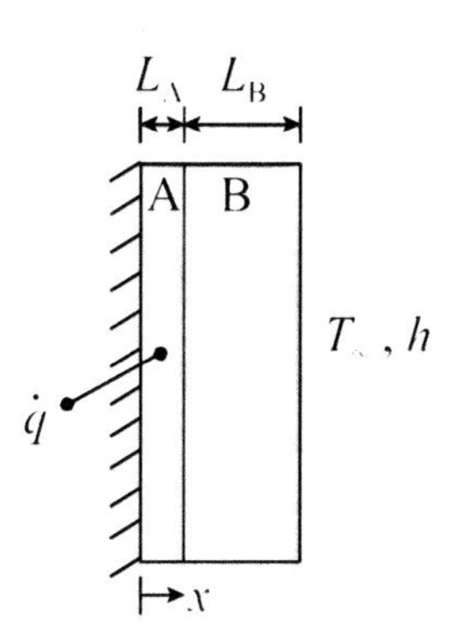

ASSUMPTIONS

1. One-dimensional steady conduction.
2. Radiation is neglected.
3. Layers are in intimate thermal contact.
4. Energy is generated uniformly in layer A.
5. Constant properties.

PROPERTIES

material A: k_A
material B: k_B

ANALYSIS

The heat equation for one-dimensional steady conduction in layer A is

$$\frac{d^2T_A}{dx^2} + \frac{\dot{q}}{k_A} = 0$$

and the heat equation for one-dimensional steady conduction in layer B is

$$\frac{d^2T_B}{dx^2} = 0$$

The boundary conditions are

$$\frac{dT_A(0)}{dx} = 0 \quad , \quad T_A(L_A) = T_B(L_A)$$

$$-k_A \frac{dT_A(L_A)}{dx} = -k_B \frac{dT_B(L_A)}{dx} \quad , \quad -k_B \frac{dT_B(L_A + L_B)}{dx} = h\left[T_B(L_A + L_B) - T_\infty\right]$$

The general solutions of the differential equations for layers A and B are

$$T_A(x) = -\frac{\dot{q}}{2k_A}x^2 + C_1 x + C_2$$

$$T_B(x) = C_3 x + C_4$$

where C_1, C_2, C_3 and C_4 are integration constants. Substituting the boundary conditions into the general solutions, the integration constants are

$$C_1 = 0$$

$$C_2 = \frac{\dot{q}L_A^2}{2k_A}\left(1 + \frac{1}{Bi_A} + \frac{Bi_B}{Bi_A}\right) + T_\infty$$

$$C_3 = -\frac{\dot{q}L_A}{2k_B}$$

$$C_4 = \frac{\dot{q}L_A L_B}{2k_B}\left[\frac{1}{Bi_B} + \left(1 + \frac{L_A}{L_B}\right)\right] + T_\infty$$

where the Biot numbers for each layer are defined as

$$Bi_A \equiv \frac{hL_A}{k_A} \quad , \quad Bi_B \equiv \frac{hL_B}{k_B}$$

Substituting the expressions for the integration constants into the general solutions and simplifying, the final solutions are

$$\frac{T_A(x) - T_\infty}{\left(\dfrac{\dot{q}L_A^2}{2k_A}\right)} = \frac{1 + Bi_A + Bi_B}{Bi_A} - \left(\frac{x}{L_A}\right)^2$$

$$\frac{T_B(x) - T_\infty}{\left(\dfrac{\dot{q}L_A L_B}{2k_B}\right)} = 1 + \frac{1}{Bi_B} + \frac{L_A}{L_B} - \frac{x}{L_B}$$

The temperature distribution is sketched below.

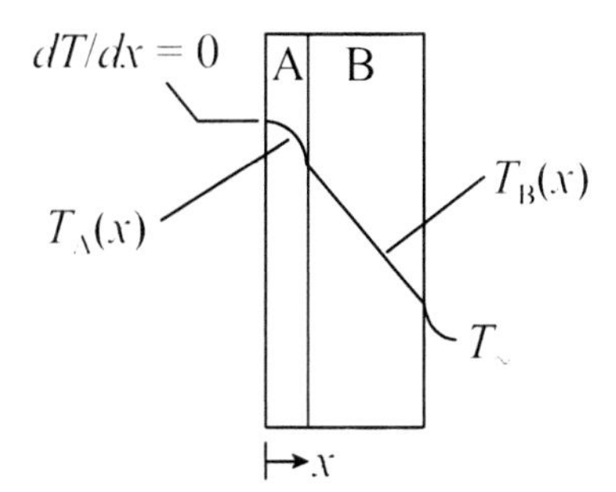

DISCUSSION

According to Fourier's law, the slope of the temperature distribution is zero at the insulated surface. In layer A where there is energy generation, the temperature distribution is parabolic, whereas in layer B where there is no energy generation, the temperature distribution is linear. At the interface, the slopes are not equal, but the products of the slopes and their respective thermal conductivities are equal, as given in the boundary conditions.

PROBLEM 3.10 Cooling Load for a Low-Temperature Laboratory Freezer

The walls of a low-temperature laboratory freezer with inside dimensions 128 cm × 76 cm × 50 cm are constructed of a 12.7-cm thick layer of polyurethane insulation sandwiched between two 3.4-mm thick sheets of galvanized steel. The thermal conductivities of the polyurethane insulation and steel are 0.025 W/m·K and 60 W/m·K, respectively. The thermal contact resistance at each interface is 2.5×10^{-4} m^2·K/W. If the freezer is designed to operate when the inside and outside surface temperatures of the freezer walls are −86°C and 25°C, respectively, what is the cooling load of the freezer?

DIAGRAM

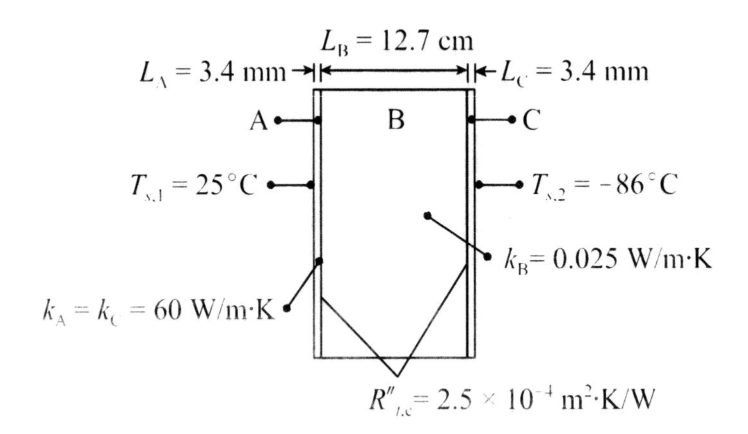

ASSUMPTIONS

1. One-dimensional steady conduction.
2. Heat transfer through the edges and corners of the freezer is neglected.
3. Constant properties.

PROPERTIES

polyurethane insulation: $k_B = 0.025$ W/m·K
galvanized steel: $k_A = k_C = 60$ W/m·K

ANALYSIS

The total surface area of the freezer walls is

$$A = 2(1.28 \times 0.50 + 1.28 \times 0.76 + 0.76 \times 0.50) \text{ m}^2$$

$= 3.986 \text{ m}^2$

The thermal resistance of a plane wall layer is given by the relation

$$R_t = \frac{L}{kA}$$

The thermal resistances of the wall layers are

$$R_{t,\text{A}} = R_{t,\text{C}} = \frac{3.4 \times 10^{-3} \text{ m}}{(60 \text{ W/m·K})(3.986 \text{ m}^2)}$$

$$= 1.422 \times 10^{-5} \text{ K/W}$$

$$R_{t,\text{B}} = \frac{0.127 \text{ m}}{(0.025 \text{ W/m·K})(3.986 \text{ m}^2)}$$

$$= 1.275 \text{ K/W}$$

The thermal resistance of the steel/insulation interfaces is given by the relation

$$R_{t,1} = R_{t,2} = \frac{R''_{t,c}}{A}$$

$$R_{t,1} = R_{t,2} = \frac{2.5 \times 10^{-4} \text{ m}^2\text{·K/W}}{3.986 \text{ m}^2}$$

$$= 6.272 \times 10^{-5} \text{ K/W}$$

There are two steel sheets and two interfaces, so the total thermal resistance is

$$R_{\text{tot}} = 2(R_{t,1} + R_{t,A}) + R_{t,B}$$

$$R_{\text{tot}} = 2(6.272 \times 10^{-5} + 1.422 \times 10^{-5}) \text{ K/W} + 1.275 \text{ K/W}$$

$$= 1.275 \text{ K/W}$$

The thermal resistances of the steel sheets and interfaces are negligibly small compared to that of the polyurethane insulation. The cooling load is the heat transfer through the wall of the freezer, given by the relation

$$q = \frac{T_{s,1} - T_{s,2}}{R_{\text{tot}}}$$

$$q = \frac{25 - (-86)°\text{C}}{1.275 \text{ K/W}}$$

$$= \underline{\underline{87.1 \text{ W}}}$$

DISCUSSION

For the freezer to maintain the specified inside wall temperature, the refrigeration system must remove 87.1 W from the contents of the freezer. Assuming the coefficient of performance (COP) of the freezer is 2.5, the electrical power required to run the freezer is given by the relation

$$P = \frac{q}{COP}$$

so the electrical power is

$$P = \frac{87.1\text{W}}{2.5}$$

$$= 34.8 \text{ W}$$

❐ ❐ ❐

PROBLEM 3.11 Cost of Heat Loss from an Insulated Steam Pipe

A long pipe of low-carbon steel (k_A = 64 W/m·K) carries 230°C steam across a room where the air temperature is 20°C and the associated heat transfer coefficient is 25 W/m^2·K. The cost of generating the steam is $5/GJ, and the steam line operates 6500 h/yr. Calculate the heat loss per meter of pipe length and the corresponding annual cost of this heat loss for various thicknesses of magnesia insulation (k_B = 0.05 W/m·K). The heat transfer coefficient for the inside surface of the pipe is 650 W/m^2·K, and the inside and outside radii of the steel pipe are 3.0 cm and 3.8 cm, respectively.

DIAGRAM

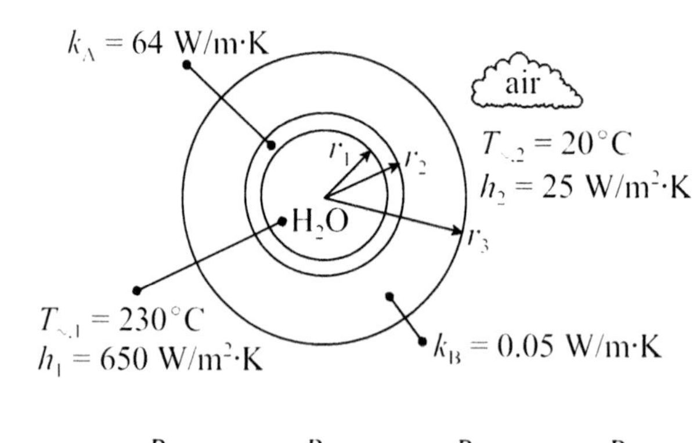

ASSUMPTIONS

1. One-dimensional steady conduction.
2. Pipe and insulation are in intimate thermal contact.
3. Radiation is neglected.
4. Constant properties.

PROPERTIES

low-carbon steel: $k_A = 64$ W/m·K
magnesia insulation: $k_B = 0.05$ W/m·K

ANALYSIS

Thermal resistance associated with convection is given by the relation

$$R_t = \frac{1}{2\pi rLh}$$

A unit length, L = 1m, is used in all calculations. The thermal resistance associated with convection on the inside surface of the pipe is

$$R_{t,1} = \frac{1}{2\pi\,(0.030\text{ m})(1\text{ m})(650\text{ W/m}^2\cdot\text{K})}$$

$$= 8.162 \times 10^{-3}\text{ K/W}$$

The thermal resistance of a cylindrical wall is given by the relation

$$R_t = \frac{\ln(r_o/r_i)}{2\pi Lk}$$

so the thermal resistance of the steel pipe is

$$R_{t,A} = \frac{\ln(0.038\text{ m}/0.030\text{ m})}{2\pi\,(1\text{ m})(64\text{ W/m}\cdot\text{K})}$$

$$= 5.879 \times 10^{-4}\text{ K/W}$$

For insulation thicknesses, we choose 1.0, 2.0, 3.0, 4.0 and 5.0 cm. The thermal resistances of the insulation and outside boundary layer are functions of the insulation thickness δ. The outside radius of the insulation is given by the relation

$$r_3 = r_2 + \delta$$

The thermal resistances of the insulation and outside boundary layer are given by the following relations, respectively,

$$R_{t,B} = \frac{\ln(r_3/r_2)}{2\pi L k_B}$$

$$R_{t,2} = \frac{1}{2\pi r_3 L h_2}$$

The total thermal resistance is given by the relation

$$R_{\text{tot}} = R_{t,1} + R_{t,A} + R_{t,B} + R_{t,2}$$

and the heat loss from the pipe is given by the relation

$$q = \frac{T_{\infty,1} - T_{\infty,2}}{R_{\text{tot}}}$$

Using the insulation thicknesses above, thermal resistances, heat losses, and corresponding costs are summarized in the table below. As an example of a calculation, the cost for the first insulation thickness is

$$\text{cost} = \frac{\$5}{10^9\ \text{J}} \times \frac{237.3\ \text{J}}{\text{s}} \times \frac{6500\ \text{h}}{\text{yr}} \times \frac{3600\ \text{s}}{\text{h}} = \$27.76$$

δ (cm)	r_3 (cm)	$R_{t,B}$ (K/W)	$R_{t,2}$ (K/W)	R_{tot} (K/W)	q (W)	cost ($)
1.0	4.8	0.7436	0.1326	0.8849	237.3	27.76
2.0	5.8	1.3460	0.1098	1.4645	143.4	16.78
3.0	6.8	1.8523	0.0936	1.9546	107.4	12.57
4.0	7.8	2.2890	0.0816	2.3793	88.3	10.33
5.0	8.8	2.6730	0.0723	2.7540	76.3	8.93

DISCUSSION

From the calculations above it is evident that for every centimeter of insulation added to the steam pipe there are diminishing cost savings. For example, increasing the thickness from 3.0 cm to 5.0 cm saves only $3.64 per year. This may not be enough to offset the higher cost of thicker magnesia insulation. Thus, an insulation thickness of approximately 3.0 cm may be the optimum as far as cost is concerned.

❒ ❒ ❒

PROBLEM 3.12 Maximum Electrical Current in an Insulated Wire

A heating element consisting of a long coil of 18-gage (1.024 mm diameter) nichrome wire is used to boil water at 180°C in a pressurized vessel. The wire is covered with a 0.2-mm thick layer of plastic insulation with a thermal conductivity of 0.16 W/m·K, and the heat transfer coefficient on the surface of the insulation is 12 kW/m²·K. According to the wire manufacturer, the insulation is rated for a maximum temperature of 300°C. What is the maximum electrical current that the wire can carry? The electrical resistance of 18-gage nichrome wire is 1.384 Ω/m.

DIAGRAM

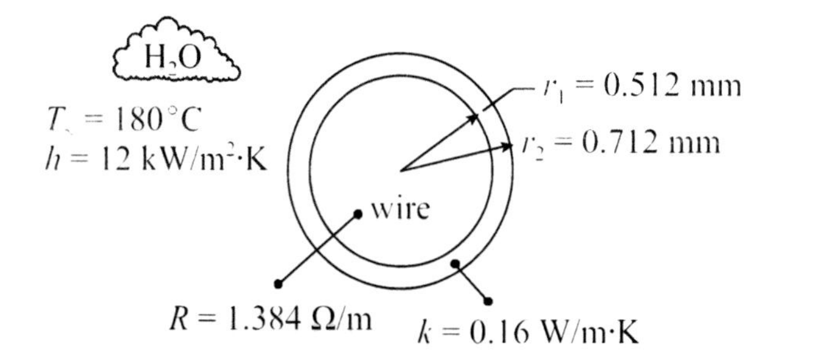

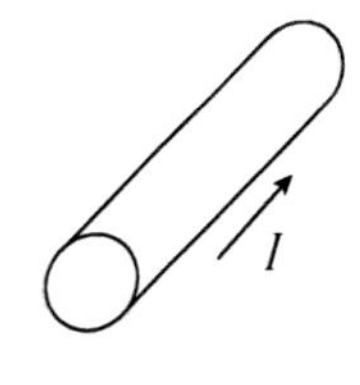

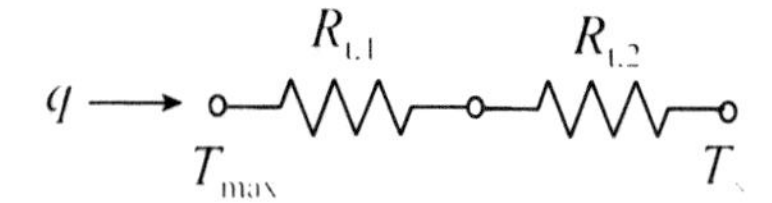

ASSUMPTIONS

1. One-dimensional steady conduction.
2. Wire and insulation are in intimate thermal contact.
3. Radiation is neglected.
4. Constant properties.

PROPERTIES

plastic insulation: $k = 0.16$ W/m·K

ANALYSIS

The thermal resistance of the insulation layer is given by the relation

$$R_{t,1} = \frac{\ln(r_2/r_1)}{2\pi kL}$$

A unit length, $L = 1$ m, will be used in all calculations. The thermal resistance of the insulation layer is

$$R_{t,1} = \frac{\ln(0.712\text{ mm}/0.512\text{ mm})}{2\pi\,(0.16\text{ W/m}\cdot\text{K})(1\text{ m})}$$

$$= 0.3280\text{ K/W}$$

The thermal resistance associated with convection is given by the relation

$$R_{t,2} = \frac{1}{2\pi r_2 Lh}$$

so the thermal resistance associated with convection is

$$R_{t,2} = \frac{1}{2\pi\,(0.712\times10^{-3}\text{ m})(1\text{ m})(12\times10^{3}\text{ W/m}^2\cdot\text{K})}$$

$$= 0.0186\text{ K/W}$$

The total thermal resistance is given by the relation

$$R_{tot} = R_{t,1} + R_{t,2}$$

so the total thermal resistance is

$$R_{tot} = (0.3280 + 0.0186)\text{ K/W}$$

$$= 0.3466\text{ K/W}$$

The heat transfer from the wire is equivalent to the rate of electrical energy dissipation. Using the electrical analogy, the heat transfer is given by the relation

$$q = I^2R = \frac{T_{max} - T_{\infty}}{R_{tot}}$$

Solving for the electrical current I,

$$I = \left[\frac{T_{max} - T_\infty}{R \, R_{tot}} \right]^{1/2}$$

The maximum electrical current is

$$I = \left[\frac{(300 - 180)°C}{(1.384 \, \Omega)(0.3466 \text{ K/W})} \right]^{1/2}$$

$$= \underline{\underline{15.8 \text{ A}}}$$

DISCUSSION

Electrical resistance of wires is a function of the metal alloys from which the wires are made. The graph below shows the maximum current that an 18-gage wire can carry as a function of wire resistance per meter of length, keeping the values of the other quantities the same as those used in this problem.

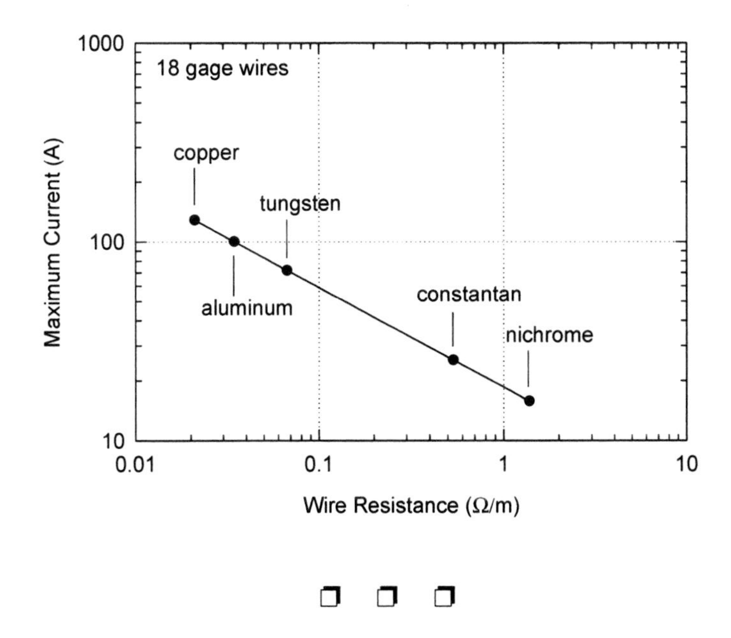

❒ ❒ ❒

PROBLEM 3.13 Roasting a Marshmallow on a Long Copper Wire over a Campfire

A boy scout constructs a holder for roasting a marshmallow over a campfire by forming a 2-gage (6.54 mm diameter) drawn copper wire ($k = 287$ W/m·K). His scoutmaster, who is a heat-transfer expert, informs the scout that if he holds the wire within 25 cm of the marshmallow, the wire will burn his hand. If the temperature of the wire at the marshmallow is 180°C, what is the temperature of the wire at the scout's hand, assuming a heat transfer coefficient of 22 W/m^2·K and an ambient air temperature of 10°C? Is the scoutmaster correct?

DIAGRAM

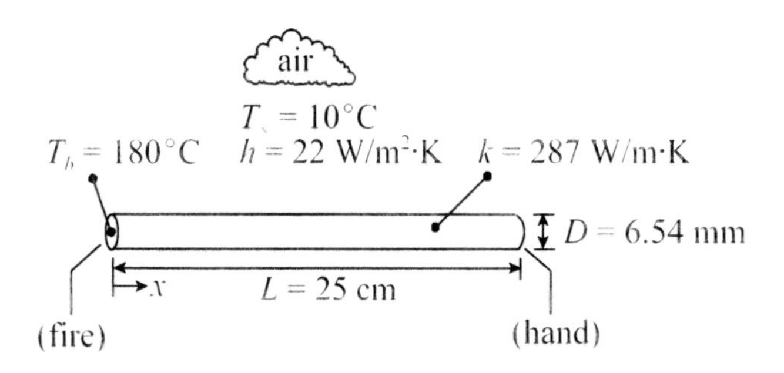

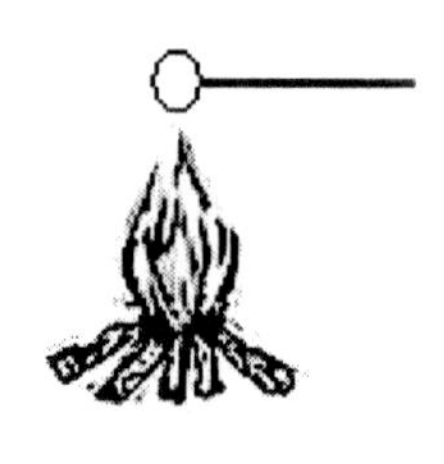

ASSUMPTIONS

1. One-dimensional steady conduction.
2. The end (at the hand) of the wire is insulated.
3. Radiation is neglected.
4. Uniform heat transfer coefficient.
5. Constant properties.

PROPERTIES

drawn copper: $k = 287$ W/m·K

ANALYSIS

The wire marshmallow holder acts as a long pin fin, conducting heat from the end near the fire to the end held by the scout. The differential equation for one-dimensional steady heat conduction in a fin is

$$\frac{d^2\theta}{dx^2} - m^2\theta = 0$$

where the quantities θ and m are defined by the relations

$$\theta \equiv T(x) - T_\infty$$

$$m \equiv \left(\frac{hP}{kA_c}\right)^{1/2}$$

The boundary conditions are

$$\theta(0) = \theta_b \quad , \quad \frac{d\theta(L)}{dx} = 0$$

The general solution of the above differential equation is

$$\theta(x) = C_1 e^{mx} + C_2 e^{-mx}$$

where C_1 and C_2 are integration constants. The temperature distribution in the fin is given by the relation

$$\frac{\theta}{\theta_b} = \frac{T(x) - T_\infty}{T_b - T_\infty} = \frac{\cosh[m(L-x)]}{\cosh(mL)}$$

The cross sectional area and perimeter of the pin fin, respectively, are

$$A_c = \frac{\pi D^2}{4}$$

$$A_c = \frac{\pi\,(6.54 \times 10^{-3}\ \text{m})^2}{4}$$

$$= 3.359 \times 10^{-5}\ \text{m}^2$$

$$P = \pi D$$

$$P = \pi\,(6.54 \times 10^{-3}\ \text{m})$$

$$= 0.0205\ \text{m}$$

and the fin parameter m is

$$m = \left[\frac{(22\ \text{W/m}^2{\cdot}\text{K})(0.0205\ \text{m})}{(287\ \text{W/m}{\cdot}\text{K})(3.359 \times 10^{-5}\ \text{m}^2)}\right]^{1/2}$$

$$= 6.840\ \text{m}^{-1}$$

Solving for $T(L)$ from the temperature distribution above, the wire temperature at the scout's hand,

$$T(L) = (T_b - T_\infty)\frac{\cosh(0)}{\cosh(mL)} + T_\infty$$

$$T(L) = \frac{(180 - 10)°\text{C}\ (1)}{\cosh\ [(6.840\ \text{m}^{-1})(0.25\ \text{m})]} + 10°\text{C}$$

$$= \underline{\underline{69.5°\text{C}}}\quad (157°\text{F})$$

DISCUSSION

The scoutmaster is correct. A wire temperature of 69.5 °C will burn the scout's hand, making it impossible for him to hold the marshmallow over the fire. To avoid being burned, the scout should hold the wire farther away from the marshmallow. For example, if the scout holds the wire at a distance of $L = 35$ cm, the temperature at his hand is only 40.8 °C (105 °F), a temperature that can be tolerated. Other solutions include using a wire with a lower thermal conductivity and/or a smaller diameter.

❐ ❐ ❐

PROBLEM 3.14 Effective Fin Lengths for a Fin Heated at Both Ends

A fin of uniform cross sectional area A_c spans the distance between two heat sources, which supply the fin with different heat transfers q_1 and q_2. The length, perimeter and thermal conductivity of the fin are L, P and k, respectively. Surrounding the fin is a fluid at temperature T_∞, and the heat transfer coefficient is h. Derive an expression for the effective fin lengths L_1 and L_2, i.e., the location on the fin where the temperature gradient is zero.

DIAGRAM

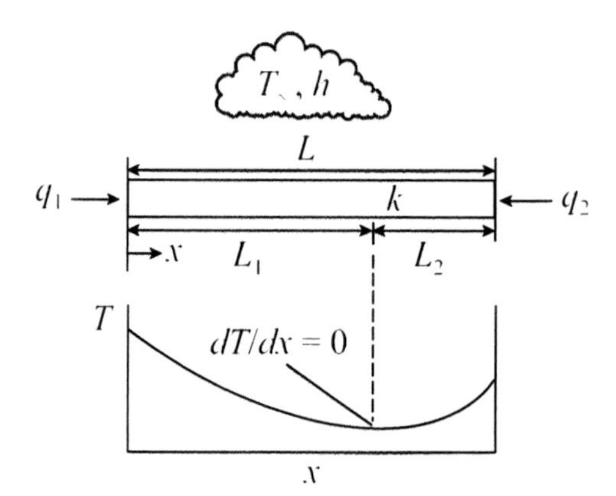

ASSUMPTIONS

1. One-dimensional steady conduction.
2. Radiation is neglected.
3. Uniform heat transfer coefficient.
4. Constant properties.

PROPERTIES

thermal conductivity: k

ANALYSIS

The differential equation for one-dimensional steady heat conduction in a fin is

$$\frac{d^2\theta}{dx^2} - m^2\theta = 0$$

where the quantities θ and m are defined by the relations

$$\theta \equiv T(x) - T_\infty$$

$$m \equiv \left(\frac{hP}{kA_c}\right)^{1/2}$$

The heat conducted into each end equals the heat flux, so the boundary conditions are

$$-k\frac{d\theta(0)}{dx} = q_1'' \quad , \quad -k\frac{d\theta(L)}{dx} = -q_2''$$

The general solution of the above differential equation is

$$\theta(x) = C_1 e^{mx} + C_2 e^{-mx}$$

where C_1 and C_2 are integration constants. Substituting the boundary conditions into the general solution, the integration constants are

$$C_1 = \frac{q_1''\left[e^{mL} - 2\sinh(mL)\right] + q_2''}{2km\sinh(mL)}$$

$$C_2 = \frac{q_1'' e^{mL} + q_2''}{2km\,\sinh(mL)}$$

To find the location on the fin where the temperature gradient is zero, we differentiate $\theta\,(x)$ with respect to x, set the result to zero, and solve for x. The result is

$$x = L_1 = -\frac{1}{2m}\ln\left(\frac{C_1}{C_2}\right)$$

Simplifying the ratio C_1/C_2 and expressing the above relation in dimensionless form,

$$\frac{L_1}{L} = -\frac{1}{2\varphi}\ln\psi$$

where,

$$\varphi = mL \quad , \quad \lambda = \frac{q_1}{q_2}$$

$$\psi = \frac{\lambda\left(e^{\varphi} - 2\sinh\varphi\right) + 1}{\lambda e^{\varphi} + 1}$$

DISCUSSION

The quantity L_1/L is graphed below as a function of q_1/q_2 for several values of mL. Note that if $q_1 = q_2$, the quantity $L_1/L = 0.5$ for all values of mL. Knowing the effective fin lengths L_1 and L_2 allows us to analyze the fin as two separate fins with insulated ends.

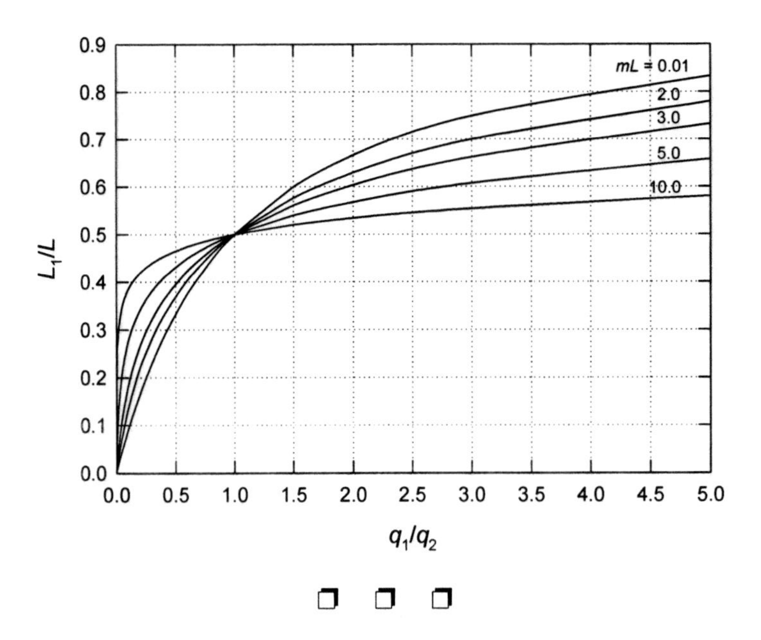

❒ ❒ ❒

PROBLEM 3.15 The Optimum Rectangular Fin

Consider a rectangular fin of length L, thickness t, width w, and thermal conductivity k. The fin is exposed to a fluid at temperature T_∞ where the heat transfer coefficient is h. Derive a relation that represents the optimum dimensions for the fin, i.e., dimensions that yield the maximum heat transfer for a minimum fin size. Construct a graph that shows the relationship between the pertinent fin dimensions, fin thermal conductivity and heat transfer coefficient.

DIAGRAM

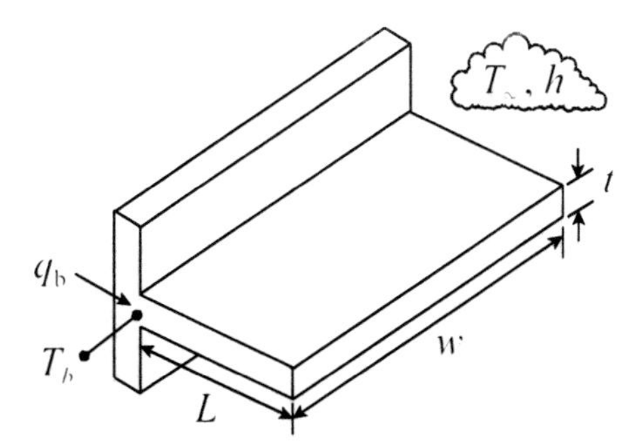

ASSUMPTIONS

1. One-dimensional steady conduction.
2. The tip of the fin is insulated.
2. Fin width w is much greater than the fin thickness t.
3. Radiation is neglected.
4. Uniform heat transfer coefficient.
5. Constant properties.

PROPERTIES

thermal conductivity: k

ANALYSIS

The heat transfer through the base of the fin is given by the relation

$$q_b = \left(hPkA_c\right)^{1/2}\theta_b \tanh(mL)$$

where $P \approx 2w$, $A_c = w\,t$, $\theta_b = T_b - T_\infty$, and

$$m = \left(\frac{hP}{kA_c}\right)^{1/2} = \left(\frac{2h}{kt}\right)^{1/2}$$

We define the fin profile area A_p as

$$A_p = Lt$$

Substituting the relation for m in to the relation for the heat transfer q_b, and replacing the fin thickness t with A_p/L, the heat transfer per unit fin width is

$$\frac{q_b}{w} = \left(\frac{2hkA_p}{L}\right)^{1/2} \theta_b \tanh\left(\frac{2hL^3}{kA_p}\right)^{1/2}$$

Differentiating the above relation with respect to fin length L, and setting the result to zero, we obtain the relation

$$\tanh\beta - 3\operatorname{sech}^2\beta = 0$$

where,

$$\beta = mL$$

There is one root relation, $\beta = 1.4192$. Expressing the optimum fin thickness in terms of the optimum fin length,

$$t_{opt} = 0.9930\frac{h\,L_{opt}^2}{k}$$

Rectangular fins are commonly made from the alloy 6061 aluminum, which has a room-temperature thermal conductivity of 180 W/m·K. For this metal alloy, the graph below shows the variation of optimum fin thickness t_{opt} with optimum fin length L_{opt} for several values of the heat transfer coefficient h.

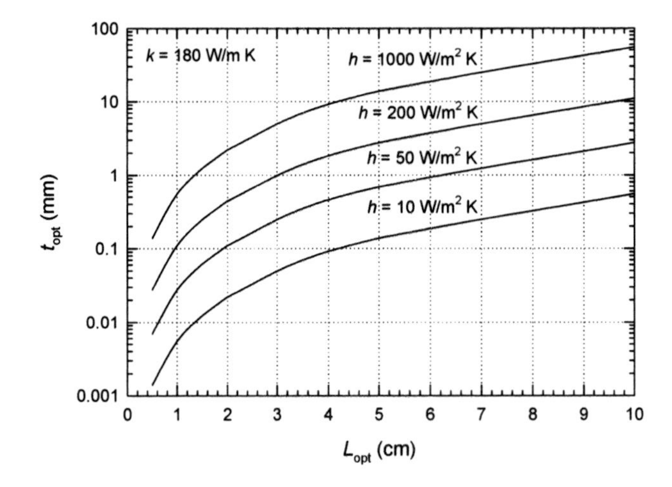

DISCUSSION

For low heat transfer coefficients the values of optimum fin thickness t_{opt} are probably too low to manufacture using conventional methods. Most heat sinks that incorporate rectangular fins are manufactured using an extrusion process. For example, if a 2-cm long optimum fin is desired for an application in which the heat transfer coefficient is 50 W/m^2·K, the optimum fin thickness is a meager 0.11 mm. But if a 10-cm long optimum fin is desired for the same application, the optimum fin thickness is about 2.8 mm, a more reasonable value from the standpoint of manufacturability.

❐ ❐ ❐

PROBLEM 3.16 Effectiveness of a Rectangular Fin

A rectangular fin constructed of 316 stainless steel (k = 13.4 W/m·K) has a length, width and thickness of 8 cm, 17 cm and 6 mm, respectively. The fin is surrounded by boiling water where the heat transfer coefficient is 8000 W/m^2·K. Should this fin be used in this application?

DIAGRAM

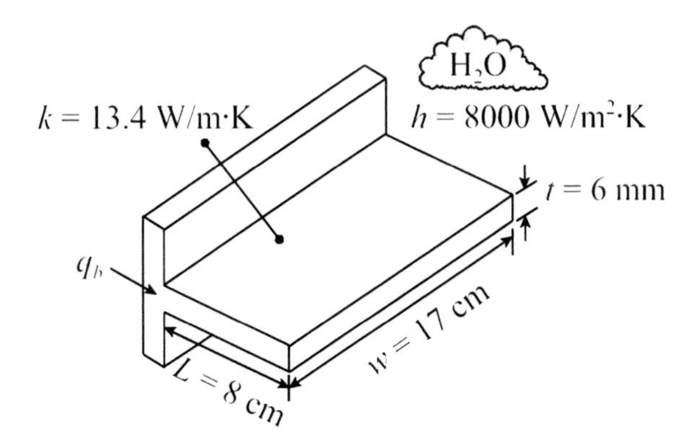

ASSUMPTIONS

1. One-dimensional steady conduction.
2. The tip of the fin is insulated.
3. Heat transfer from the fin edges is neglected.
4. Radiation is neglected.
5. Uniform heat transfer coefficient.
6. Constant properties.

PROPERTIES

316 stainless steel: k = 13.4 W/m·K

ANALYSIS

Fin effectiveness is defined as the ratio of the fin heat transfer to the heat transfer that would exist without the fin.

$$\varepsilon_f = \frac{q_b}{hA_c\theta_b}$$

The heat transfer through the base of the fin is

$$q_b = (hPkA_c)^{1/2}\theta_b \tanh(mL)$$

Thus, the fin effectiveness is

$$\varepsilon_f = \left(\frac{kP}{hA_c}\right)^{1/2} \tanh(mL)$$

where, $P \approx 2w$, $A_c = w\,t$ and

$$m = \left(\frac{hP}{kA_c}\right)^{1/2}$$

$$m = \left[\frac{(8000\ \text{W/m}^2\cdot\text{K})(2 \times 0.17\ \text{m})}{(13.4\ \text{W/m}\cdot\text{K})(1.02\times 10^{-3}\ \text{m}^2)}\right]^{1/2}$$

$$= 446\ \text{m}^{-1}$$

The fin effectiveness is

$$\varepsilon_f = \left[\frac{(13.4\ \text{W/m}\cdot\text{K})(2 \times 0.17\ \text{m})}{(8000\ \text{W/m}^2\cdot\text{K})(1.02 \times 10^{-3}\ \text{m}^2)}\right]^{1/2} \tanh\left[(446\ \text{m}^{-1})(0.08\ \text{m})\right]$$

$$= \underline{\underline{0.747}}$$

Because $\varepsilon_f < 1$, a fin should not be used in this application. Adding a fin decreases the heat transfer by approximately 25 percent.

DISCUSSION

If the application involved a gas instead of boiling water, the heat transfer coefficient would be much lower, say $h = 250$ W/m²·K. For this case the fin effectiveness is $\varepsilon_f = 4.23$, indicating that a fin should be used.

❒ ❒ ❒

PROBLEM 3.17 A Transmission Line as an Infinitely Long Fin

A transformer at a power substation experiences a short circuit, resulting in a steady power dissipation of 300 W into a transmission line, which acts as a long fin emanating from the transformer. The transmission line has a diameter of 2.0-cm and is made of an aluminum alloy (k = 190 W/m·K). The air temperature and heat transfer coefficient are 20°C and 40 W/m²·K, respectively. What is the temperature of the transmission line at its connection point with the transformer? Is the transmission line at risk of experiencing a failure?

DIAGRAM

air

$T_\infty = 20°C$

$h = 40$ W/m²·K

$q_b = 300$ W

$D = 2.0$ cm

x

T_b

$k = 190$ W/m·K

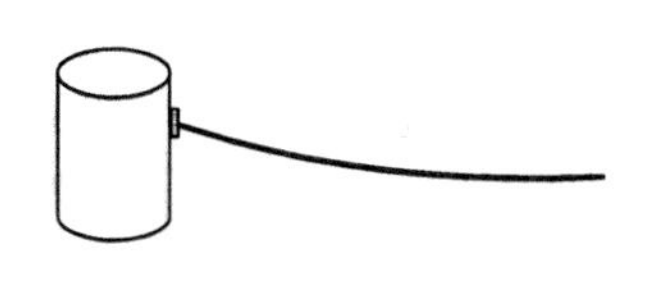

ASSUMPTIONS

1. One-dimensional steady conduction.
2. The fin is infinitely long.
3. Radiation is neglected.
4. Uniform heat transfer coefficient.
5. Constant properties.

PROPERTIES

aluminum alloy: $k = 190$ W/m·K

ANALYSIS

The differential equation for one-dimensional steady heat conduction in a fin is

$$\frac{d^2\theta}{dx^2} - m^2\theta = 0$$

where the quantities θ and m are defined by the relations

$$\theta \equiv T(x) - T_\infty$$

$$m \equiv \left(\frac{hP}{kA_c}\right)^{1/2}$$

The boundary conditions are

$$\theta(0) = \theta_b \quad , \quad \theta(\infty) = 0$$

The general solution of the above differential equation is

$$\theta(x) = C_1 e^{mx} + C_2 e^{-mx}$$

where C_1 and C_2 are integration constants. Substituting the boundary conditions into the general solution, the integration constants are

$$C_1 = 0 \quad , \quad C_2 = \theta_b$$

Thus, the temperature distribution in the fin is given by the relation

$$\frac{\theta}{\theta_b} = \frac{T(x) - T_\infty}{T_b - T_\infty} = e^{-mx}$$

The cross sectional area and perimeter of the pin fin, respectively, are

$$A_c = \frac{\pi D^2}{4}$$

$$A_c = \frac{\pi\,(0.02 \text{ m})^2}{4}$$

$$= 3.142 \times 10^{-4} \text{ m}^2$$

$$P = \pi D$$

$$P = \pi\,(0.02 \text{ m})$$

$$= 0.0628 \text{ m}$$

From Fourier's law,

$$q_b = -kA_c \frac{d\theta(0)}{dx} = kA_c m\theta_b = \left(hPkA_c\right)^{1/2} \theta_b$$

Solving for the fin base temperature T_b, the temperature of the transmission line at the transformer,

$$T_b = \frac{q_b}{\left(hPkA_c\right)^{1/2}} + T_\infty$$

$$T_b = \frac{300 \text{ W}}{[(40 \text{ W/m}^2\cdot\text{K})(0.0628 \text{ m})(190 \text{ W/m}\cdot\text{K})(3.142 \times 10^{-4} \text{ m}^2)]^{1/2}} + 20°\text{C}$$

$$= \underline{\underline{795°\text{C}}}$$

DISCUSSION

Pure aluminum has a melting point of 660°C. If the alloying metals in the aluminum transmission line have a higher melting point, the line may still be at risk of a thermal failure. Even if the base temperature of the fin is lower than the melting point, the mechanical strength may be significantly reduced, thereby causing a structural failure.

❒ ❒ ❒

PROBLEM 3.18 Power Transistor Leads as Fins

A power transistor with three 0.8 mm-diameter leads of a copper-silver alloy (k = 400 W/m·K) dissipates 3.5 W of power. The printed circuit board on which the transistor is mounted is maintained at 30°C. Show that the predicted case temperature of the transistor is lower if the leads are analyzed as fins rather than simple thermal resistors with no convection from their surface. The air temperature is 20°C, and the heat transfer coefficient for the leads is 60 W/m^2·K. The length of the leads from the transistor case to the printed circuit board is 9.0 mm.

DIAGRAM

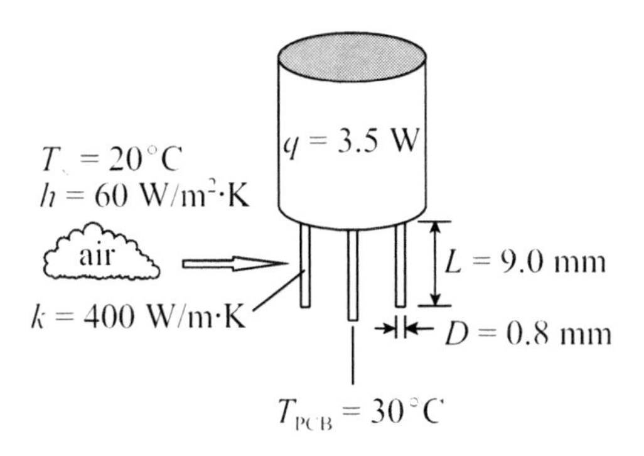

ASSUMPTIONS

1. One-dimensional steady conduction.
2. All the power dissipated by the transistor is conducted into the leads.
3. The three leads equally share the dissipated power.
4. Radiation is neglected.
5. Uniform heat transfer coefficient.
6. Constant properties.

PROPERTIES

copper-silver alloy: $k = 400$ W/m·K

ANALYSIS

To show that the predicted case temperature of the transistor is lower by analyzing the leads as fins, we first analyze the leads as simple resistors with no convection from their surface. By dividing the dissipated power equally among the three leads, we perform the analysis for a single lead. The cross sectional area of a lead is

$$A_c = \frac{\pi D^2}{4}$$

$$A_c = \frac{\pi (0.8 \times 10^{-3}\ \text{m})^2}{4}$$

$$= 5.027 \times 10^{-7}\ \text{m}^2$$

The thermal resistance of a lead is given by the relation

$$R_t = \frac{L}{kA_c}$$

$$R_t = \frac{(9.0 \times 10^{-3}\ \text{m})}{(400\ \text{W/m·K})(5.027 \times 10^{-7}\ \text{m}^2)}$$

$$= 44.76\ \text{K/W}$$

The heat conducted into the base of a lead is given by the relation

$$q_b = q/3 = \frac{T_c - T_{\text{PCB}}}{R_t}$$

where T_c is the case temperature of the transistor. Thus, the case temperature is

$$T_c = q_b R_t + T_{\text{PCB}}$$

$$T_c = (1.167\ \text{W})(44.76\ \text{K/W}) + 30°\text{C}$$

$$= \underline{\underline{82.2°\text{C}}}$$

Now we determine the case temperature by treating the leads as fins. The differential equation for one-dimensional steady heat conduction in a fin is

$$\frac{d^2\theta}{dx^2} - m^2\theta = 0$$

where the quantities θ and m are defined by the relations

$$\theta \equiv T(x) - T_\infty$$

$$m \equiv \left(\frac{hP}{kA_c}\right)^{1/2}$$

The boundary conditions are

$$\theta(0) = \theta_b \quad , \quad \theta(L) = \theta_L$$

where $\theta_b = T_c - T_\infty$ and $\theta_L = T_{PCB} - T_\infty$. The general solution of the above differential equation is

$$\theta(x) = C_1 e^{mx} + C_2 e^{-mx}$$

where C_1 and C_2 are integration constants. Substituting the boundary conditions into the differential equation and solving for the integration constants, the temperature distribution in the fin is given by the relation

$$\frac{\theta}{\theta_b} = \frac{(\theta_L/\theta_b)\sinh(mx) + \sinh[m(L-x)]}{\sinh(mL)}$$

From Fourier's law,

$$q_b = -kA_c\frac{d\theta(0)}{dx} = M\frac{\cosh(mL) - \theta_L/\theta_b}{\sinh(mL)}$$

where,

$$M = (hPkA_c)^{1/2}\theta_b$$

$$M = [(60 \text{ W/m}^2\cdot\text{K})(2.513 \times 10^{-3} \text{ m})(400 \text{ W/m}\cdot\text{K})(5.027 \times 10^{-7} \text{ m}^2)]^{1/2}(T_c - T_\infty)$$

$$= 5.507 \times 10^{-3}(T_c - T_\infty)$$

We have

$$m = \left[\frac{(60\ \text{W/m}^2\cdot\text{K})(2.513\times10^{-3}\ \text{m})}{(400\ \text{W/m}\cdot\text{K})(5.027\times10^{-7}\ \text{m}^2)}\right]^{1/2}$$

$$= 27.38\ \text{m}^{-1}$$

Rearranging the expression above for q_b and solving for the case temperature T_c,

$$T_c = \frac{q_b \sinh(mL) + (hPkA_c)^{1/2}(T_{PCB} - T_\infty)}{(hPkA_c)^{1/2}\cosh(mL)} + T_\infty$$

$$T_c = \frac{(1.167\ \text{W})(0.2489) + (5.507\times10^{-3}\ \text{K}^{-1})(30-20)^\circ\text{C}}{(5.507\times10^{-3}\ \text{K}^{-1})(1.0305)} + 20^\circ\text{C}$$

$$= \underline{\underline{80.9^\circ\text{C}}}$$

Thus, the predicted case temperature is 1.3 °C lower when the leads are analyzed as fins.

DISCUSSION

By modeling the leads as fins, the predicted case temperature of the transistor is 1.3 °C lower than if the leads were modeled as simple thermal resistors with no convection from their surface. This temperature reduction would probably not be significant in most applications. However, if the transistor operates at or very near its maximum case temperature, the second approach to the analysis would at least suggest that the transistor will not fail due to overheating.

❒ ❒ ❒

PROBLEM 3.19 Air Cooled Dual In-Line Device with and without a Heat Sink

An air cooled dual in-line (DIP) electronic device dissipates 0.6 W to 20 °C air. The maximum allowable case temperature is 85 °C. Compare the predicted case temperatures of the device with and without a heat sink. For each case, use a heat transfer coefficient of 35 W/m²·K. The DIP package measures 12 mm × 8.0 mm, and the fins on the heat sink have a length and thickness of 3.8 mm and 0.25 mm, respectively. The heat sink is made of 1100 aluminum alloy with a thermal conductivity of 222 W/m·K.

DIAGRAM

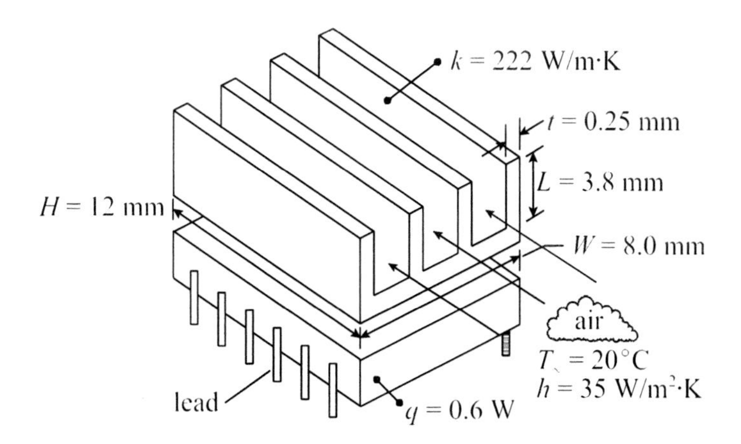

ASSUMPTIONS

1. One-dimensional steady conduction in the fins.
2. Heat transfer coefficient is the same with and without the heat sink.
3. Uniform heat transfer coefficient.
4. Radiation is neglected.
5. Heat transfer from the sides and bottom of the device and through the leads is neglected.
6. Heat is uniformly dissipated from the top surface of the device.
7. Thermal contact resistance between the device and heat sink is neglected.
8. The tips of the fins are insulated.
9. The heat sink base has negligible thermal resistance.
10. Constant properties.

PROPERTIES

1100 aluminum: $k = 222$ W/m·K

ANALYSIS

First, we find the case temperature of the device without a heat sink. The surface area of the top of the device is

$$A = WH$$

$$= (8.0 \times 10^{-3}\ \text{m})(0.012\ \text{m})$$

$$= 9.60 \times 10^{-5}\ \text{m}^2$$

Newton's law of cooling is given by the relation

$$q = hA(T_s - T_\infty)$$

so the case (surface) temperature of the device is

$$T_s = \frac{q}{hA} + T_\infty$$

$$T_s = \frac{0.6\ \text{W}}{(35\ \text{W/m}^2\cdot\text{K})(9.60 \times 10^{-5}\ \text{m}^2)} + 20°\text{C}$$

$$= \underline{\underline{199°\text{C}}}$$

Under the given assumptions, the device requires a heat sink. The cross sectional area and perimeter of a fin are given by the relations

$$A_c = Ht$$

$$A_c = (0.012\ \text{m})(0.25 \times 10^{-3})$$

$$= 3.00 \times 10^{-6}\ \text{m}^2$$

$$P \approx 2H$$

$$P \approx 2(0.012\ \text{m})$$

$$= 0.024\ \text{m}$$

A fin parameter needed in the analysis is defined by the relation

$$m \equiv \left(\frac{hP}{kA_c}\right)^{1/2}$$

$$m = \left[\frac{(35\ \text{W/m}^2\cdot\text{K})(0.024\ \text{m})}{(222\ \text{W/m}\cdot\text{K})(3.00 \times 10^{-6}\ \text{m}^2)}\right]^{1/2}$$

$$= 35.5\ \text{m}^{-1}$$

Fin efficiency η_f is given by the relation

$$\eta_f = \frac{\tanh(mL)}{mL}$$

$$\eta_f = \frac{\tanh[(35.5\ \text{m}^{-1})(3.8 \times 10^{-3}\ \text{m})]}{(35.5\ \text{m}^{-1})(3.8 \times 10^{-3}\ \text{m})}$$

$= 0.9940$

In order to find the case temperature, we must first find the overall surface efficiency η_o for the heat sink. The fin surface area A_f and total heat sink surface area A_t are required. We have

$$A_f = 2N_f HL$$

where $N_f = 4$, the number of fins. Thus,

$$A_f = 2(4)(0.012\text{ m})(3.8 \times 10^{-3}\text{ m})$$

$$= 3.648 \times 10^{-4}\text{ m}^2$$

The total heat sink surface area is the area of the fins plus the interfin areas minus the areas occupied by the footprints of the fins.

$$A_t = A_f + HW - N_f Ht$$

$$A_t = 3.648 \times 10^{-4}\text{ m}^2 + (0.012\text{ m})(8.0 \times 10^{-3}\text{ m}) - (4)(0.012\text{ m})(0.25 \times 10^{-3}\text{ m})$$

$$= 4.488 \times 10^{-4}\text{ m}^2$$

The surface efficiency is given by the relation

$$\eta_o = 1 - \frac{A_f}{A_t}(1 - \eta_f)$$

$$\eta_o = 1 - (3.648 \times 10^{-4}\text{ m}^2)/(4.488 \times 10^{-4}\text{ m}^2)(1 - 0.9940)$$

$$= 0.9951$$

The heat sink base temperature, which we assume is equal to the surface temperature of the device, is given by the relation

$$T_s = \frac{q}{\eta_o h A_t} + T_\infty$$

$$T_s = \frac{0.6\text{ W}}{(0.9951)(35\text{ W/m}^2\cdot\text{K})(4.488 \times 10^{-4}\text{ m}^2)} + 20°\text{C}$$

$$= \underline{\underline{58.4°\text{C}}}$$

DISCUSSION

Given the assumptions of the problem, the analysis shows that the device will overheat without a heat sink. By attaching a miniature heat sink to the device, the case temperature is below the maximum allowable value of 85°C. However, if assumptions 4 and 5 were removed, the predicted case temperature might be lower than 85°C without a heat sink. A separate analysis would be required to show this.

❐ ❐ ❐

PROBLEM 3.20 Design of a Pin Fin Heat Sink

A 10 cm × 16 cm electronic device is air cooled using a stainless steel (k = 15 W/m·K) heat sink with circular pin fins. The surface of the device must be maintained at 56°C in an environment where the air temperature is 10°C. The device dissipates 75 W, and the length and diameter of the pin fins are 3.0 cm and 4.2 mm, respectively. If the heat transfer coefficient is 71 W/m²·K, how many fins are required?

DIAGRAM

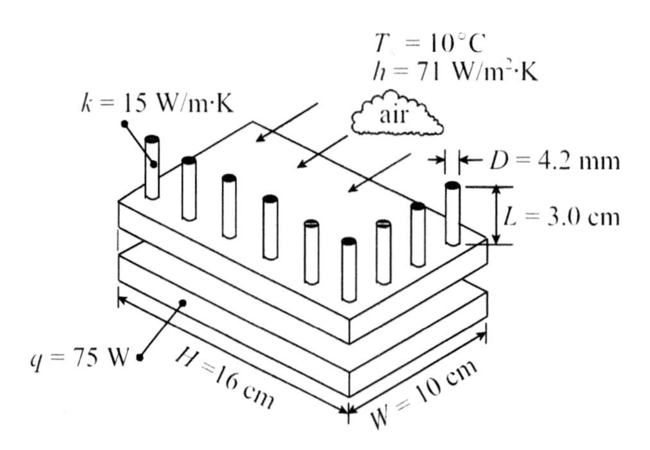

ASSUMPTIONS

1. One-dimensional steady conduction in the fins.
2. Uniform heat transfer coefficient.
3. Radiation is neglected.
4. All dissipated power is uniformly distributed across the top surface of the device.
5. Thermal contact resistance between the device and heat sink is neglected.
6. The tips of the fins are insulated.
7. The heat sink base has negligible thermal resistance.
8. Constant properties.

PROPERTIES

stainless steel: $k = 15$ W/m·K

ANALYSIS

The heat transfer for the heat sink is given by the relation

$$q = hA_t\theta_b\eta_o$$

where the total heat transfer surface area of the heat sink is given by the relation

$$A_t = A_f + HW - N_f A_c$$

where A_c is the fin cross sectional area. The fin surface area A_f is given by the relation

$$A_f = N_f PL$$

where N_f is the number of fins and P is the fin perimeter. The heat sink surface efficiency η_o is given by the relation

$$\eta_o = 1 - \frac{A_f}{A_t}\left(1 - \eta_f\right)$$

where fin efficiency η_f is given by the relation

$$\eta_f = \frac{\tanh(mL)}{mL}$$

After substituting the relations for A_t, A_f and η_o into the relation above for heat transfer, and after some simplification, we obtain a relation for the number of fins as

$$N_f = \frac{q/h\theta_b - HW}{PL\eta_f - A_c}$$

We have,

$$\theta_b = T_b - T_\infty$$

$\theta_b = (56 - 10)$°C

$= 46$°C

$$P = \pi D$$

$P = \pi\,(4.2 \times 10^{-3}$ m)

$= 0.0132 \text{ m}$

$$A_c = \frac{\pi D^2}{4}$$

$$A_c = \frac{\pi (4.2 \times 10^{-3} \text{ m})^2}{4}$$

$= 1.385 \times 10^{-5} \text{ m}^2$

$$m \equiv \left(\frac{hP}{kA_c} \right)^{1/2}$$

$$m = \left[\frac{(71 \text{ W/m}^2\cdot\text{K})(0.0132 \text{ m})}{15 \text{ W/m}\cdot\text{K})(1.385 \times 10^{-5} \text{ m}^2)} \right]^{1/2}$$

$= 67.2 \text{ m}^{-1}$

$$\eta_f = \frac{\tanh [(67.2 \text{ m}^{-1})(0.030 \text{ m})]}{(67.2 \text{ m}^{-1})(0.030 \text{ m})}$$

$= 0.4787$

Thus, the number of fins required is

$$N_f = \frac{(75 \text{ W})/[(71 \text{ W/m}^2\cdot\text{K})(46\,°\text{C})] - (0.16 \text{ m})(0.10 \text{ m})}{(0.0132 \text{ m})(0.030 \text{ m})(0.4787) - 1.385 \times 10^{-5} \text{ m}^2}$$

$= 39.6 \approx \underline{\underline{40}}$

DISCUSSION

Forty fins fit nicely on a heat sink measuring 10 cm × 16 cm. The aspect ratio of the heat sink base is 5:8, so a 5 × 8 array (5 rows and 8 columns) of pin fins works well for this application. We note that if $q/h\theta_b \le HW$, no fins are required to maintain the surface of the device at its maximum allowable temperature. In this case we set $N_f = 0$, and we have

$$N_f = \frac{q/h\theta_b - HW}{PL\eta_f - A_c}$$

which reduces to Newton's law of cooling for an unfinned surface of area HW,

$$q = hHW(T_b - T_\infty)$$

❒ ❒ ❒

Chapter 4

Two-Dimensional Steady Conduction

PROBLEM 4.1 Conduction in a Rectangular Region with Prescribed Temperatures at the Boundaries

A two-dimensional rectangular region with dimensions $L = 4.0$ m and $W = 1.0$ m is subjected to prescribed temperatures at the boundaries, as shown in the diagram. Using the infinite series solution generated by the separation of variables method, find the temperature at (2.0, 0.5) m, the midpoint of the region. Use only the first six nonzero terms of the infinite series. Assess the error if only three terms are used. Plot the temperature distributions $T(x, 0.5)$ and $T(2.0, y)$.

DIAGRAM

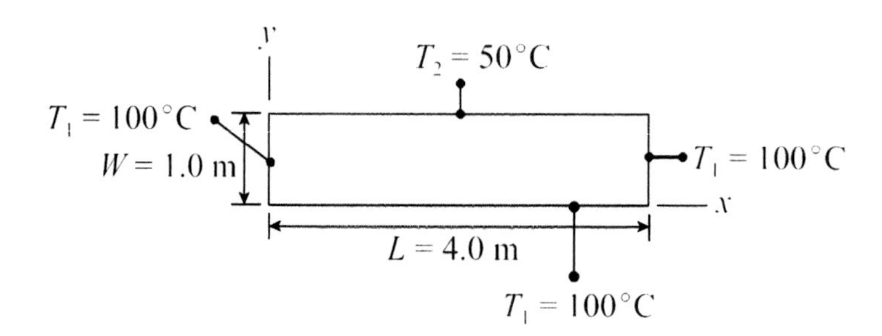

ASSUMPTIONS

1. Steady two-dimensional conduction.
2. Constant temperatures prescribed at the boundaries.

PROPERTIES

not applicable

ANALYSIS

The temperature distribution is given by the relation

$$\theta(x,y) = \frac{2}{\pi}\sum_{n=1}^{\infty}\frac{(-1)^{n+1}+1}{n}\sin\left(\frac{n\pi x}{L}\right)\frac{\sinh(n\pi y/L)}{\sinh(n\pi W/L)}$$

where,

$$\theta(x,y) \equiv \frac{T(x,y)-T_1}{T_2-T_1}$$

Quantities in the relation for the temperature distribution have the following values:

$x = 2.0$ m , $y = 0.5$ m , $L = 4.0$ m , $W = 1.0$ m

The first six nonzero terms of the infinite series and the solution for the temperature distribution are summarized in the table below.

n	term	$\theta\,(2.0,0.5)$
1	0.5905	0.5905
3	−0.1193	0.4712
5	0.03505	0.5063
7	−0.01159	0.4947
9	0.004125	0.4988
11	−0.001540	0.4972

The temperature at the midpoint of the region is

$$T(2.0,0.5) = \theta\,(2.0,0.5)\,(T_2 - T_1) + T_1$$

$$= (0.4972)(50 - 100) + 100$$

$$= \underline{\underline{75.1°\text{C}}}$$

A graph of $T(x,0.5)$ is shown below.

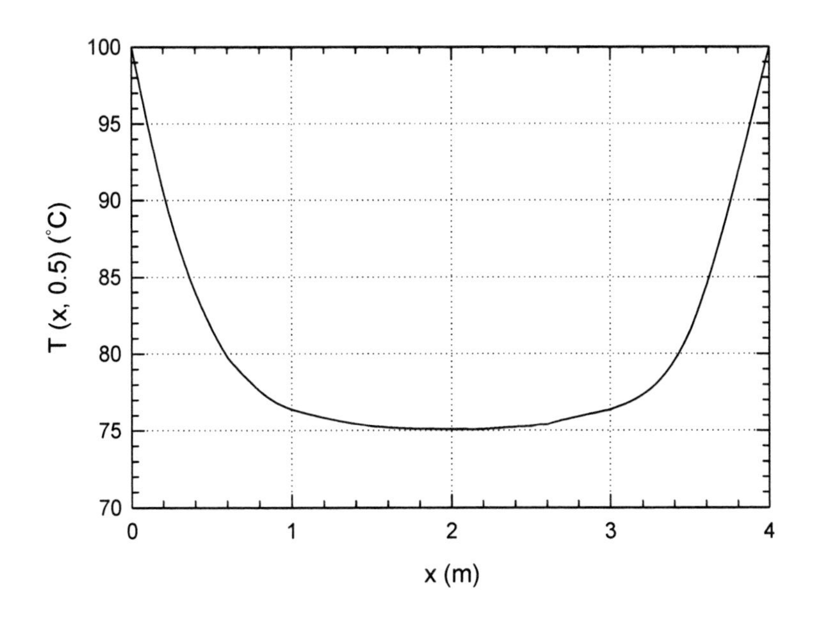

A graph of $T(2.0,y)$ is shown below.

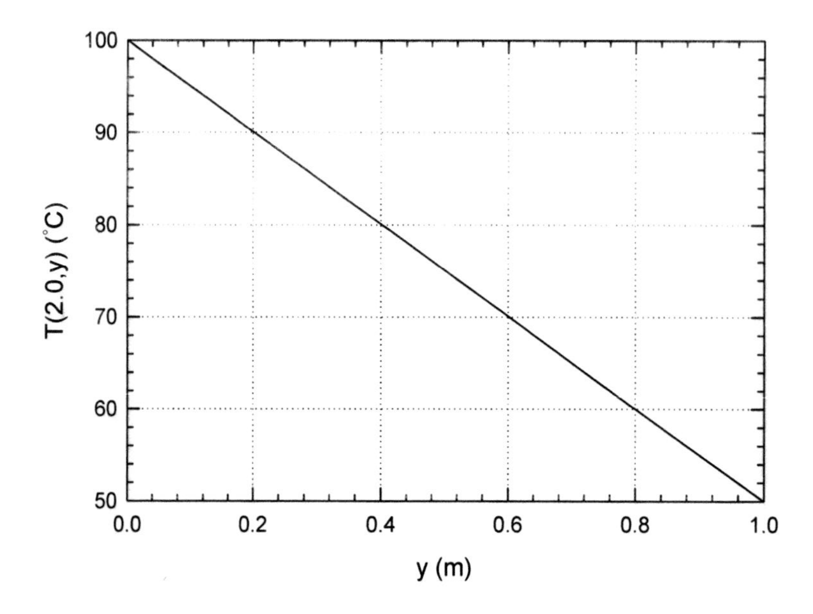

If only three terms are taken in the series, the temperature of the midpoint of the region is

$$T(2.0,0.5) = \theta(2.0,0.5)(T_2 - T_1) + T_1$$

$$= (0.5063)(50 - 100) + 100$$

$$= \underline{\underline{74.7°C}}$$

The percent error is

$$\text{error} = \frac{75.1 - 74.7}{75.1} \times 100 = \underline{\underline{0.53 \text{ percent}}}$$

DISCUSSION

The error associated with taking only three terms instead of six in the series would be acceptable for most applications. As expected, the graph of $T(x,0.5)$ is symmetric about $x = 2.0$ m, whereas the graph of $T(2.0,y)$ is a straight line from $T = 100°C$ at $y = 0$ to $T = 50°C$ at $y = 1.0$ m, in accordance with the boundary conditions.

PROBLEM 4.2 **Conduction in a Rectangular Region with Energy Generation and Prescribed Temperatures at the Boundaries**

A long electrical heater with a rectangular cross section of 6 cm × 4 cm generates energy uniformly at $\dot{q} = 30$ MW/m^3. The heater is a solid with thermal conductivity $k = 12$ W/m·K, and all sides of the heater are held at 0°C. Using the infinite series solution generated by the separation of variables method, find the temperature at (0,0) cm, the center of the heater, and (0,1) cm. Plot the temperature distribution $T(x,1)$.

DIAGRAM

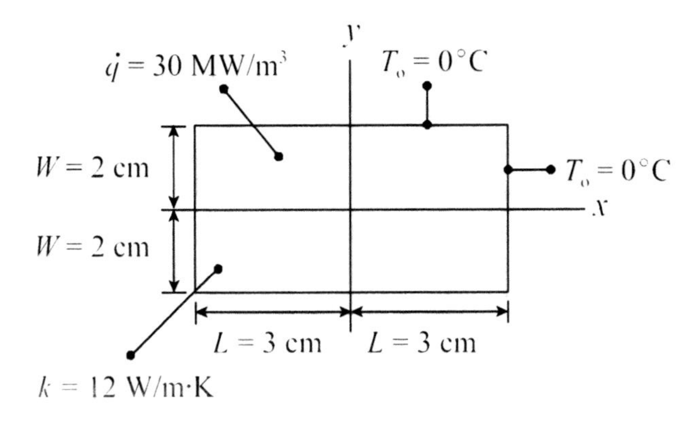

ASSUMPTIONS

1. Steady two-dimensional conduction.
2. Constant temperatures prescribed at the boundaries.
3. Uniform generation.
4. Constant properties.

PROPERTIES

thermal conductivity: $k = 12$ W/m·K

ANALYSIS

Because the heater has thermal symmetry in both the x and y directions, the origin is placed at the center of the rectangular cross section, as shown in the diagram. The temperature distribution is given by the relation

$$\frac{\theta(x,y)}{\dot{q}L^2/k} = \frac{1}{2}\left[1-\left(\frac{x}{L}\right)^2\right] - 2\sum_{n=0}^{\infty}\frac{(-1)^n}{(\lambda_n L)^3}\left[\frac{\cosh(\lambda_n y)}{\cosh(\lambda_n W)}\right]\cos(\lambda_n x)$$

where,

$$\theta(x,y) \equiv T(x,y) - T_o$$

$$\lambda_n = \frac{(2n+1)\pi}{2L}$$

For $x = y = 0$, the infinite series above converges rapidly, giving a sum of 0.3208 after only four terms. The temperature at the center of the heater is

$$T(0,0) = \frac{(30 \times 10^6 \text{ W/m}^3)(0.03 \text{ m})^2}{12 \text{ W/m·K}}(0.5 - 0.3208) + 0°\text{C}$$

$$= \underline{\underline{403°\text{C}}}$$

For $x = 0$, $y = 1$, the infinite series converges to 0.3638. The temperature at these coordinates is

$$T(0,1) = \frac{(30 \times 10^6 \text{ W/m}^3)(0.03 \text{ m})^2}{12 \text{ W/m·K}}(0.5 - 0.3638) + 0°\text{C}$$

$$= \underline{\underline{306°\text{C}}}$$

A graph of $T(x,1)$ is shown below.

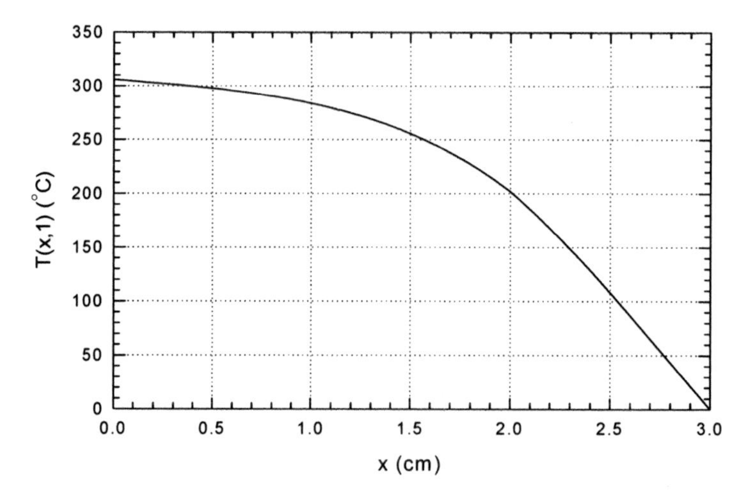

DISCUSSION

The point (0,0) has the highest temperature, 403°C, in the region.

❐ ❐ ❐

PROBLEM 4.3 **Conduction in a Cylinder with Prescribed Temperatures at the Boundaries**

The end of a solid cylinder of length $L = 24$ cm and radius $R = 8$ cm is maintained at temperature $T_o = 100°C$, while the other surfaces are maintained at temperature $T_1 = 0°C$, as shown in the diagram. Using the infinite series solution generated by the separation of variables method, find the temperature at (0,20) cm, and plot the temperature distributions $T(r,18)$, $T(r,20)$ and $T(r,22)$ on the same graph.

DIAGRAM

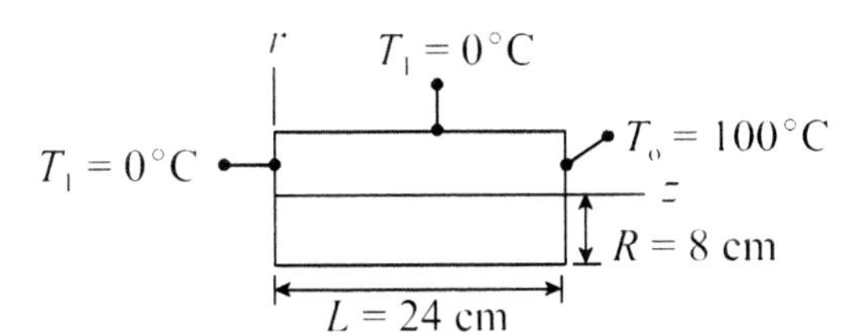

ASSUMPTIONS

1. Steady two-dimensional conduction.
2. Constant temperatures prescribed at the boundaries.

PROPERTIES

not applicable

ANALYSIS

For mathematical convenience and due to the symmetry of the problem, the axes are defined as shown in the diagram. The temperature distribution is given by the relation

$$\frac{\theta(r,z)}{\theta_o} = 2\sum_{n=1}^{\infty}\left[\frac{\sinh(\lambda_n z)}{\sinh(\lambda_n L)}\right]\frac{J_o(\lambda_n r)}{(\lambda_n R)J_1(\lambda_n R)}$$

where,

$$\theta(r,z) \equiv T(r,z) - T_1$$

$$\theta_o = T_o - T_1$$

and J_o and J_1 are Bessel functions of the first kind of zero and first order, respectively. The parameters λ_n are zeros of the relation

$$J_o(\lambda_n R) = 0$$

For the temperature at (0,20) cm, the infinite series above converges rapidly, giving a sum of 0.4235 after only five terms. The temperature at this location is

$$T(0,20) = 0.4235\theta_o + T_1$$

$$= (0.4235)(100 - 0) + 0$$

$$= \underline{\underline{42.4°C}}$$

A graph of $T(r,18)$, $T(r,20)$ and $T(r,22)$ is shown below.

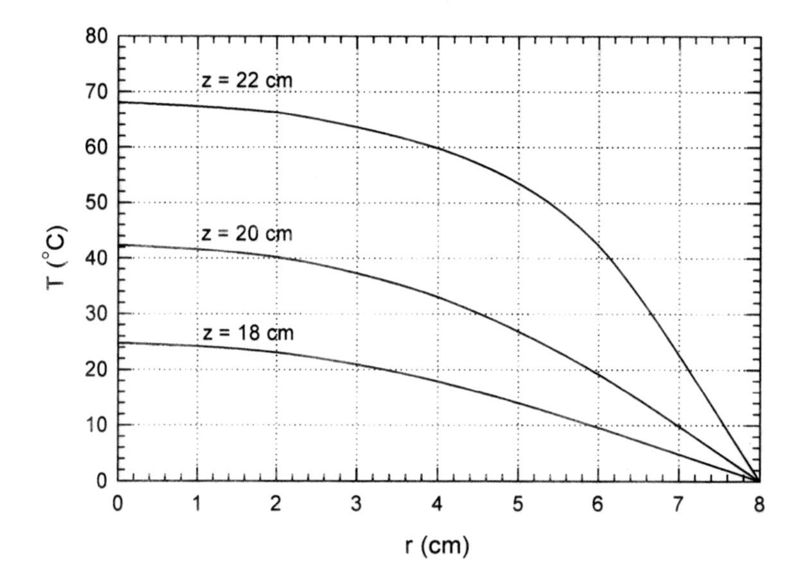

DISCUSSION

As shown in the graph, the radial temperature change is more pronounced near the end where the surface temperature is 100°C. The radial temperature change decreases toward the other end of the cylinder where the surface temperature is 0°C. We also note from the graph that for all values of z the slope of the temperature is zero at $r = 0$, i.e., $\partial T(0,z)/\partial r = 0$. This is one of the boundary conditions used in the solution to the partial differential equation for this problem,

$$\frac{\partial}{\partial r}\left(r\frac{\partial\theta}{\partial r}\right) + r\frac{\partial^2\theta}{\partial z^2} = 0$$

❐ ❐ ❐

PROBLEM 4.4 Heat Loss from a Buried Pipe Carrying Warm Oil

A long steel pipe carrying warm oil is buried in the earth such that its center line is 60 cm below the surface of the ground. The oil temperature is 50°C, and the surface temperature of the ground is 15°C. If the outside radius of the pipe is 17 cm, what is the heat loss per meter of pipe length? For the thermal conductivity of soil, use $k = 1$ W/m·K.

DIAGRAM

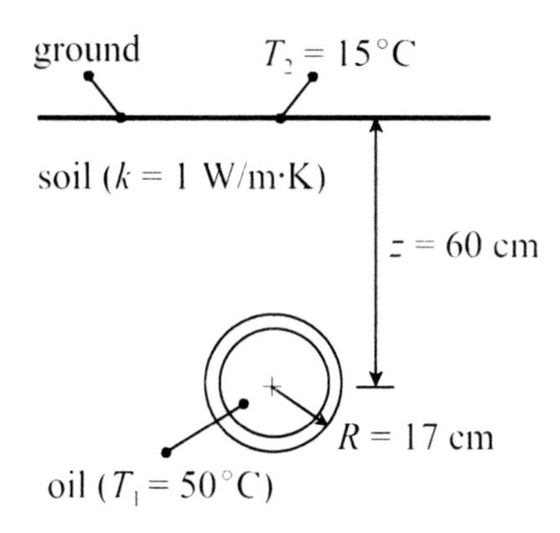

ASSUMPTIONS

1. Steady two-dimensional conduction.
2. Neglect thermal resistances of the thermal boundary layer and pipe wall.
3. The pipe is parallel to the ground.
4. Constant oil and ground temperatures.
5. Constant properties.

PROPERTIES

soil: $k = 1$ W/m·K

ANALYSIS

The relation for two-dimensional conduction is

$$q = Sk(T_1 - T_2)$$

where S is conduction shape factor and k is thermal conductivity. For a cylinder of length L buried in a semi-infinite region and parallel to its surface, the relation for S is

$$S = \frac{2\pi L}{\cosh^{-1}(2z/D)}$$

This shape factor relation is valid for $L >> D$. For a unit length of pipe, the shape factor is

$$S = \frac{2\pi\,(1\text{m})}{\cosh^{-1}[2(0.60\text{ m})/(0.34\text{ m})]}$$

$$= 3.249\text{ m}$$

so the heat loss for a unit length of pipe is

$$q = S\,k\,(T_1 - T_2)$$

$$= (3.249\text{ m})(1\text{ W/m·K})(50 - 15)°\text{C}$$

$$= \underline{\underline{114\text{ W}}}$$

DISCUSSION

The heat loss could be reduced by burying the pipe deeper in the ground. For example, if $z = 1.20$ m, twice the depth used above, the shape factor is

$$S = \frac{2\pi\,(1\text{m})}{\cosh^{-1}[2(1.20\text{ m})/(0.34\text{ m})]}$$

$$= 2.378\text{ m}$$

and the heat loss per meter of pipe length is

$$q = (2.378\text{ m})(1\text{ W/m·K})(50 - 15)°\text{C}$$

$$= 83.2\text{ W}$$

An alternative relation for the shape factor is

$$S = \frac{2\pi L}{\ln(4z/D)}$$

This relation yields a value nearly identical to the first,

$$S = \frac{2\pi\,(1\text{ m})}{\ln[4(1.20\text{ m})/(0.34\text{ m})]}$$

$$= 2.373\text{ m}$$

❒ ❒ ❒

PROBLEM 4.5 **Temperature of a Current-Carrying Wire Encased in a Porcelain Tube**

A long 14-gage nichrome wire (D = 1.628 mm) is encased within the center of a long porcelain rod (k = 2.2 W/m·K) of square cross section measuring 25 mm × 25 mm. The wire carries a steady current of I = 30 A, and the resistance of the wire is 0.54 Ω/m. If the outside surface of the porcelain rod is maintained at10°C, what is the wire temperature? Plot the wire temperature for the range of current I = 20 A to I = 50 A.

DIAGRAM

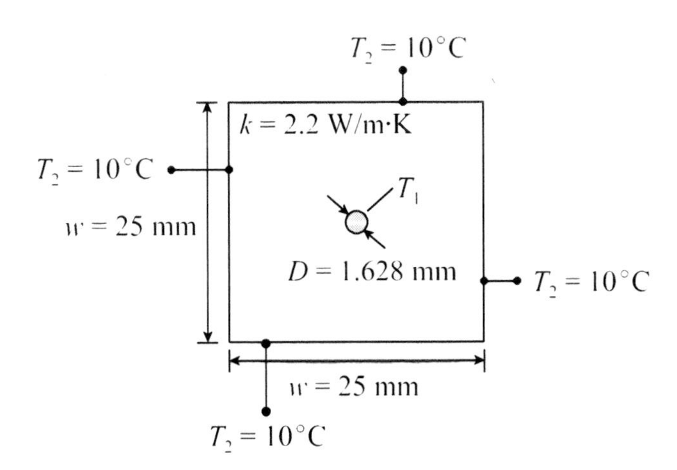

ASSUMPTIONS

1. Steady two-dimensional conduction.
2. Neglect contact resistance between the wire and rod.
3. Constant surface temperature.
4. Steady current.
5. Constant properties.

PROPERTIES

porcelain: k = 2.2 W/m·K

ANALYSIS

The heat transfer is the electrical power dissipated by the wire.

$$P = q = I^2 R$$

For a unit length of wire, the heat transfer is

$$q = (30\text{ A})^2\,(0.54\ \Omega)$$

$$= 486\text{ W}$$

The conduction shape factor for a cylinder of length L centered in a square solid of equal length is given by the relation

$$S = \frac{2\pi L}{\ln(1.08w/D)}$$

$$S = \frac{2\pi\,(1\text{ m})}{\ln[1.08(0.025\text{ m})/(1.628 \times 10^{-3}\text{ m})]}$$

$$= 2.237\text{ m}$$

The heat transfer in terms of the shape factor is given by the relation

$$q = Sk(T_1 - T_2)$$

Solving for T_1, the wire temperature,

$$T_1 = \frac{q}{Sk} + T_2$$

$$T_1 = \frac{486\text{ W}}{(2.237\text{ m})(2.2\text{ W/m·K})} + 10°\text{C}$$

$$= \underline{\underline{109°\text{C}}}$$

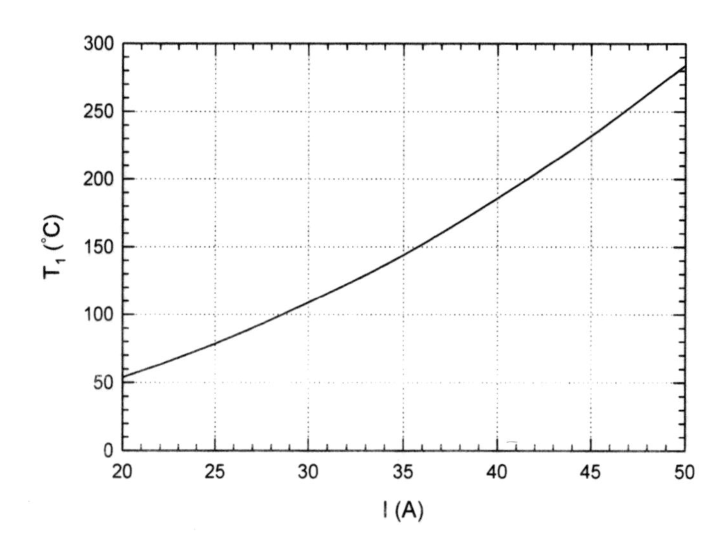

DISCUSSION

Nichrome consists of 80 percent nickel and 20 percent chromium. The melting point of nichrome is about 1400°C, which is one of the reasons this alloy is used for making electrical resistance heating elements. The maximum current the wire can carry before melting occurs is 113 A. Strictly speaking, this current corresponds to the surface temperature of the wire. There is a small radial temperature gradient in the wire, where the maximum temperature occurs at the wire's center. Thus, a current slightly less than 113 A is required to melt the wire.

❐ ❐ ❐

PROBLEM 4.6 Design of a Driveway Deicing System

A driveway deicing system consists of an array of polyethylene pipes ($k_p = 0.33$ W/m·K) buried beneath and parallel to a concrete slab (k_c = 1.4 W/m·K). The pipes, which have an outside diameter and wall thickness of 4.0 cm and 2.0 mm, respectively, carry a water/ethylene glycol solution that has an average temperature of 50°C. The heat transfer coefficient for the inside surface of the pipes is 3000 W/m^2·K. Through laboratory tests, it has been determined that ice will not form on the driveway under typical winter conditions if the heat transfer per unit length of pipe is at least 40 W/m and the surface temperature of the concrete is 10°C. Determine combinations of pipe depth z and center-to-center pipe spacing w to satisfy these design specifications. For the thermal conductivity of soil, use k_s = 1 W/m·K.

DIAGRAM

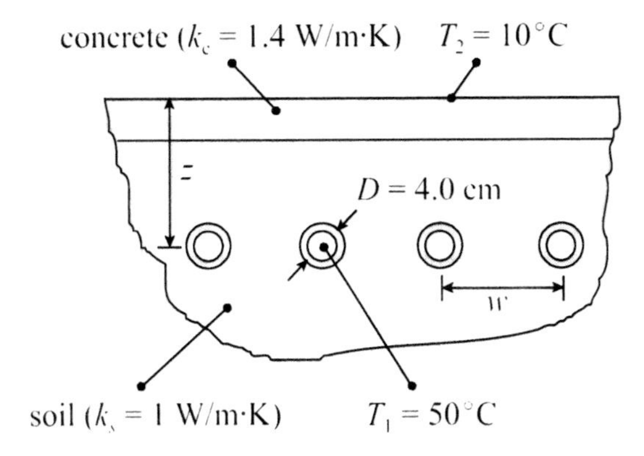

ASSUMPTIONS

1. Two-dimensional steady conduction.
2. Neglect thermal contact resistance between the pipes and soil.
3. Thermal conductivity of the concrete slab is equivalent to that of the soil.
4. Constant properties.

PROPERTIES

concrete: $k_c = 1.4$ W/m·K
soil: $k_s = 1$ W/m·K
polyethylene: $k_p = 0.33$ W/m·K

ANALYSIS

There are three thermal resistances in the deicing system: the thermal boundary layer, the pipe wall and the soil. For a unit length of pipe, these are given by the relations

$$R_{t,1} = \frac{1}{2\pi r_1 L h}$$

$$R_{t,1} = \frac{1}{2\pi\,(0.0180\text{ m})(1\text{ m})(3000\text{ W/m}^2\cdot\text{K})}$$

$$= 2.947 \times 10^{-3}\text{ K/W}$$

$$R_{t,2} = \frac{\ln(r_2/r_1)}{2\pi L k_p}$$

$$R_{t,2} = \frac{\ln(0.020\text{ m}/0.0180\text{ m})}{2\pi\,(1\text{ m})(0.33\text{ W/m}\cdot\text{K})}$$

$$= 0.0508\text{ K/W}$$

$$R_{t,3} = \frac{1}{Sk_s}$$

where the conduction shape factor S is given by the relation

$$S = \frac{2\pi L}{\ln\left[\frac{2w}{\pi D}\sinh(2\pi z/w\right]}$$

This relation is valid for $z > D$. Consistent with the third assumption and as a worst case, the entire semi-infinite region is assumed to have the same thermal conductivity as the soil, as indicated in the relation for $R_{t,3}$ above. The conduction shape factor contains the design variables w and z. The total thermal resistance is given by the relation

$$R_{tot} = R_{t,1} + R_{t,2} + R_{t,3}$$

Using the thermal analogy of Ohm's law, the heat transfer can be written as

$$q = \frac{T_1 - T_2}{R_{tot}}$$

Setting the pipe spacing to $w = 10$ cm, 20 cm and 30 cm, and solving the equations above, we plot the heat transfer per unit pipe length q' as a function of pipe depth z.

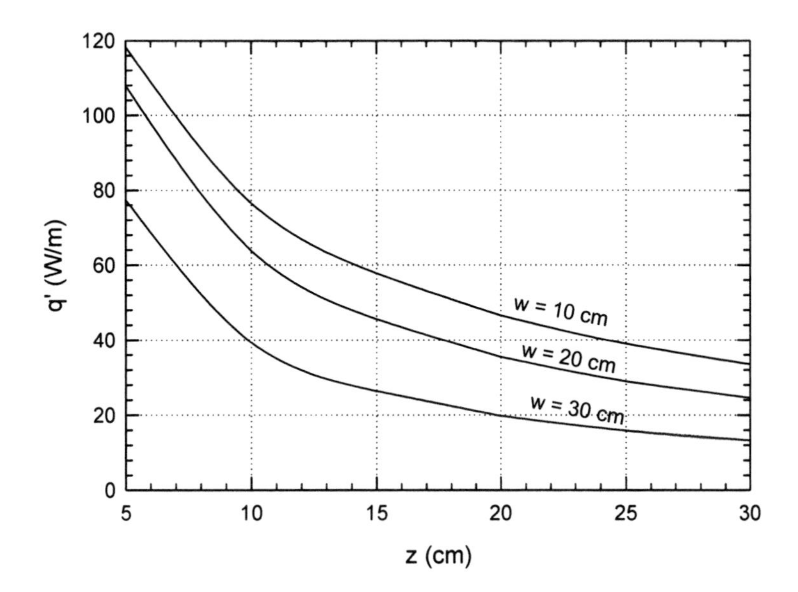

DISCUSSION

The heat transfer $q' = 40$ W/m in the graph represents a threshold below which the deicing system will not meet the design specifications. Thus, every combination of pipe spacing w and pipe depth z that lies above the threshold will meet the design specifications. For pipe spacings of $w = 10$ cm, 20 cm and 30 cm, the maximum pipe depths are $z = 25$ cm, 18 cm and 10 cm, respectively. Of course, these values would change if the temperature of the water/ethylene-glycol solution or the temperature of the concrete surface were different. Changing the diameter of the pipes and the material from which they are made would also affect the results.

The fluid flow and convective heat transfer characteristics of a similar deicing system are illustrated in Problem 8.12.

❒ ❒ ❒

PROBLEM 4.7 Design of a Cubical Heat Treating Oven

A small cubical oven for heat treating metal specimens in a materials laboratory is to be designed. The maximum electrical power available to the oven is 16 kW, and space considerations limit the outside dimensions of the oven to 80 cm. To heat treat a variety of metals under different conditions, the oven must be capable of sustaining an inside air temperature of 1200°C. The walls of the oven are to be made of fire clay brick (k = 1.5 W/m·K). The heat transfer coefficients for the inside and outside surfaces of the oven walls are 17 W/m^2·K and 10 W/m^2·K, respectively, and the temperature of the air in the laboratory is 25°C. Find the minimum wall thickness for this oven.

DIAGRAM

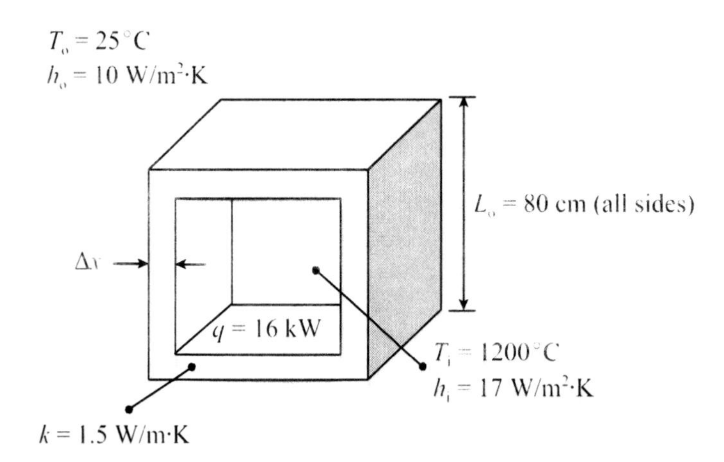

ASSUMPTIONS

1. Two-dimensional steady conduction.
2. Constant heat transfer coefficients and temperatures.
3. Radiation is neglected.
4. Constant properties.

PROPERTIES

fire clay brick: k = 1.5 W/m·K

ANALYSIS

The inside surface area for one wall is given by the relation

$$A_i = L_i^2$$

where the quantity L_i is given by

$$L_i = L_o - 2\Delta x$$

and Δx is the wall thickness. The conduction shape factors for a wall, edge and corner, respectively, for the cubical oven are given by the relations

$$S_w = A_i / \Delta x$$

$$S_e = 0.54 L_i$$

$$S_c = 0.15 \Delta x$$

There are six walls, twelve edges and eight corners, so the conduction shape factor for the cubical enclosure is

$$S = 6S_w + 12S_e + 8S_c$$

The thermal resistances for the inside thermal boundary layer, the oven enclosure and the outside thermal boundary layer, respectively, are given by the relations

$$R_{t,1} = \frac{1}{6A_i h_i}$$

$$R_{t,2} = \frac{1}{Sk}$$

$$R_{t,3} = \frac{1}{6A_o h_o}$$

where the outside surface area of one wall is given by

$$A_o = L_o^2$$

The total thermal resistance is the sum of the resistances above,

$$R_{tot} = R_{t,1} + R_{t,2} + R_{t,3}$$

and the heat transfer is given by the relation

$$q = \frac{T_i - T_o}{R_{tot}}$$

The above equations can be solved iteratively to obtain a value for the wall thickness Δx. We obtain,

$\Delta x = \underline{\underline{8.90 \text{ cm}}}$

DISCUSSION

It is instructive to examine how electrical power affects oven wall thickness. The graph below shows the variation of wall thickness with electrical power for three different oven air temperatures T_i. For a given oven temperature, the wall thickness decreases with electrical power, which is equivalent to the heat transfer. For a given electrical power, thicker walls are required to sustain higher oven air temperatures, as expected.

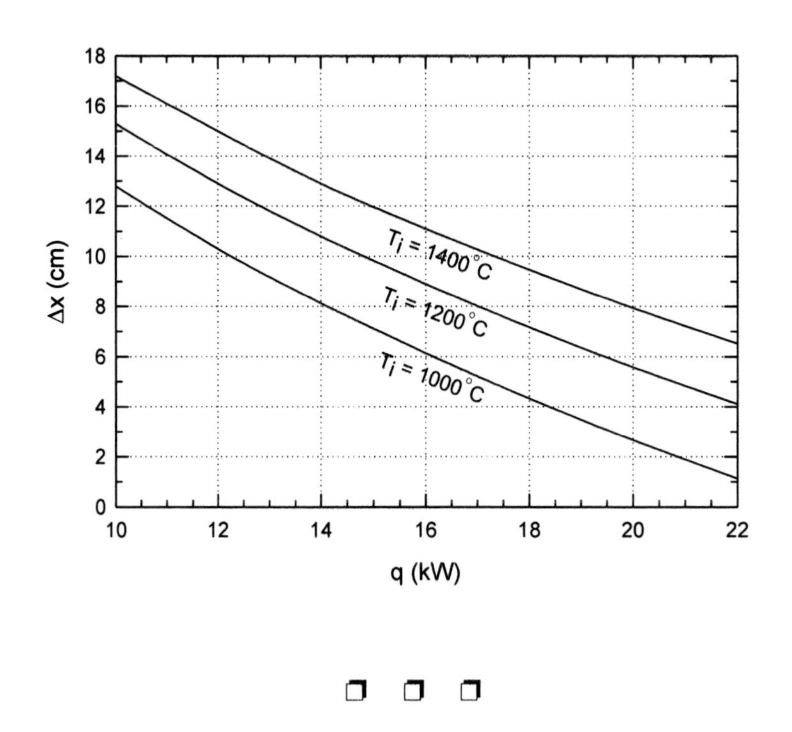

❐ ❐ ❐

PROBLEM 4.8 Heat Transfer Between Buried Hot and Cold Water Pipes

Two long, parallel copper pipes, one carrying hot water at 60°C and the other carrying cold water at 8°C, are buried deep in the ground. Each pipe has an outside diameter of 3.5 cm, and their centerlines are spaced 10 cm apart. If the thermal conductivity of the soil is 1.2 W/m·K, find the heat transfer between the pipes. Investigate the variation of heat transfer with pipe diameter and spacing.

DIAGRAM

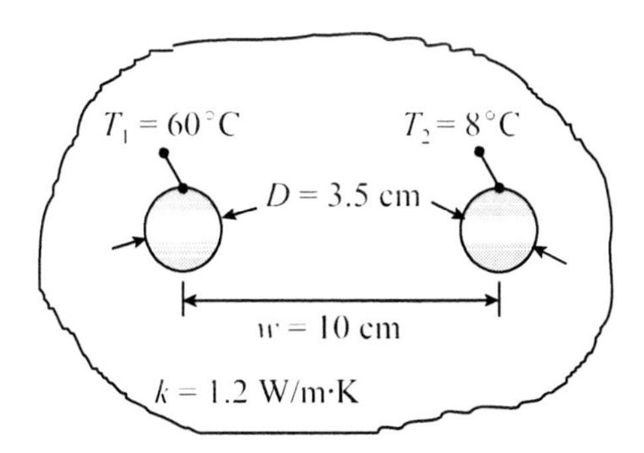

ASSUMPTIONS

1. Steady two-dimensional conduction.
2. Constant temperatures.
3. Neglect thermal resistances of the thermal boundary layer and pipe wall.
4. Neglect thermal contact resistance between the pipes and soil.
5. Constant properties.

PROPERTIES

soil: $k = 1.2$ W/m·K

ANALYSIS

The conduction shape factor for two parallel cylinders of length L buried in an infinite medium is given by the relation

$$S = \frac{2\pi L}{\cosh^{-1}\left(\dfrac{4w^2 - D_1^2 - D_2^2}{2D_1D_2}\right)}$$

The restrictions for this relation are $L >> D_1$, D_2 and $L >> w$. In this problem, $D_1 = D_2$, so for a unit length of pipe we have

$$S = \frac{2\pi\,(1\text{m})}{\cosh^{-1}\left(\dfrac{4(0.10\text{ m})^2 - 2(0.035\text{ m})^2}{2(0.035\text{ m})^2}\right)}$$

$$= 1.836\text{ m}$$

The heat transfer between the pipes is given by

$$q = Sk(T_1 - T_2)$$

q = (1.836 m)(1.2 W/m·K)(60 − 8)°C

= <u>115 W</u>

DISCUSSION

To examine the effect of pipe diameter and spacing on heat transfer, the graph below shows the variation of q with w for three values of pipe diameter D. Heat transfer is clearly greatest for close spacing of large pipes and least for distant spacing of small pipes.

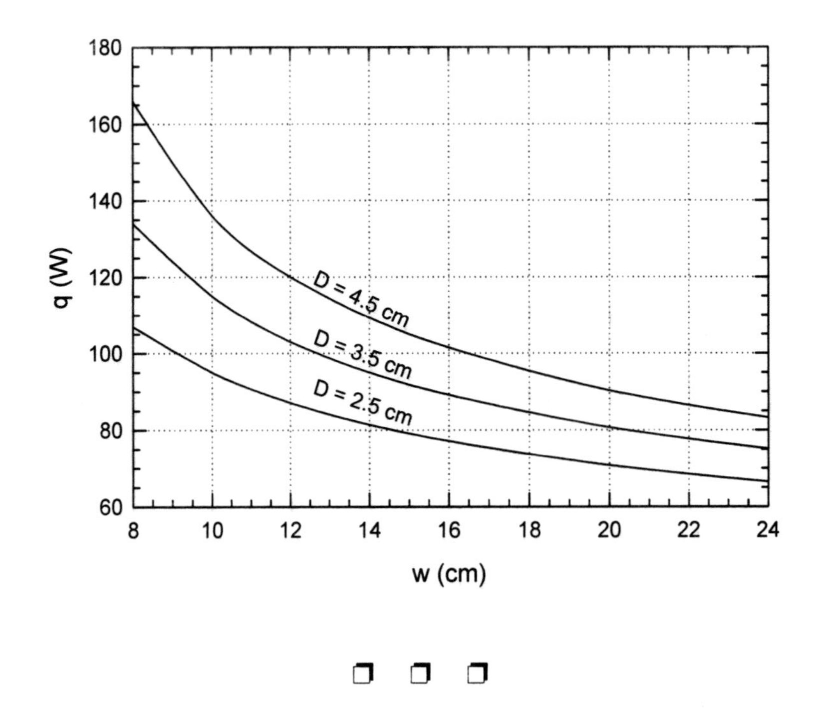

❒ ❒ ❒

PROBLEM 4.9 Derivation of a Finite Difference Equation for an Interior Node with Generation, Point Heat Source and Convection

Derive a finite difference equation for the temperature of an interior node in a two-dimensional region. In the derivation, include energy generation, a point heat source at the node and convection from the surface surrounding the node. Let the spacing between nodes be equal in the x and y directions.

DIAGRAM

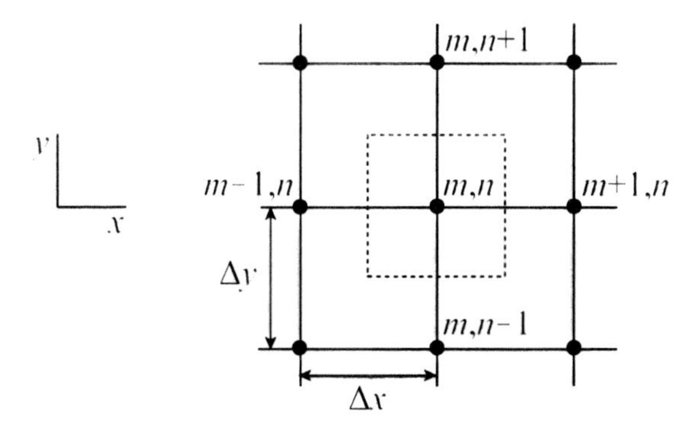

ASSUMPTIONS

1. Steady two-dimensional conduction.
2. Constant properties.

PROPERTIES

thermal conductivity: k

ANALYSIS

The finite difference equation may be derived by applying conservation of energy to the control volume surrounding the interior node, as indicated by the shaded area in the diagram. In the energy balance we may assume that all the heat flows are into the node. For steady conditions, the conservation of energy relation is

$$\dot{E}_{in} + \dot{E}_g = 0$$

The rate of heat transfer into node m,n is given by the relation

$$\dot{E}_{in} = k\Delta y\delta\left(\frac{T_{m-1,n} - T_{m,n}}{\Delta x}\right) + k\Delta y\delta\left(\frac{T_{m+1,n} - T_{m,n}}{\Delta x}\right) + k\Delta x\delta\left(\frac{T_{m,n+1} - T_{m,n}}{\Delta y}\right) + k\Delta x\delta\left(\frac{T_{m,n-1} - T_{m,n}}{\Delta y}\right)$$

$$+ q_s - h\Delta x\Delta y\left(T_{m,n} - T_\infty\right)$$

The first four terms on the right side of the relation above are the finite difference versions of Fourier's law, the fifth term represents the point heat source, and the sixth term represents the convection to one surface of the control volume. The parameters δ and k represent the region thickness and thermal conductivity, respectively, and T_∞ is the free stream fluid temperature. The rate of energy generation is given by the relation

$$\dot{E}_g = \dot{q}\Delta x \Delta y \delta$$

where $\dot{q}$ represents the rate of energy generation per unit volume. Setting $\Delta x = \Delta y = \Delta$, and substituting the above relations into the energy balance,

$$k\delta\left[\left(T_{m-1,n} - T_{m,n}\right) + \left(T_{m+1,n} - T_{m,n}\right) + \left(T_{m,n+1} - T_{m,n}\right) + \left(T_{m,n-1} - T_{m,n}\right)\right]$$

$$+ q_s + \dot{q}\Delta^2\delta - h\Delta^2\left(T_{m,n} - T_\infty\right) = 0$$

Simplifying this relation and solving for $T_{m,n}$, the temperature of the interior node,

$$T_{m,n} = \frac{T_{m-1,n} + T_{m+1,n} + T_{m,n+1} + T_{m,n-1} + \dfrac{1}{k}\left(\dfrac{q_s}{\delta} + \dot{q}\Delta^2\right) + Bi\, T_\infty}{4 + Bi}$$

where the Biot number Bi is given by

$$Bi = \frac{h\Delta^2}{k\delta}$$

DISCUSSION

As an example calculation, let's find the temperature of the interior node given the following values:

$T_{m-1,n} = 100°\text{C}$ $\quad k = 40$ W/m·K $\quad \dot{q} = 50$ kW/m^3

$T_{m+1,n} = 50°\text{C}$ $\quad \delta = 2.5$ mm $\quad T_\infty = 25°\text{C}$

$T_{m,n+1} = 80°\text{C}$ $\quad \Delta = 2.0$ cm $\quad h = 130$ W/m^2·K

$T_{m,n-1} = 60°\text{C}$ $\quad q_s = 3$ W

$$Bi = \frac{(130\ \text{W/m}^2\cdot\text{K})(0.020\ \text{m})^2}{(40\ \text{W/m}\cdot\text{K})(0.0025\ \text{m})}$$

$$= 0.520$$

The temperature of the node is

$$T_{m,n} = \frac{100 + 50 + 80 + 60 + (1/40)[3/0.0025 + (50 \times 10^3)(0.020)^2] + (0.520)(25)}{4 + 0.520}$$

$$= 73.8°\text{C}$$

❒ ❒ ❒

PROBLEM 4.10 **Derivation of a Finite Difference Equation for an Interior Node with Generation, Heat Flux and Radiation**

Derive a finite difference equation for the temperature of an interior node in a two-dimensional region. In the derivation, include energy generation, a uniform heat flux and radiation from the surface surrounding the node. Let the spacing between nodes be equal in the x and y directions.

DIAGRAM

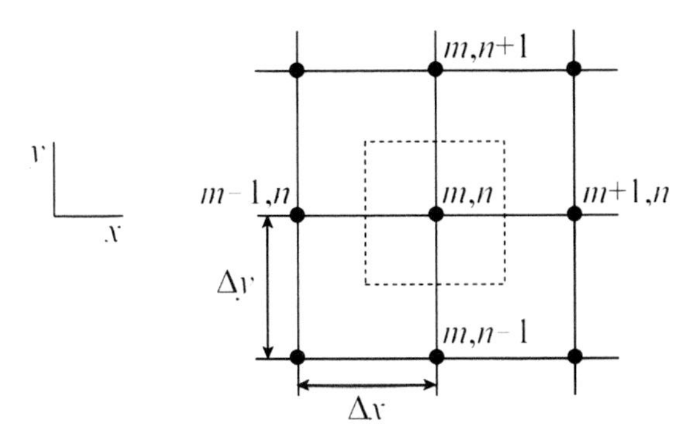

ASSUMPTIONS

1. Steady two-dimensional conduction.
2. Constant properties.

PROPERTIES

thermal conductivity: k
emissivity: ε

ANALYSIS

The finite difference equation may be derived by applying conservation of energy to the control volume surrounding the interior node, as indicated by the shaded area in the diagram. In the energy balance we may assume that all the heat flows are into the node. For steady conditions, the conservation of energy relation is

$$\dot{E}_{in} + \dot{E}_g = 0$$

The rate of heat transfer into node m,n is given by the relation

$$\dot{E}_{in} = k\Delta y\delta\left(\frac{T_{m-1,n} - T_{m,n}}{\Delta x}\right) + k\Delta y\delta\left(\frac{T_{m+1,n} - T_{m,n}}{\Delta x}\right) + k\Delta x\delta\left(\frac{T_{m,n+1} - T_{m,n}}{\Delta y}\right) + k\Delta x\delta\left(\frac{T_{m,n-1} - T_{m,n}}{\Delta y}\right)$$

$$+ q''\Delta x\Delta y - \Delta x\Delta y\sigma\varepsilon\left(T_{m,n}^4 - T_{sur}^4\right)$$

The first four terms on the right side of the relation above are the finite difference versions of Fourier's law, the fifth term represents heat flow into one surface of the control volume by a heat flux, and the sixth term represents the radiation exchange between one surface of the control volume and the surroundings. The parameters δ and k represent the region thickness and thermal conductivity, respectively, and T_{sur} is the temperature of the surroundings. The rate of energy generation is given by the relation

$$\dot{E}_g = \dot{q}\Delta x \Delta y \delta$$

where $\dot{q}$ represents the rate of energy generation per unit volume. Setting $\Delta x = \Delta y = \Delta$, and substituting the above relations into the energy balance,

$$k\delta\left[\left(T_{m-1,n} - T_{m,n}\right) + \left(T_{m+1,n} - T_{m,n}\right) + \left(T_{m,n+1} - T_{m,n}\right) + \left(T_{m,n-1} - T_{m,n}\right)\right]$$

$$+ q''\Delta^2 - \Delta^2\sigma\varepsilon\left(T_{m,n}^4 - T_{sur}^4\right) + \dot{q}\Delta^2\delta = 0$$

Simplifying,

$$4T_{m,n} = T_{m-1,n} + T_{m+1,n} + T_{m,n+1} + T_{m,n-1} + \frac{\Delta^2}{k}\left[\frac{q''}{\delta} - \frac{\sigma\varepsilon}{\delta}\left(T_{m,n}^4 - T_{sur}^4\right) + \dot{q}\right]$$

This relation may be solved iteratively for $T_{m,n}$.

DISCUSSION

As an example calculation, let's find the temperature of the interior node given the following values:

$T_{m-1,n} = 120°C = 393$ K
$T_{m+1,n} = 70°C = 343$ K
$T_{m,n+1} = 90°C = 363$ K
$T_{m,n-1} = 60°C = 333$ K

$k = 30$ W/m·K
$\delta = 2.5$ mm
$\Delta = 2.0$ cm
$\dot{q} = 50$ kW/m^3

$q'' = 20$ kW/m^2
$\varepsilon = 0.9$
$T_{sur} = 10°C = 283$ K

Iteratively solving for $T_{m,n}$, and noting that $\sigma = 5.669 \times 10^{-8}$ W/m^2·K^4,

$T_{m,n} = 384$ K $= 111$ °C

❒ ❒ ❒

PROBLEM 4.11 Steady Temperatures of a Rectangular Plate with Specified Edge Temperatures

The edges of a rectangular plate measuring 25 cm × 15 cm are maintained at temperatures 100°C, 500°C, 300°C and 50°C, as shown in the diagram, and the plate is made of a stainless steel alloy (k = 15 W/m·K). There is no energy generation in the plate, no heat flux at the surface, and no convection or radiation from the surface. Subdivide the plate into a square finite difference mesh with eight nodes, and find the steady temperatures of the plate.

DIAGRAM

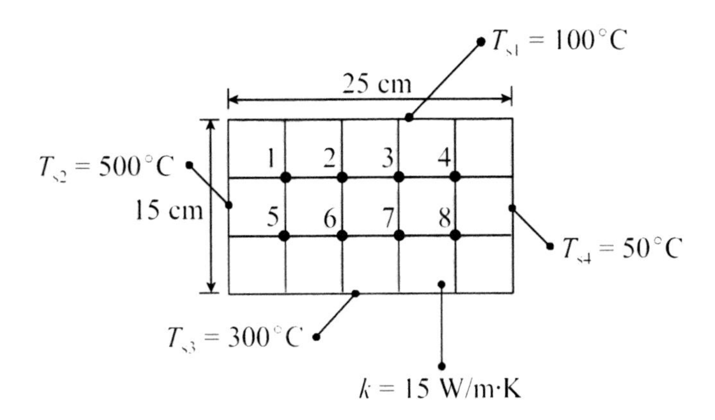

ASSUMPTIONS

1. Steady two-dimensional conduction.
2. Edge temperatures are constant.
3. No energy generation.
4. No heat flux, convection or radiation at the surface.

PROPERTIES

stainless steel: k = 15 W/m·K

ANALYSIS

The finite difference equation for an interior node with no generation, heat flux, convection or radiation is

$$T_{m,n} = \left(T_{m,n+1} + T_{m-1,n} + T_{m,n-1} + T_{m+1,n}\right)/4$$

where the subscripts m and n correspond to the x and y coordinates of the nodes, respectively. (See Problem 4.9). Thus, the temperature of an interior node is simply the arithmetic average of the temperatures of the four surrounding nodes. Writing the finite difference equations for each node,

node 1	$T_1 = (T_{s1} + T_{s2} + T_5 + T_2)/4$
node 2	$T_2 = (T_{s1} + T_1 + T_6 + T_3)/4$
node 3	$T_3 = (T_{s1} + T_2 + T_7 + T_4)/4$
node 4	$T_4 = (T_{s1} + T_3 + T_8 + T_{s4})/4$
node 5	$T_5 = (T_1 + T_{s2} + T_{s3} + T_6)/4$
node 6	$T_6 = (T_2 + T_5 + T_{s3} + T_7)/4$
node 7	$T_7 = (T_3 + T_6 + T_{s3} + T_8)/4$
node 8	$T_8 = (T_4 + T_7 + T_{s3} + T_{s4})/4$

where the edge temperatures T_{s1}, T_{s2}, T_{s3} and T_{s4} have the values shown in the diagram. Solving this system of equations iteratively,

$T_1 = 285.5°C$
$T_2 = 203.9°C$
$T_3 = 163.0°C$
$T_4 = 121.9°C$
$T_5 = 338.1°C$
$T_6 = 267.0°C$
$T_7 = 226.1°C$
$T_8 = 174.5°C$

DISCUSSION

For a unit thickness, the heat transfer from the right edge of the plate may be expressed using Fourier's law,

$$q = k(1)\left[(T_4 - T_{s4}) + (T_8 - T_{s4}) + (T_{s1} - T_{s4})/2 + (T_{s3} - T_{s4})/2\right]$$

where the node spacing Δ has divided out. The third and fourth temperature differences are divided by 2 because the conduction area is half the node spacing. The heat transfer is

q = (15 W/m·K)(1 m)[(121.9−50)°C + (174.5−50)°C + (100−50)/2°C + (300−50)/2°C]

= 5196 W

❒ ❒ ❒

PROBLEM 4.12 Temperature Distribution of a Thick Fin

A rectangular fin of length 12.0 cm and thickness 3.0 cm is exposed to 25°C air where the heat transfer coefficient is 32 W/m^2·K. The base of the fin is maintained at 90°C, and the fin has a thermal conductivity of 16 W/m·K. Using a square mesh with a node spacing of 1.5 cm, use the finite difference method to find the temperature distribution of the fin.

DIAGRAM

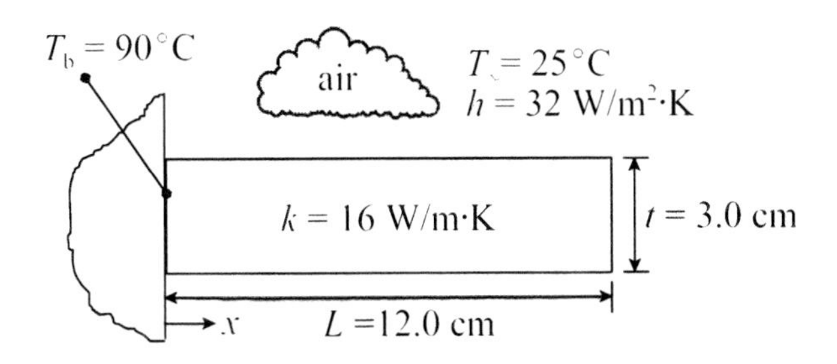

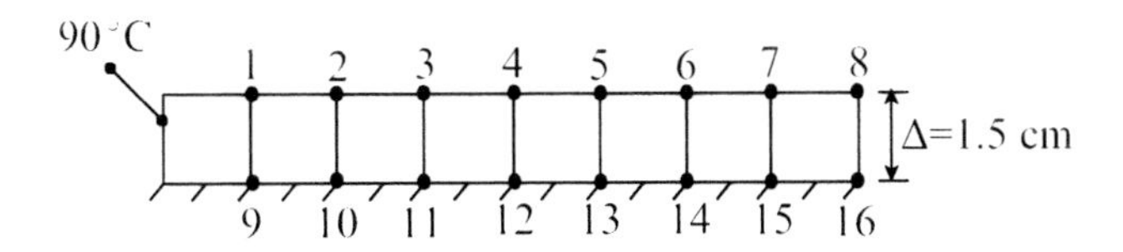

ASSUMPTIONS

1. Two-dimensional steady conduction.
2. Tip of fin loses heat by convection.
3. Radiation is neglected.
4. Constant properties.

PROPERTIES

thermal conductivity: $k = 16$ W/m·K

ANALYSIS

The fin has thermal symmetry, so we analyze the upper half of the fin, placing an adiabatic boundary condition on the fin's center plane, as shown. Nodes 1 through 7 are located on a plane surface with convection. The finite difference equation for these nodes is of the form

$$2T_{m,n-1} + T_{m-1,n} + T_{m+1,n} + 2BiT_\infty - 2(Bi + 2)T_{m,n} = 0$$

where m and n correspond to the x and y coordinates, respectively, and Bi is the Biot number given by

$$Bi = \frac{h\Delta}{k}$$

$$Bi = \frac{(32 \text{ W/m}^2\cdot\text{K})(0.015 \text{ m})}{16 \text{ W/m}\cdot\text{K}}$$

$= 0.03$

Nodes 9 through 15 are located on a plane insulated surface. The finite difference equation for these nodes is of the form given above, but with $Bi = 0$.

$$2T_{m,n+1} + T_{m-1,n} + T_{m+1,n} - 4T_{m,n} = 0$$

Node 8 is an external corner node with convection. The finite difference equation for this node is of the form

$$T_{m-1,n} + T_{m,n-1} + 2BiT_{\infty} - 2(Bi+1)T_{m,n} = 0$$

Half of the control surface for node 16 has convection, while the other half is insulated. Thus, the finite difference equation for this node is of the form above with Bi replaced with $Bi/2$.

$$T_{m-1,n} + T_{m,n+1} + BiT_{\infty} - (Bi+2)T_{m,n} = 0$$

Note that in all the finite difference equations, a unit depth for the fin is assumed. The finite difference equations for the nodes are as follows:

node 1	$2T_9 + T_b + T_2 + 2BiT_{\infty} - 2(Bi+2)T_1 = 0$
node 2	$2T_{10} + T_1 + T_3 + 2BiT_{\infty} - 2(Bi+2)T_2 = 0$
node 3	$2T_{11} + T_2 + T_4 + 2BiT_{\infty} - 2(Bi+2)T_3 = 0$
node 4	$2T_{12} + T_3 + T_5 + 2BiT_{\infty} - 2(Bi+2)T_4 = 0$
node 5	$2T_{13} + T_4 + T_6 + 2BiT_{\infty} - 2(Bi+2)T_5 = 0$
node 6	$2T_{14} + T_5 + T_7 + 2BiT_{\infty} - 2(Bi+2)T_6 = 0$
node 7	$2T_{15} + T_6 + T_8 + 2BiT_{\infty} - 2(Bi+2)T_7 = 0$
node 8	$T_7 + T_{16} + 2BiT_{\infty} - 2(Bi+1)T_8 = 0$
node 9	$2T_1 + T_b + T_{10} - 4T_9 = 0$
node 10	$2T_2 + T_9 + T_{11} - 4T_{10} = 0$
node 11	$2T_3 + T_{10} + T_{12} - 4T_{11} = 0$
node 12	$2T_4 + T_{11} + T_{13} - 4T_{12} = 0$
node 13	$2T_5 + T_{12} + T_{14} - 4T_{13} = 0$
node 14	$2T_6 + T_{13} + T_{15} - 4T_{14} = 0$
node 15	$2T_7 + T_{14} + T_{16} - 4T_{15} = 0$
node 16	$T_{15} + T_8 + BiT_{\infty} - (Bi+2)T_{16} = 0$

Solving this system of equations iteratively, the following table of temperatures are obtained:

node	1	2	3	4	5	6	7	8
temperature (°C)	80.3	72.7	66.5	61.5	57.6	54.7	52.7	51.5
node	9	10	11	12	13	14	15	16
temperature (°C)	81.0	73.3	67.1	62.0	58.1	55.2	53.1	51.9

DISCUSSION

The temperature difference from the fin's center plane to its surface is less than 1 °C, indicating that conduction in the fin is nearly one-dimensional. It is instructive to compare the temperature distribution obtained from the finite difference solution with that obtained analytically for one-dimensional conduction in a fin. The temperature distribution in a fin with convection from its tip is given by the relation

$$\frac{\theta}{\theta_b} = \frac{T(x) - T_\infty}{T_b - T_\infty} = \frac{\cosh[m(L-x)] + (h/mk)\sinh[m(L-x)]}{\cosh(mL) + (h/mk)\sinh(mL)}$$

where,

$$m = \left(\frac{hP}{kA_c}\right)^{1/2}$$

$m = 11.72 \text{ m}^{-1}$

The graph below shows the variation of θ/θ_b with x/L for the finite difference and analytical solutions, and the agreement is excellent.

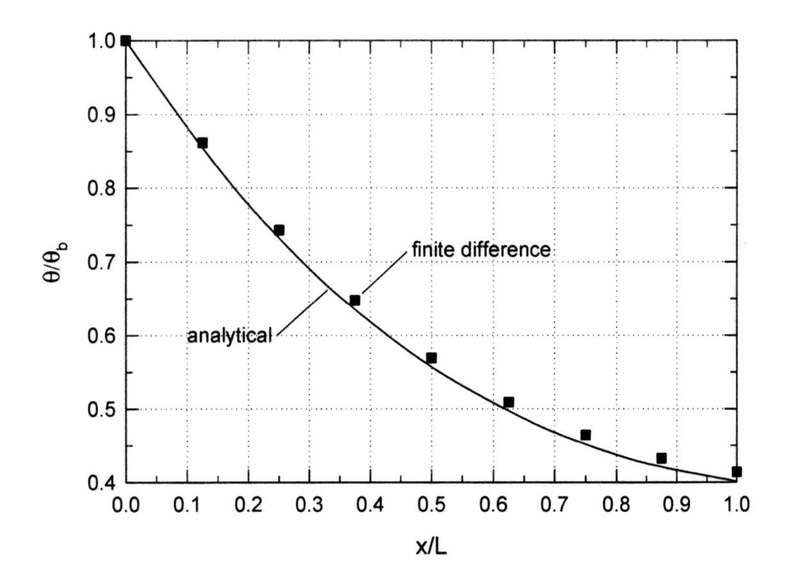

❒ ❒ ❒

PROBLEM 4.13 **Temperature Distribution of a Bracket Holding Two Power Transistors**

A 1.5-mm thick bracket of an aluminum alloy ($k = 200$ W/m·K) holds a pair of power transistors that dissipate 6 W each, as shown in the diagram. The bottom edge of the bracket is attached to a cold plate maintained at 10°C. Using the finite difference method, find the steady temperature distribution of the bracket by treating each power dissipation as a point heat source. Use a square mesh with a spacing of 2 cm, and neglect convection and radiation from the bracket.

DIAGRAM

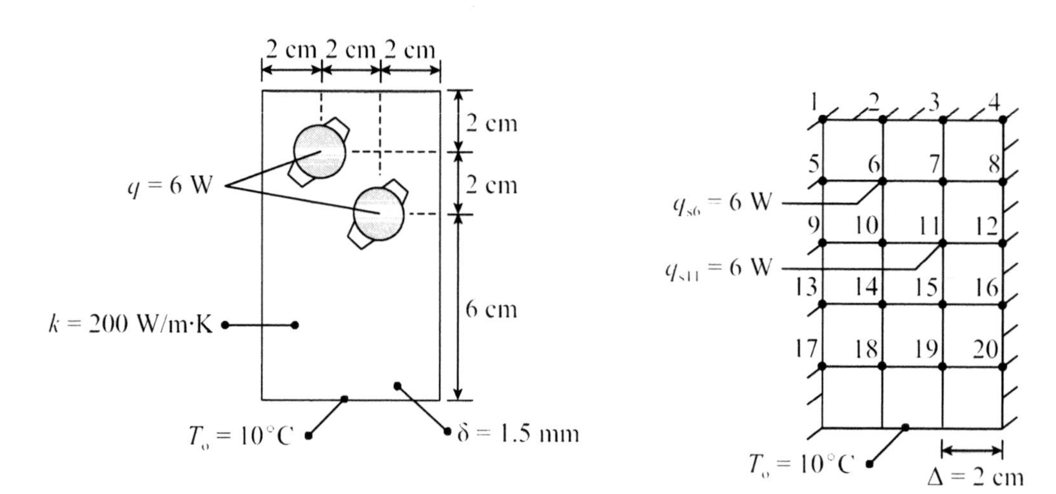

ASSUMPTIONS

1. Two-dimensional steady conduction.
2. Convection and radiation from the bracket are neglected.
3. Heat transfer from the edges of the bracket is neglected.
4. Constant properties.

PROPERTIES

aluminum alloy: $k = 200$ W/m·K

ANALYSIS

Nodes 7, 10, 11, 14, 15, 18 and 19 are interior nodes, and nodes 6 and 11 are interior nodes with point heat sources. Nodes 1 and 4 are insulated external corner nodes, and the rest of the boundary nodes are insulated edge nodes. The finite difference equations for the nodes are as follows:

node 1	$T_2 + T_5 - 2T_1 = 0$
node 2	$2T_6 + T_1 + T_3 - 4T_2 = 0$
node 3	$2T_7 + T_2 + T_4 - 4T_3 = 0$
node 4	$T_3 + T_8 - 2T_4 = 0$
node 5	$2T_6 + T_1 + T_9 - 4T_5 = 0$
node 6	$T_2 + T_5 + T_{10} + T_7 - 4T_6 + q_{s6}/(k\delta) = 0$
node 7	$T_3 + T_6 + T_{11} + T_8 - 4T_7 = 0$
node 8	$2T_7 + T_4 + T_{12} - 4T_8 = 0$
node 9	$2T_{10} + T_5 + T_{13} - 4T_9 = 0$
node 10	$T_6 + T_9 + T_{14} + T_{11} - 4T_{10} = 0$
node 11	$T_7 + T_{10} + T_{15} + T_{12} - 4T_{11} + q_{s11}/(k\delta) = 0$
node 12	$2T_{11} + T_8 + T_{16} - 4T_{12} = 0$
node 13	$2T_{14} + T_9 + T_{17} - 4T_{13} = 0$
node 14	$T_{10} + T_{13} + T_{18} + T_{15} - 4T_{14} = 0$
node 15	$T_{11} + T_{14} + T_{19} + T_{16} - 4T_{15} = 0$
node 16	$2T_{15} + T_{12} + T_{20} - 4T_{16} = 0$
node 17	$2T_{18} + T_{13} + T_o - 4T_{17} = 0$
node 18	$T_{14} + T_{17} + T_o + T_{19} - 4T_{18} = 0$
node 19	$T_{15} + T_{18} + T_o + T_{20} - 4T_{19} = 0$
node 20	$2T_{19} + T_{16} + T_o - 4T_{20} = 0$

This system of equations is solved iteratively. The steady temperature distribution of the bracket is shown schematically below. All temperatures are expressed in units of °C.

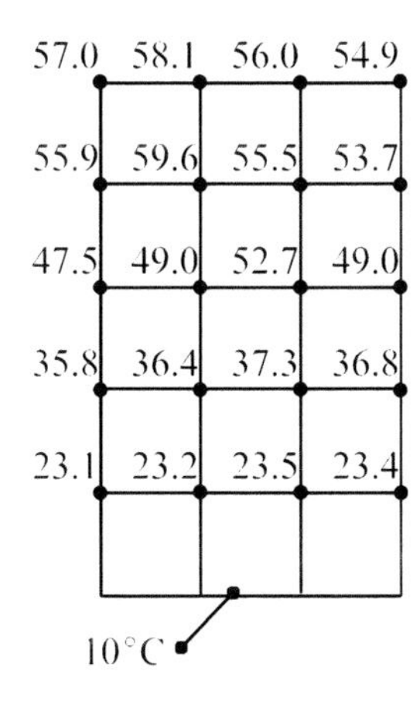

DISCUSSION

The highest temperature occurs at node 6, which corresponds to the position of one of the power transistors. As expected, temperatures decrease toward the cold plate, and conduction becomes nearly one-dimensional near the bottom of the bracket. As a check of the analysis, we calculate the heat transfer from the bracket to the cold plate. Because the bracket does not transfer heat to the surroundings by convection or radiation, all power dissipated by the transistors must be conducted through the bracket to the cold plate. Thus, the conduction at the bottom of the bracket must equal 12 W. The heat transfer from the bracket to the cold plate is given by the finite difference form of Fourier's law,

$$q = k\delta\left[\tfrac{1}{2}(T_{17} - T_o) + (T_{18} - T_o) + (T_{19} - T_o) + \tfrac{1}{2}(T_{20} - T_o)\right]$$

q = (200 W/m·K)(0.0015 m)[½ (23.1 - 10) + (23.2 - 10) + (23.5 - 10) + ½ (23.4 - 10)]°C

= 11.99 W ≈ 12 W

❒ ❒ ❒

PROBLEM 4.14 Temperature Distribution of a Square Duct

A square duct made of phenolic (k = 0.02 W/m·K) carries a 300°C gas across a room where the air temperature is 25°C. The inside and outside dimensions of the duct are 20 cm and 40 cm, respectively, and the heat transfer coefficients for the inside and outside duct surfaces are 75 W/m²·K and 18 W/m²·K, respectively. Using the finite difference method, find the steady temperatures of the duct and the heat loss per unit length of duct. Use a square mesh of 5 cm.

DIAGRAM

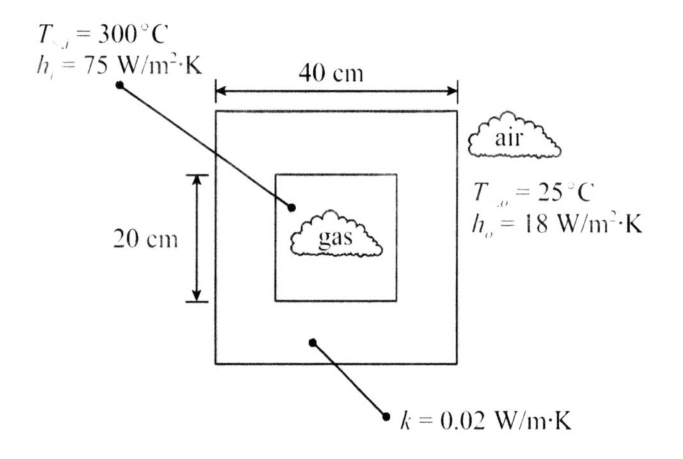

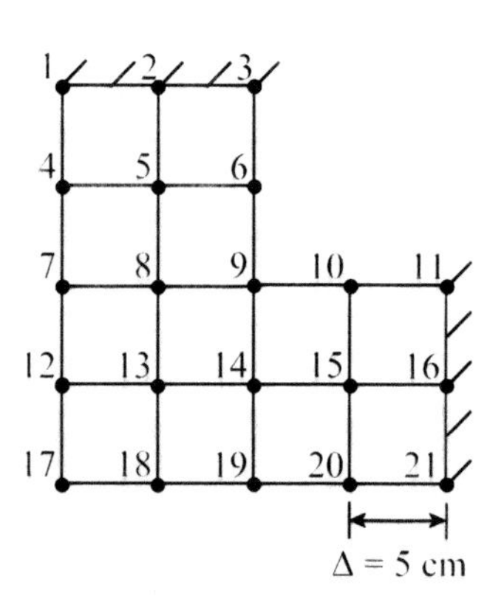

ASSUMPTIONS

1. Two-dimensional steady conduction.
2. Radiation is neglected.
3. Constant properties.

PROPERTIES

phenolic: k = 0.02 W/m·K

ANALYSIS

Because the duct has thermal symmetry, we can analyze a quarter-section of the duct wall, as shown in the diagram above. The two planes of symmetry, indicated by nodes 1, 2 and 3 and 11, 16 and 21, respectively, are insulated. The finite difference equations for all the nodes are as follows:

node 1	$T_2 + T_4 + Bi_o T_{\infty,o} - 2(Bi_o/2 + 1)T_1 = 0$
node 2	$2T_5 + T_1 + T_3 - 4T_2 = 0$
node 3	$T_2 + T_6 + Bi_i T_{\infty,i} - 2(Bi_i/2 + 1)T_3 = 0$
node 4	$2T_5 + T_1 + T_7 + 2Bi_o T_{\infty,o} - 2(Bi_o + 2)T_4 = 0$
node 5	$T_2 + T_4 + T_8 + T_6 - 4T_5 = 0$
node 6	$2T_5 + T_3 + T_9 + 2Bi_i T_{\infty,i} - 2(Bi_i + 2)T_6 = 0$
node 7	$2T_8 + T_4 + T_{12} + 2Bi_o T_{\infty,o} - 2(Bi_o + 2)T_7 = 0$
node 8	$T_5 + T_7 + T_{13} + T_9 - 4T_8 = 0$
node 9	$2(T_8 + T_{14}) + T_6 + T_{10} + 2Bi_i T_{\infty,i} - 2(3 + Bi_i)T_9 = 0$
node 10	$2T_{15} + T_9 + T_{11} + 2Bi_i T_{\infty,i} - 2(Bi_i + 2)T_{10} = 0$
node 11	$T_{10} + T_{16} + Bi_i T_{\infty,i} - 2(Bi_i/2 + 1)T_{11} = 0$
node 12	$2T_{13} + T_7 + T_{17} + 2Bi_o T_{\infty,o} - 2(Bi_o + 2)T_{12} = 0$
node 13	$T_8 + T_{12} + T_{18} + T_{14} - 4T_{13} = 0$
node 14	$T_9 + T_{13} + T_{19} + T_{15} - 4T_{14} = 0$
node 15	$T_{10} + T_{14} + T_{20} + T_{16} - 4T_{15} = 0$
node 16	$2T_{15} + T_{11} + T_{21} - 4T_{16} = 0$
node 17	$T_{12} + T_{18} + 2Bi_o T_{\infty,o} - 2(Bi_o + 1)T_{17} = 0$
node 18	$2T_{13} + T_{17} + T_{19} + 2Bi_o T_{\infty,o} - 2(Bi_o + 2)T_{18} = 0$
node 19	$2T_{14} + T_{18} + T_{20} + 2Bi_o T_{\infty,o} - 2(Bi_o + 2)T_{19} = 0$
node 20	$2T_{15} + T_{19} + T_{21} + 2Bi_o T_{\infty,o} - 2(Bi_o + 2)T_{20} = 0$
node 21	$T_{16} + T_{20} + Bi_o T_{\infty,o} - 2(Bi_o/2 + 1)T_{21} = 0$

The quantities Bi_i and Bi_o are defined as the Biot numbers for the inside and outside surfaces, respectively.

$$Bi_i \equiv \frac{h_i \Delta}{k}$$

$$Bi_i = \frac{(75\ \text{W/m}^2\cdot\text{K})(0.05\ \text{m})}{0.02\ \text{W/m}\cdot\text{K}}$$

$$= 187.5$$

$$Bi_o \equiv \frac{h_o \Delta}{k}$$

$$Bi_o = \frac{(18\ \text{W/m}^2\cdot\text{K})(0.05\ \text{m})}{0.02\ \text{W/m}\cdot\text{K}}$$

$$= 45.0$$

The system of finite difference equations is solved iteratively. The steady temperature distribution of the duct is shown schematically below. All temperatures are expressed in units of °C.

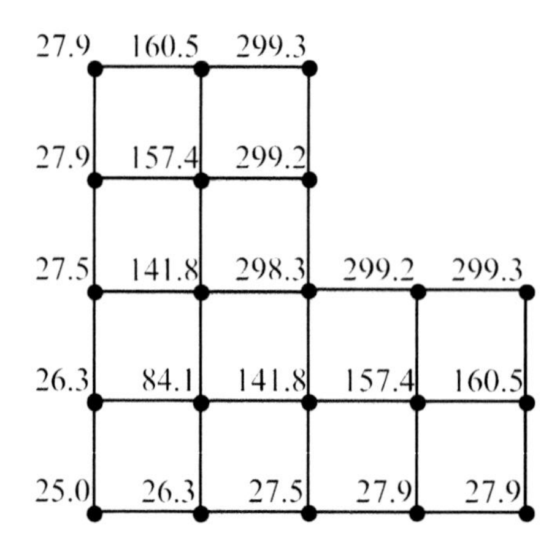

Note the symmetry of the temperatures about the corner of the duct. The heat loss per unit length of duct can be calculated by using Newton's law of cooling at either the inside or outside surface of the duct. Since the inside surface involves fewer nodes, the inside surface is used. Noting the symmetry of the temperatures, the heat loss for a quarter section of a unit length of duct is

$$q = 2h_i\Delta(1)\left[\tfrac{1}{2}(T_{\infty,i} - T_3) + (T_{\infty,i} - T_6) + \tfrac{1}{2}(T_{\infty,i} - T_9)\right]$$

$$q = 2(75\ \text{W/m}^2\cdot\text{K})(0.05\ \text{m})(1\ \text{m})[\tfrac{1}{2}(300 - 299.3) + (300 - 299.2) + \tfrac{1}{2}(300 - 298.3)]°\text{C}$$

$$= 15.0\ \text{W}$$

Thus, the total heat loss for a unit length of duct is

$$q = 4(15.0\ \text{W})$$

$$= \underline{\underline{60.0\ \text{W}}}$$

DISCUSSION

It is instructive to compare this heat loss with that obtained using the conduction shape factor method. For this problem, the conduction shape factor for two-dimensional conduction in a square solid with a centered square hole is given by the relation

$$S = \frac{2\pi L}{0.931\ln(a/b) - 0.0502}$$

where a and b are the outside and inside dimensions, respectively, and L is length. The conduction shape factor is

$$S = \frac{2\pi(1\ \text{m})}{0.931\ \ln(0.40\ \text{m}/0.20\ \text{m}) - 0.0502}$$

$$= 10.56\ \text{m}$$

The thermal resistances for the inside thermal boundary layer, the duct wall and the outside thermal boundary layer, respectively, are given by

$$R_{t,1} = \frac{1}{h_i A_i}$$

$$R_{t,1} = \frac{1}{(75\ \text{W/m}^2\cdot\text{K})(0.20\ \text{m})(1\ \text{m})}$$

$$= 0.0667\ \text{K/W}$$

$$R_{t,2} = \frac{1}{Sk}$$

$$R_{t,2} = \frac{1}{(10.56\ \text{m})(0.02\ \text{W/m}\cdot\text{K})}$$

$$= 4.735\ \text{K/W}$$

$$R_{t,3} = \frac{1}{h_o A_o}$$

$$R_{t,3} = \frac{1}{(18\ \text{W/m}^2\cdot\text{K})(0.40\ \text{m})(1\ \text{m})}$$

$$= 0.1389\ \text{K/W}$$

The total thermal resistance is

$$R_{tot} = R_{t,1} + R_{t,2} + R_{t,3}$$

$$R_{tot} = (0.0667 + 4.735 + 0.1389)\ \text{K/W}$$

$$= 4.941\ \text{K/W}$$

Thus, the heat loss from the duct is

$$q = \frac{T_{\infty,i} - T_{\infty,o}}{R_{tot}}$$

$$q = \frac{(300 - 25)°C}{4.941 \text{ K/W}}$$

$$= 55.7 \text{ W}$$

This heat loss is approximately equal to the value found using the finite difference method. The difference is about 7.2 percent, which is reasonable given the crude finite difference mesh used.

❐ ❐ ❐

PROBLEM 4.15 Temperature Distribution of a Slag Chute

In a copper smelting operation, hot slag flows into a special furnace where it is burned. The slag flows along a chute, a cross section of which is shown in the diagram below, constructed of a refractory material (k = 2 W/m·K). At a certain location along the chute, the slag has a temperature of 1100°C. All sides of the chute not in direct contact with the slag lose heat to the surrounding 35°C air by convection, where the heat transfer coefficient is 20 W/m²·K. Using the finite difference method, find the steady temperature distribution of the chute. Use a square mesh of 4 cm. Assume that the heat transfer coefficient between the slag and chute is very large.

DIAGRAM

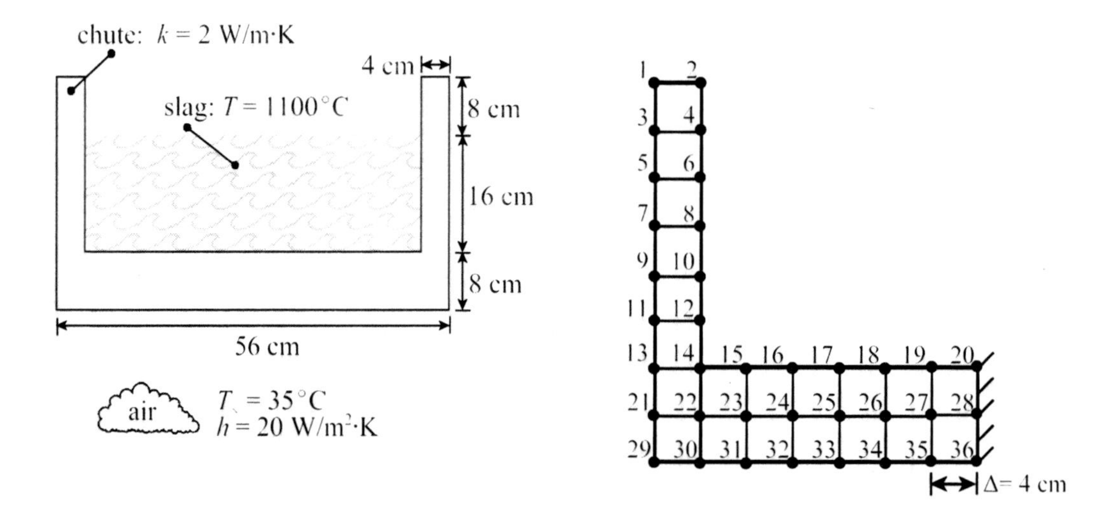

ASSUMPTIONS

1. Two-dimensional steady conduction.
2. Radiation is neglected.
3. Constant properties.

PROPERTIES

phenolic: $k = 2$ W/m·K

ANALYSIS

Because the chute has thermal symmetry, we can analyze a half-section of the chute wall, as shown in the diagram above. The plane of symmetry, indicated by nodes 20, 28 and 36, are insulated. The heat transfer coefficient for the slag is very large, so the temperature of the nodes on the inside surface of the chute in contact with the slag is 1100°C. All the other surface nodes have a convective boundary condition.

A convenient way to solve the finite difference equations for a large number of nodes is to use a spreadsheet. In a spreadsheet each cell represents a node in the finite difference model and contains a finite difference equation for that node. The table below summarizes the spreadsheet versions of the finite difference equations for selected nodes. In the spreadsheet used here, nodes 1 and 2 correspond to cells A1 and B1, respectively. Nodes 29 to 36 correspond to cells A9 to H9, respectively. Given these locations, the rest of the nodes can be readily located in the spreadsheet. Cell C12 contains the air temperature, $T_\infty = 35$°C, and cell E12 contains the Biot number, $Bi = h\Delta/k = 0.40$.

Node	Cell	Spreadsheet Finite Difference Equation
1	A1	(0.5*A2+0.5*B1+E12*C12)/(1+E12)
3	A2	(0.5*A1+0.5*A3+B2+E12*C12)/(2+E12)
6	B3	1100
14	B7	1100
20	H7	1100
22	B8	(B7+A8+B9+C8)/4
28	H8	(0.5*H7+0.5*H9+G8)/2
29	A9	1100
36	H9	(0.5*H8+0.5*G9+0.5*E12*C12)/(1+0.5*E12)

Specifying finite difference equations for the rest of the nodes in a similar fashion, the spreadsheet solves the system of equations iteratively. The slag chute temperatures in units of °C are shown in the spreadsheet output below.

	A	B	C	D	E	F	G	H
1	242.1	257.9						
2	392.0	452.1						
3	707.5	1100						
4	775.9	1100						
5	789.0	1100						
6	783.3	1100						
7	743.0	1100	1100	1100	1100	1100	1100	1100
8	554.8	749.0	819.5	846.3	856.6	860.6	862.1	862.5
9	394.4	521.5	582.8	608.9	619.6	623.8	625.4	625.8
10								
11	Δ	k	T_∞	h	Bi			
12	0.04	2	35	20	0.40			

DISCUSSION

From the results shown above, the chute temperature varies from 242.1°C at the top outside corner to 1100°C at the inside surface of the chute that is in contact with the hot slag. As expected, the outside surface temperatures of the bottom of the chute are lower than the outside surface temperatures of the side of the chute because the chute bottom is twice as thick as the side. The exceptions to this are the temperatures at nodes 1 and 2 because the free surface of the hot slag is 8 cm below the top of the chute.

❒ ❒ ❒

PROBLEM 4.16 Temperature Distribution of a Metal Core Circuit Board

A 15 cm × 9 cm circuit board is populated with heat-producing devices, as shown in the diagram. The circuit board has a 0.8 mm thick metal core (k = 165 W/m·K) that conducts heat to a 20°C liquid-cooled heat sink that is clamped to one edge of the metal core. Using the finite difference method, determine the steady temperature distribution of the metal core. Use a square mesh of 1 cm.

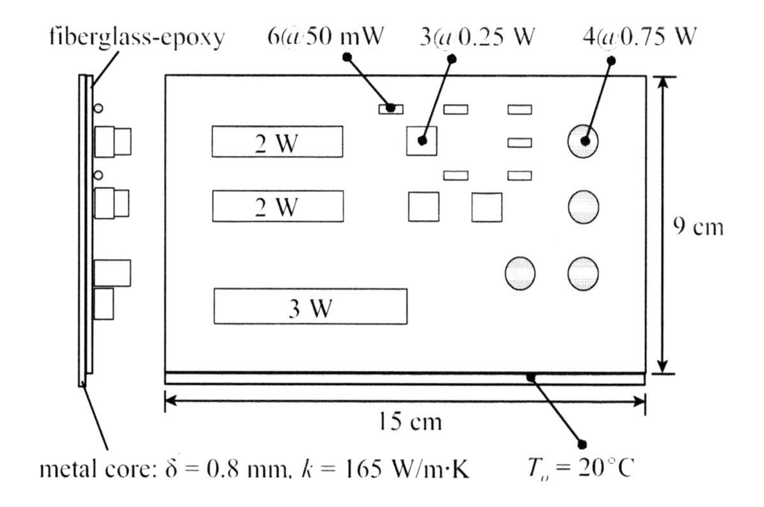

DIAGRAM

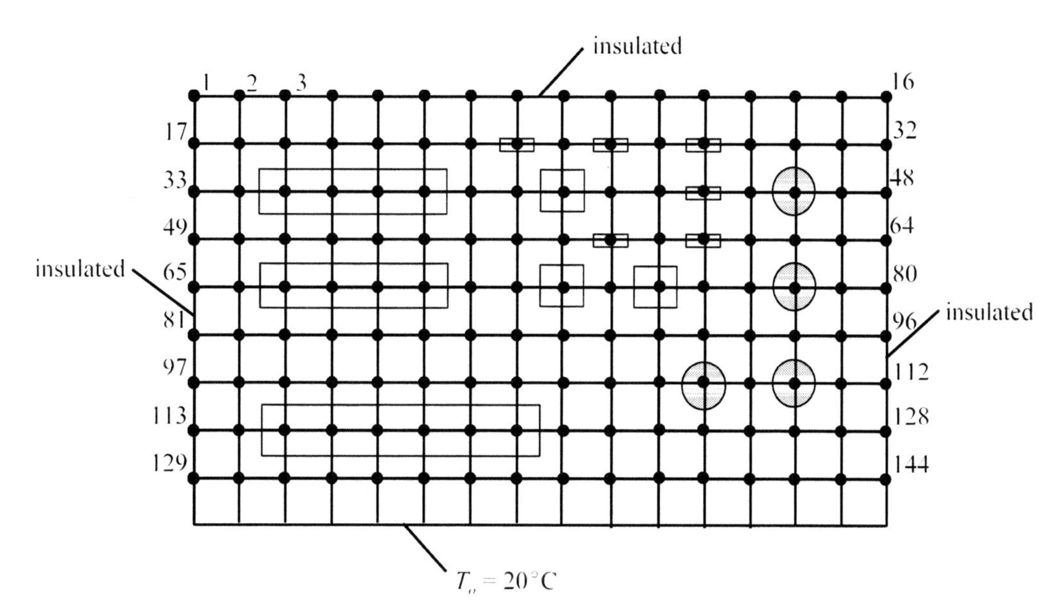

ASSUMPTIONS

1. Two-dimensional steady conduction.
2. Three edges of the metal core are insulated.
3. The fourth edge of the metal core is maintained at 20°C.
4. All the heat dissipated by the devices is transferred through the fiberglass-epoxy layer into the metal core.
5. Heat dissipations are approximated as point heat sources.
6. Constant properties.

PROPERTIES

metal core: $k = 165$ W/m·K

ANALYSIS

A convenient way to solve the finite difference equations for a large number of nodes is to use a spreadsheet. In a spreadsheet each cell represents a node in the finite difference model and contains a finite difference equation for that node. The table below summarizes the spreadsheet versions of the finite difference equations for selected nodes. In the spreadsheet used here, nodes 1 and 2 correspond to cells A1 and B1, respectively, and nodes 129 and 144 correspond to cells A9 and P9, respectively. Given these locations, the rest of the nodes can be readily located in the spreadsheet. Cells A14 and B14 contain the thermal conductivity, k, and thickness of the metal core, δ, respectively.

Node	Cell	Spreadsheet Finite Difference Equation
1	A1	0.5*A2+0.5*B1
2	B1	(0.5*A1+0.5*C1+B2)/2
17	A2	(0.5*A1+0.5*A3+B2)/2
18	B2	(B1+A2+B3+C2)/4
24	H2	(H1+G2+H3+I2+0.05/(A14*B14))/4
35	C3	(C2+B3+C4+D3+0.5/(A14*B14))/4
110	N7	(N6+M7+N8+O7+0.75/(A14*B14))/4
130	B9	(B8+A9+20+C9)/4
144	P9	(0.5*P8+0.5*20+O9)/2

Specifying the finite difference equations for the rest of the nodes in a similar fashion, the spreadsheet solves the system of equations iteratively. The metal core temperatures in units of °C are shown in the spreadsheet output below.

	A	B	C	D	E	F	G	H	I	J	K	L	M	N	O	P
1	47.4	47.6	48.0	48.1	47.8	47.0	45.9	44.9	44.2	43.6	43.2	43.1	43.2	43.3	43.1	43.0
2	47.2	47.5	48.1	48.0	47.1	45.8	44.9	44.1	43.5	43.0	43.1	43.1	43.4	43.1	43.1	42.9
3	46.6	47.0	48.6	49.1	48.8	47.7	45.4	44.2	43.8	42.8	42.4	42.5	42.9	44.3	42.8	42.4
4	45.0	45.4	46.4	46.9	46.6	45.6	44.0	42.8	42.2	41.6	41.3	41.4	41.6	42.2	41.5	41.2
5	42.7	43.1	44.8	45.4	45.2	44.2	42.0	40.8	40.5	39.7	39.9	39.7	40.0	41.4	39.9	39.4
6	39.4	39.7	40.5	40.9	40.8	40.1	39.0	38.0	37.4	36.9	37.0	37.4	37.4	37.8	37.1	36.7
7	35.4	35.7	36.5	36.9	37.0	36.5	35.8	34.9	34.0	33.6	33.9	35.3	34.5	35.4	33.8	33.4
8	30.9	31.3	32.8	33.4	33.5	33.3	32.8	31.9	30.2	29.5	29.5	29.9	29.8	29.9	29.4	29.2
9	25.6	25.8	26.3	26.6	26.7	26.6	26.3	25.9	25.2	24.9	24.8	24.9	24.9	24.9	24.7	24.7
10	20.0	20.0	20.0	20.0	20.0	20.0	20.0	20.0	20.0	20.0	20.0	20.0	20.0	20.0	20.0	20.0
11																
12																
13	*k*	δ														
14	165	8E-4														

DISCUSSION

As shown in the results above, the highest metal core temperature occurs at node 36, cell D3 in the spreadsheet finite difference model. If the maximum operating temperatures of all devices were equal, the device whose foot print covers this node (the 2-W device in the upper left corner) would be the most critical device on the circuit board. To find the actual case temperature of this device, we would find the total thermal resistance of the interfacial materials between the metal core and the bottom surface of the device. We would then use this parameter to calculate the temperature rise between the device and the metal core and add this result to the metal core temperature. The total heat dissipated by all the devices on the circuit board is 11.05 W. This total can also be found by applying Fourier's law of conduction to nodes 129 to 144 in the finite difference model.

❒ ❒ ❒

PROBLEM 4.17 **Temperature of a Traveling Wave Tube on a Ceramic Heat Spreader**

In an attempt to reduce the temperature of a traveling wave tube (TWT) in a communications system, the tube is mounted on the surface of a ceramic spreader plate ($k = 7$ W/m·K), as shown in the diagram. The heat dissipated by the tube results in a steady heat flux at its bottom surface of $q'' = 2$ W/cm^2. The problem is rendered two-dimensional by assuming that the system is very long in the direction perpendicular to the page. The top surface of the plate is convectively cooled by air, while the other surfaces are assumed to be insulated. The TWT is centered on the ceramic heat spreader. Using the finite difference method, determine the steady temperature of the bottom surface of the TWT. Use a square mesh of 1 cm.

DIAGRAM

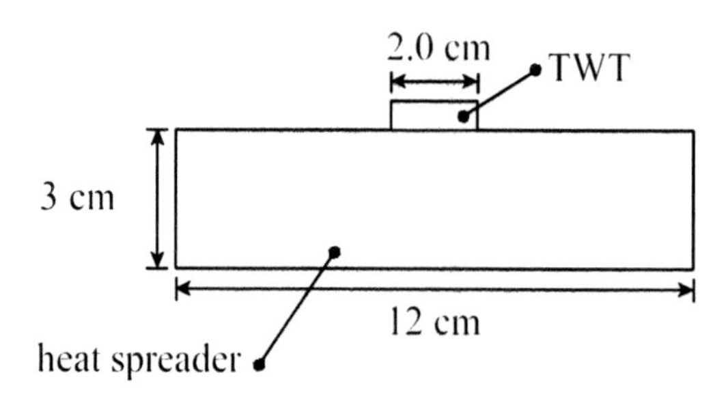

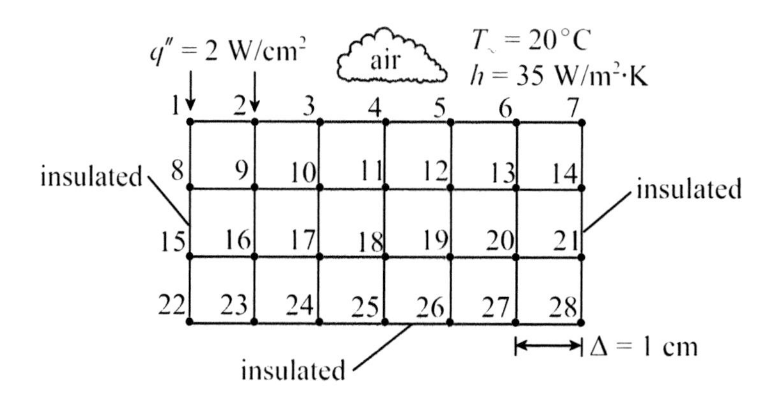

ASSUMPTIONS

1. Two-dimensional steady conduction.
2. Bottom and sides of the spreader plate are insulated.
3. Radiation from the top surface of the spreader plate is neglected.
4. Thermal contact resistance between the TWT and spreader plate is neglected.
5. Constant properties.

PROPERTIES

ceramic: $k = 7$ W/m·K

ANALYSIS

The finite difference equation for an edge node (node 2) with an applied heat flux and convection is of the form

$$T_{m,n} = \frac{T_{m,n-1} + \frac{1}{2}T_{m-1,n} + \frac{1}{2}T_{m+1,n} + BiT_\infty + q''\Delta/k}{2 + Bi}$$

For an exterior corner node with an applied heat flux (node 1), the finite difference equation is of the form

$$T_{m,n} = \frac{1}{2}T_{m,n-1} + \frac{1}{2}T_{m+1,n} + q''\Delta/k$$

A convenient way to solve the finite difference equations for a large number of nodes is to use a spreadsheet. In a spreadsheet each cell represents a node in the finite difference model and contains a finite difference equation for that node. The table below summarizes the spreadsheet versions of the finite difference equations for selected nodes. Note that only half the region is modeled due to thermal symmetry about the center plane of the system. In the spreadsheet used here, nodes 1 to 7 correspond to cells A1 to G1, respectively, and nodes 22 to 28 correspond to cells A4 to G4, respectively. Given these locations, the rest of the nodes can be readily located in the spreadsheet. Cells A8 to F8 contain the heat flux, q'', thermal conductivity, k, air temperature, T_∞, heat transfer coefficient, h, node spacing, Δ, and Biot number, $Bi = h\Delta/k$, respectively. Note that the heat flux is halved on node 1 because this node uses only half the control surface. Also note that the heat flux and Biot number are halved on node 2 to reflect the transition from a heat flux to a convective boundary condition on the top surface of the ceramic plate.

Node	Cell	Spreadsheet Finite Difference Equation
1	A1	0.5*A2 + 0.5*B1 + 0.5*A8*E8/B8
2	B1	(B2+0.5*A1+0.5*C1+0.5*F8*C8 +0.5*A8*E8/B8)/(2+0.5*F8)
3	C1	(C2 + 0.5*B1 + 0.5*D1 + F8*C8)/(2 + F8)
7	G1	(0.5*F1 + 0.5*G2 + 0.5*F8*E8)/(1 + 0.5*F8)
8	A2	(B2 + 0.5*A1 + 0.5*A3)/2
9	B2	(B1 + A2 + B3 + C2)/4
22	A4	0.5*A3 + 0.5*B4

Specifying the finite difference equations for the rest of the nodes in a similar fashion, the spreadsheet solves the system of equations iteratively. The temperatures of the ceramic heat spreader in units of °C are shown in the spreadsheet output below.

	A	B	C	D	E	F	G
1	171.5	158.9	140.6	131.4	126.1	123.1	121.7
2	155.6	150.9	142.1	135.1	130.3	127.4	126.4
3	149.2	146.9	141.8	136.6	132.5	129.9	129.0
4	147.4	145.7	141.5	136.9	133.1	130.7	129.8
5							
6							
7	q''	k	T_∞	h	Δ	Bi	
8	2E4	7	20	35	0.01	0.05	

DISCUSSION

As shown in the output above, the temperature of the bottom surface of the traveling wave tube is 171.5°C at its center and 158.9°C at its corner. If the 171.5°C surface temperature is higher than that recommended by the TWT manufacturer, the cooling design would have to be modified. For example, a spreader plate with a higher thermal conductivity would reduce the TWT surface temperature. Replacing the ceramic with 302 stainless steel (k = 15.1 W/m·K), the TWT surface temperature is 150.7°C. Replacing the ceramic with 6061 aluminum (k = 180 W/m·K), the TWT surface temperature is only 133.9°C. Other methods for reducing the TWT temperature include using a larger and thicker spreader plate and increasing the air velocity over the top surface of the plate.

❒ ❒ ❒

Chapter 5

Unsteady Conduction

PROBLEM 5.1 Air Cooling a Large Steel Plate in a Heat Treating Operation

A heat treatment is performed on a large plate of 4-mm thick steel (ρ = 7840 kg/m^3, c_p = 460 J/kg·K, k = 52 W/m·K). The plate is heated in an oven until the plate attains a uniform temperature of 575°C. The plate is then removed from the oven and allowed to cool in air at 22°C, where the heat transfer coefficient is 14 W/m^2·K. What time is required for the plate to cool such that an operator can handle the plate if the maximum handling temperature is 60°C?

DIAGRAM

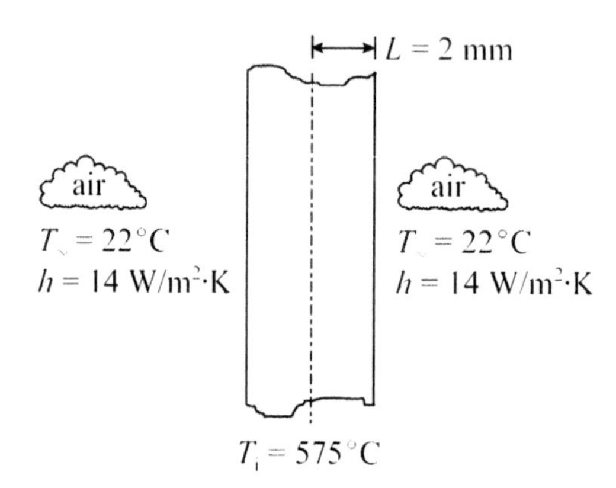

ASSUMPTIONS

1. The temperature of the plate is uniform at any instant of time.
2. The plate has a uniform temperature prior to cooling.
3. The plate is surrounded by the cooling air.
4. Heat transfer from the edges of the plate is neglected.
5. Radiation is neglected.
6. Constant properties.

PROPERTIES

steel: $\rho = 7840$ kg/m^3 , $c_p = 460$ J/kg·K , $k = 52$ W/m·K

ANALYSIS

Before proceeding with the analysis, the first assumption must be checked. If the temperature of a substance is uniform at any instant of time, the substance is said to be *lumped.* The criterion for this condition is that the Biot number is less than 0.1. For a plate of half-thickness L, the Biot number is defined as

$$Bi \equiv \frac{hL}{k}$$

The Biot number is

$$Bi = \frac{(14 \text{ W/m}^2\cdot\text{K})(0.002 \text{ m})}{52 \text{ W/m}\cdot\text{K}}$$

$$= 5.38 \times 10^{-4} < 0.1$$

Thus, the plate is lumped. The temperature history of the plate is given by the relation

$$\frac{\theta}{\theta_i} = \frac{T - T_\infty}{T_i - T_\infty} = \exp\left[-\left(\frac{hA}{\rho V c_p}\right)t\right]$$

where T is temperature at time t, A is surface area and V is volume. Noting that $A/V = 1/L$, and solving for t,

$$t = -\frac{\rho L c_p}{h}\ln\left(\frac{T - T_\infty}{T_i - T_\infty}\right)$$

$$t = -\frac{(7840 \text{ kg/m}^3)(0.002 \text{ m})(460 \text{ J/kg}\cdot\text{K})}{14 \text{ W/m}^2\cdot\text{K}} \ln\left(\frac{60 - 22}{575 - 22}\right)$$

$$= 1380 \text{ s} = \underline{\underline{23.0 \text{ min}}}$$

DISCUSSION

The plate may be safely handled by an operator after it has cooled for 23 minutes. We note that the cooling time is directly proportional to the density, thickness and specific heat of the plate, as indicated in the relation above. We also note that the cooling time is inversely proportional to the heat transfer coefficient. These mathematical relationships are consistent with the expected physical behavior of the system. If radiation was included in the analysis, the cooling time would decrease. This effect is illustrated in Problem 5.2.

PROBLEM 5.2 Convective and Radiative Cooling of a Copper Sphere

A 5-cm radius sphere of pure copper (ρ = 8933 kg/m^3, c_p = 385 J/kg·K, k = 401 W/m·K) has a uniform initial temperature of 300°C. The sphere is then exposed to 10°C air, where the heat transfer coefficient is 40 W/m^2·K. Surrounding the sphere is a 0°C enclosure, and the surface of the sphere has an emissivity of 0.8 due to high oxidation. Construct a graph showing the temperature history of the sphere for the case when it convects and radiates and for the case when it convects only.

DIAGRAM

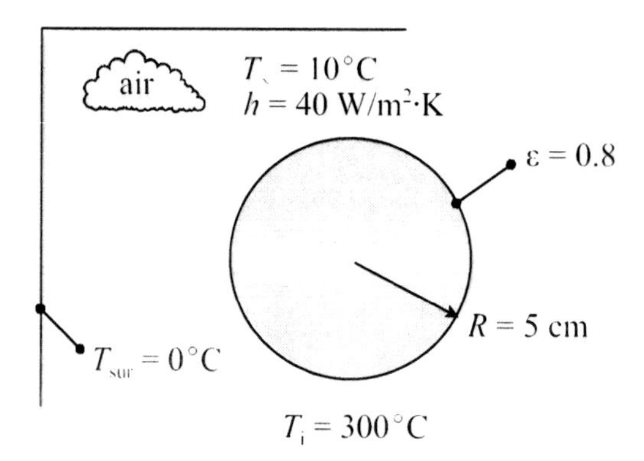

ASSUMPTIONS

1. The temperature of the sphere is uniform at any instant of time.
2. The sphere has a uniform temperature prior to cooling.
3. The sphere is surrounded by the cooling air.
4. Constant properties.

PROPERTIES

copper: $\rho = 8933$ kg/m^3 , $c_p = 385$ J/kg·K , $k = 401$ W/m·K , $\varepsilon = 0.8$

ANALYSIS

The first assumption must be checked to determine whether the sphere is lumped. The Biot number for a sphere is defined as

$$Bi \equiv \frac{h(R/3)}{k}$$

The Biot number is

$$Bi = \frac{(40\ \text{W/m}^2\cdot\text{K})(0.05\ \text{m}/3)}{401\ \text{W/m}\cdot\text{K}}$$

$$= 1.66 \times 10^{-3} < 0.1$$

Thus, the sphere is lumped. An energy balance for the sphere yields the relation

$$-\dot{E}_{out} = \frac{dE_{st}}{dt}$$

The rate of heat transfer from the sphere includes convection and radiation,

$$\dot{E}_{out} = hA(T - T_\infty) + \varepsilon\sigma A(T^4 - T_{sur}^4)$$

and the rate of energy stored by the sphere is written as

$$\frac{dE_{st}}{dt} = \rho V c_p \frac{dT}{dt}$$

Substituting these relations into the energy balance and simplifying, we obtain the first order nonlinear differential equation,

$$-\frac{dT}{dt} = C_1 T + C_2 T^4 + C_3$$

where,

$$C_1 = \frac{hA}{\rho V c_p}$$

$$C_2 = \frac{\varepsilon\sigma A}{\rho V c_p}$$

$$C_3 = \frac{-1}{\rho V c_p}\left(hAT_\infty + \varepsilon\sigma A T_{sur}^4\right)$$

The values of these constants are

$$C_1 = 6.979 \times 10^{-4}\ \text{s}^{-1},\ C_2 = 7.913 \times 10^{-13}\ \text{K}^{-3}\cdot\text{s}^{-1},\ C_3 = -0.2019\ \text{K/s}$$

Because radiation is involved, units of K must be used for temperatures. The differential equation may be solved numerically by writing it in the finite difference form,

$$-\left(\frac{T^{p+1}-T^{p}}{\Delta t}\right)=C_1T^p+C_2\left(T^p\right)^4+C_3$$

where the p and $p+1$ superscripts on temperature refer to present and future values, respectively, and Δt is the time step. The initial condition is $T(0) = T_i$. Choosing a time step of $\Delta t = 1$ s, the solution for temperature T is marched out in time to an arbitrary value of 3600 s for the case when the sphere convects and radiates and for the case when the sphere convects only. The second case is easily handled by setting $\varepsilon = 0$. The graph below shows the temperature history of the sphere for both cases.

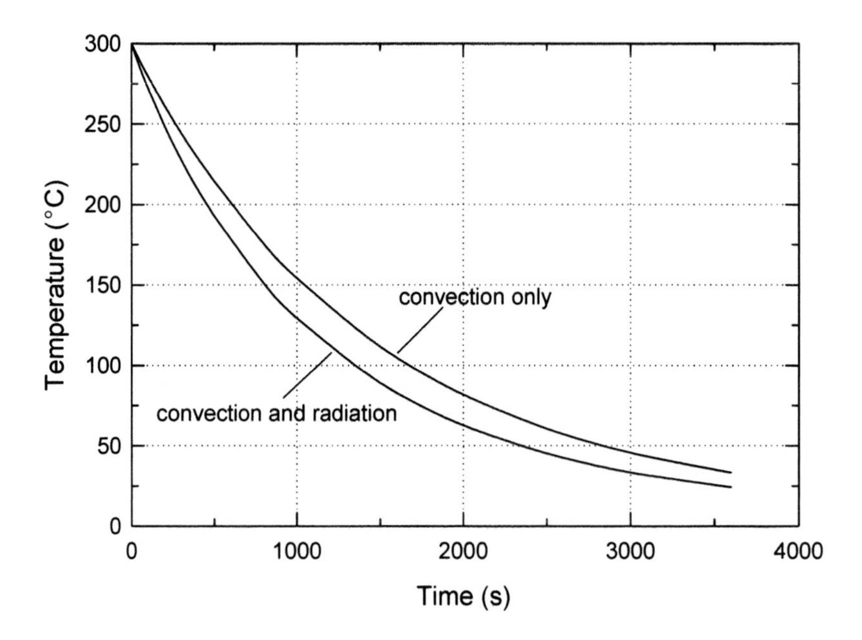

DISCUSSION

As shown in the graph, the sphere cools more rapidly when radiation is included in the analysis. After one hour, the sphere cools to 24 °C when radiation is included, whereas the sphere cools to 33 °C if convection only is included. The accuracy of the numerical solution for the convection only case may be checked by solving the problem using the equation given in Problem 5.1. A time step of 1 s is small enough to yield reasonable accuracy.

❒ ❒ ❒

PROBLEM 5.3 Heating Time of a Lumped Mass with Energy Generation

A solid cylinder of length 30 cm and radius 1.5 cm has a density, specific heat and thermal conductivity of $\rho = 11{,}000$ kg/m^3, $c_p = 120$ J/kg·K and $k = 54$ W/m·K, respectively. When the cylinder is immersed in a large container of 25°C water, where the heat transfer coefficient is 550 W/m^2·K, the cylinder begins to generate energy uniformly within it at a rate of 0.6 MW/m^3. If the initial temperature of the cylinder is equal to that of the water, how much time passes before the temperature of the cylinder reaches 400°C?

DIAGRAM

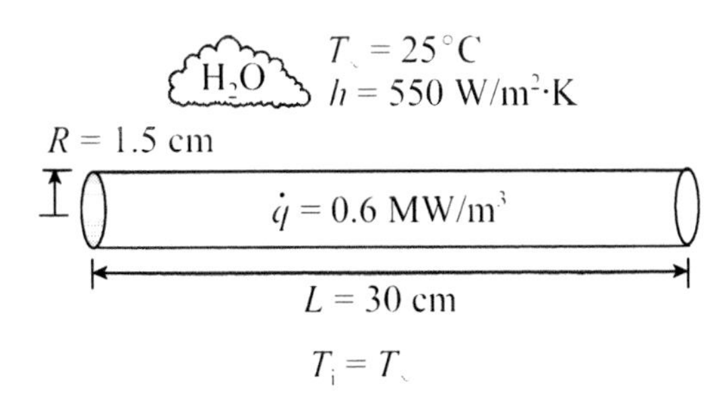

ASSUMPTIONS

1. The temperature of the cylinder is uniform at any instant of time.
2. The cylinder has a uniform temperature prior to the start of energy generation.
3. The heat transfer coefficient is constant.
4. Convection from the ends of the cylinder is neglected.
5. Radiation is neglected.
6. Constant properties.

PROPERTIES

$\rho = 11{,}000$ kg/m^3 , $c_p = 120$ J/kg·K , $k = 54$ W/m·K

ANALYSIS

The first assumption must be checked to determine whether the cylinder is lumped. The Biot number for a cylinder is defined as

$$Bi \equiv \frac{h(R/2)}{k}$$

The Biot number is

$$Bi = \frac{(550 \text{ W/m}^2\cdot\text{K})(0.015 \text{ m}/2)}{401 \text{ W/m}\cdot\text{K}}$$

$$= 0.0764 < 0.1$$

Thus, the cylinder is lumped. An energy balance for the cylinder yields the relation

$$\dot{E}_{in} - \dot{E}_{out} + \dot{E}_g = \frac{dE_{st}}{dt}$$

The rate of heat transfer to the cylinder is zero, and the rate of heat transfer from the cylinder is

$$\dot{E}_{out} = hA(T - T_\infty)$$

and the rate of energy stored by the cylinder is

$$\frac{dE_{st}}{dt} = \rho V c_p \frac{dT}{dt}$$

Substituting these relations into the energy balance and simplifying,

$$\frac{d\theta}{dt} = -\left(\frac{hA}{\rho V c_p}\right)\theta + \frac{\dot{q}V}{\rho c_p}$$

where,

$$\theta = T - T_\infty$$

Separating variables and integrating,

$$\int_{\theta_i}^{\theta} \frac{d\theta}{a\theta + b} = \int_0^t dt$$

where,

$$a = -\frac{hA}{\rho V c_p} \quad , \quad b = \frac{\dot{q}}{\rho c_p}$$

Noting that $A = 2\pi RL$ and $V = \pi R^2 L$, the values of the two constants above are

$a = 0.05554 \text{ s}^{-1}$, $b = 0.45455$ K/s

Integrating the relation above, and solving for time,

$$t = \frac{1}{a}\ln\left(\frac{a\theta + b}{a\theta_i + b}\right)$$

where,

$$\theta_i = T_i - T_\infty$$

The time required for the cylinder to reach a temperature of 400°C is

$$t = \frac{1}{0.05554\ \text{s}^{-1}} \ln\left[\frac{(0.05554\ \text{s}^{-1})(375\ \text{K}) + 0.45455\ \text{K/s}}{(0.05554\ \text{s}^{-1})(0\ \text{K}) + 0.45455\ \text{K/s}}\right]$$

$$= 69.25\ \text{s} = \underline{\underline{1.15\ \text{min}}}$$

DISCUSSION

If the fourth assumption were relaxed, thereby allowing convection from the ends of the cylinder, the time required for the cylinder to reach 400°C increases only slightly to 70.1 s. A graph of cylinder temperature as a function of time is shown below. The temperature of the cylinder increases exponentially with time, so within a few minutes the cylinder will melt or otherwise destroy itself thermally due to its own energy generation.

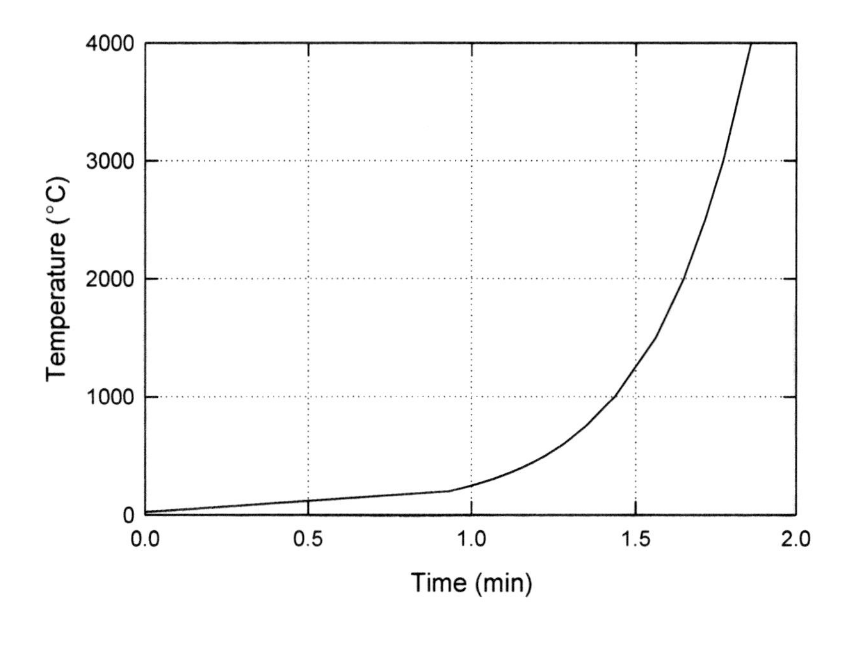

❒ ❒ ❒

PROBLEM 5.4 Cooling of a Rod During Two Convective Regimes

A long 3-mm diameter rod of steel (ρ = 7840 kg/m^3, c_p = 460 J/kg·K, k = 43 W/m·K) has a uniform initial temperature of 650 °C. The rod is then suddenly immersed in a large container of 25 °C water. The hot rod causes the water immediately surrounding the rod to boil, giving rise to a heat transfer coefficient of 5600 W/m^2·K. The boiling stops when the rod reaches a temperature of 100 °C, at which time the heat transfer coefficient becomes 750 W/m^2·K. Find the time period during which boiling occurs and the total time required for the rod to reach a temperature of 40 °C.

DIAGRAM

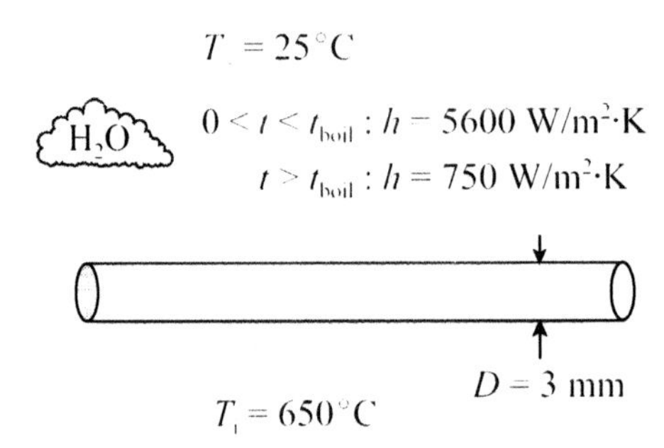

ASSUMPTIONS

1. The temperature of the rod is uniform at any instant of time.
2. There are two convective regimes, one during boiling and one after boiling ends.
3. The heat transfer coefficients are constant during the two regimes.
4. Convection from the ends of the rod is neglected.
5. Radiation is neglected.
6. Constant properties.

PROPERTIES

steel: ρ = 7840 kg/m^3, c_p = 460 J/kg·K, k = 43 W/m·K

ANALYSIS

The first assumption must be checked to determine whether the rod is lumped. The Biot number for a cylinder is defined as

$$Bi \equiv \frac{h(R/2)}{k}$$

If the rod is lumped for the boiling regime, it is also lumped for the second regime since the heat transfer coefficient for boiling is the larger of the two values. The Biot number for the boiling regime is

$$Bi = \frac{(5600\ \text{W/m}^2\cdot\text{K})(0.0015\ \text{m}/2)}{43\ \text{W/m}\cdot\text{K}}$$

$$= 0.0977 < 0.1$$

Thus, the rod is lumped for both convective regimes. The temperature history of the rod is given by the relation

$$\frac{\theta}{\theta_i} = \frac{T - T_\infty}{T_i - T_\infty} = \exp\left[-\left(\frac{hA}{\rho V c_p}\right)t\right]$$

where T is temperature at time t, A is surface area and V is volume. Noting that $V/A = R/2$, and solving for t,

$$t = -\frac{\rho R c_p}{2h}\ln\left(\frac{T - T_\infty}{T_i - T_\infty}\right)$$

The time required for the rod to reach 100°C, i.e., for the water to cease boiling, is

$$t = -\frac{(7840\ \text{kg/m}^3)(0.0015\ \text{m})(460\ \text{J/kg}\cdot\text{K})}{2(5600\ \text{W/m}^2\cdot\text{K})}\ln\left(\frac{100°\text{C} - 25°\text{C}}{650°\text{C} - 25°\text{C}}\right)$$

$$= \underline{\underline{1.024\ \text{s}}}$$

For the second regime we set $T_i = 100°\text{C}$ and $T = 40°\text{C}$. The time required for the temperature of the rod to change from 100°C to 40°C is

$$t = -\frac{(7840\ \text{kg/m}^3)(0.0015\ \text{m})(460\ \text{J/kg}\cdot\text{K})}{2(750\ \text{W/m}^2\cdot\text{K})}\ln\left(\frac{40°\text{C} - 25°\text{C}}{100°\text{C} - 25°\text{C}}\right)$$

$$= 5.804\ \text{s}$$

Thus, the total time required for the rod to achieve a temperature of 40°C is

$$t_{\text{total}} = 1.024\ \text{s} + 5.804\ \text{s}$$

$$= \underline{\underline{6.83\ \text{s}}}$$

DISCUSSION

A graph of the temperature history of the rod distinctly shows the two convective regimes. Because the heat transfer coefficient is much larger during the first regime than the second, the first part of the temperature history is steeper than the second. The third assumption, that the heat transfer coefficients during each regime are constant, is only approximate, particularly during the boiling regime. The heat transfer coefficient for boiling is a strong function of temperature, so the value used in the analysis should be regarded as an average value for the time interval over which boiling occurs.

During the first regime, radiation would be significant due to the high temperature of the rod, but to simplify the problem, radiation was neglected.

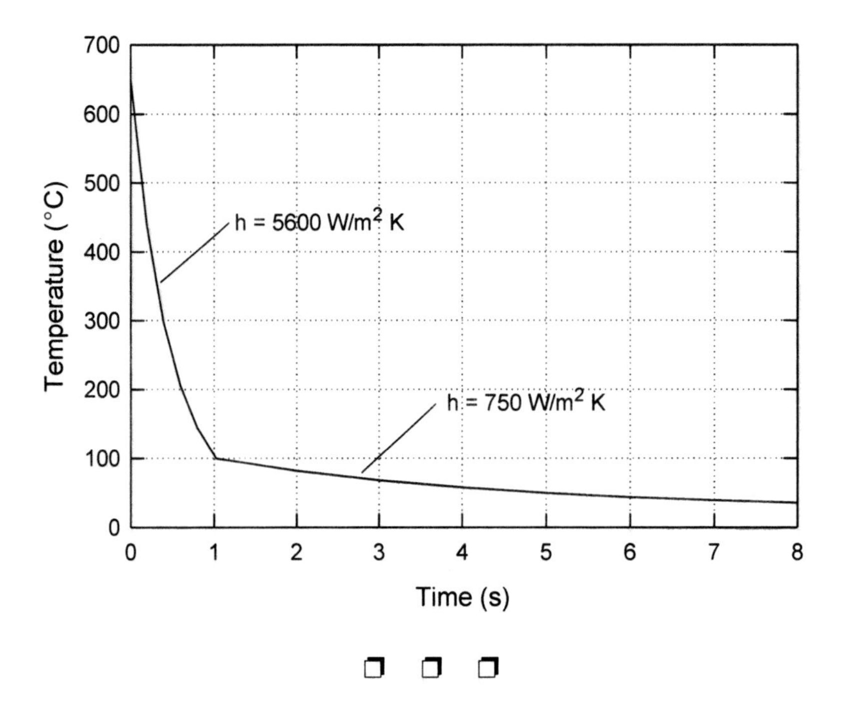

❒ ❒ ❒

PROBLEM 5.5 Heat Treating Time for a Moving Sheet of Steel

A 6-mm thick sheet of mild steel (ρ = 7830 kg/m^3·K, c_p = 434 J/kg·K, k = 64 W/m·K) is heat treated by passing it through a large furnace operating at a steady temperature of 800°C. The sheet is to be heated from 30°C to 750°C. If the heat transfer coefficient in the furnace is 90 W/m^2·K, find the required heating time. If the sheet is moving through the furnace at 0.3 m/s, how long must the furnace be?

DIAGRAM

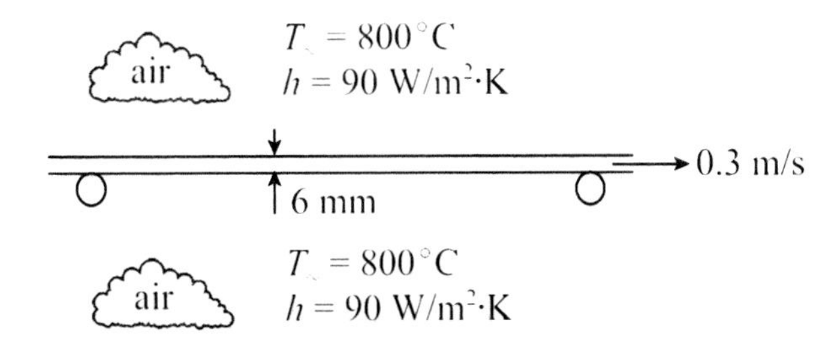

ASSUMPTIONS

1. The temperature of the sheet is uniform at any instant of time.
2. Convection from the edges of the sheet is neglected.
3. Sheet is moving with a constant velocity.
4. Radiation is neglected.
5. Constant properties.

PROPERTIES

mild steel: $\rho = 7830$ kg/m³·K, $c_p = 434$ J/kg·K, $k = 64$ W/m·K

ANALYSIS

The first assumption must be checked to determine whether the sheet of steel is lumped. The Biot number for a plate is defined as

$$Bi \equiv \frac{hL}{k}$$

where L is the half-thickness of the plate. The Biot number is

$$Bi = \frac{(90 \text{ W/m}^2\cdot\text{K})(0.003 \text{ m})}{64 \text{ W/m}\cdot\text{K}}$$

$$= 4.22 \times 10^{-3} < 0.1$$

Thus, the plate is lumped. The temperature history of the plate is given by the relation

$$\frac{\theta}{\theta_i} = \frac{T - T_\infty}{T_i - T_\infty} = \exp\left[-\left(\frac{hA}{\rho V c_p}\right)t\right]$$

where T is temperature at time t, A is surface area and V is volume. Noting that $A/V = 1/L$, and solving for t,

$$t = -\frac{\rho L c_p}{h} \ln\left(\frac{T - T_\infty}{T_i - T_\infty}\right)$$

$$t = -\frac{(7830 \text{ kg/m}^3)(0.003 \text{ m})(434 \text{ J/kg·K})}{90 \text{ W/m}^2\text{·K}} \ln\left(\frac{750 - 800}{30 - 800}\right)$$

$$= 309.7 \text{ s} = \underline{\underline{5.16 \text{ min}}}$$

Thus, the "residence" time in the furnace for the sheet of steel is 5.16 min. The length of furnace required to achieve this time is the velocity of the sheet multiplied by the residence time,

$$l = vt$$

$$l = (0.3 \text{ m/s})(309.7 \text{ s})$$

$$= \underline{\underline{92.9 \text{ m}}}$$

DISCUSSION

To achieve the residence time required for the sheet of steel to reach 750°C, the heat treating furnace must be about 93 m long, which may not be very realistic in practice due to manufacturing facility limitations. The furnace length can be decreased by increasing the furnace operating temperature and decreasing the velocity of the sheet. The graph below shows these relationships.

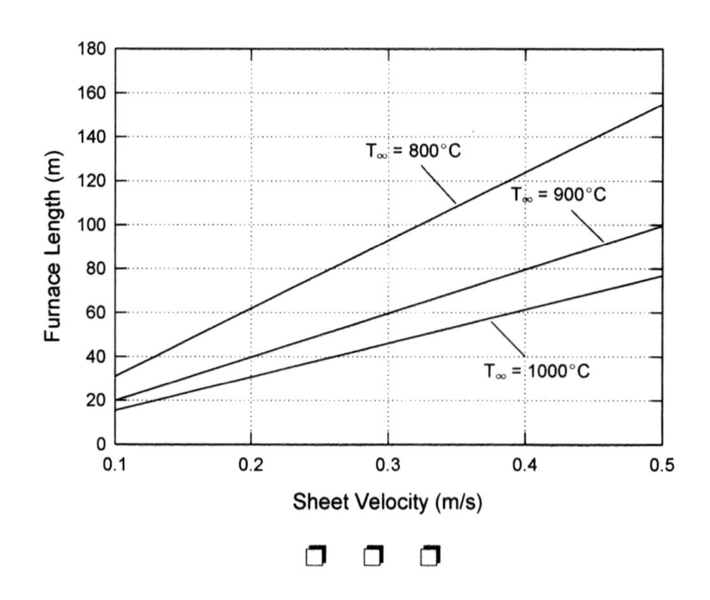

❒ ❒ ❒

PROBLEM 5.6 Heating Time for a Thick Slab of Concrete

The surface of a thick slab of concrete ($\alpha = 6.9 \times 10^{-7}$ m^2/s) is suddenly changed to 50°C and maintained at that temperature for a long time. If the initial temperature of the concrete slab is 15°C, how much time is required for a point 2 cm deep in the slab to reach 40°C?

DIAGRAM

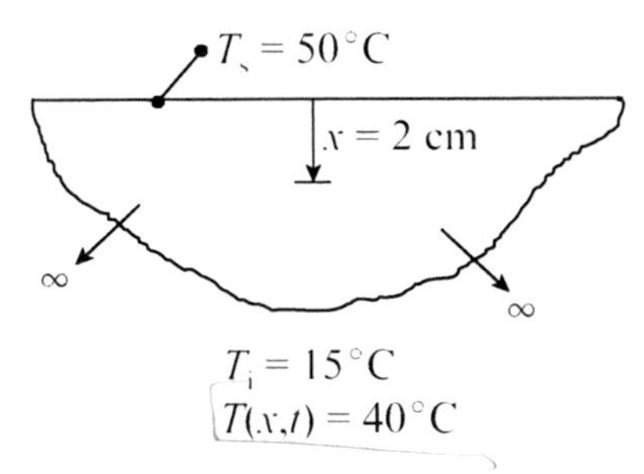

ASSUMPTIONS

1. The slab can be analyzed as a semi-infinite solid.
2. One-dimensional conduction.
3. Constant properties.

PROPERTIES

concrete: $\alpha = 6.9 \times 10^{-7}$ m^2/s

ANALYSIS

For a semi-infinite solid initially at temperature T_i whose surface is suddenly changed to T_s, the temperature at a specified depth x and time t is given by the relation

$$\frac{T(x,t) - T_s}{T_i - T_s} = erf\left(\frac{x}{2\sqrt{\alpha t}}\right)$$

where α is the thermal diffusivity of the solid, and *erf* is the Gaussian error function. This equation may be solved for time t by recognizing that the value of the left side is known and using a table of the error function to obtain a value for the argument of the function. Alternatively, time may be found by iteration using a computer-based equation solver. Using the second approach, the time required for a point 2 cm deep in the slab to reach 40°C is found to be

$t = 2163$ s $=$ <u>36.1 min</u>

DISCUSSION

The primary criterion underlying the first assumption is that the thermal penetration does not reach the back side of the concrete slab in the specified time. If the temperature of the back side of the slab were to significantly change, an analysis that included the properties of the material in contact with the back side of the concrete (soil, most likely) would have to be done to obtain an accurate result. Most concrete slabs used as sidewalks, patios and other flat work in the construction industry are approximately 9 cm thick. Setting x = 0.09 m and t = 2163 s, the temperature of the back side of the slab is calculated as 18.5 °C, only 3.5 °C higher than the initial temperature of the slab. Thus, our answer is not exceptionally precise, but since the temperature of the back side of the slab does not change appreciably, our answer is a reasonable approximation.

❒ ❒ ❒

PROBLEM 5.7 Using a Semi-infinite Solid Model to Determine Thermal Conductivity

A test procedure for determining the thermal conductivity of a plastic material involves imbedding a small thermocouple in a thick slab of the plastic at a known depth and measuring the temperature change due to a sudden change in temperature of one of the surfaces. In the test, the thermocouple is imbedded at a depth of 6 mm, and the surface of the slab is suddenly changed to 100 °C by exposing it to a steady flow of boiling water. After 5 minutes of exposure, the thermocouple indicates a temperature of 60 °C. If the initial temperature of the slab is 20 °C, what is the thermal conductivity of the plastic? The density and specific heat of the plastic are known to be $\rho = 1150\ \text{kg/m}^3$ and $c_p = 1700\ \text{J/kg·K}$.

DIAGRAM

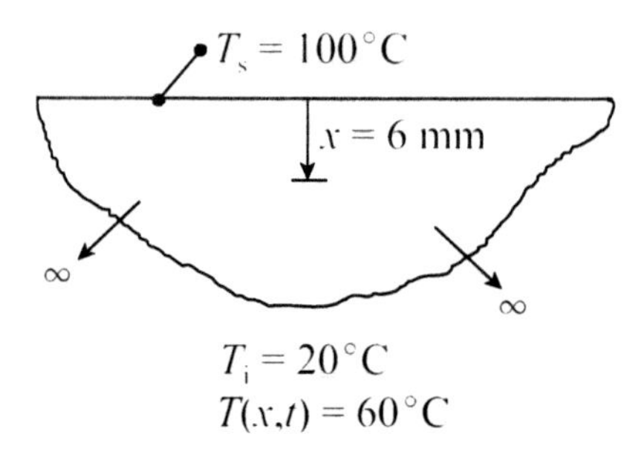

ASSUMPTIONS

1. The slab can be analyzed as a semi-infinite solid.
2. One-dimensional conduction.
3. Constant properties.

PROPERTIES

plastic: $\rho = 1150$ kg/m^3, $c_p = 1700$ J/kg·K

ANALYSIS

For a semi-infinite solid initially at temperature T_i whose surface is suddenly changed to T_s, the temperature at a specified depth x and time t is given by the relation

$$\frac{T(x,t) - T_s}{T_i - T_s} = erf\left(\frac{x}{2\sqrt{\alpha t}}\right)$$

where α is the thermal diffusivity of the material, defined as

$$\alpha \equiv \frac{k}{\rho c_p}$$

Iteratively solving for α,

$\alpha = 1.32 \times 10^{-7}$ m^2/s

Thus, the thermal conductivity k of the plastic is

$k = \alpha \rho c_p$

$= (1.32 \times 10^{-7}$ m^2/s$)(1150$ kg/m$^3)(1700$ J/kg·K$)$

$=$ 0.258 W/m·K

DISCUSSION

The known values of density and specific heat and experimentally determined value of thermal conductivity are typical for the nylon family of plastics. For example, at room temperature nylon 6/6 has a density, specific heat and thermal conductivity of $\rho = 1140$ kg/m^3, $c_p = 1680$ J/kg·K and $k = 0.25$ W/m·K. There are more accurate experimental methods for determining the thermal conductivity of an unknown solid material, but this problem illustrates how a semi-infinite solid model may be used.

PROBLEM 5.8 Firewall Evaluation using a Semi-infinite Solid Model

A 25-cm thick wall of concrete (ρ = 2250 kg/m^3, c_p = 875 J/kg·K, k = 1.4 W/m·K) serves as a firewall in a commercial building. A test specification in the fire code states that a uniform radiant heat flux of 10^4 W/m^2 applied to one surface of this wall must not induce temperatures on the irradiated side and back side of the wall that are higher than 300°C and 20°C, respectively, after 30 minutes of exposure. The initial temperature of the wall is 20°C. Does this firewall satisfy the code?

DIAGRAM

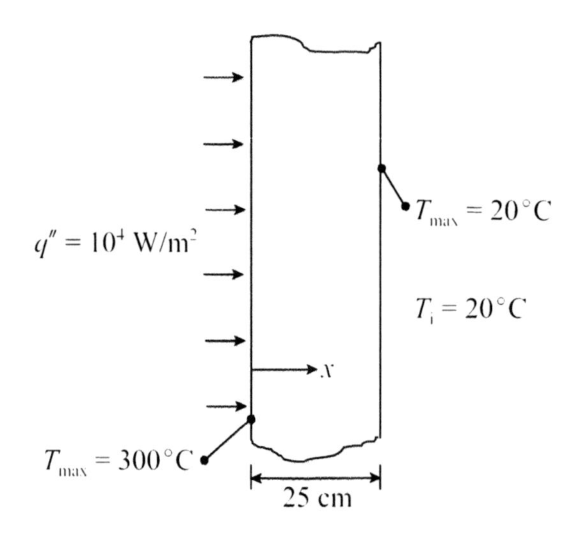

ASSUMPTIONS

1. The firewall can be analyzed as a semi-infinite solid.
2. One-dimensional conduction.
3. All incident heat flux is absorbed by the surface (α_s = 1).
4. Convection and radiation from the surface are neglected.
5. Constant properties.

PROPERTIES

concrete: ρ = 2250 kg/m^3, c_p = 875 J/kg·K, k = 1.4 W/m·K, α_s = 1

ANALYSIS

Because the initial temperature of the firewall is equivalent to the maximum temperature of the back side of the wall after the specified time interval, we can model the wall as a semi-infinite solid. The temperature of a semi-infinite solid at depth x after time t, exposed to a heat flux q'', is given by the relation

$$T(x,t) - T_i = \frac{2q''(\alpha t/\pi)^{1/2}}{k}\exp\left(\frac{-x^2}{4\alpha t}\right) - \frac{q''x}{k} erfc\left(\frac{x}{2\sqrt{\alpha t}}\right)$$

where T_i is the initial temperature of the solid, and *erfc* is the complementary error function defined by

$$erfc(\eta) \equiv 1 - erf(\eta)$$

The thermal diffusivity of the concrete is

$$\alpha \equiv \frac{k}{\rho c_p}$$

$$\alpha = \frac{(1.4 \text{ W/m·K})}{(2250 \text{ kg/m}^3)(875 \text{ J/kg·K})}$$

$$= 7.111 \times 10^{-7} \text{ m}^2\text{/s}$$

We must calculate the temperature of both surfaces of the wall, for $x = 0$ m and $x = 0.25$ m. For $x = 0$ m the relation above reduces to

$$T(x,t) - T_i = \frac{2q''(\alpha t/\pi)^{1/2}}{k}$$

Thus, the temperature of the irradiated surface after an exposure time of 30 min is

$$T(x,t) = \frac{2(10^4 \text{ W/m}^2)[(7.111 \times 10^{-7} \text{ m}^2\text{/s})(30 \text{ min} \times 60 \text{ s/min})/\pi]^{1/2}}{1.4 \text{ W/m·K}} + 20°\text{C}$$

$$= \underline{\underline{308°\text{C}}}$$

and the temperature of the back surface of the wall after an exposure time of 30 min is

$$T(x,t) = \underline{\underline{20.0°\text{C}}}$$

DISCUSSION

This firewall does not meet the code because the temperature of the irradiated surface exceeds the maximum temperature by 8°C. Increasing the wall thickness does not remedy the situation, but a slight adjustment in concrete properties may be in order. Increasing the thermal conductivity of the concrete to $k = 1.44$ W/m·K results in a surface temperature of 300°C.

PROBLEM 5.9 Semi-infinite Solid Model of Laser Heating a Retinal Tear

An argon or krypton laser may be used to repair a retinal tear to prevent the retina from separating from the back wall of the eye. A focused pulse of laser light is directed on the tear, producing an inflammation reaction that fuses the tissues together. If the laser pulse has a magnitude of 0.8 MJ/m^2, calculate the temperature of the surface of the tear after 1 s. For thermal properties of retinal tissue, use $\rho = 998$ kg/m^3, $c_p = 4182$ J/kg·K and $k = 0.6$ W/m·K. The initial temperature of the retinal tissue is 30°C.

DIAGRAM

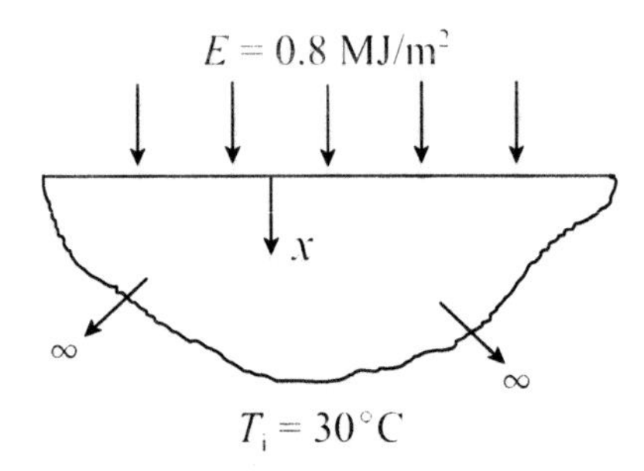

ASSUMPTIONS

1. The retinal tissue can be analyzed as a semi-infinite solid.
2. One-dimensional conduction.
3. The laser pulse is instantaneous.
4. Convection and radiation from the surface are neglected.
5. Constant properties.

PROPERTIES

retinal tissue: $\rho = 998$ kg/m^3, $c_p = 4182$ J/kg·K, $k = 0.6$ W/m·K

ANALYSIS

The temperature of a semi-infinite solid at depth x after time t, exposed to an instantaneous surface energy pulse E, is given by the relation

$$\frac{T - T_i}{E / (\rho c_p \sqrt{\pi \alpha t})} = \exp\left(\frac{-x^2}{4\alpha t}\right)$$

where T_i is the initial temperature of the solid and E is the instantaneous energy pulse at the surface in units of J/m^2. The thermal diffusivity of the retinal tissue is

$$\alpha \equiv \frac{k}{\rho c_p}$$

$$\alpha = \frac{(0.6 \text{ W/m·K})}{(998 \text{ kg/m}^3)(4182 \text{ J/kg·K})}$$

$$= 1.438 \times 10^{-7} \text{ m}^2\text{/s}$$

The temperature of the surface of the tear after a 1 s time period is

$$T = \frac{E}{\rho c_p \sqrt{\pi \alpha t}} + T_i$$

$$T = \frac{0.8 \times 10^6 \text{ J/m}^2}{(998 \text{ kg/m}^3)(4182 \text{ J/kg·K})[\pi(1.438 \times 10^{-7} \text{ m}^2\text{/s})(1 \text{ s})]^{1/2}} + 30°\text{C}$$

$$= \underline{\underline{315°\text{C}}}$$

DISCUSSION

Theoretically, the surface temperature of a semi-infinite solid exposed to an instantaneous energy pulse is infinite at $t = 0$. The surface temperature of the retinal tissue for a time period of about 10 s after the laser pulse in shown in the graph below.

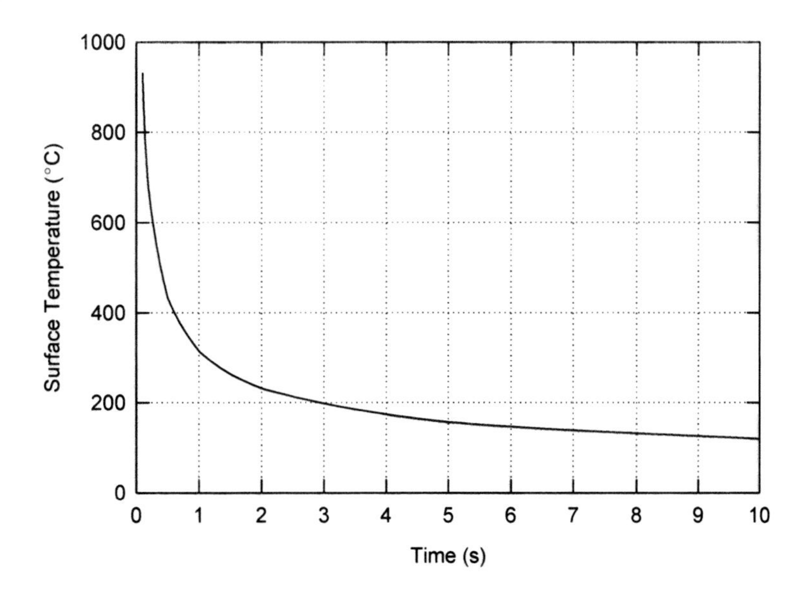

❒ ❒ ❒

PROBLEM 5.10 Rocket Motor Insulation Design using a Semi-infinite Solid Model

During the test firing of a solid-propellant rocket motor, the aft dome is radiantly heated by the plume of exhaust gases that emanates from the nozzle, as shown in the diagram. An insulating material covers the aft dome to assure that, during the motor firing, the underlying composite motor case does not exceed a maximum allowable temperature. Using a semi-infinite solid model of the insulation-case-propellant system, estimate the thickness of cork insulation ($\rho = 117$ kg/m^3, $c_p = 1880$ J/kg·K, $k = 0.046$ W/m·K) required to limit the case to a temperature of 95°C for a motor firing time of 80 s. The radiative heat flux from the plume is 7000 W/m^2, and the initial temperature of the motor is 20°C.

DIAGRAM

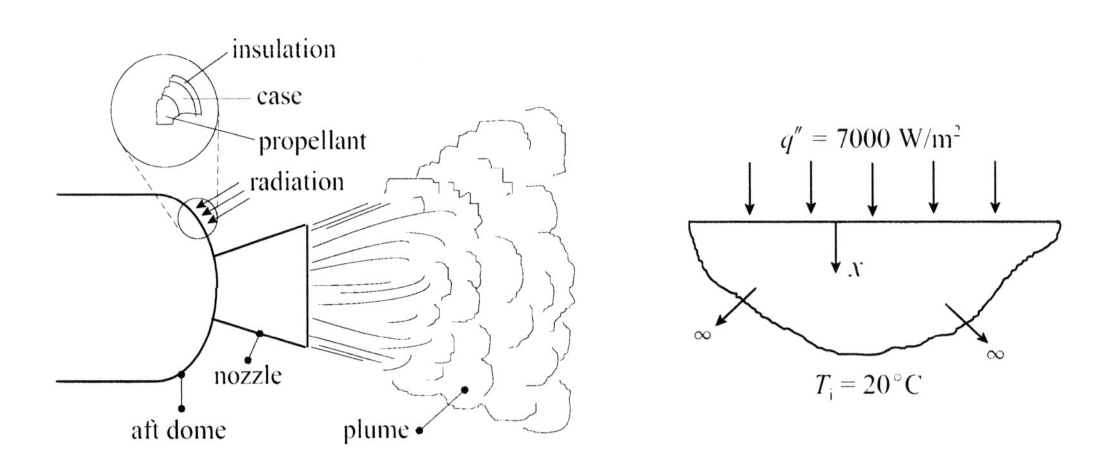

ASSUMPTIONS

1. The insulation-case-propellant system can be analyzed as a semi-infinite solid with thermal properties of cork.
2. One-dimensional conduction.
3. All incident radiant heat flux is absorbed by the insulation surface ($\alpha_s = 1$).
4. Convection and radiation from the insulation surface are neglected.
5. Constant properties.

PROPERTIES

cork: $\rho = 117$ kg/m^3, $c_p = 1880$ J/kg·K, $k = 0.046$ W/m·K, $\alpha_s = 1$

ANALYSIS

The temperature of a semi-infinite solid at depth x after time t, exposed to a heat flux q'', is given by the relation

$$T(x,t) - T_i = \frac{2q''(\alpha t/\pi)^{1/2}}{k}\exp\left(\frac{-x^2}{4\alpha t}\right) - \frac{q''x}{k}erfc\left(\frac{x}{2\sqrt{\alpha t}}\right)$$

where T_i is the initial temperature of the solid, and *erfc* is the complementary error function. We treat the insulation-case-propellant system as a single semi-infinite solid with thermal properties of cork. This is a reasonable approximation as long as the thermal properties of these three materials are somewhat similar. The thermal diffusivity of the cork, and by assumption the case and propellant, is

$$\alpha \equiv \frac{k}{\rho c_p}$$

$$\alpha = \frac{(0.046 \text{ W/m·K})}{(117 \text{ kg/m}^3)(1880 \text{ J/kg·K})}$$

$$= 2.091 \times 10^{-7} \text{ m}^2\text{/s}$$

We wish to find the depth in the semi-infinite solid that attains a temperature of 95°C after 80 s of exposure to the radiant heat flux. This depth is equivalent to the required thickness of cork insulation. Setting $T(x,t) = 95°C$, $t = 80$ s, and $q'' = 7000$ W/m^2 and iteratively solving the above relation for x,

$$x = 7.70 \times 10^{-3} \text{ m} = \underline{\underline{7.70 \text{ mm}}}$$

DISCUSSION

The cork insulation must be at least 7.70 mm thick in order to keep the composite case below 95°C during the 80-s firing time. If the cork insulation is 7.70 mm thick, the outer surface of the rocket motor case will attain a temperature of 95°C after 80 s, the duration of the motor firing.

It is instructive to determine the penetration depth, i.e., the depth to which the temperature changes during the 80-s time period. Setting $T(x,t) = 20°C$, the initial temperature of the rocket motor, and solving for x, we obtain $x = 43.1$ mm (1.70 in). This depth corresponds to a point well into the rocket motor case and probably into the solid propellant. Our primary assumption, that the thermal properties of the insulation, case and propellant are similar, is therefore an important one. The thermal properties of cork, graphite or kevlar composites, and solid rocket propellants are within an order of magnitude of one another, so modeling the insulation-case-propellant system as a semi-infinite solid with thermal properties of cork is reasonable.

PROBLEM 5.11 Air Cooling a Thick Stainless Steel Plate

Upon removal from a 400°C heat treating oven, a large plate of stainless steel ($\rho = 7900$ kg/m^3, $c_p = 480$ J/kg·K, $k = 15$ W/m·K) is allowed to cool in 20°C air, where the heat transfer coefficient is 18 W/m^2·K. If the plate is 25 cm thick, find the temperature of the center and surface of the plate after cooling for 2 h.

DIAGRAM

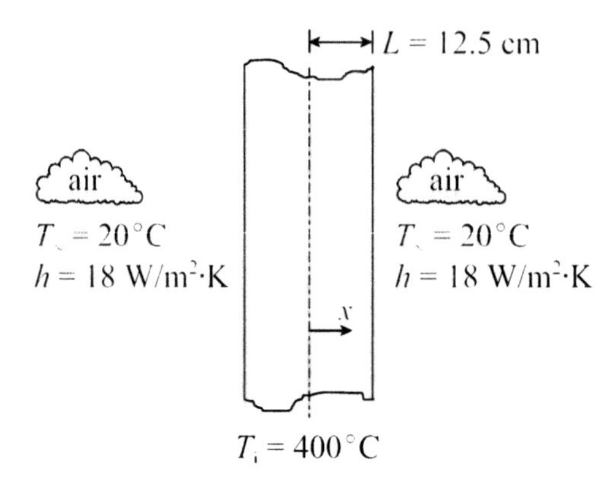

ASSUMPTIONS

1. The plate has a uniform temperature prior to cooling.
2. One-dimensional conduction in the x direction.
3. The plate is surrounded by the cooling air.
4. Heat transfer from the edges of the plate is neglected.
5. Radiation is neglected.
6. Constant properties.

PROPERTIES

stainless steel: $\rho = 7900$ kg/m^3, $c_p = 480$ J/kg·K, $k = 15$ W/m·K

ANALYSIS

The method of analysis depends on whether the plate is lumped. The Biot number for a plate of half-thickness L is defined as

$$Bi \equiv \frac{hL}{k}$$

Thus, the Biot number for the stainless steel plate is

$$Bi = \frac{(18\ \text{W/m}^2\cdot\text{K})(0.125\ \text{m})}{15\ \text{W/m}\cdot\text{K}}$$

$$= 0.150 > 0.1$$

The plate is not lumped, so there are spatial temperature effects in the plate. Because the plate is large, we assume one-dimensional conduction in the x direction. Thermal diffusivity is defined as

$$\alpha \equiv \frac{k}{\rho c_p}$$

Thus,

$$\alpha = \frac{(15\ \text{W/m}\cdot\text{K})}{(7900\ \text{kg/m}^3)(480\ \text{J/kg}\cdot\text{K})}$$

$$= 3.956 \times 10^{-6}\ \text{m}^2/\text{s}$$

The Fourier number is defined as

$$Fo \equiv \frac{\alpha t}{L^2}$$

$$Fo = \frac{(3.956 \times 10^{-6}\ \text{m}^2/\text{s})(2\ \text{h} \times 3600\ \text{s/h})}{(0.125\ \text{m})^2}$$

$$= 1.8229$$

For a large plate, with uniform initial temperature, that is suddenly exposed to a convective environment, the temperature may be found by a one-term approximation of an infinite series,

$$\frac{\theta}{\theta_i} = \frac{T(x,t) - T_\infty}{T_i - T_\infty} = C_1 \exp(-\zeta_1^2 Fo)\cos(\zeta_1 x/L)$$

where the parameters C_1 and ζ_1 are obtained from the relations

$$\zeta_1 \tan\zeta_1 = Bi$$

$$C_1 = \frac{4\sin\zeta_1}{2\zeta_1 + \sin(2\zeta_1)}$$

The values of these parameters are

$$C_1 = 1.0237 \quad , \quad \zeta_1 = 0.3779$$

Substituting numerical values into the one-term approximation for $x = 0$,

$$\frac{T(x,t) - T_\infty}{T_i - T_\infty} = 1.0237 \exp[-(0.3779)^2(1.8229)] \cos(0)$$

$$= 0.7891$$

Thus,

$$T(x,t) = 0.7891(400 - 20)°\text{C} + 20°\text{C}$$

$$= \underline{\underline{320°\text{C}}}$$

Substituting numerical values into the one-term approximation for $x = L$,

$$\frac{T(x,t) - T_\infty}{T_i - T_\infty} = 1.0237 \exp[-(0.3779)^2(1.8229)] \cos[(0.3779)(1)]$$

$$= 0.7164$$

Thus,

$$T(x,t) = 0.7164(400 - 20)°\text{C} + 20°\text{C}$$

$$= \underline{\underline{292°\text{C}}}$$

DISCUSSION

As shown in the analysis, the center of the plate has a higher temperature after 2 h of cooling than does the surface since the surface is in direct contact with the air. The difference in temperatures between the center and surface is not great, however, because the material is metallic. The Biot number for the plate is not much greater than 0.1, which suggests that a reasonable answer could be obtained by treating the plate as a lumped mass. The relation for a lumped mass is

$$\frac{T - T_\infty}{T_i - T_\infty} = \exp\left[-\left(\frac{hA}{\rho V c_p}\right)t\right]$$

After substituting values,

$$T = 309°\text{C}$$

which lies between the center and surface temperatures.

❒ ❒ ❒

PROBLEM 5.12 Predicting the Baking Time of a Potato

A potato is to be baked in a conventional oven set at a temperature of 175 °C. The initial temperature of the potato is 25 °C, and the heat transfer coefficient when the potato is in the oven is 14 W/m^2·K. If the potato is considered completely baked when its center reaches a temperature of 115 °C, what baking time is required? The potato can be approximated as a sphere with a radius of 4 cm. For thermal properties, use $k = 0.51$ W/m·K and $\alpha = 1.33 \times 10^{-7}$ m^2/s.

DIAGRAM

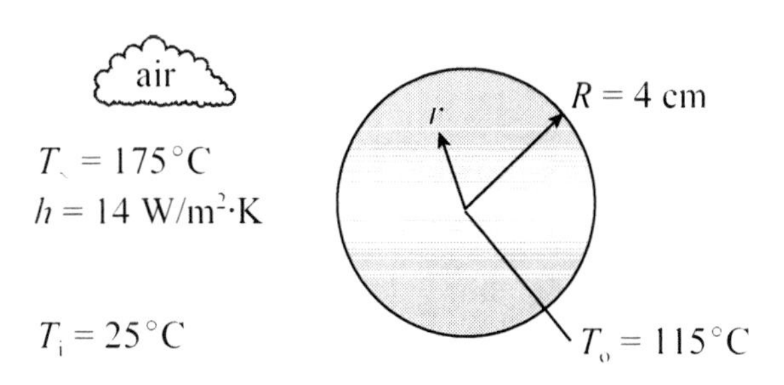

ASSUMPTIONS

1. The potato has a uniform temperature prior to baking.
2. One-dimensional conduction in the r direction.
3. The potato is surrounded by air in the oven.
4. The potato is spherical.
5. Radiation is neglected.
6. Constant properties.

PROPERTIES

potato: $k = 0.51$ W/m·K and $\alpha = 1.33 \times 10^{-7}$ m^2/s

ANALYSIS

For a sphere, with uniform initial temperature, that is suddenly exposed to a convective environment, the temperature may be found by a one-term approximation of an infinite series,

$$\frac{\theta}{\theta_i} = \frac{T(r,t) - T_\infty}{T_i - T_\infty} = C_1 \exp(-\zeta_1^2 Fo) \frac{1}{\zeta_1 (r/R)} \sin(\zeta_1 r/R)$$

where the parameters C_1 and ζ_1 may be obtained from a knowledge of the Biot number,

$$Bi \equiv \frac{hR}{k}$$

$$Bi = \frac{(14\ \text{W/m}^2\cdot\text{K})(0.04\ \text{m})}{0.51\ \text{W/m}\cdot\text{K}}$$

$$= 1.098$$

The values of C_1 and ζ_1 (consult a heat transfer text for a table of Bi, C_1, and ζ_1) are

$$C_1 = 1.2934 \quad , \quad \zeta_1 = 1.6157$$

Noting that the Fourier number is $Fo \equiv \alpha t/R^2$, setting $r = 0$, and solving the one-term approximation for time t,

$$t = \frac{-R^2}{\zeta_1^2 \alpha} \ln\left(\frac{1}{C_1} \frac{\theta}{\theta_i} \right)$$

$$t = \frac{-(0.04\ \text{m})^2}{(1.6157)^2\,(1.33 \times 10^{-7}\ \text{m}^2/\text{s})} \ln\left(\frac{1}{1.2934} \frac{115°\text{C} - 175°\text{C}}{25°\text{C} - 175°\text{C}} \right)$$

$$= 5408\ \text{s} = \underline{\underline{1.50\ \text{h}}}$$

DISCUSSION

Radiation is an important heat transfer mechanism for cooking foods in conventional ovens because the heat transfer coefficients are typically low, so the radiative portion of the total heat transfer may be comparable to the convective portion. If radiation had been included in the analysis, the baking time would have decreased significantly. Most people are not willing to wait an hour or so to bake a potato, or to cook anything else for that matter. Our result illustrates why a microwave oven, which cooks food in a relatively short period of time, is the preferred way to bake a potato!

❒ ❒ ❒

PROBLEM 5.13 Oven Temperature for a Plastic Rod in a Manufacturing Operation

A long 18-mm diameter rod of plastic ($\rho = 1210\ \text{kg/m}^3$, $c_p = 1375\text{J/kg}\cdot\text{K}$, $k = 0.23\ \text{W/m}\cdot\text{K}$) is uniformly heated in a conveyor oven in preparation for forming the rod into a useful shape for a manufactured product. For optimum forming characteristics, the temperature of the rod should not be less than 210°C. If the rod spends 4 min in the oven, where the heat transfer coefficient is 10 $\text{W/m}^2\cdot\text{K}$, what should be the operating temperature of the oven? The initial temperature of the rod is 20°C.

DIAGRAM

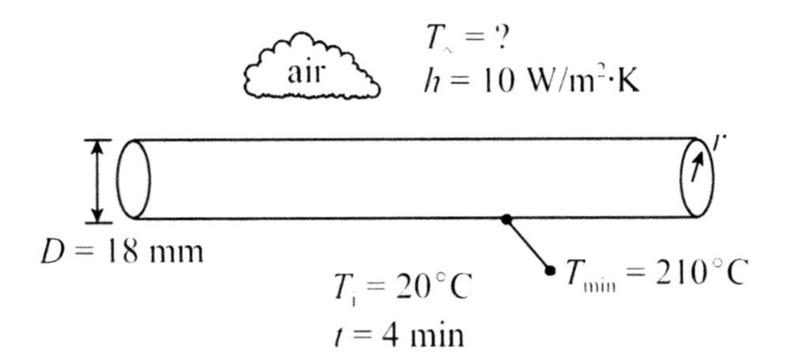

ASSUMPTIONS

1. The rod has a uniform initial temperature.
2. One-dimensional conduction in the r direction.
3. The rod is surrounded by air in the oven.
4. Radiation is neglected.
5. Constant properties.

PROPERTIES

plastic: $\rho = 1210$ kg/m^3, $c_p = 1375$J/kg·K, $k = 0.23$ W/m·K

ANALYSIS

The Biot number for a long cylinder of radius R is defined as

$$Bi \equiv \frac{hR}{k}$$

Thus, the Biot number for the rod is

$$Bi = \frac{(10 \text{ W/m}^2\cdot\text{K})(0.009 \text{ m})}{0.23 \text{ W/m}\cdot\text{K}}$$

$$= 0.391$$

Thermal diffusivity is defined as

$$\alpha \equiv \frac{k}{\rho c_p}$$

Thus,

$$\alpha = \frac{(0.23 \text{ W/m·K})}{(1210 \text{ kg/m}^3)(1375 \text{ J/kg·K})}$$

$$= 1.382 \times 10^{-7} \text{ m}^2\text{/s}$$

The Fourier number is defined as

$$Fo \equiv \frac{\alpha t}{R^2}$$

$$Fo = \frac{(1.382 \times 10^{-7} \text{ m}^2\text{/s})(4 \text{ min} \times 60 \text{ s/min})}{(0.009 \text{ m})^2}$$

$$= 0.4095$$

For a long cylinder, with uniform initial temperature, that is suddenly exposed to a convective environment, the temperature may be found by a one-term approximation of an infinite series,

$$\frac{\theta}{\theta_i} = \frac{T(r,t) - T_\infty}{T_i - T_\infty} = C_1 \exp(-\zeta_1^2 Fo) J_o(\zeta_1 r/R)$$

where J_0 is the Bessel function of the first kind of zero order, and C_1 and ζ_1 are constants determined from a knowledge of the Biot number Bi. The values of C_1 and ζ_1 (consult a heat transfer text for a table of Bi, C_1, and ζ_1) are

$$C_1 = 1.0912 \quad , \quad \zeta_1 = 0.8421$$

The minimum temperature occurs at $r = 0$, the center of the rod. Noting that $J_0(0) = 1$, and solving the one-term approximation for the air (oven) temperature T_∞,

$$T_\infty = \frac{T(0,t) - KT_i}{1 - K}$$

where, $T(0,t) = 210°\text{C}$ and

$$K = C_1 \exp(-\zeta_1^2 Fo)$$

$$K = (1.0912) \exp[-(0.8421)^2(0.4095)]$$

$$= 0.8162$$

Thus, the oven temperature is

$$T_\infty = \frac{210°\text{C} - (0.8162)(20°\text{C})}{1 - 0.8162}$$

$= \underline{\underline{1054°C}}$

DISCUSSION

The oven must operate at a temperature of 1054°C for the center of the rod to reach 210°C in 4 min. It is instructive to examine the effect of time on the solution. If the rod is left in the oven for a longer period of time, the operating temperature of the oven need not be as high. A lower oven temperature results in lower fuel or power requirements to run the oven, resulting in a cost savings to the company. However, a longer heating time means a lower production rate, resulting in a lower profit for the company. Thus, there is a tradeoff between heating temperature and time. The graph below shows the required oven temperature for a range of heating times. To determine whether there is an optimum heating time and temperature, relationships for the cost of oven operation and production rate would have to be included in the analysis.

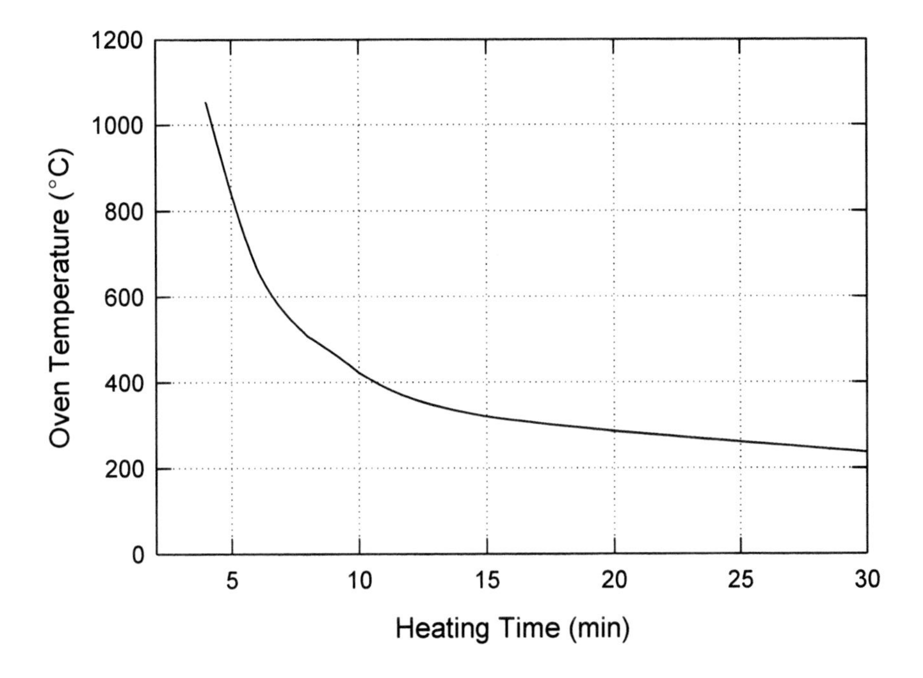

❒ ❒ ❒

PROBLEM 5.14 Heat Transfer Coefficient Prediction for a Windshield Defrosting System

While a car is parked outside during a winter night when the ambient air temperature is −20°C, a layer of ice forms on the 5-mm thick windshield. In the morning when the defrosting system is turned on, the inside surface of the windshield is suddenly exposed to a 35°C flow of air. Assuming that the ice layer acts as a perfect insulator, what is the value of the inside heat transfer coefficient that would allow the outside surface of the windshield to reach 0°C in 70 s? For thermal properties of the windshield, use $\rho = 2100$ kg/m^3, $c_p = 850$ J/kg·K and $k = 1.1$ W/m·K.

DIAGRAM

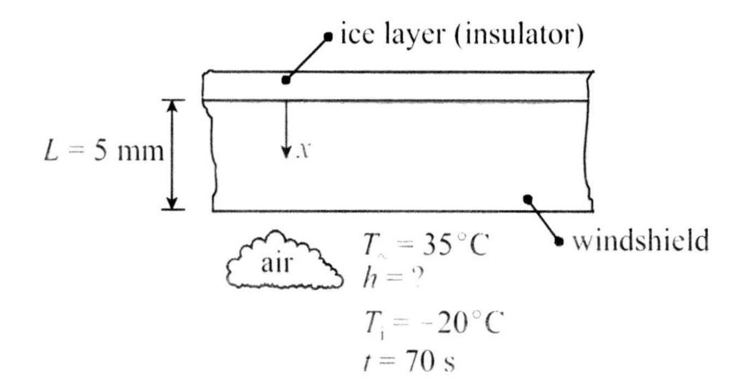

ASSUMPTIONS

1. The windshield is a large flat plate.
2. One-dimensional conduction in the x direction.
3. The outside surface of the windshield is perfectly insulated by the ice layer.
4. The entire inside surface of the windshield is exposed to the defrosting air.
5. Radiation is neglected.
6. Constant properties.

PROPERTIES

windshield: $\rho = 2100$ kg/m^3, $c_p = 850$ J/kg·K, $k = 1.1$ W/m·K

ANALYSIS

The thermal diffusivity of the windshield is

$$\alpha \equiv \frac{k}{\rho c_p}$$

$$\alpha = \frac{1.1 \text{ W/m·K}}{(2100 \text{ kg/m}^3)(850 \text{ J/kg·K})}$$

$= 6.163 \times 10^{-7}$ m²/s

The Fourier number is defined as

$$Fo \equiv \frac{\alpha t}{L^2}$$

where L is the windshield thickness. Thus,

$$Fo = \frac{(6.163 \times 10^{-7}\ \text{m}^2/\text{s})(70\ \text{s})}{(0.005\ \text{m})^2}$$

$= 1.726$

When a large plate, with uniform initial temperature, of thickness $2L$ is suddenly exposed to a convective environment, there is thermal symmetry about the center plane of the plate. Thus, half of the plate can be analyzed, placing an adiabatic (insulated) boundary condition on the center plane. In this problem, the ice layer provides that insulation. The temperature may be found by a one-term approximation of an infinite series,

$$\frac{\theta}{\theta_i} = \frac{T(x,t) - T_\infty}{T_i - T_\infty} = C_1 \exp(-\zeta_1^2 Fo)\cos(\zeta_1 x/L)$$

where the parameters C_1 and ζ_1 are obtained from the relations

$$\zeta_1 \tan\zeta_1 = Bi$$

$$C_1 = \frac{4\sin\zeta_1}{2\zeta_1 + \sin(2\zeta_1)}$$

Since the heat transfer coefficient h is unknown, we do not know the Biot number, defined as

$$Bi \equiv \frac{hL}{k}$$

but we know all temperatures and the Fourier number Fo, so we can solve the above four equations for C_1, ζ_1, Bi and h. Setting $x = 0$ and solving these equations iteratively,

$C_1 = 1.0478$, $\zeta_1 = 0.5375$, $Bi = 0.320$, $h = 70.5$ W/m²·K

DISCUSSION

Let's examine the effect of initial temperature (ambient air temperature) on the solution. If the ambient air temperature during the night is lower than −20°C, a larger heat transfer coefficient is required to clear the frost in the same time period. On the other hand, if the ambient air temperature during the night is higher than −20°C, a smaller heat transfer coefficient is required. The graph below shows the relationship between initial temperature and the required heat transfer coefficient. As expected, the heat transfer coefficient decreases with increasing initial temperature. From a practical point of view, this means that the air flow rate from the fan in the car's dashboard must increase with decreasing ambient air temperature if the frost is to be removed during a given period of time.

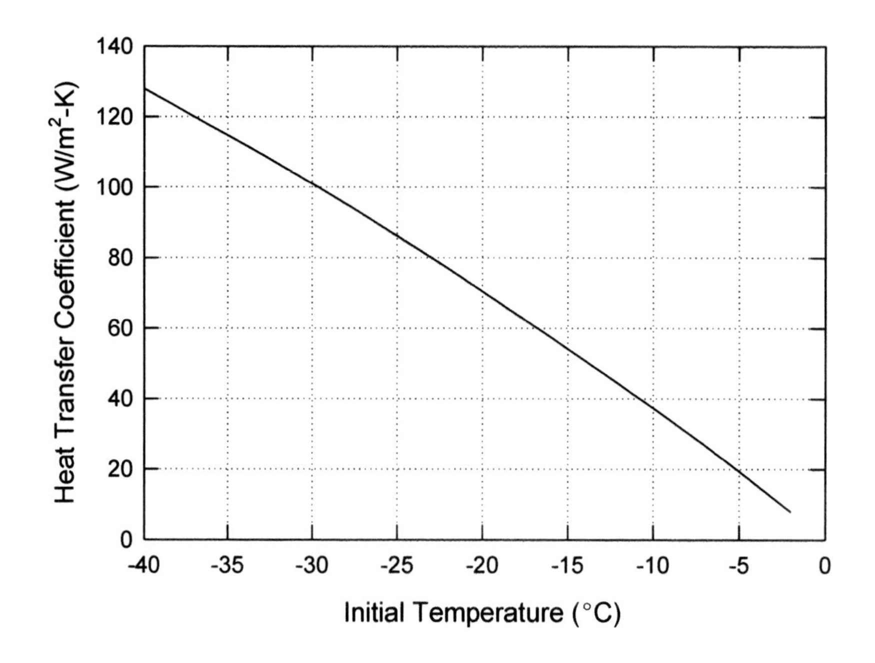

❒ ❒ ❒

PROBLEM 5.15 Cooling a Short Ceramic Cylinder

A cylinder of ceramic (ρ = 2300 kg/m^3, c_p = 750 J/kg·K, k = 2.2 W/m·K) with a length and diameter of 16 cm and 8 cm, respectively, has an initial temperature of 450°C. The cylinder is then suddenly exposed to 20°C air, where the heat transfer coefficient is 30 W/m^2·K. For a cooling time of 30 min, find the temperature of the center of the cylinder and the center of one end of the cylinder.

DIAGRAM

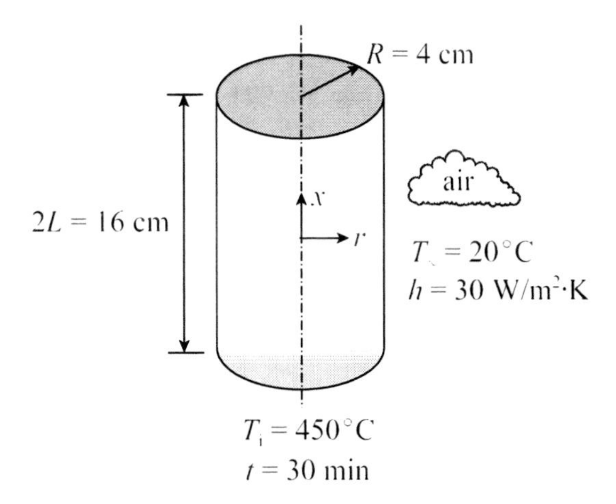

ASSUMPTIONS

1. The cylinder has a uniform initial temperature.
2. Two-dimensional conduction in the x and r directions.
3. The entire surface of the cylinder is exposed to the cooling air.
4. Radiation is neglected.
5. Constant properties.

PROPERTIES

ceramic: $\rho = 2300$ kg/m^3, $c_p = 750$ J/kg·K, $k = 2.2$ W/m·K

ANALYSIS

A short cylinder is formed by the intersection of a large plate of thickness $2L$ and a long cylinder of radius R. The Biot numbers for the plate and cylinder, respectively, are

$$Bi \equiv \frac{hL}{k}$$

$$Bi = \frac{(30 \text{ W/m}^2\cdot\text{K})(0.08 \text{ m})}{2.2 \text{ W/m}\cdot\text{K}}$$

$$= 1.091$$

$$Bi \equiv \frac{hR}{k}$$

$$Bi = \frac{(30 \text{ W/m}^2\cdot\text{K})(0.04 \text{ m})}{2.2 \text{ W/m}\cdot\text{K}}$$

$$= 0.5455$$

Thus, the cylinder is not lumped in either direction. The thermal diffusivity of the ceramic is

$$\alpha \equiv \frac{k}{\rho c_p}$$

$$\alpha = \frac{2.2 \text{ W/m}\cdot\text{K}}{(2300 \text{ kg/m}^3)(750 \text{ J/kg}\cdot\text{K})}$$

$$= 1.275 \times 10^{-6} \text{ m}^2\text{/s}$$

The Fourier numbers for the plate and cylinder, respectively, are

$$Fo \equiv \frac{\alpha t}{L^2}$$

$$Fo = \frac{(1.275 \times 10^{-6} \text{ m}^2\text{/s})(30 \text{ min} \times 60 \text{ s/min})}{(0.08 \text{ m})^2}$$

$$= 0.3586$$

$$Fo \equiv \frac{\alpha t}{R^2}$$

$$Fo = \frac{(1.275 \times 10^{-6} \text{ m}^2\text{/s})(30 \text{ min} \times 60 \text{ s/min})}{(0.04 \text{ m})^2}$$

$$= 1.434$$

For a large plate, with uniform initial temperature, that is suddenly exposed to a convective environment, the temperature may be found by a one-term approximation of an infinite series,

$$P(x,t) = \frac{\theta}{\theta_i} = \frac{T(x,t) - T_\infty}{T_i - T_\infty} = C_1 \exp(-\zeta_1^2 Fo)\cos(\zeta_1 x/L)$$

where, by knowing that $Bi = 1.091$, the parameters C_1 and ζ_1 are

$C_1 = 1.1246$, $\zeta_1 = 0.8800$

For a long cylinder, with uniform initial temperature, that is suddenly exposed to a convective environment, the temperature may be found by a one-term approximation of an infinite series,

$$C(r,t) = \frac{\theta}{\theta_i} = \frac{T(r,t) - T_\infty}{T_i - T_\infty} = C_1 \exp(-\zeta_1^2 Fo) J_o(\zeta_1 r/R)$$

where J_0 is the Bessel function of the first kind of zero order, and C_1 and ζ_1 are

$$C_1 = 1.1235 \quad , \quad \zeta_1 = 0.8922$$

For the center of the cylinder, $x = 0$ and $r = 0$, giving the plate and cylinder solutions

$$P(0,t) = 0.8519 \quad , \quad C(0,t) = 0.3588$$

The product of these two solutions yields the solution for the short cylinder.

$$\frac{T(0,0,t) - T_\infty}{T_i - T_\infty} = P(0,t)C(0,t)$$

Thus, the temperature of the center of the cylinder after 30 min is

$$T(0,0,t) = (0.8519)(0.3588)(450°C - 20°C) + 20°C$$

$$= \underline{\underline{151°C}}$$

For the center of one end of the cylinder, $x = L$ and $r = 0$, so the cylinder solution is the same as before, but the plate solution is

$$P(L,t) = 0.5428$$

Thus, the temperature of the center of one end of the cylinder after 30 min is

$$T(L,0,t) = (0.5428)(0.3588)(450°C - 20°C) + 20°C$$

$$= \underline{\underline{104°C}}$$

DISCUSSION

As shown by the calculations, the temperature of the center of the cylinder must be higher than that of the surface since the surface is in direct contact with the cooling air. It turns out that the temperature at $x = L$ and $r = R$ is

$$T(L,R,t) = (0.5428)(0.2908)(450°C - 20°C) + 20°C$$

$$= 87.9°C$$

❐ ❐ ❐

PROBLEM 5.16 Initiation of Melting for an Ice Cube

A 40-cm cube of ice ($\rho = 920$ kg/m^3, $c_p = 2040$ J/kg·K, $k = 1.88$ W/m·K) at a uniform temperature of $-18\,°C$ is suddenly brought into a 15 °C room in an ice-making plant where the cube is carried along a conveyor to a delivery truck. The heat transfer coefficient for the surfaces of the ice cube is 24 W/m^2·K. Neglecting conduction into the conveyor, how long will it take for the ice cube to begin melting?

DIAGRAM

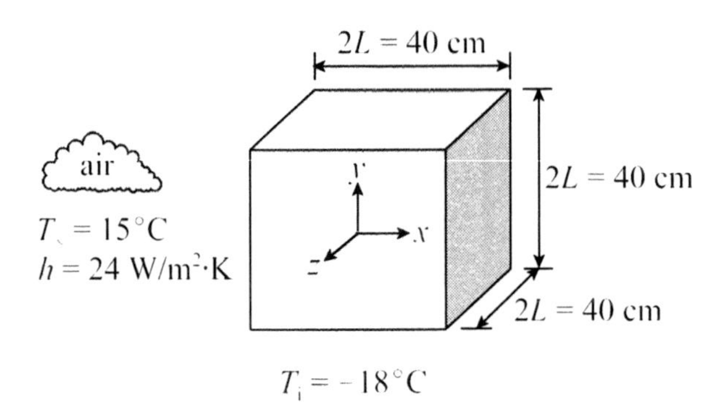

ASSUMPTIONS

1. The cube has a uniform initial temperature.
2. Three-dimensional conduction in the *x*, *y* and *z* directions.
3. All surfaces of the cube are exposed to the surrounding air.
4. Conduction into the conveyor is neglected.
5. The time required for the solid-liquid phase change is neglected.
6. Radiation is neglected.
7. Constant properties.

PROPERTIES

ice: $\rho = 920$ kg/m^3, $c_p = 2040$ J/kg·K, $k = 1.88$ W/m·K

ANALYSIS

The thermal diffusivity of the ice is

$$\alpha \equiv \frac{k}{\rho c_p}$$

$$\alpha = \frac{1.88\ \text{W/m·K}}{(920\ \text{kg/m}^3)(2040\ \text{J/kg·K})}$$

$= 1.002 \times 10^{-6}\ \text{m}^2/\text{s}$

A cube is formed by the orthogonal intersection of three large plates of thickness $2L$. The Biot number for the plate is

$$Bi \equiv \frac{hL}{k}$$

$$Bi = \frac{(24\ \text{W/m}^2\cdot\text{K})(0.20\ \text{m})}{1.88\ \text{W/m}\cdot\text{K}}$$

$= 2.553$

For a large plate, with uniform initial temperature, that is suddenly exposed to a convective environment, the temperature may be found by a one-term approximation of an infinite series,

$$P(x,t) = \frac{\theta}{\theta_i} = \frac{T(x,t) - T_\infty}{T_i - T_\infty} = C_1 \exp(-\zeta_1^2 Fo)\cos(\zeta_1 x/L)$$

where, by knowing that $Bi = 2.553$, the parameters C_1 and ζ_1 are

$C_1 = 1.1965 \quad , \quad \zeta_1 = 1.1408$

The corners of the ice cube, which are farthest from the center of the cube, will warm the fastest, so the ice cube begins to melt when the corners reach a temperature of 0°C. Upon substituting temperatures, the solution for the plate is

$P(L,t) = 0.4545$

For a corner of the cube, $x = y = z = L$, so the solution for the cube is the product of the three plate solutions

$$\frac{T(L,L,L,t) - T_\infty}{T_i - T_\infty} = [P(L,t)]^3$$

Knowing that $Fo = \alpha t/L^2$, and solving the one-term approximation for time t,

$$t = -\frac{L^2}{\alpha\zeta_1^2}\ln\left(\frac{1}{C_1\cos\zeta_1}\frac{\theta}{\theta_i}\right)$$

$$t = \frac{-(0.20\ \text{m})^2}{(1.002 \times 10^{-6}\ \text{m}^2/\text{s})(1.1408)^2}\ln\left(\frac{1}{1.1965\cos(1.1408)}\frac{0°\text{C} - 15°\text{C}}{-18°\text{C} - 15°\text{C}}\right)$$

$= 2849\ \text{s} = \underline{\underline{47.5\ \text{min}}}$

DISCUSSION

The ice cube will begin to melt after 47.5 min of exposure to the 15°C air, and because the corners are farthest from the center of the cube, liquid water droplets will appear at the corners first. Note that an additional amount of time is required for the ice cube to absorb the heat of fusion to cause 0°C solid water to change to 0°C liquid water, so our result does not reflect the phase change.

PROBLEM 5.17 Cooking Time for a Roast

A beef expert recommends that, in order to cook a roast so that it is "medium done" and that any bacteria in the meat is killed, the roast should have a minimum temperature of 70°C. If the roast has an initial temperature of 10°C, and the oven temperature is 175°C, where the heat transfer coefficient is 20 W/m^2·K, how long will it take to cook a 3.0-kg roast? Treat the roast as a cylinder with a diameter equal to its length, and use the properties of liquid water at room temperature (ρ = 998 kg/m^3, c_p = 4182 J/kg·K, k = 0.602 W/m·K).

DIAGRAM

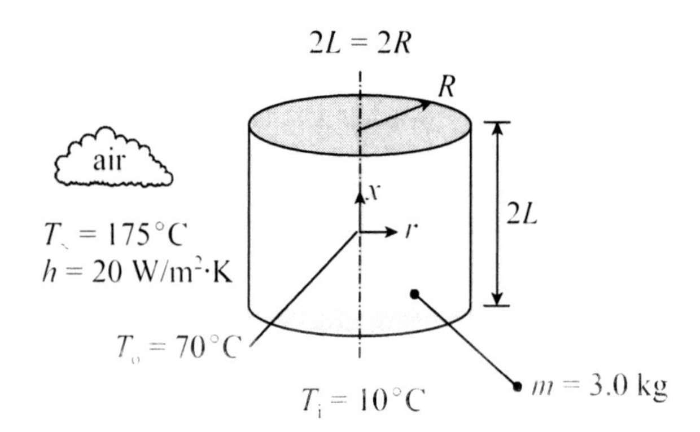

ASSUMPTIONS

1. The roast has a uniform initial temperature.
2. Two-dimensional conduction in the x and r directions.
3. All surfaces of the roast are exposed to the surrounding air in the oven.
4. The roast is approximated as a cylinder with a length equal to its diameter.
5. The thermal properties of beef are approximated with those of liquid water.
6. Radiation is neglected.
7. Constant properties.

PROPERTIES

liquid water: $\rho = 998$ kg/m^3, $c_p = 4182$ J/kg·K, $k = 0.602$ W/m·K

ANALYSIS

The volume of the roast is given by

$$V = \pi R^2 2L$$

and density is

$$\rho = \frac{m}{V}$$

Knowing that $L = R$ and combining these relations,

$$L = R = \left(\frac{m}{2\pi\rho}\right)^{1/3}$$

$$L = R = \left(\frac{3.0\ \text{kg}}{2\pi\ (998\ \text{kg/m}^3)}\right)^{1/3}$$

$$= 0.0782\ \text{m}$$

The Biot number is defined as

$$Bi \equiv \frac{hL}{k}$$

$$Bi = \frac{(20\ \text{W/m}^2\cdot\text{K})(0.0782\ \text{m})}{0.602\ \text{W/m}\cdot\text{K}}$$

$$= 2.598$$

The thermal diffusivity of the roast is

$$\alpha \equiv \frac{k}{\rho c_p}$$

$$\alpha = \frac{0.602\ \text{W/m}\cdot\text{K}}{(998\ \text{kg/m}^3)(4182\ \text{J/kg}\cdot\text{K})}$$

$$= 1.442 \times 10^{-7}\ \text{m}^2/\text{s}$$

A short cylinder is formed by the intersection of a large plate of thickness $2L$ and a long cylinder of radius R. For the roast to be thoroughly cooked, the center must be at least 70°C. For a large plate, with uniform initial temperature, that is suddenly exposed to a convective environment, the temperature of the center may be found by a one-term approximation of an infinite series,

$$P(0,t) = \frac{\theta_0}{\theta_i} = \frac{T(0,t) - T_\infty}{T_i - T_\infty} = C_1 \exp(-\zeta_1^2 Fo)$$

where, by knowing that $Bi = 2.598$, the parameters C_1 and ζ_1 for a plate are

$C_{1,p} = 1.1979$, $\zeta_{1,p} = 1.1460$

The one-term approximation for a long cylinder is identical in form to that for a large plate,

$$C(0,t) = \frac{\theta_0}{\theta_i} = \frac{T(0,t) - T_\infty}{T_i - T_\infty} = C_1 \exp(-\zeta_1^2 Fo)$$

where, by knowing that $Bi = 2.598$, the parameters C_1 and ζ_1 for a cylinder are

$C_{1,c} = 1.3867$, $\zeta_{1,c} = 1.7126$

The solution for the short cylinder is the product of the plate and cylinder solutions,

$$\left(\frac{\theta_0}{\theta_i}\right)_{short\ cylinder} = P(0,t)C(0,t) = \frac{T(0,0,t) - T_\infty}{T_i - T_\infty}$$

Substituting temperatures,

$$\left(\frac{\theta_0}{\theta_i}\right)_{short\ cylinder} = \frac{70°C - 175°C}{10°C - 175°C} = 0.6364$$

Recognizing that $Fo = \alpha t/L^2 = \alpha t/R^2$, and solving the product solution above for time t,

$$t = \frac{\ln\left(\dfrac{C_{1,p}C_{1,c}}{P(0,t)C(0,t)}\right)}{\dfrac{\alpha}{L^2}\left(\zeta_{1,p}^2 + \zeta_{1,c}^2\right)}$$

$t = 9582$ s = 2.66 h

from which the Fourier number is $Fo = 0.2259$.

DISCUSSION

Radiation is a significant heat transfer mechanism in an oven where the heat transfer coefficient is low due to natural convection. Thus, the actual cooking time would be less than 2.66 h. Furthermore, a more accurate cooking time may be obtained by using the actual thermal properties of beef. One source gives the thermal diffusivity of beef as $\alpha = 1.56 \times 10^{-7}$ m^2/s. Using this value instead of that for water gives a cooking time of 2.46 h.

❒ ❒ ❒

PROBLEM 5.18 Estimating the Time of Death for a Human

To estimate the time of death for a human being, forensics experts may employ a method that is based on unsteady conduction in multidimensional systems. Approximate an adult human as a cylinder with a length and diameter of 1.8 m and 0.35 m, respectively. The temperature of the body at the time of death is 37°C, and the temperature of the surrounding air is 21°C. If the temperature at the center of the body is measured to be 26°C, how long has the person been dead? Use a heat transfer coefficient of 14 W/m^2·K, and neglect conduction from the body into any surrounding structures. Use the thermal properties of liquid water at room temperature ($\rho = 998$ kg/m^3, $c_p = 4182$ J/kg·K, $k = 0.602$ W/m·K).

DIAGRAM

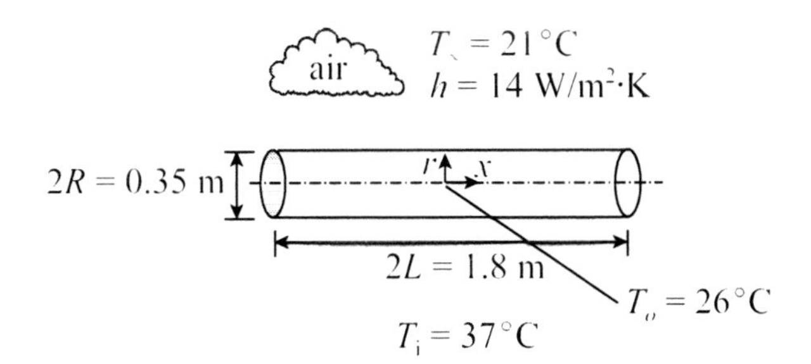

ASSUMPTIONS

1. The body has a uniform initial temperature.
2. The body is approximated as a cylinder.
3. Two-dimensional conduction in the x and r directions.
4. All surfaces of the body are exposed to the surrounding air.
5. The thermal properties of the body are approximated with those of liquid water.
6. Radiation is neglected.
7. Constant properties.

PROPERTIES

liquid water: ρ = 998 kg/m^3, c_p = 4182 J/kg·K, k = 0.602 W/m·K

ANALYSIS

Thermal diffusivity is defined as

$$\alpha \equiv \frac{k}{\rho c_p}$$

$$\alpha = \frac{0.602 \text{ W/m·K}}{(998 \text{ kg/m}^3)(4182 \text{ J/kg·K})}$$

$$= 1.442 \times 10^{-7} \text{ m}^2\text{/s}$$

The human body is modeled as a short cylinder, which is formed by the orthogonal intersection of a large plate of half-thickness L and a long cylinder of radius R. The Biot numbers for the plate and cylinder, respectively, are

$$Bi_p \equiv \frac{hL}{k}$$

$$Bi_p = \frac{(14 \text{ W/m}^2\text{·K})(0.9 \text{ m})}{0.602 \text{ W/m·K}}$$

$$= 20.93$$

$$Bi_c \equiv \frac{hR}{k}$$

$$Bi_c = \frac{(14 \text{ W/m}^2\text{·K})(0.175 \text{ m})}{0.602 \text{ W/m·K}}$$

$$= 4.070$$

For a large plate, with uniform initial temperature, that is suddenly exposed to a convective environment, the temperature of the center may be found by a one-term approximation of an infinite series,

$$P(0,t) = \frac{\theta_0}{\theta_i} = \frac{T(0,t) - T_\infty}{T_i - T_\infty} = C_1 \exp(-\zeta_1^2 Fo)$$

where, by knowing that Bi = 20.93, the parameters C_1 and ζ_1 for a plate are

$C_{1,p} = 1.2701$, $\zeta_{1,p} = 1.4983$

The one-term approximation for a long cylinder is identical in form to that for a large plate,

$$C(0,t) = \frac{\theta_0}{\theta_i} = \frac{T(0,t) - T_\infty}{T_i - T_\infty} = C_1 \exp(-\zeta_1^2 Fo)$$

where, by knowing that $Bi = 4.070$, the parameters C_1 and ζ_1 for a cylinder are

$C_{1,c} = 1.4721$, $\zeta_{1,c} = 1.9138$

The solution for the short cylinder is the product of the plate and cylinder solutions,

$$\left(\frac{\theta_0}{\theta_i}\right)_{short\ cylinder} = P(0,t)C(0,t) = \frac{T(0,0,t) - T_\infty}{T_i - T_\infty}$$

Substituting temperatures,

$$\left(\frac{\theta_0}{\theta_i}\right)_{short\ cylinder} = \frac{26°C - 21°C}{37°C - 21°C} = 0.3125$$

Recognizing that $Fo = \alpha t/L^2$ and $Fo = \alpha t/R^2$ for the plate and cylinder, respectively, and solving the product solution above for time t,

$$t = \frac{\ln\left(\dfrac{C_{1,p}C_{1,c}}{P(0,t)C(0,t)}\right)}{\alpha\left[\left(\dfrac{\zeta_{1,p}}{L}\right)^2 + \left(\dfrac{\zeta_{1,c}}{R}\right)^2\right]}$$

$t = 1.0138 \times 10^5$ s = 28.2 h

DISCUSSION

The estimated time of death is 28.2 h before the deep body temperature measurement is made. Body cooling, referred to as algor mortis, is regarded as the most accurate method for estimating the time of death for a human being during the first 24 hours post mortem. For longer periods of time, forensics experts may use other methods such as the degree of rigor mortis and body decomposition.

PROBLEM 5.19 **Temperature Prediction for a Thick Slab of Iron using the Finite Difference Method**

A thick slab of iron (ρ = 7870 kg/m^3, c_p = 447 J/kg·K, k = 80.2 W/m·K) has a uniform initial temperature of 0°C. One surface of the slab is suddenly changed to 60°C. What is the temperature at a depth of 3 cm in the slab after 1 min? Approximate the thick slab as a semi-infinite region, and use the explicit finite difference method.

DIAGRAM

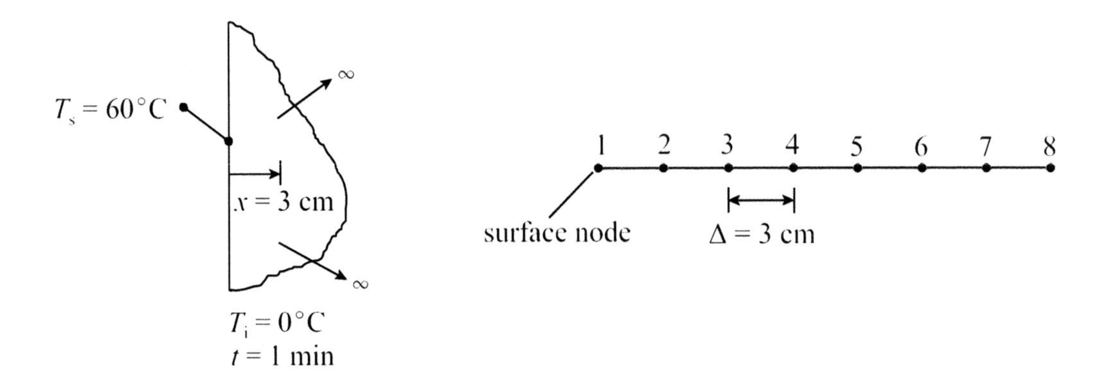

ASSUMPTIONS

1. The thick slab is treated as a semi-infinite region.
2. One-dimensional conduction.
3. Convection and radiation from the surface are neglected.
4. Constant properties.

PROPERTIES

iron: ρ = 7870 kg/m^3, c_p = 447 J/kg·K, k = 80.2 W/m·K

ANALYSIS

The thermal diffusivity is

$$\alpha \equiv \frac{k}{\rho c_p}$$

$$\alpha = \frac{80.2 \text{ W/m·K}}{(7870 \text{ kg/m}^3)(447 \text{ J/kg·K})}$$

$$= 2.280 \times 10^{-5} \text{ m}^2/\text{s}$$

The stability criterion for a one-dimensional finite difference problem is

$$Fo(1+Bi) \le \tfrac{1}{2}$$

where the Biot number is defined as

$$Bi = \frac{h\Delta}{k}$$

and the Fourier number is defined as

$$Fo = \frac{\alpha \Delta t}{\Delta^2}$$

where Δt is the time step and Δ is the node spacing. Convection is absent, so $Bi = 0$. Thus, the stability criterion is

$$Fo \le \tfrac{1}{2}$$

Hence, the largest value of time step Δt that yields a stable solution is

$$\Delta t = \frac{\Delta^2}{2\alpha}$$

$$\Delta t = \frac{(0.03 \text{ m})^2}{2(2.280 \times 10^{-5} \text{ m}^2/\text{s})}$$

$$= 19.74 \text{ s}$$

The solution will be more accurate for a smaller time step, so we choose a value of $\Delta t = 1$ s, giving a Fourier number of $Fo = 0.02533$. We anticipate that eight nodes with a spacing of 3 cm is sufficient to effectively model the thick slab as a semi-infinite region. The surface (node 1) is maintained at 60°C, while the back side (node 8) is treated as an adiabatic boundary. The finite difference equations for the nodes are

node 1	$T_1^{p+1} = T_s$
node 2	$T_2^{p+1} = Fo(T_1^p + T_3^p) + (1-2Fo)T_2^p$
node 3	$T_3^{p+1} = Fo(T_2^p + T_4^p) + (1-2Fo)T_3^p$
node 4	$T_4^{p+1} = Fo(T_3^p + T_5^p) + (1-2Fo)T_4^p$
node 5	$T_5^{p+1} = Fo(T_4^p + T_6^p) + (1-2Fo)T_5^p$
node 6	$T_6^{p+1} = Fo(T_5^p + T_7^p) + (1-2Fo)T_6^p$
node 7	$T_7^{p+1} = Fo(T_6^p + T_8^p) + (1-2Fo)T_7^p$
node 8	$T_8^{p+1} = 2FoT_7^p + (1-2Fo)T_8^p$

where the superscripts p and p+1 refer to the previous time and new time, respectively. This system of equations is solved by initially setting all nodes to T_i, the initial temperature of the slab, and then marching out in time, using intervals of Δt =1 s, to the end time of 60 s. The temperatures, in units of °C, of the nodes at 10-s intervals are summarized in the table below.

	Node							
Time (s)	1	2	3	4	5	6	7	8
0	60	0	0	0	0	0	0	0
10	60	12.3	1.6	0.2	0.0	0.0	0.0	0.0
20	60	20.1	4.5	0.8	0.1	0.0	0.0	0.0
30	60	25.3	7.7	1.8	0.3	0.1	0.0	0.0
40	60	29.1	10.6	3.1	0.7	0.2	0.0	0.0
50	60	31.9	13.3	4.5	1.3	0.3	0.1	0.0
60	60	34.1	15.7	6.0	1.9	0.5	0.1	0.1

Thus, the temperature of the slab at a depth of 3 cm after 1 min is 34.1°C.

DISCUSSION

Note that the temperature of node 8, the back side of the slab, changed by only 0.1°C after 60 s, indicating that the finite difference model reasonably approximates a semi-infinite region. Our result can be checked by solving the problem analytically. For a semi-infinite solid initially at temperature T_i whose surface is suddenly changed to T_s, the temperature at a specified depth x and time t is given by the relation

$$\frac{T(x,t)-T_s}{T_i-T_s} = erf\left(\frac{x}{2\sqrt{\alpha t}}\right)$$

where α is the thermal diffusivity of the solid. Upon substituting values and solving for $T(x,t)$,

$$T(x,t) = (0°C - 60°C)\ \mathrm{erf}\left(\frac{0.03\ \mathrm{m}}{2[(2.280 \times 10^{-5}\ \mathrm{m^2/s})(60\ \mathrm{s})]^{1/2}}\right) + 60°C$$

$$= 34.0°C$$

which is in excellent agreement with our first result.

❐ ❐ ❐

PROBLEM 5.20 Heating Time for a Plastic Sheet in a Manufacturing Operation

An 8–mm thick sheet of polycarbonate plastic (ρ = 1205 kg/m^3, c_p = 1260 J/kg·K, k = 0.199 W/m·K) is heated in an oven prior to a forming operation. The initial temperature of the sheet is 18°C, and the oven is set to 200°C. If convective conditions in the oven create a heat transfer coefficient of 35 W/m^2·K for the surfaces of the plastic sheet, how long should the sheet be in the oven to achieve a minimum forming temperature of 150°C? Use the explicit finite difference method with a node spacing of 1 mm.

DIAGRAM

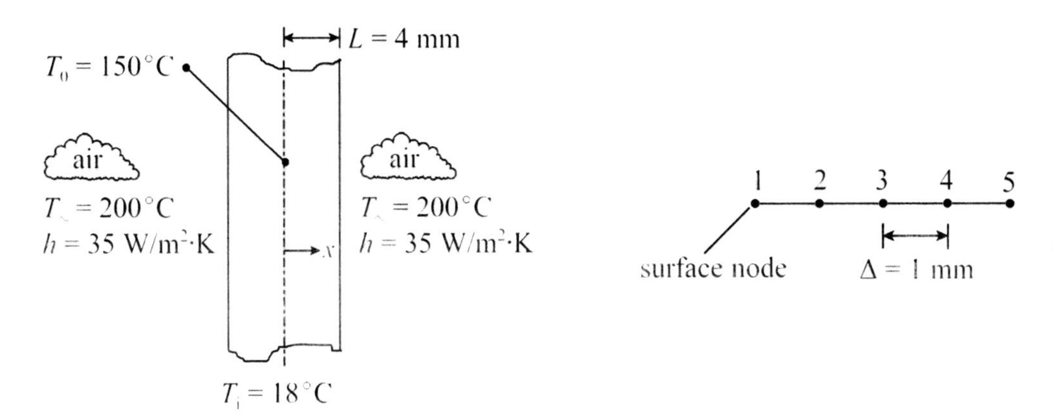

ASSUMPTIONS

1. The sheet has a uniform initial temperature.
2. One-dimensional conduction in the x direction.
3. The sheet is surrounded by air in the oven.
4. Radiation is neglected.
5. Constant properties.

PROPERTIES

polycarbonate plastic: ρ = 1205 kg/m^3, c_p = 1260 J/kg·K, k = 0.199 W/m·K

ANALYSIS

The thermal diffusivity is

$$\alpha \equiv \frac{k}{\rho c_p}$$

$$\alpha = \frac{0.199 \text{ W/m·K}}{(1205 \text{ kg/m}^3)(1260 \text{ J/kg·K})}$$

$= 1.311 \times 10^{-7}$ m²/s

The stability criterion for a one-dimensional finite difference problem is

$$Fo(1 + Bi) \le \tfrac{1}{2}$$

where the Biot number is

$$Bi = \frac{h\Delta}{k}$$

$$Bi = \frac{(35 \text{ W/m}^2\cdot\text{K})(0.001 \text{ m})}{0.199 \text{ W/m}^2\cdot\text{K}}$$

$= 0.1759$

and the Fourier number is

$$Fo = \frac{\alpha \Delta t}{\Delta^2}$$

where Δt is the time step and Δ is the node spacing. For node 1, the surface of the plastic sheet, the largest value of time step Δt for a stable solution is

$$\Delta t = \frac{\Delta^2}{2\alpha} \frac{1}{1 + Bi}$$

$$\Delta t = \frac{(0.001 \text{ m})^2}{2(1.311 \times 10^{-7} \text{ m}^2\text{/s})(1 + 0.1759)}$$

$= 3.24$ s

For the other nodes, $Bi = 0$, and the largest value of time step Δt for a stable solution is

$$\Delta t = \frac{\Delta^2}{2\alpha}$$

$$\Delta t = \frac{(0.001 \text{ m})^2}{2(1.311 \times 10^{-7} \text{ m}^2\text{/s})}$$

$= 3.81$ s

The smaller of the two time steps governs the problem. We choose a value of $\Delta t = 0.1$ s, giving a Fourier number of $Fo = 0.0131$. The surface of the plastic sheet (node 1) has a convective boundary condition, and the center plane (node 5) of the plastic sheet has an adiabatic boundary condition. The finite difference equations for the nodes are

node 1 $T_1^{p+1} = 2Fo(T_2^p + BiT_\infty) + (1 - 2Fo - 2FoBi)T_1^p$
node 2 $T_2^{p+1} = Fo(T_1^p + T_3^p) + (1-2Fo)T_2^p$
node 3 $T_3^{p+1} = Fo(T_2^p + T_4^p) + (1-2Fo)T_3^p$
node 4 $T_4^{p+1} = Fo(T_3^p + T_5^p) + (1-2Fo)T_4^p$
node 5 $T_5^{p+1} = 2FoT_4^p + (1-2Fo)T_5^p$

where the superscripts p and $p+1$ refer to the previous time and new time, respectively. This system of equations is solved by initially setting all nodes to T_i, the initial temperature of the sheet, and then marching out in time, using intervals of $\Delta t = 0.1$ s, to a value of time at which node 5 reaches a temperature of 150°C. The temperatures, in units of °C, of the nodes at 60-s intervals are summarized in the table below.

	Node				
Time (s)	1	2	3	4	5
0	18	18	18	18	18
60	90.4	73.1	60.4	52.6	50.0
120	117.4	104.4	94.7	88.8	86.8
180	137.7	127.9	120.6	116.2	114.7
240	153.1	145.6	140.2	136.8	135.7
293.4	163.5	157.7	153.5	150.8	150.0

Thus, the time it takes for the center of the plastic sheet to reach 150°C is 293.4 s (4.89 min).

DISCUSSION

Our result can be checked by solving the problem analytically. For a large plate, with uniform initial temperature, that is suddenly exposed to a convective environment, the temperature of the center may be found by a one-term approximation of an infinite series,

$$P(0,t) = \frac{\theta_0}{\theta_i} = \frac{T(0,t) - T_\infty}{T_i - T_\infty} = C_1 \exp(-\zeta_1^2 Fo)$$

where, by knowing that the Biot number for a plate of half-thickness L is $Bi = hL/k = 0.7035$, the parameters C_1 and ζ_1 are

$$C_1 = 1.0922 \quad , \quad \zeta_{1,p} = 0.7520$$

Noting that the Fourier number is $Fo \equiv \alpha t/L^2$, and solving the one-term approximation for time t,

$$t = \frac{-L^2}{\zeta_1^2 \alpha} \ln\left(\frac{1}{C_1}\frac{\theta_0}{\theta_i}\right)$$

$$t = \frac{-(0.004 \text{ m})^2}{(0.7520)^2 (1.311 \times 10^{-7} \text{ m}^2/\text{s})} \ln \left(\frac{1}{1.0922} \frac{150°\text{C} - 200°\text{C}}{18°\text{C} - 200°\text{C}} \right)$$

$$= 297.9 \text{ s} = \underline{\underline{4.96 \text{ min}}}$$

which is in excellent agreement with our first result.

❒ ❒ ❒

PROBLEM 5.21 Warmup Time Prediction for Power Transistors

A 1.5-mm thick bracket of an aluminum alloy ($\rho = 2700$ kg/m^3, $c_p = 920$ J/kg·K, $k = 200$ W/m·K) holds a pair of power transistors, as shown in the diagram. The bottom edge of the bracket is attached to a cold plate maintained at 10°C. Just prior to powering on the transistors, the initial temperature of the bracket and transistors is 20°C. When the power is turned on, each transistor dissipates 6 W. Using the explicit finite difference method with a node spacing of 2 cm, find the time required for the transistors to achieve steady temperatures. What are the steady case temperatures of the transistors?

DIAGRAM

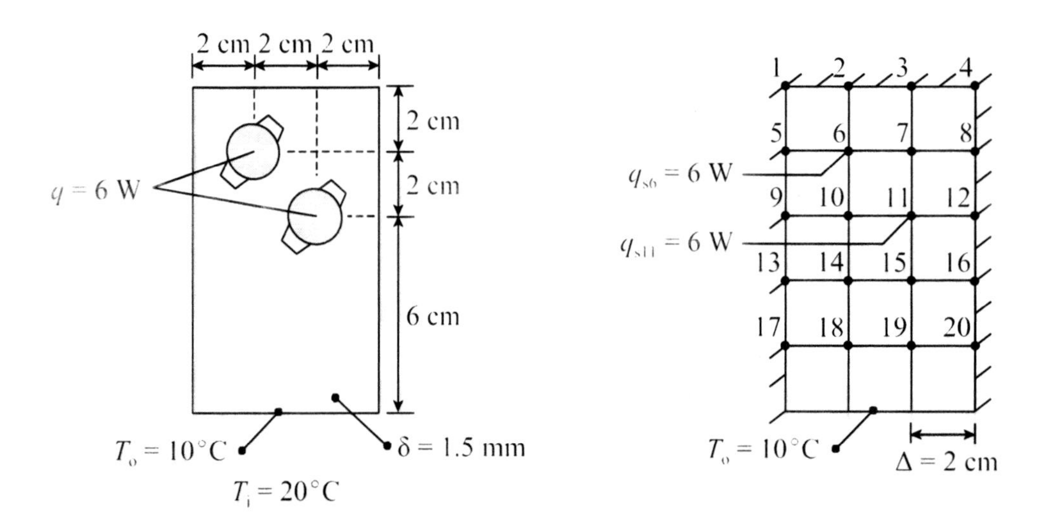

ASSUMPTIONS

1. Two-dimensional conduction.
2. Convection and radiation from the bracket and transistors are neglected.
3. Heat transfer from the edges of the bracket is neglected.
4. Power dissipations are steady.
5. Constant properties.

PROPERTIES

aluminum alloy: $\rho = 2700$ kg/m^3, $c_p = 920$ J/kg·K, $k = 200$ W/m·K

ANALYSIS

Nodes 7, 10, 11, 14, 15, 18 and 19 are interior nodes with no heat sources. Power dissipations of the transistors are treated as point heat sources on nodes 6 and 11. Nodes 1 and 4 are insulated exterior corner nodes, and the rest of the boundary nodes are insulated edge nodes. The Fourier number is

$$Fo = \frac{\alpha \Delta t}{\Delta^2}$$

where Δt is the time step, and thermal diffusivity is defined as

$$\alpha \equiv \frac{k}{\rho c_p}$$

$$\alpha = \frac{(200 \text{ W/m·K})}{(2700 \text{ kg/m}^3)(920 \text{ J/kg·K})}$$

$$= 8.052 \times 10^{-5} \text{ m}^2\text{/s}$$

The stability criterion for two-dimensional conduction is

$$Fo(4 + C\,Bi) \leq 1$$

where the value of the constant C depends on the location of the node in the two-dimensional region. Because convection is absent, $Bi = 0$, so the stability criterion is

$$Fo \leq 1/4$$

Thus, the largest time step for a stable solution is

$$\Delta t = \frac{\Delta^2}{4\alpha}$$

$$\Delta t = \frac{(0.02 \text{ m})^2}{4(8.052 \times 10^{-5} \text{ m}^2\text{/s})}$$

$$= 1.24 \text{ s}$$

We choose a time step of $\Delta t = 0.1$ s, giving a Fourier number of $Fo = 0.0201$. The finite difference equations for the nodes are as follows:

node 1	$T_1^{p+1} = 2Fo(T_2^p + T_5^p) + (1 - 4Fo)T_1^p$
node 2	$T_2^{p+1} = Fo(2T_6^p + T_1^p + T_3^p) + (1 - 4Fo)T_2^p$
node 3	$T_3^{p+1} = Fo(2T_7^p + T_2^p + T_4^p) + (1 - 4Fo)T_3^p$
node 4	$T_4^{p+1} = 2Fo(T_3^p + T_8^p) + (1 - 4Fo)T_4^p$
node 5	$T_5^{p+1} = Fo(2T_6^p + T_1^p + T_9^p) + (1 - 4Fo)T_5^p$
node 6	$T_6^{p+1} = Fo(T_2^p + T_5^p + T_7^p + T_{10}^p) + (1 - 4Fo)T_6^p + Foq_{s6}/(k\delta)$
node 7	$T_7^{p+1} = Fo(T_3^p + T_6^p + T_8^p + T_{11}^p) + (1 - 4Fo)T_7^p$
node 8	$T_8^{p+1} = Fo(2T_7^p + T_4^p + T_{12}^p) + (1 - 4Fo)T_8^p$
node 9	$T_9^{p+1} = Fo(2T_{10}^p + T_5^p + T_{13}^p) + (1 - 4Fo)T_9^p$
node 10	$T_{10}^{p+1} = Fo(T_6^p + T_9^p + T_{11}^p + T_{14}^p) + (1 - 4Fo)T_{10}^p$
node 11	$T_{11}^{p+1} = Fo(T_7^p + T_{10}^p + T_{12}^p + T_{15}^p) + (1 - 4Fo)T_{11}^p + Foq_{s11}/(k\delta)$
node 12	$T_{12}^{p+1} = Fo(2T_{11}^p + T_8^p + T_{16}^p) + (1 - 4Fo)T_{12}^p$
node 13	$T_{13}^{p+1} = Fo(2T_{14}^p + T_9^p + T_{17}^p) + (1 - 4Fo)T_{13}^p$
node 14	$T_{14}^{p+1} = Fo(T_{10}^p + T_{13}^p + T_{15}^p + T_{18}^p) + (1 - 4Fo)T_{14}^p$
node 15	$T_{15}^{p+1} = Fo(T_{11}^p + T_{14}^p + T_{16}^p + T_{19}^p) + (1 - 4Fo)T_{15}^p$
node 16	$T_{16}^{p+1} = Fo(2T_{15}^p + T_{12}^p + T_{20}^p) + (1 - 4Fo)T_{16}^p$
node 17	$T_{17}^{p+1} = Fo(2T_{18}^p + T_{13}^p + T_o) + (1 - 4Fo)T_{17}^p$
node 18	$T_{18}^{p+1} = Fo(T_{14}^p + T_{17}^p + T_{19}^p + T_o) + (1 - 4Fo)T_{18}^p$
node 19	$T_{19}^{p+1} = Fo(T_{15}^p + T_{18}^p + T_{20}^p + T_o) + (1 - 4Fo)T_{19}^p$
node 20	$T_{20}^{p+1} = Fo(2T_{19}^p + T_{16}^p + T_o^p) + (1 - 4Fo)T_{20}^p$

where the superscripts p and p+1 refer to the previous time and new time, respectively. The parameter δ is the bracket thickness, k is thermal conductivity of the aluminum alloy, and q_{s6} and q_{s11} are the point heat sources (6 W) for nodes 6 and 11, respectively. This system of equations is solved by initially setting all nodes to T_i, the initial temperature of the bracket, and then marching out in time, using intervals of $\Delta t = 0.1$ s, to a value of time at which the temperatures of the nodes stop changing, indicating that a steady thermal condition has been reached.

A graph of bracket temperatures at nodes 6 and 11 is shown below. As illustrated in the graph, it takes approximately 410 s (6.83 min) for the bracket to achieve a steady thermal condition. This is the time required for the temperatures to stop changing in the finite difference calculations in the first place past the decimal point. The resulting steady temperatures are therefore accurate to within 0.1 °C.

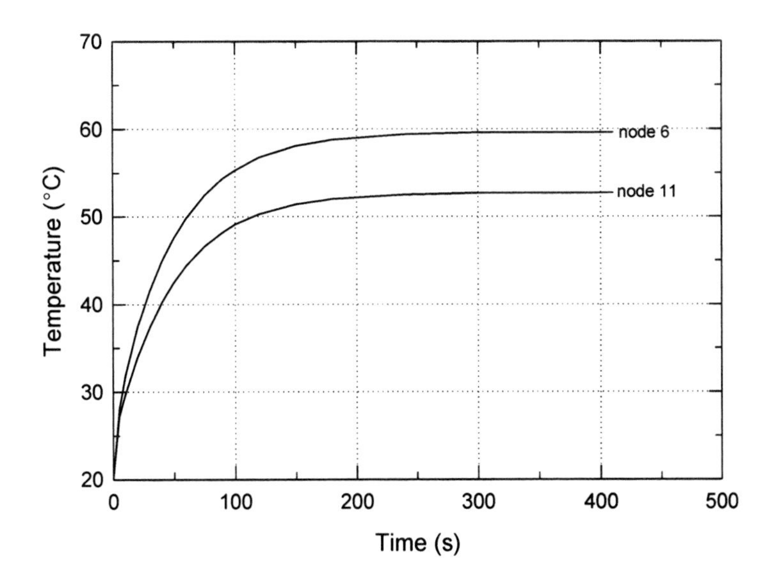

The steady temperatures at each node are shown schematically below. As shown in the schematic, The steady case temperatures of the two transistors at nodes 6 and 11are 59.6°C and 52.7°C, respectively.

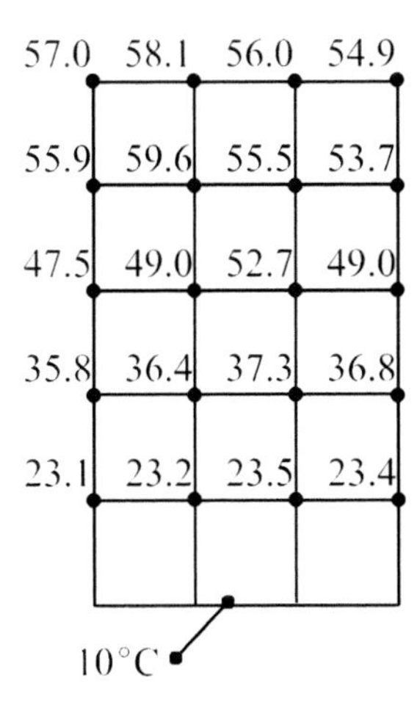

DISCUSSION

The steady version of this problem was worked in Problem 4.13. Comparing the two solutions, we see that they are identical.

PROBLEM 5.22 **Temperatures of a Stainless Steel Tube Carrying Liquid Sodium in a Nuclear Reactor**

A square tube of stainless steel ($\rho = 7770$ kg/m^3, $c_p = 460$ J/kg·K, $k = 25$ W/m·K) for transporting liquid sodium in a nuclear reactor has a uniform initial temperature of 10°C. When the liquid sodium enters the tube, a heat transfer coefficient of 1800 W/m^2·K is established on the inside surface of the tube. The outside surface of the tube is in contact with 10°C air, where the heat transfer coefficient is 20 W/m^2·K. The inside and outside dimensions of the tube are 20 mm and 40 mm, respectively, as shown in the diagram. Find the tube temperatures at times 10, 30,60, 90 and 120 s after the liquid sodium enters the tube. Use an explicit finite difference method with a node spacing of 5 mm.

DIAGRAM

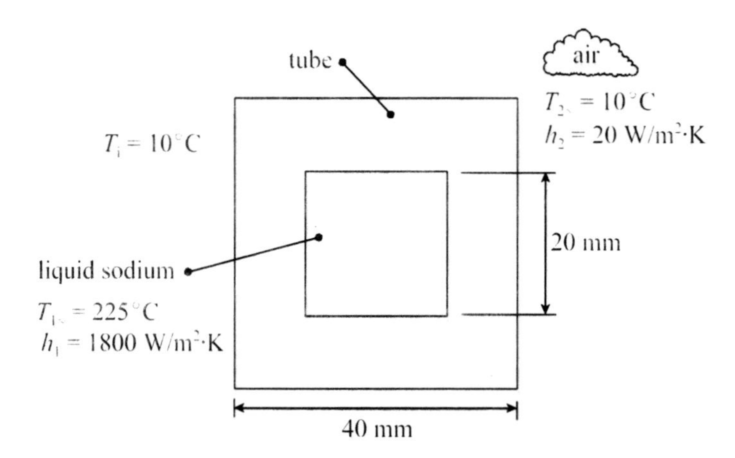

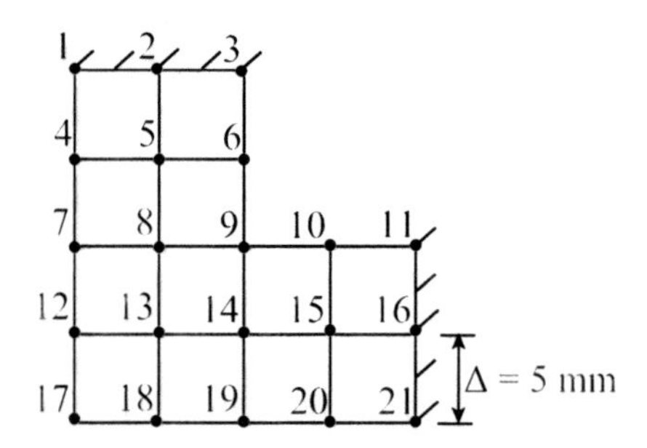

ASSUMPTIONS

1. Two-dimensional conduction.
2. Radiation is neglected.
3. Constant properties.

PROPERTIES

stainless steel: $\rho = 7770$ kg/m^3, $c_p = 460$ J/kg·K, $k = 25$ W/m·K

ANALYSIS

The tube has thermal symmetry about the horizontal and vertical center planes, so only a quarter of the tube cross section is modeled. An insulated boundary condition is applied to the surfaces thus created. The Fourier number is

$$Fo = \frac{\alpha \Delta t}{\Delta^2}$$

where Δt is the time step, and thermal diffusivity is defined as

$$\alpha \equiv \frac{k}{\rho c_p}$$

$$\alpha = \frac{(25\ \text{W/m·K})}{(7770\ \text{kg/m}^3)(460\ \text{J/kg·K})}$$

$$= 6.995 \times 10^{-6}\ \text{m}^2\text{/s}$$

The stability criterion for two-dimensional conduction is

$$Fo(4 + C\,Bi) \leq 1$$

where the value of the constant C depends on the location of the node in the two-dimensional region. For interior, edge, exterior corner, and interior corner nodes, the values of C are 0, 2, 4 and 4/3, respectively. The Biot number for the inside tube surface is

$$Bi_1 = \frac{h_1 \Delta}{k}$$

$$Bi_1 = \frac{(1800\ \text{W/m}^2\text{·K})(0.005\ \text{m})}{25\ \text{W/m·K}}$$

$$= 0.360$$

and the Biot number for the outside tube surface is

$$Bi_2 = \frac{h_2 \Delta}{k}$$

$$Bi_2 = \frac{(20\ \text{W/m}^2\cdot\text{K})(0.005\ \text{m})}{25\ \text{W/m}\cdot\text{K}}$$

$$= 4.00 \times 10^{-3}$$

By checking the stability criterion for each type of node, we see that the exterior corner nodes for the larger Biot number (nodes 3 and 11) govern the stability of the problem. Thus, the largest time step for a stable solution is

$$\Delta t = \frac{\Delta^2}{4\alpha(1 + Bi_1)}$$

$$\Delta t = \frac{(0.005\ \text{m})^2}{4(6.995 \times 10^{-6}\ \text{m}^2/\text{s})(1 + 0.360)}$$

$$= 0.657\text{s}$$

We choose a time step of $\Delta t = 0.01$ s, which yields a Fourier number of $Fo = 2.798 \times 10^{-3}$. The finite difference equations for the nodes are as follows:

node 1	$T_1^{p+1} = Fo(2T_2^p + 2T_4^p + 2Bi_2T_{\infty 2}) + (1 - 4Fo - 2Bi_2Fo)T_1^p$
node 2	$T_2^{p+1} = Fo(2T_5^p + T_1^p + T_3^p) + (1 - 4Fo)T_2^p$
node 3	$T_3^{p+1} = Fo(2T_2^p + 2T_6^p + 2Bi_1T_{\infty 1}) + (1 - 4Fo - 2Bi_1Fo)T_3^p$
node 4	$T_4^{p+1} = Fo(2T_5^p + T_1^p + T_7 + 2Bi_2T_{\infty 2}) + (1 - 4Fo - 2Bi_2Fo)T_4^p$
node 5	$T_5^{p+1} = Fo(T_2 + T_4 + T_6 + T_8) + (1 - 4Fo)T_5^p$
node 6	$T_6^{p+1} = Fo(2T_5^p + T_3^p + T_9 + 2Bi_1T_{\infty 1}) + (1 - 4Fo - 2Bi_1Fo)T_6^p$
node 7	$T_7^{p+1} = Fo(2T_8^p + T_4^p + T_{12} + 2Bi_2T_{\infty 2}) + (1 - 4Fo - 2Bi_2Fo)T_7^p$
node 8	$T_8^{p+1} = Fo(T_5 + T_7 + T_9 + T_{13}) + (1 - 4Fo)T_8^p$
node 9	$T_9^{p+1} = 2Fo(\tfrac{2}{3}T_8^p + \tfrac{2}{3}T_{14}^p + \tfrac{1}{3}T_6^p + \tfrac{1}{3}T_{10}^p + \tfrac{2}{3}Bi_1T_{\infty 1})$ $+ (1 - 4Fo - \tfrac{4}{3}Bi_1Fo)T_9^p$
node 10	$T_{10}^{p+1} = Fo(2T_{15}^p + T_9^p + T + 2Bi_1T_{\infty 1}) + (1 - 4Fo - 2Bi_1Fo)T_{10}^p$
node 11	$T_{11}^{p+1} = Fo(2T_{10}^p + 2T_{16}^p + 2Bi_1T_{\infty 1}) + (1 - 4Fo - 2Bi_1Fo)T_{11}^p$
node 12	$T_{12}^{p+1} = Fo(2T_{13}^p + T_7^p + T_{17} + 2Bi_2T_{\infty 2}) + (1 - 4Fo - 2Bi_2Fo)T_{12}^p$
node 13	$T_{13}^{p+1} = Fo(T_8 + T_{12} + T_{14} + T_{18}) + (1 - 4Fo)T_{13}^p$
node 14	$T_{14}^{p+1} = Fo(T_9 + T_{13} + T_{15} + T_{19}) + (1 - 4Fo)T_{14}^p$
node 15	$T_{15}^{p+1} = Fo(T_{10} + T_{14} + T_{16} + T_{20}) + (1 - 4Fo)T_{15}^p$
node 16	$T_{16}^{p+1} = Fo(2T_{15}^p + T_{11}^p + T_{21}) + (1 - 4Fo)T_{16}^p$
node 17	$T_{17}^{p+1} = Fo(2T_{12}^p + 2T_{18}^p + 4Bi_2T_{\infty 2}) + (1 - 4Fo - 4Bi_2Fo)T_{17}^p$
node 18	$T_{18}^{p+1} = Fo(2T_{13}^p + T_{17}^p + T_{19} + 2Bi_2T_{\infty 2}) + (1 - 4Fo - 2Bi_2Fo)T_{18}^p$
node 19	$T_{19}^{p+1} = Fo(2T_{14}^p + T_{18}^p + T_{20} + 2Bi_2T_{\infty 2}) + (1 - 4Fo - 2Bi_2Fo)T_{19}^p$
node 20	$T_{20}^{p+1} = Fo(2T_{15}^p + T_{19}^p + T_{21} + 2Bi_2T_{\infty 2}) + (1 - 4Fo - 2Bi_2Fo)T_{20}^p$
node 21	$T_{21}^{p+1} = Fo(2T_{16}^p + 2T_{20}^p + 2Bi_2T_{\infty 2}) + (1 - 4Fo - 2Bi_2Fo)T_{21}^p$

where the superscripts p and p+1 refer to the previous time and new time, respectively. This system of equations is solved by initially setting all nodes to T_i, the initial temperature of the tube, and then marching out in time, using intervals of Δt = 0.01 s, to a final time value of 120 s. The temperatures of the tube at 10, 30, 60, 90 and 120 s are shown in the table below. Also shown in the table is the steady solution, which is reached at a time of approximately 310 s.

	Temperature (°C)					
Node	$t = 10$ s	$t = 30$ s	$t = 60$ s	$t = 90$ s	$t = 120$ s	$t = 310$ s
1	57.5	126.6	178.2	200.9	210.8	218.6
2	68.9	133.6	181.8	203.0	212.3	219.6
3	102.8	153.5	191.3	207.9	215.2	220.9
4	54.8	124.5	177.2	200.4	210.5	218.4
5	65.8	131.3	180.7	202.4	212.0	219.4
6	99.1	151.1	190.1	207.3	214.8	220.7
7	47.3	119.0	174.5	198.9	209.6	217.9
8	56.2	124.7	177.5	200.7	210.9	218.8
9	84.1	141.6	185.6	205.0	213.4	220.1
10	99.1	151.1	190.2	207.3	214.8	220.7
11	102.9	153.6	191.3	207.9	215.2	220.9
12	38.1	112.4	171.2	197.0	208.4	217.2
13	43.3	116.1	173.3	198.4	209.4	218.0
14	56.3	124.7	177.5	200.7	210.9	218.8
15	65.9	131.4	180.8	202.4	212.0	219.4
16	69.0	133.7	181.9	203.0	212.3	219.6
17	34.3	109.4	169.5	195.8	207.4	216.4
18	38.1	112.4	171.2	197.0	208.4	217.2
19	47.3	119.0	174.5	198.9	209.6	217.9
20	54.9	124.6	177.3	200.4	210.5	218.4
21	57.6	126.7	178.3	200.9	210.8	218.6

DISCUSSION

At any given time, the lowest temperature of the tube occurs at the outside corner (node 17), and the highest temperature of the tube occurs at the center of the inside surface (nodes 3 and 11). Due to thermal symmetry, the temperatures at corresponding nodes across a diagonal drawn through the inside and outside corners of the tube are equal. For example, the temperatures at nodes 1 and 21, 4 and 20, 6 and 10, 5 and 15, and 8 and 14, are equal.

❐ ❐ ❐

Chapter 6

Principles of Convection

PROBLEM 6.1 **Ratio of Average Heat Transfer Coefficient to Local Heat Transfer Coefficient for a Flat Plate**

For flow over a flat plate, the local heat transfer coefficient h_x varies as $x^{-1/n}$, where x is the distance from the leading edge of the plate, and n is a constant such that $n > 1$. Find the ratio of the average heat transfer coefficient to the local heat transfer coefficient for any location x on the surface of the plate.

DIAGRAM

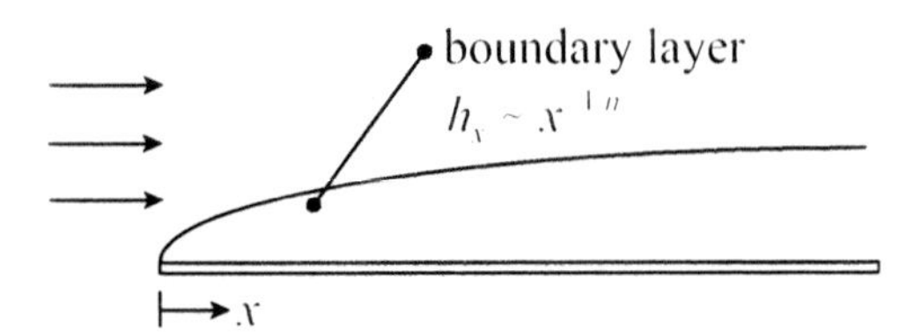

ASSUMPTIONS

1. $n > 1$.

PROPERTIES

Not applicable.

ANALYSIS

For flow over a flat plate, the average heat transfer coefficient is obtained from the local heat transfer coefficient by performing an integrated average,

$$\bar{h}_x(x) = \frac{1}{x}\int_0^x h_x(x)dx$$

The local heat transfer coefficient may be expressed as

$$h_x(x) = ax^{-1/n}$$

where a is a constant. Substituting this expression and integrating,

$$\bar{h}_x(x) = \frac{ax^{-1/n}}{1 - 1/n}$$

Thus, the ratio of the average heat transfer coefficient to the local heat transfer coefficient is

$$\frac{\bar{h}_x(x)}{h_x(x)} = \frac{\left(\dfrac{ax^{-1/n}}{1-\frac{1}{n}}\right)}{ax^{-1/n}} = \frac{n}{n-1}$$

DISCUSSION

Note that the result is independent of the constant of proportionality a. It is instructive to mention a few special cases. For forced laminar flow over a flat plate, $n = 2$, and for laminar flow over a vertical flat plate experiencing natural convection, $n = 4$. Hence, for these two special cases, the average heat transfer coefficient is greater than the local value by a factor of 2 and 4/3, respectively.

PROBLEM 6.2 Heat Transfer for a Section of Flat Plate with Laminar Flow

For laminar flow over a flat plate, the local heat transfer coefficient varies as $h_x = ax^{-1/2}$, where x is the distance from the leading edge of the plate, and $a = 100$ W/m$^{3/2}$·K. The surface of the plate is isothermal, with a temperature of 80°C, and the free stream temperature of the fluid flowing over the plate is 20°C. If the plate is 1 m wide, find the convective heat transfer for the section of the plate defined by 0.75 m < x < 2.0 m.

DIAGRAM

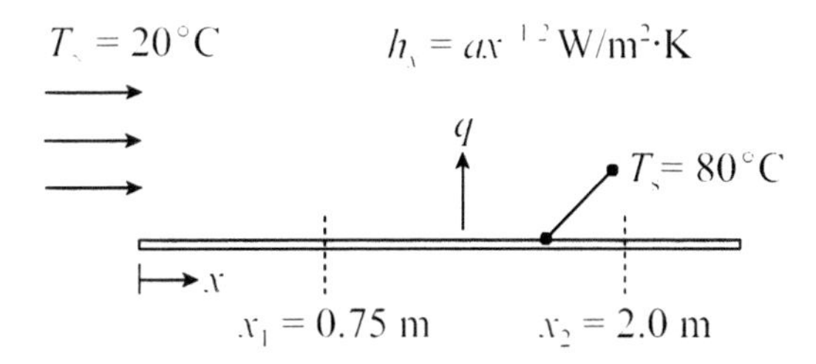

ASSUMPTIONS

1. Laminar flow.
2. Isothermal surface.
3. Constant free stream fluid temperature.

PROPERTIES

Not applicable.

ANALYSIS

For flow over a flat plate, the average heat transfer coefficient is obtained from the local heat transfer coefficient by performing an integrated average,

$$\bar{h}_x = \frac{1}{x_2 - x_1}\int_{x_1}^{x_2} h_x(x)dx$$

where the local heat transfer coefficient varies as

$$h_x(x) = ax^{-1/2}$$

and $a = 100$ W/m$^{3/2}$·K. Substituting this expression and integrating,

$$\bar{h}_x = \frac{2a(x_2^{1/2} - x_1^{1/2})}{x_2 - x_1}$$

Thus, the average heat transfer coefficient for the section 0.75 m < x < 2.0 m is

$$\bar{h}_x = \frac{2(100 \text{ W/m}^2{\cdot}\text{K})[(2.0 \text{ m})^{1/2} - (0.75 \text{ m})^{1/2}]}{2.0 \text{ m} - 0.75 \text{ m}}$$

$$= 87.7 \text{ W/m}^2{\cdot}\text{K}$$

The convective heat transfer for this section of plate is

$$q = \bar{h}_x A(T_s - T_\infty) = \bar{h}_x W(x_2 - x_1)(T_s - T_\infty)$$

$$q = (87.7 \text{ W/m}^2{\cdot}\text{K})(1 \text{ m})(2.0 \text{ m} - 0.75 \text{ m})(80°\text{C} - 20°\text{C})$$

$$= \underline{\underline{6578 \text{ W}}}$$

DISCUSSION

The average heat transfer coefficient for the entire plate from the leading edge to x = 2.0 m is found by setting $x_1 = 0$, which yields

$$\bar{h}_x = 141 \text{ W/m}^2 \cdot \text{K}$$

$$q = 16{,}920 \text{ W}$$

PROBLEM 6.3 Derivation of the Friction and Heat Transfer Coefficients from Boundary Layer Profiles

For flow over a surface, the velocity boundary layer is described by the velocity profile

$$u(y) = A + By + Cy^2 + Dy^3$$

and the thermal boundary layer is described by the temperature profile

$$T(y) = E + Fy + Gy^2 + Hy^3$$

where the coefficients A through H are constants. Derive expressions for the friction coefficient C_f and the heat transfer coefficient h in terms of the free stream velocity and temperature u_∞ and T_∞, respectively, fluid properties and the profile constants.

DIAGRAM

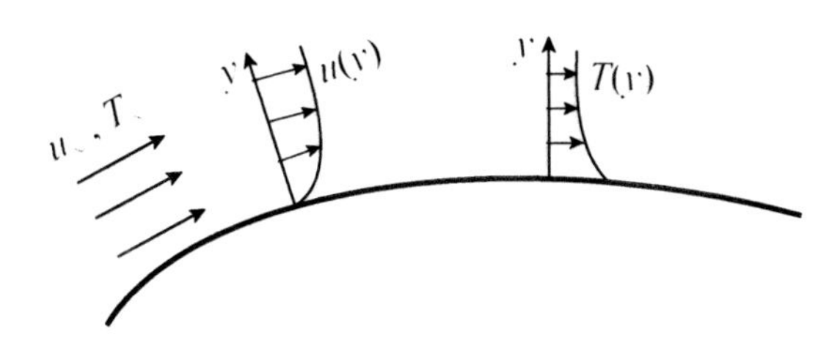

ASSUMPTIONS

1. Constant free stream fluid conditions.
2. Constant fluid properties.

PROPERTIES

density: ρ
dynamic viscosity: μ
thermal conductivity: k_f

ANALYSIS

The friction coefficient is defined as

$$C_f \equiv \frac{\tau_s}{\rho u_\infty^2 / 2}$$

where u_∞ is the free stream fluid velocity, and τ_s is the shear stress at the surface, defined by

$$\tau_s \equiv \mu \frac{\partial u}{\partial y}\bigg|_{y=0}$$

which applies to Newtonian fluids only. Differentiating the velocity profile,

$$\frac{\partial u}{\partial y}\bigg|_{y=0} = B + 2Cy + 3Dy^2\Big|_{y=0} = B$$

which yields a shear stress

$$\tau_s = \mu B$$

Thus, the expression for the friction coefficient is

$$C_f = \frac{2\mu B}{\rho u_\infty^2}$$

The heat flux at the surface may be written in terms of Fourier's law and Newton's law of cooling,

$$q_s'' = -k_f \frac{\partial T}{\partial y}\bigg|_{y=0} = h(T_s - T_\infty)$$

Differentiating the temperature profile,

$$\frac{\partial T}{\partial y}\bigg|_{y=0} = F + 2Gy + 3Hy^2\Big|_{y=0} = F$$

and the temperature at the surface is

$$T_s = T(0) = E$$

Rearranging the expression for the heat flux, and solving for the heat transfer coefficient,

$$h = \frac{-k_f F}{E - T_\infty}$$

DISCUSSION

The difficulty in obtaining expressions for C_f and h lies in finding the boundary layer profiles, which may require a great deal of theoretical and experimental work.

PROBLEM 6.4 **Boundary Layer Transition Length for the Flow of Three Fluids on a Flat Plate**

A fluid at 300 K flows across a flat plate with a constant velocity. The plate is in thermal equilibrium with the fluid. Find the distance from the leading edge of the plate at which transition occurs for the following fluids and free stream velocities:

atmospheric air	$u_\infty = 15$ m/s
saturated liquid water	$u_\infty = 3$ m/s
mercury	$u_\infty = 1$ m/s

DIAGRAM

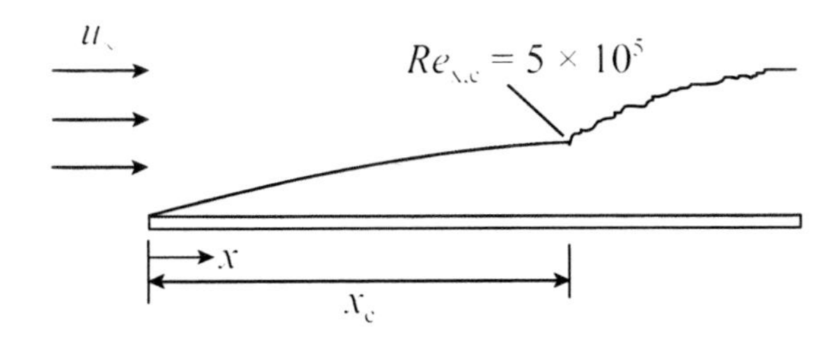

ASSUMPTIONS

1. Constant fluid velocity.
2. Constant fluid properties.

PROPERTIES

$T_f = 300$ K

atmospheric air:	$\nu = 15.89 \times 10^{-6}$ m^2/s
saturated liquid water:	$\nu = 8.576 \times 10^{-7}$ m^2/s
mercury:	$\nu = 1.125 \times 10^{-7}$ m^2/s

ANALYSIS

For flow on a flat plate, Reynolds number is defined as

$$Re_x \equiv \frac{u_\infty x}{\nu}$$

where u_∞ is the free stream fluid velocity and ν is kinematic viscosity. The Reynolds number at which the velocity boundary layer undergoes transition from a laminar to a turbulent condition is called the *critical* Reynolds number, and it has the value $Re_{x,c} = 5 \times 10^5$. Thus, the distance from the leading edge of the plate at which transition occurs is

$$x_c = \frac{Re_{x,c}\nu}{u_\infty}$$

For atmospheric air,

$$x_c = \frac{(5 \times 10^5)(15.89 \times 10^{-6}\ \text{m}^2/\text{s})}{15\ \text{m/s}}$$

$$= 0.530\ \text{m} = \underline{\underline{53.0\ \text{cm}}}$$

For saturated liquid water,

$$x_c = \frac{(5 \times 10^5)(8.576 \times 10^{-7}\ \text{m}^2/\text{s})}{3\ \text{m/s}}$$

$$= 0.143\ \text{m} = \underline{\underline{14.3\ \text{cm}}}$$

For mercury,

$$x_c = \frac{(5 \times 10^5)(1.125 \times 10^{-7}\ \text{m}^2/\text{s})}{1\ \text{m/s}}$$

$$= 0.0563\ \text{m} = \underline{\underline{5.63\ \text{cm}}}$$

DISCUSSION

Through extensive experiments, it is has been determined that the critical Reynolds number for flow over a flat plate varies from about 10^5 to 3×10^6, depending on surface roughness and the turbulence level of the free stream. The value $Re_{x,c} = 5 \times 10^5$ is representative of this range, and is the value typically used in heat transfer calculations.

The distance from the leading edge of the plate required to achieve transition increases with kinematic viscosity because viscous forces attenuate the instabilities that give rise to transition. On the other hand, this distance decreases with free stream velocity because velocity has the effect of shifting or delaying the occurrence of transition to locations farther downstream.

PROBLEM 6.5 Velocity and Temperature Profiles in Couette Flow

The one-dimensional flow of a viscous fluid between two large parallel plates with one plate stationary and other plate moving with constant velocity is called *Couette* flow. Consider a lubricating oil ($\mu = 6.41 \times 10^{-3}$ Pa·s, $k = 0.138$ W/m·K) between two large parallel plates with a 2-mm gap between them. The stationary plate has a temperature of 77°C, and the other plate, which is has a constant velocity of 25 m/s, is insulated. Find the temperature of the moving plate and the heat flux at the stationary plate.

DIAGRAM

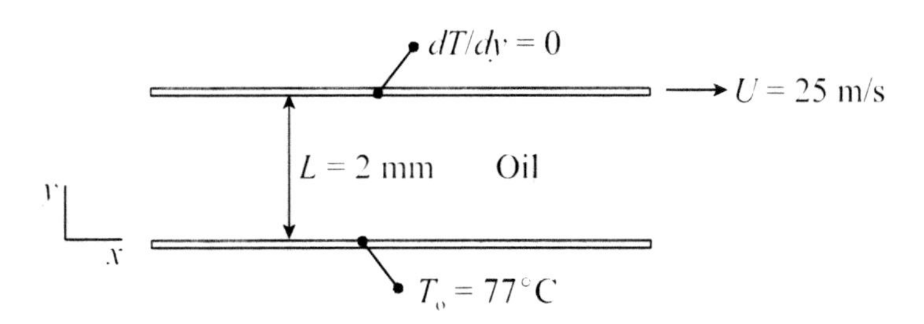

ASSUMPTIONS

1. One-dimensional steady, parallel flow.
2. Incompressible fluid.
3. Body forces are neglected.
4. No energy generation.
5. Constant fluid properties.

PROPERTIES

$T = 77°C$

lubricating oil: $\mu = 6.41 \times 10^{-3}$ Pa·s, $k = 0.138$ W/m·K

ANALYSIS

For an incompressible fluid with constant properties, the x-momentum equation for the boundary layer reduces to

$$\frac{\partial^2 u}{\partial y^2} = 0$$

where u is the x-component of velocity. The velocity profile may be obtained by solving this equation. Integrating twice,

$$u(y) = C_1 y + C_2$$

where C_1 and C_2 are constants of integration, which are determined from the boundary conditions

$$u(0) = 0 \text{ , } u(L) = U$$

Applying the boundary conditions, the velocity profile is

$$u(y) = U\frac{y}{L}$$

Under the assumptions given, the energy equation is

$$k\frac{\partial^2 T}{\partial y^2} + \mu\left(\frac{\partial u}{\partial y}\right)^2 = 0$$

The first term is the conduction term, and the second term is the viscous dissipation term. Substituting the velocity profile, and integrating twice, the temperature distribution in the fluid is

$$T(y) = -\frac{\mu}{2k}\left(\frac{U}{L}\right)^2 y^2 + C_3 y + C_4$$

where C_3 and C_4 are constants of integration, which are determined from the boundary conditions

$$T(0) = T_o \text{ , } \frac{dT(L)}{dy} = 0$$

Applying the boundary conditions and simplifying, the temperature distribution is

$$T(y) = T_o + \frac{\mu U^2}{2k}\left[2\left(\frac{y}{L}\right) - \left(\frac{y}{L}\right)^2\right]$$

The temperature of the moving plate is

$$T(L) = T_o + \frac{\mu U^2}{2k}$$

$$T(L) = 77°\text{C} + \frac{(6.41 \times 10^{-3}\ \text{Pa·s})(25\ \text{m/s})^2}{2(0.138\ \text{W/m·K})}$$

$$= \underline{\underline{91.5°\text{C}}}$$

The heat flux in the fluid is found by invoking Fourier's law,

$$q'' = -k\frac{dT}{dy} = -\frac{\mu U^2}{L}\left(1 - \frac{y}{L}\right)$$

Thus, the heat flux at the stationary plate is

$$q'' = -\frac{\mu U^2}{L}$$

$$q'' = -\frac{(6.41 \times 10^{-3}\ \text{Pa·s})(25\ \text{m/s})^2}{0.002\ \text{m}}$$

$$= -3.13 \times 10^5\ \text{W/m}^2 = \underline{\underline{-313\ \text{kW/m}^2}}$$

DISCUSSION

Due to viscous dissipation in the oil, the moving plate has a higher temperature than the stationary plate. The significance of the minus sign on the heat flux is that the direction of heat transfer is in the negative *y* direction, i.e., from the fluid into the stationary plate, which means that the fluid is at a higher temperature than the plate. The graph below shows the temperature variation in the oil for several values of velocity *U*. Note that the slope of the temperature gradient at the moving plate is zero, consistent with an insulated boundary condition.

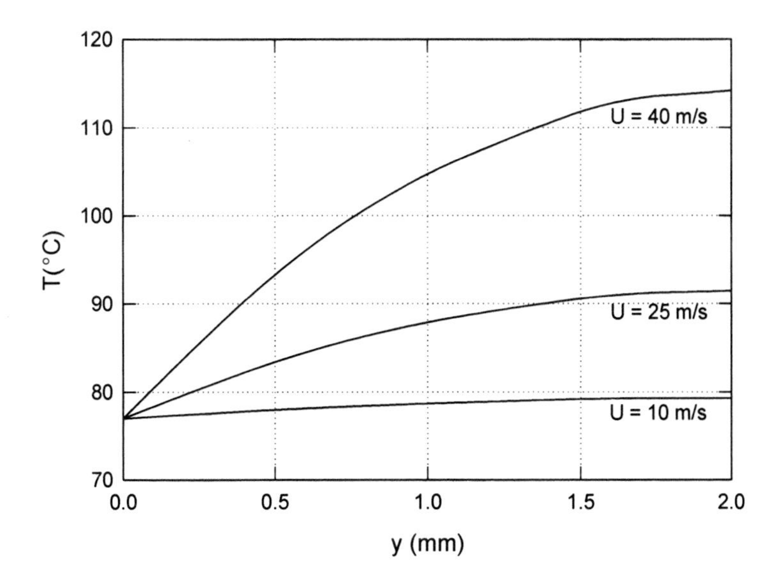

Thermal properties of the lubricating oil were based on 77°C, the temperature of the stationary plate. This is a good approximation for low plate velocities because the oil temperature does not change significantly due to low viscous dissipation. But as plate velocity increases, the oil temperature increases significantly toward the moving plate, as shown in the graph. A better approximation is to base oil thermal properties on an average temperature of the two plates, which involves an iterative approach.

❒ ❒ ❒

PROBLEM 6.6 Couette Flow with Uniform Energy Generation

The one-dimensional flow of a viscous fluid between two large parallel plates with one plate stationary and other plate moving with constant velocity is called *Couette* flow. The stationary and moving plates are maintained at temperatures T_o and T_L, respectively. For the case where the fluid generates energy uniformly, derive an expression for the temperature distribution in the fluid.

DIAGRAM

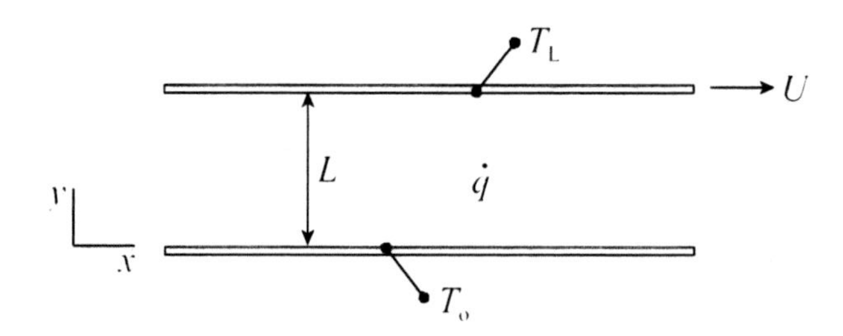

ASSUMPTIONS

1. One-dimensional steady, parallel flow.
2. Incompressible fluid.
3. Body forces are neglected.
4. Uniform energy generation.
5. Constant fluid properties.

PROPERTIES

dynamic viscosity: μ
thermal conductivity: k

ANALYSIS

For an incompressible fluid with constant properties, the x-momentum equation for the boundary layer reduces to

$$\frac{\partial^2 u}{\partial y^2} = 0$$

where u is the x-component of velocity. The velocity profile may be obtained by solving this equation. Integrating twice,

$$u(y) = C_1 y + C_2$$

where C_1 and C_2 are constants of integration, which are determined from the boundary conditions

$$u(0) = 0 \, , \, u(L) = U$$

Applying the boundary conditions, the velocity profile is

$$u(y) = U \frac{y}{L}$$

Under the assumptions given, the energy equation is

$$k \frac{\partial^2 T}{\partial y^2} + \mu \left(\frac{\partial u}{\partial y} \right)^2 + \dot{q} = 0$$

The first term is the conduction term, the second term is the viscous dissipation term, and the third term is the energy generation. Substituting the velocity profile, and integrating twice, the temperature distribution in the fluid is

$$T(y) = -\left[\frac{\mu}{k} \left(\frac{U}{L} \right)^2 + \frac{\dot{q}}{k} \right] \frac{y^2}{2} + C_3 y + C_4$$

where C_3 and C_4 are constants of integration, which are determined from the boundary conditions

$$T(0) = T_o \, , \, T(L) = T_L$$

Applying the boundary conditions and simplifying, the temperature distribution is

$$T(y) = T_o + \frac{1}{2k} \left(\mu U^2 + \dot{q} L^2 \right) \left[\frac{y}{L} - \left(\frac{y}{L} \right)^2 \right] + \left(T_L - T_o \right) \frac{y}{L}$$

DISCUSSION

The heat flux at any point in the fluid is given by

$$q'' = -k\frac{dT(y)}{dy}$$

from which the heat flux at the surface of either plate may be obtained.

❒ ❒ ❒

PROBLEM 6.7 Velocity and Thermal Boundary Layer Thicknesses for Laminar Flow on a Flat Plate

Saturated liquid water at 34°C flows across an isothermal flat plate at a velocity of 2 m/s. The temperature of the plate is 60°C, and the plate is 10 cm long in the direction of the flow. Verify that the flow is laminar everywhere on the plate, and find the thickness of the velocity and thermal boundary layers at the trailing edge of the plate.

DIAGRAM

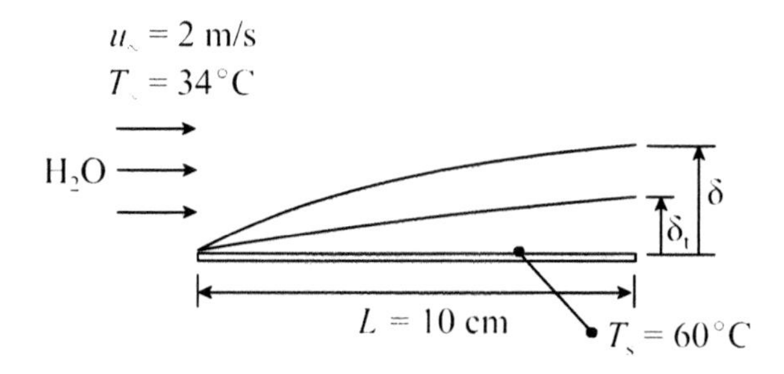

ASSUMPTIONS

1. Steady conditions.
2. Isothermal surface.
3. Constant properties.

PROPERTIES

$T_f = (T_s + T_\infty)/2 = (60°C + 34°C)/2 = 47°C = 320$ K

saturated liquid water: $\rho = 989.1$ kg/m^3, $c_p = 4180$ J/kg·K, $\mu = 577 \times 10^{-6}$ Pa·s,
$k = 0.640$ W/m·K, $Pr = 3.77$

ANALYSIS

Kinematic viscosity is

$$\nu \equiv \frac{\mu}{\rho}$$

$$\nu = \frac{577 \times 10^{-6}\ \text{Pa·s}}{989.1\ \text{kg/m}^3}$$

$$= 5.834 \times 10^{-7}\ \text{m}^2\text{/s}$$

and thermal diffusivity is

$$\alpha \equiv \frac{k}{\rho c_p}$$

$$\alpha = \frac{0.640\ \text{W/m·K}}{(989.1\ \text{kg/m}^3)(4180\ \text{J/kg·K})}$$

$$= 1.548 \times 10^{-7}\ \text{m}^2\text{/s}$$

The Reynolds number for the trailing edge of the plate is

$$Re_L = \frac{u_\infty L}{\nu}$$

$$Re_L = \frac{(2\ \text{m/s})(0.10\ \text{m})}{5.834 \times 10^{-7}\ \text{m}^2\text{/s}}$$

$$= 3.428 \times 10^5$$

Since $Re_L < 5 \times 10^5$, the flow is laminar over the entire plate. The velocity boundary layer thickness for laminar flow on a flat plate is

$$\delta = \frac{5.0L}{\sqrt{Re_L}}$$

$$\delta = \frac{5.0(0.10\ \text{m})}{(3.428 \times 10^5)^{1/2}}$$

$$= 8.54 \times 10^{-4}\ \text{m} = \underline{\underline{0.854\ \text{mm}}}$$

The thermal boundary layer thickness for laminar flow on a flat plate is

$$\delta_t = \frac{\delta}{Pr^{1/3}}$$

$$\delta_t = \frac{8.54 \times 10^{-4}\ \text{m}}{(3.77)^{1/3}}$$

$$= 5.49 \times 10^{-4}\ \text{m} = \underline{\underline{0.549\ \text{mm}}}$$

DISCUSSION

At the trailing edge of the plate, the velocity boundary layer is thicker than the thermal boundary layer. We already knew this because the Prandtl number, a dimensionless parameter that reflects the relative thicknesses of the velocity and thermal boundary layers, is defined as

$$Pr \equiv \frac{\nu}{\alpha}$$

and $Pr = 3.77$ for saturated liquid water at 320 K.

❒ ❒ ❒

PROBLEM 6.8 Dimensionless Numbers for Four Different Fluids

A fluid flows across a flat plate with a length of 1 m at a free stream velocity of 1 m/s. If the average heat transfer coefficient for the surface is 150 W/m^2·K, find the values of the Reynolds, Nusselt and Stanton numbers and Colburn j factor for atmospheric air, saturated liquid water, glycerin, and mercury. Base all thermal properties on a film temperature of 300 K.

DIAGRAM

$u_\infty = 1$ m/s

$T_f = 300$ K

$h = 150$ W/m^2·K

$L = 1$ m

ASSUMPTIONS

1. Constant properties.

PROPERTIES

$T_f = 300$ K

Fluid	ν (m²/s)	*k* (W/m·K)	α (m²/s)	*Pr*
Atmospheric air	15.89×10^{-6}	0.0263	22.5×10^{-6}	0.707
Sat. liquid water	8.576×10^{-7}	0.613	1.471×10^{-7}	5.83
Glycerin	634×10^{-6}	0.286	0.935×10^{-7}	6780
Mercury	0.1125×10^{-6}	8.540	45.30×10^{-7}	0.0248

ANALYSIS

The relations for the dimensionless parameters are given below. The Reynolds number is

$$Re_L = \frac{u_\infty L}{\nu}$$

The Nusselt number is

$$Nu_L = \frac{hL}{k}$$

The Stanton number is

$$St = \frac{Nu}{Re_L\, Pr}$$

and the Colburn *j* factor is

$$j_H = St\,\mathrm{Pr}^{2/3}$$

The dimensionless parameters are summarized in the table below.

Fluid	Re_L	Nu_L	*St*	j_H
Atmospheric air	6.29×10^{4}	5703	0.128	0.102
Sat. liquid water	1.17×10^{6}	244.7	3.59×10^{-5}	1.16×10^{-4}
Glycerin	1580	524.5	4.90×10^{-5}	0.0176
Mercury	8.89×10^{6}	17.56	7.96×10^{-5}	6.77×10^{-6}

DISCUSSION

The Prandtl numbers for the glycerin and mercury are typical values for highly viscous liquids and liquid metals, respectively. From the Reynolds numbers we see that the air and glycerin flows are laminar, but the water and mercury flows are turbulent.

❒ ❒ ❒

PROBLEM 6.9 Predicting Convective Heat Transfer from Fluid Friction on a Plate

Both surfaces of a 1 m × 1 m flat plate are subject to the parallel flow of 20°C liquid water. The velocity of the water is 4 m/s, and the temperature of the plate is 75°C. Find the drag force on the plate, the heat transfer coefficient, and the convective heat transfer.

DIAGRAM

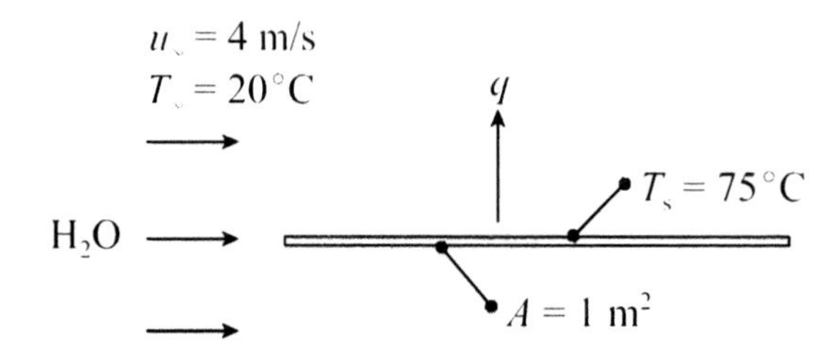

ASSUMPTIONS

1. Steady conditions.
2. Isothermal surfaces.
3. Parallel flow.
4. Constant properties.

PROPERTIES

$T_f = (T_s + T_\infty)/2 = (75°C + 20°C)/2 = 47.5°C = 321$ K

liquid water: $\rho = 988.7$ kg/m^3, $c_p = 4180$ J/kg·K, $\nu = 5.737 \times 10^{-7}$ m^2/s, $Pr = 3.70$

ANALYSIS

The Reynolds number is

$$Re_L = \frac{u_\infty L}{\nu}$$

$$Re_L = \frac{(4 \text{ m/s})(1 \text{ m})}{5.737 \times 10^{-7} \text{ m}^2/\text{s}}$$

$$= 6.972 \times 10^6$$

The Reynolds number shows that the flow condition is mixed, i.e., that the boundary layer is laminar over part of the plate and turbulent over the rest. The friction coefficient for this flow condition is

$$C_f = \frac{0.074}{Re_L^{1/5}} - \frac{1742}{Re_L}$$

$$C_f = \frac{0.074}{(6.972 \times 10^6)^{1/5}} - \frac{1742}{6.972 \times 10^6}$$

$$= 2.917 \times 10^{-3}$$

The Stanton number is defined as

$$St \equiv \frac{Nu_L}{Re_L Pr} = \frac{h}{\rho c_p u_\infty}$$

and the *Reynolds analogy* states that

$$\frac{C_f}{2} = St$$

Accounting for flow on both sides of the plate, the drag force on the plate is

$$F = 2C_f A \frac{\rho u_\infty^2}{2}$$

$$F = 2(2.917 \times 10^{-3})\frac{(1\ \text{m}^2)(988.7\ \text{kg/m}^3)(4\ \text{m/s})^2}{2}$$

$$= \underline{\underline{46.1\ \text{N}}}$$

Combining the Stanton number relation with the Reynolds analogy, the heat transfer coefficient is

$$h = \frac{C_f \rho c_p u_\infty}{2}$$

$$h = \frac{(2.917 \times 10^{-3})(988.7\ \text{kg/m}^3)(4180\ \text{J/kg·K})(4\ \text{m/s})}{2}$$

$= \underline{\underline{2.411 \times 10^4 \text{ W/m}^2\cdot\text{K}}}$

so the convective heat transfer is

$$q = 2hA(T_s - T_\infty)$$

$$q = 2(2.411 \times 10^4 \text{ W/m}^2\cdot\text{K})(1 \text{ m}^2)(75\,°\text{C} - 20\,°\text{C})$$

$$= 2.65 \times 10^6 \text{ W} = \underline{\underline{2.65 \text{ MW}}}$$

DISCUSSION

The Reynolds analogy given above is strictly valid for $Pr \approx 1$, but in our problem, $Pr = 3.70$. It has been shown that the analogy may be applied over a wide range of Pr if a Prandtl number correction is used. This correction is incorporated in the *modified Reynolds*, or *Chilton-Colburn*, analogy,

$$\frac{C_f}{2} = St\,Pr^{2/3} = j_H$$

where j_H is the Colburn j factor. Using the modified analogy, more accurate values for the heat transfer coefficient and convective heat transfer, respectively, are

$$h = \underline{\underline{1.008 \times 10^4 \text{ W/m}^2\cdot\text{K}}}$$

$$q = \underline{\underline{1.11 \times 10^6 \text{ W} = 1.11 \text{ MW}}}$$

❐ ❐ ❐

PROBLEM 6.10 Using Boundary Layer Similarity to Find Heat Transfer Coefficients

A body of irregular shape has a characteristic length of 1 m and is maintained at a uniform surface temperature of 50 °C. When the body is placed in 20 °C atmospheric air moving at a velocity of 60 m/s, the average heat flux at the surface is 15 kW/m^2. If a second body of the same shape, but with a characteristic length of 4 m, is maintained at the same surface temperature as the first body, find the average heat transfer coefficient for the second body if it is placed in 20 °C atmosphericair moving at a velocity of 15 m/s.

DIAGRAM

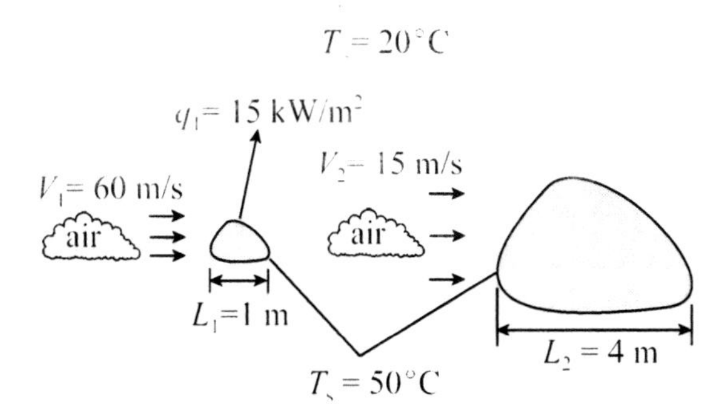

ASSUMPTIONS

1. Steady conditions.
2. Constant properties.

PROPERTIES

$T_{f1} = T_{f2} = T_f = (T_s + T_\infty)/2$

atmospheric air: $k_1 = k_2, \nu_1 = \nu_2, Pr_1 = Pr_2$

ANALYSIS

Reynolds number is given by

$$Re_L = \frac{VL}{\nu}$$

so the Reynolds number for each body is

$$Re_{L,1} = \frac{(60\text{ m/s})(1\text{ m})}{\nu_1} = \frac{60\text{ m/s}^2}{\nu_1}$$

$$Re_{L,2} = \frac{(15\text{ m/s})(4\text{ m})}{\nu_2} = \frac{60\text{ m/s}^2}{\nu_2}$$

Because each body is exposed to the same fluid, $\nu_1 = \nu_2$, so $Re_{L,1} = Re_{L,2}$. Also, $Pr_1 = Pr_2$. For a prescribed geometry, the Nusselt number is a function of the Reynolds and Prandtl numbers,

$$Nu = f(Re_L, Pr)$$

Hence,

$$Nu_1 = Nu_2$$

so we have

$$\frac{h_1 L_1}{k_1} = \frac{h_2 L_2}{k_2}$$

Since $k_1 = k_2$,

$$h_2 = \frac{L_1}{L_2} h_1 = 0.25 h_1$$

Knowing the heat flux at the surface of body 1,

$$h_1 = \frac{q_1''}{T_s - T_\infty}$$

$$h_1 = \frac{15 \times 10^3 \text{ W/m}^2}{50°\text{C} - 20°\text{C}}$$

$$= 500 \text{ W/m}^2\cdot\text{K}$$

Thus, the heat transfer coefficient for the second body is

$$h_2 = 0.25(500 \text{ W/m}^2\cdot\text{K})$$

$$= \underline{\underline{125 \text{ W/m}^2\cdot\text{K}}}$$

DISCUSSION

The crucial factor for this problem is knowing that $Re_{L,1} = Re_{L,2}$. If the Reynolds numbers for the two bodies were not equal, we would have to know the specific form of the function $f(Re_L, Pr)$ in order to determine h_2.

❒ ❒ ❒

Chapter 7

External Forced Convection

PROBLEM 7.1 **Velocity Boundary Layer Thickness for Three Gases Flowing Parallel to a Flat Plate**

Consider the following atmospheric gases at a film temperature of 27°C flowing parallel to a flat plate at a velocity of 5 m/s: air, hydrogen, and argon. On the same graph, plot the variation of velocity boundary layer thickness with distance from the leading edge of the plate.

DIAGRAM

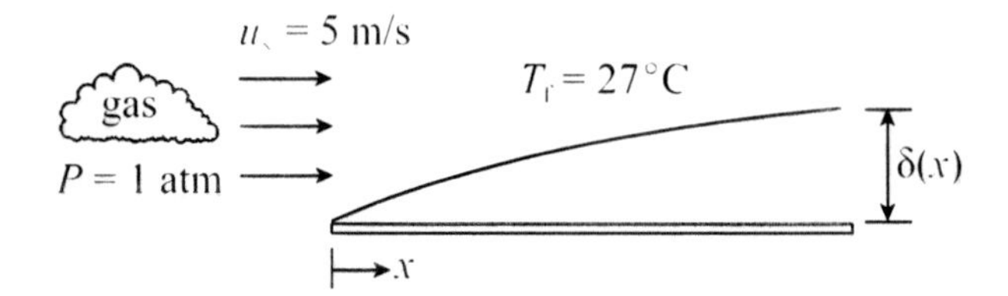

ASSUMPTIONS

1. Steady conditions.
2. Parallel flow.
3. Constant properties.

PROPERTIES

$T_f = 27°C = 300$ K

atmospheric air: $\nu = 15.89 \times 10^{-6}$ m²/s
atmospheric H_2: $\nu = 111 \times 10^{-6}$ m²/s
atmospheric Ar: $\nu = 142 \times 10^{-6}$ m²/s

ANALYSIS

For laminar boundary layers, the boundary layer thickness is given by the relation

$$\delta(x) = \frac{5.0x}{\sqrt{Re_x}}$$

where the Reynolds number is

$$Re_x \frac{u_\infty x}{\nu}$$

As long as $Re_x < 5 \times 10^5$, the transition Reynolds number, we can easily find $\delta(x)$ for the three gases, from which a graph can be constructed. This graph is shown below for the range of distances $0 < x < 50$ cm.

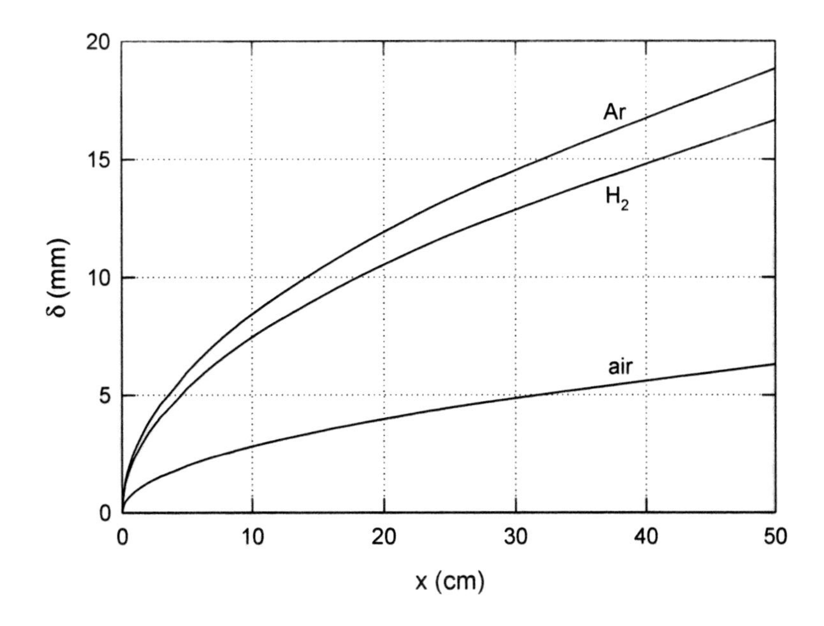

DISCUSSION

It is clear from the graph that, at a given distance x, the boundary layer thickness increases with kinematic viscosity ν. From the relation for boundary layer thickness, it is also evident that δ is directly proportional to $(\nu x)^{1/2}$ and inversely proportional to $u_\infty^{1/2}$.

❐ ❐ ❐

PROBLEM 7.2 **Convective Heat Transfer for Parallel Laminar Flow on a Flat Plate**

One side of a flat plate measuring 1 m × 1 m is exposed to the parallel flow of 19 °C glycerin at a velocity of 1.5 m/s. The surface of the plate is maintained at a uniform temperature of 75 °C. Verify that the flow is laminar over the entire plate, and find the convective heat transfer.

DIAGRAM

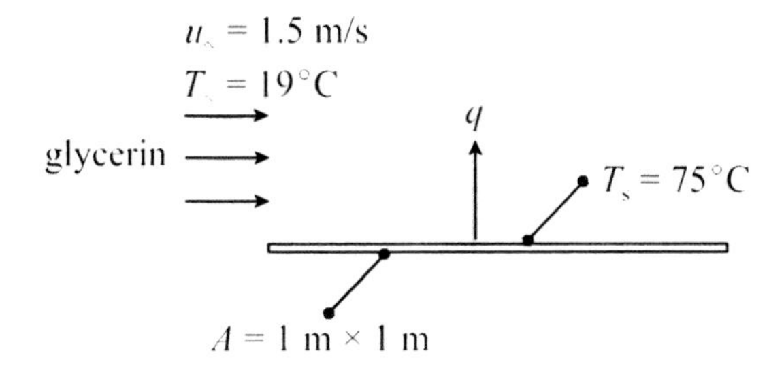

ASSUMPTIONS

1. Steady conditions.
2. Parallel flow.
3. Constant properties.

PROPERTIES

$T_f = (T_s + T_\infty)/2 = (75°C + 19°C)/2 = 47°C = 320$ K

glycerin: $k = 0.287$ W/m·K, $\nu = 168 \times 10^{-6}$ m^2/s, $Pr = 1870$

ANALYSIS

The Reynolds number is

$$Re_L = \frac{u_\infty L}{\nu}$$

$$Re_L = \frac{(1.5 \text{ m/s})(1 \text{ m})}{168 \times 10^{-6} \text{ m}^2/\text{s}}$$

$$= 8929$$

Because $Re_L < 5 \times 10^5$, the flow is laminar over the entire plate. The average Nusselt number for laminar flow is

$$Nu_L = \frac{hL}{k} = 0.664 Re_L^{1/2} Pr^{1/3}$$

$$Nu_L = 0.664(8929)^{1/2}(1870)^{1/3}$$

$$= 773$$

Thus, the average heat transfer coefficient is

$$h = \frac{(0.287 \text{ W/m·K})(773)}{1 \text{ m}}$$

$$= 222 \text{ W/m}^2\text{·K}$$

so the convective heat transfer is

$$q = hA(T_s - T_\infty)$$

$$q = (222 \text{ W/m}^2\cdot\text{K})(1 \text{ m}^2)(75°\text{C} - 19°\text{C})$$

$$= 1.243 \times 10^4 \text{ W} = \underline{\underline{1.24 \text{ kW}}}$$

DISCUSSION

The relation above for average Nusselt number comes from an analytical solution to the energy equation for a laminar boundary layer. The local Nusselt number is exactly half the average value, so

$$Nu_x = \frac{h_x x}{k} = 0.332 Re_x^{1/2} Pr^{1/3}$$

where x is the distance from the leading edge of the plate. From this relation we see that

$$h_x = Cx^{-1/2}$$

where C is a constant that incorporates thermal properties and other constants. This result implies that the heat transfer coefficient is theoretically infinite at the leading edge of the plate and decreases as $x^{-½}$ in the flow direction

❒ ❒ ❒

PROBLEM 7.3 Convective Heat Transfer for Parallel Mixed Flow on a Flat Plate

One side of a flat plate that measures 5 m in the flow direction and 1 m wide is exposed to the parallel flow of 15°C liquid water at a velocity of 3 m/s. If the surface of the plate is maintained at a uniform temperature of 60°C, find the convective heat transfer.

DIAGRAM

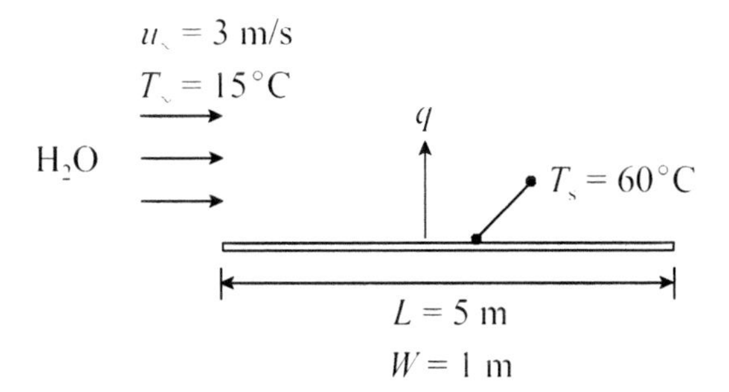

ASSUMPTIONS

1. Steady conditions.
2. Parallel flow.
3. Constant properties.

PROPERTIES

$T_f = (T_s + T_\infty)/2 = (60°C + 15°C)/2 = 37.5°C = 311$ K

liquid water: $k = 0.629$ W/m·K, $\nu = 6.870 \times 10^{-7}$ m²/s, $Pr = 4.53$

ANALYSIS

The Reynolds number is

$$Re_L = \frac{u_\infty L}{\nu}$$

$$Re_L = \frac{(3 \text{ m/s})(5 \text{ m})}{6.870 \times 10^{-7} \text{ m}^2\text{/s}}$$

$$= 2.183 \times 10^7$$

Because $Re_L > 5 \times 10^5$, the flow is mixed, i.e., laminar from the leading edge to the point at which the boundary layer undergoes transition and turbulent downstream from that point. The average Nusselt number for mixed flow is

$$Nu_L = \frac{hL}{k} = (0.037 Re_L^{4/5} - 871) Pr^{1/3}$$

$$Nu_L = (0.037(2.183 \times 10^7)^{4/5} - 871)(4.53)^{1/3}$$

$$= 4.407 \times 10^4$$

Thus, the average heat transfer coefficient is

$$h = \frac{(0.629 \text{ W/m·K})(4.407 \times 10^4)}{5 \text{ m}}$$

$$= 5544 \text{ W/m}^2\text{·K}$$

so the convective heat transfer is

$$q = hA(T_s - T_\infty)$$

$$q = (5544 \text{ W/m}^2\text{·K})(5 \text{ m}^2)(60°C - 15°C)$$

$$= 1.25 \times 10^6 \text{ W} = \underline{\underline{1.25 \text{ MW}}}$$

DISCUSSION

The distance downstream from the leading edge of the plate where the boundary layer undergoes transition is called the *critical* distance x_c, and is related to the critical Reynolds number by

$$Re_{L,c} = \frac{u_\infty x_c}{\nu} = 5 \times 10^5$$

Thus, the critical distance is

$$x_c = \frac{(5 \times 10^5)(6.870 \times 10^{-7}\ \text{m}^2/\text{s})}{3\ \text{m/s}}$$

$$= 0.115\ \text{m} = 11.5\ \text{cm}$$

Hence, only the first 11.5 cm of the plate experiences laminar flow. For the rest of the plate the boundary layer is turbulent. In situations for which $L >> x_c$, turbulent flow exists over most of the plate, and the Nusselt number relation reduces to

$$Nu_L = 0.037 Re_L^{4/5} Pr^{1/3}$$

from which we obtain

$$Nu_L = 4.551 \times 10^4$$

$$h = 5726\ \text{W/m}^2\cdot\text{K}$$

$$q = 1.29\ \text{MW}$$

The heat transfer is within 4 percent of the value obtained using the relation for mixed flow. Hence, we could have assumed turbulent flow for the entire plate with little error.

❐ ❐ ❐

PROBLEM 7.4 Convective Heat Transfer for Parallel Laminar Flow on a Flat Plate with an Unheated Starting Length

One side of a square flat plate measuring 1 m × 1 m experiences the parallel flow of 20°C ethylene glycol. The first 20 cm of the plate from the leading edge is unheated, but the rest of the plate has a uniform surface temperature of 90°C. If the free stream velocity of the ethylene glycol is 2.5 m/s, what is the convective heat transfer from the plate?

DIAGRAM

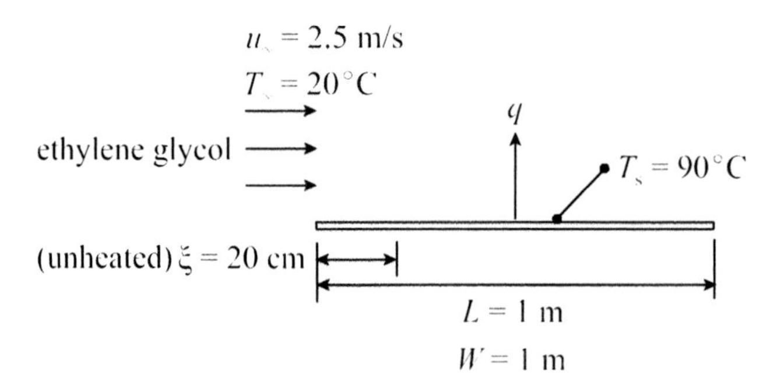

ASSUMPTIONS

1. Steady conditions.
2. Parallel flow.
3. Constant properties.

PROPERTIES

$T_f = (T_s + T_\infty)/2 = (90°C + 20°C)/2 = 55°C = 328$ K

ethylene glycol: $k = 0.260$ W/m·K, $\nu = 5.50 \times 10^{-6}$ m²/s, $Pr = 58.7$

ANALYSIS

The Reynolds number is

$$Re_L = \frac{u_\infty L}{\nu}$$

$$Re_L = \frac{(2.5 \text{ m/s})(1 \text{ m})}{5.50 \times 10^{-6} \text{ m}^2/\text{s}}$$

$$= 4.545 \times 10^5$$

Because $Re_L < 5 \times 10^5$, the flow is laminar over the entire plate. The average Nusselt number for this configuration is

$$\overline{Nu}_L = \frac{2[1-(\xi/L)^{3/4}]}{1-\xi/L} Nu_x$$

where ξ is the unheated starting length and Nu_x is the local Nusselt number at $x = L$ measured from the leading edge of the plate,

$$Nu_x = 0.332 Re_x^{1/2} Pr^{1/3}$$

$$Nu_x = 0.332(4.545 \times 10^5)^{1/2}(58.7)^{1/3}$$

$$= 869.9$$

Thus, the average Nusselt number is

$$Nu_L = \frac{2[1 - (0.20\text{ m}/1\text{ m})^{3/4}]}{1 - (0.20\text{ m}/1\text{ m})}(869.9)$$

$$= 1524$$

so the average heat transfer coefficient is

$$h = \frac{Nu_L k}{L}$$

$$h = \frac{(1524)(0.260\text{ W/m·K})}{1\text{ m}}$$

$$= 396\text{ W/m}^2\text{·K}$$

and the convective heat transfer is

$$q = hA(T_s - T_\infty)$$

$$q = (396\text{ W/m}^2\text{·K})(1\text{ m}^2)(90°\text{C} - 20°\text{C})$$

$$= 2.772 \times 10^4\text{ W} = \underline{\underline{27.7\text{ kW}}}$$

DISCUSSION

Downstream of the unheated starting length, the temperature of the plate is uniform. Another common type of thermal condition is a uniform heat flux. For this case, the local Nusselt number is

$$Nu_x = 0.453 Re^{1/2} Pr^{1/3}$$

which yields a value for Nu_x that is 36 percent higher than for the isothermal surface case. Consequently, $Nu_L = 2080$, $h = 541$ W/m²·K, and $q = 37.9$ kW.

❐ ❐ ❐

PROBLEM 7.5 Heat Loss from a Window

A window in a tall building experiences a horizontal wind flow parallel to the surface at a velocity of 8 m/s. The window measures 2 m high, 1.5 m wide and 6.5 mm thick. The window casement, which surrounds the edge of the glass pane, acts as a turbulence promoter that trips the boundary layer at the leading edge. Inside the building, the air temperature and heat transfer coefficient for the interior surface of the window are 21 °C and 9 W/m^2·K, respectively. If the outdoor air temperature is 5 °C, what is the convective heat loss from the window?

DIAGRAM

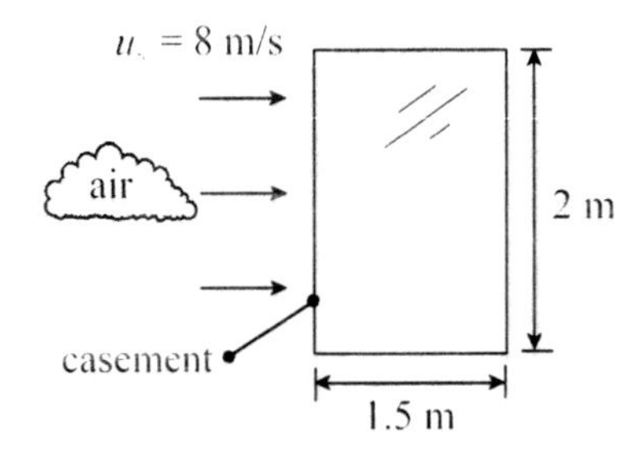

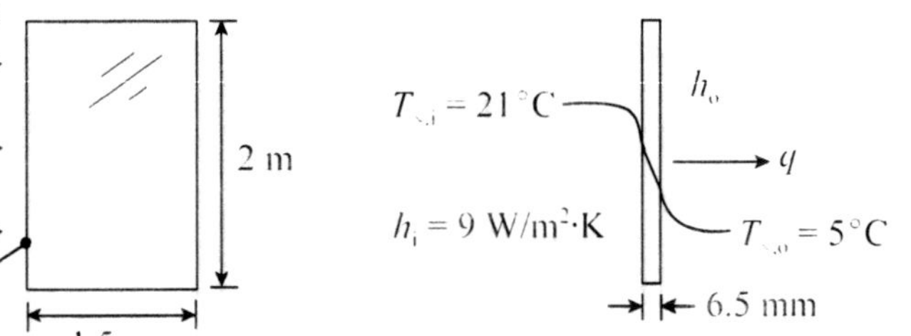

ASSUMPTIONS

1. Steady flow.
2. Steady one-dimensional conduction.
3. Radiation is neglected.
4. Turbulence exists on entire window.
5. Constant properties.

PROPERTIES

window glass: k_g = 1.4 W/m·K

Assume T_s = 10°C. $T_f = (T_s + T_{\infty,o})/2$ = (10°C + 5°C)/2 = 7.5°C = 281 K

atmospheric air: k = 0.0248 W/m·K, $\nu = 14.20 \times 10^{-6}$ m^2/s, Pr = 0.712

ANALYSIS

The Reynolds number is

$$Re_L = \frac{u_\infty L}{\nu}$$

$$Re_L = \frac{(8 \text{ m/s})(1.5 \text{ m})}{14.20 \times 10^{-6} \text{ m}^2/\text{s}}$$

$$= 8.451 \times 10^5$$

This Reynolds number would normally indicate mixed flow, but the window casement trips the boundary layer at the leading edge of the window, leading to complete turbulent flow. The average Nusselt number for turbulent flow on the window is

$$Nu_L = \frac{h_o L}{k} = 0.037 Re^{4/5} Pr^{1/3}$$

$$Nu_L = 0.037(8.451 \times 10^5)^{4/5}(0.712)^{1/3}$$

$$= 1822$$

which yields the heat transfer coefficient for the outside surface of the window,

$$h_o = \frac{Nu_L k}{L}$$

$$h_o = \frac{(1822)(0.0248 \text{ W/m·K})}{1.5 \text{ m}}$$

$$= 30.1 \text{ W/m}^2\text{·K}$$

The total thermal resistance includes the conduction resistance of the glass plus the resistance of the inside and outside boundary layers,

$$R_{tot} = \frac{1}{h_i A} + \frac{\Delta x}{k_g A} + \frac{1}{h_o A}$$

where A is window surface area and Δx is window thickness. Thus,

$$R_{tot} = \frac{1}{(9 \text{ W/m}^2\text{·K})(3 \text{ m}^2)} + \frac{0.0065 \text{ m}}{(1.4 \text{ W/m·K})(3 \text{ m}^2)} + \frac{1}{(30.1 \text{ W/m}^2\text{·K})(3 \text{ m}^2)}$$

$$= 0.0497 \text{ K/W}$$

The heat loss is

$$q = \frac{T_{\infty,i} - T_{\infty,o}}{R_{tot}}$$

$$q = \frac{(21 - 5)°C}{0.0497 \text{ K/W}}$$

$$= \underline{\underline{322 \text{ W}}}$$

DISCUSSION

Thermal properties were based on an assumed value of 10°C for the outside surface of the glass. Now that the heat transfer coefficient has been calculated, we can check the accuracy of this assumption. Using Newton's law of cooling,

$$q = h_o A(T_s - T_{\infty,o})$$

and solving for T_s,

$$T_s = \frac{(322 \text{ W})}{(30.1 \text{ W/m}^2\cdot\text{K})(3 \text{ m}^2)} + 5°C$$

$$= 8.6°C$$

This temperature is close to the assumed value, so a property correction is not warranted.

❒ ❒ ❒

PROBLEM 7.6 Surface Temperatures of an Array of Solid State Devices

Atmospheric air at 55°C is forced over a circuit board populated with a closely spaced array of solid state devices that dissipate 300 mW each. The top surface of each device is square, measuring 2.5 cm × 2.5 cm, and the free stream air velocity is 7 m/s. Assuming that each device loses all dissipated heat through its top surface to the forced air, find the surface temperature of the first through fifth devices from the leading edge of the circuit board.

DIAGRAM

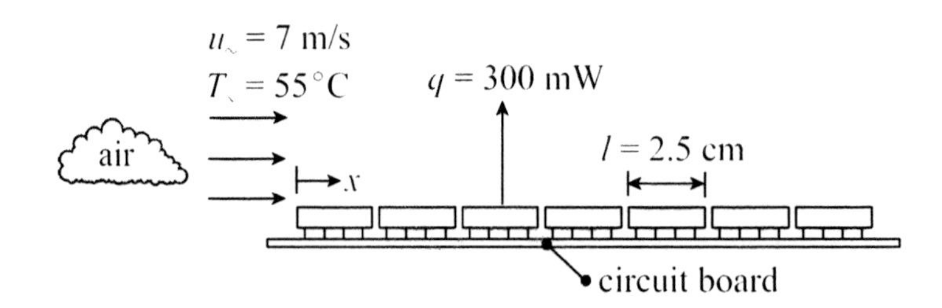

ASSUMPTIONS

1. Steady conditions.
2. Spaces between the devices are neglected.
3. All the heat dissipated by each device is lost from its top surface.
4. The top surface of each device is flat and smooth.
5. The top surface of each device is isothermal.
6. Radiation is neglected.
7. Constant properties.

PROPERTIES

Assume $T_s = 75°C$. $T_f = (T_s + T_{\infty,o})/2 = (75°C + 55°C)/2 = 65°C = 338$ K

atmospheric air: $k = 0.0291$ W/m·K, $\nu = 19.71 \times 10^{-6}$ m²/s, $Pr = 0.702$

ANALYSIS

The assumed surface temperature given above is an estimate of the average surface temperature of the first five devices. As the boundary layer develops downstream from the leading edge of the first device, the heat transfer coefficient, and hence the surface temperature, will change. Based on the assumed surface temperature, the Reynolds number for the trailing edge of the fifth device is

$$Re_L = \frac{u_\infty L}{\nu}$$

$$Re_L = \frac{(7 \text{ m/s})(5 \times 0.025 \text{ m})}{19.71 \times 10^{-6} \text{ m}^2/\text{s}}$$

$$= 4.439 \times 10^4$$

Because $Re_L < 5 \times 10^5$, the boundary layer is laminar for the first five devices. We base the remaining calculations on the center of each device. The local Reynolds and Nusselt numbers for the first five devices are given by the relations

$$Re_{x,n} = \frac{u_\infty x_n}{\nu}$$

$$Nu_{x,n} = \frac{h_{x,n} x_n}{k} = 0.332 Re_{x,n}^{1/2} Pr^{1/3}$$

where n is the number of each device ($n = 1, 2, 3,..., n$), and x_n is the location of the center of each device,

$$x_n = l(n - 1/2)$$

and l = 0.025 m, the length of each device. Once the heat transfer coefficient for each device is known, the surface temperatures may be determined from Newton's law of cooling

$$q = h_{x,n} A(T_{s,n} - T_\infty)$$

where q = 0.300 W and A = (0.025 m)^2. The results for the first five devices are summarized in the table below.

n	x (cm)	Re_x	Nu_x	h_x (W/m^2·K)	T_s (°C)
1	1.25	4439	19.7	45.8	65.5
2	3.75	1.33×10^4	34.1	26.4	73.2
3	6.25	2.22×10^4	44.0	20.5	78.5
4	8.75	3.11×10^4	52.0	17.3	82.7
5	11.25	4.00×10^4	59.0	15.3	86.5

DISCUSSION

The laminar boundary layer thickness increases with x, the distance from the leading edge of the circuit board, which means that the heat transfer coefficient decreases with x. Hence, the surface temperature of the devices increases with x, as shown in the table. Many solid state devices have maximum allowable case temperatures of 125°C. If we let T_s = 125°C and solve for n, we see that the circuit board could be populated with 22 devices in the flow direction without exceeding this temperature, and the boundary layer would still be laminar ($Re_x = 1.98 \times 10^5$).

The accuracy of the results could be improved somewhat by basing the thermal properties of air on the film temperature of each device rather than an average film temperature for all devices.

Assumption 2 suggests a continuous boundary layer along the entire circuit board, but if there are even small spaces between the devices, a new boundary layer would likely develop at the leading edge of each device, yielding about the same heat transfer coefficient for all devices.

PROBLEM 7.7 **Temperature of a Base Plate Common to Heated and Unheated Electronic Compartments**

An electronic communications system consists of two adjacent compartments, as shown in the diagram. The compartments share a common base plate that functions as an air cooled heat sink. The compartment that experiences the air flow first contains passive devices that dissipate a negligible amount of heat. The other compartment contains devices that contribute to a total heat dissipation of 40 W, all of which transfers into the heat sink. Atmospheric air at 35°C is forced over the base plate at a velocity of 4 m/s. The compartments are the same size, and the overall size of the base plate is 36 cm × 36 cm. Find the average surface temperature of the base plate.

DIAGRAM

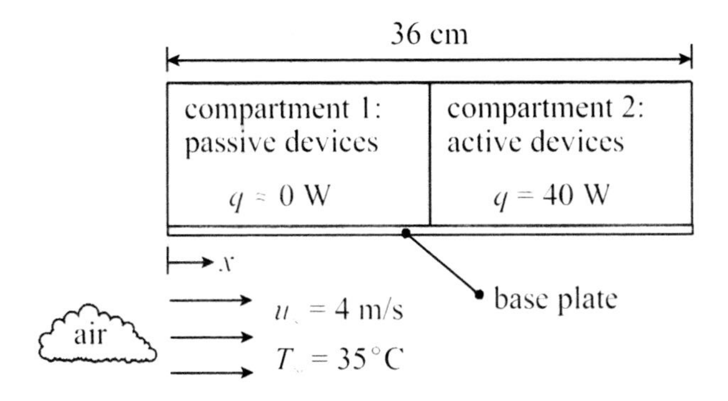

ASSUMPTIONS

1. Steady conditions.
2. All the heat dissipated in compartment 2 is transferred into the base plate.
3. In-plane conduction in the base plate from compartment 2 to compartment 1 is neglected.
4. The base plate for compartment 2 is subjected to a uniform heat flux.
5. Radiation is neglected.
6. Constant properties.

PROPERTIES

Assume $T_s = 150°C$. $T_f = (T_s + T_\infty)/2 = (150°C + 35°C)/2 = 92.5°C = 366$ K

atmospheric air: $k = 0.0312$ W/m·K, $\nu = 22.68 \times 10^{-6}$ m²/s, $Pr = 0.697$

ANALYSIS

The Reynolds number is

$$Re_L = \frac{u_\infty L}{\nu}$$

$$Re_L = \frac{(4 \text{ m/s})(0.36 \text{ m})}{22.68 \times 10^{-6} \text{ m}^2\text{/s}}$$

$$= 6.349 \times 10^4$$

Because $Re_L < 5 \times 10^5$, the flow is laminar over the entire base plate. The average Nusselt number for this configuration is

$$Nu_L = \frac{2[1-(\xi/L)^{3/4}]}{1-\xi/L} Nu_x$$

where ξ is the unheated starting length of compartment 1and Nu_x is the local Nusselt number at $x = L$ measured from the leading edge of the base plate,

$$Nu_x = 0.332 Re_x^{1/2} Pr^{1/3}$$

$$Nu_x = 0.332(6.349 \times 10^4)^{1/2}(0.697)^{1/3}$$

$$= 74.17$$

The compartments are the same size, so the average Nusselt number is

$$Nu_L = \frac{2[1 - (0.18 \text{ m}/0.36 \text{ m})^{3/4}]}{1 - (0.18 \text{ m}/0.36 \text{ m})} (74.17)$$

$$= 120.3$$

so the average heat transfer coefficient is

$$h = \frac{Nu_L k}{L}$$

$$h = \frac{(120.3)(0.0312 \text{ W/m·K})}{0.36 \text{ m}}$$

$$= 10.4 \text{ W/m}^2\text{·K}$$

Knowing the average heat transfer coefficient and the heat dissipation, we may use Newton's law of cooling to find the surface temperature of the base plate.

$$q = hA(T_s - T_\infty)$$

Solving for T_s,

$$T_s = \frac{(40\ \text{W})}{(10.4\ \text{W/m}^2\cdot\text{K})(0.18\ \text{m})^2} + 35°\text{C}$$

$$= \underline{\underline{154°\text{C}}}$$

DISCUSSION

The electronic devices in compartment 2 are attached in some way to the base plate, which the analysis shows has an average temperature of 154°C. Heat dissipating devices always have a higher temperature than the heat sink to which they are attached, so the case temperatures of the devices must be greater than 154°C. Most electronic devices cannot operate at this temperature, so the analysis suggests that something in the cooling system must be changed. One solution is to switch the direction of the air flow so that compartment 2 experiences the air flow first. With this approach, the portion of the base plate for compartment 1can be ignored. The boundary layer thickness will be smaller, so the average heat transfer coefficient will be larger, thereby giving a lower surface temperature. Noting that the Reynolds number is now based on half the base plate length, we have the following results:

$$Nu_L = \frac{hL}{k} = 0.664 Re_L^{1/2} Pr^{1/3}$$

$$Nu_L = (0.664)(3.175 \times 10^4)(0.697)^{1/3}$$

$$= 104.9$$

$$h = \frac{(104.9)(0.0312\ \text{W/m}\cdot\text{K})}{0.18\ \text{m}}$$

$$= 18.2\ \text{W/m}^2\cdot\text{K}$$

$$T_s = \frac{(40\ \text{W})}{(18.2\ \text{W/m}^2\cdot\text{K})(0.18\ \text{m})^2} + 35°\text{C}$$

$$= 103°\text{C}$$

Many electronic devices can operate reliably with this heat sink temperature.

❒ ❒ ❒

PROBLEM 7.8 Cooling Load for a Refrigerated Truck

A refrigerated truck transporting food products travels at 28 m/s during a hot afternoon when the outdoor air temperature is 36°C. The right, top and left sides of the box-shaped refrigerated cargo compartment measure 2.3 m wide × 8.6 m long and consist of a 2.5-cm thick layer of polyurethane foam insulation sandwiched between two thin stainless steel panels. To prevent product spoilage, the inside surface temperature of the cargo compartment must be maintained at 4°C. Neglecting heat transfer through the bottom and ends of the cargo compartment, how many tons of refrigeration are required to offset the heat gain to the compartment? One ton of refrigeration equals 12,000 Btu/h .

DIAGRAM

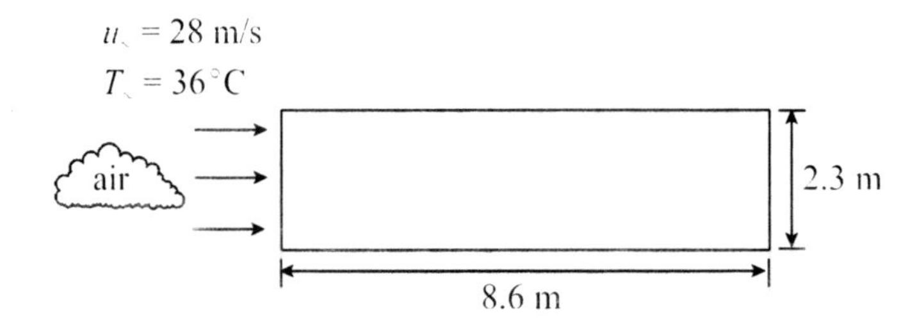

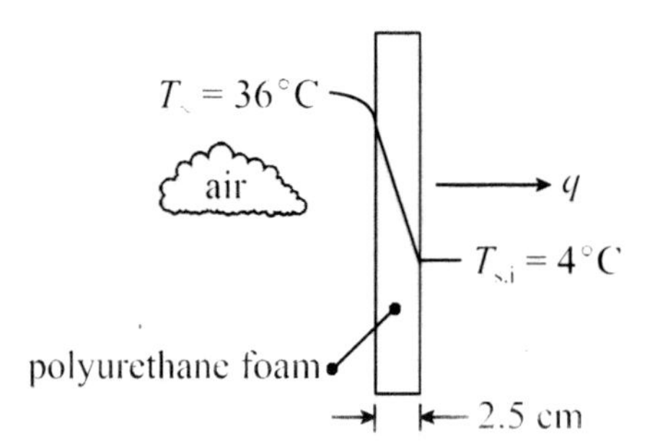

ASSUMPTIONS

1. Steady conditions.
2. Steady one-dimensional conduction.
3. Thermal resistance of the stainless steel panels is neglected.
4. Thermal contact resistance is neglected.
5. Outside surfaces of the cargo compartment are flat and smooth.
6. Radiation, including solar, is neglected.
7. Constant properties.

PROPERTIES

polyurethane foam: $k_p = 0.026$ W/m·K

Assume $T_{s,o} = 34°C$. $T_f = (T_{s,o} + T_\infty)/2 = (34°C + 36°C)/2 = 35°C = 308$ K

atmospheric air: $k = 0.0269$ W/m·K, $\nu = 16.69 \times 10^{-6}$ m²/s, $Pr = 0.706$

ANALYSIS

The Reynolds number is

$$Re_L = \frac{u_\infty L}{\nu}$$

$$Re_L = \frac{(28 \text{ m/s})(8.6 \text{ m})}{16.69 \times 10^{-6} \text{ m}^2/\text{s}}$$

$$= 1.443 \times 10^7$$

which indicates a mixed flow condition. Thus, the average Nusselt number is

$$Nu_L = \frac{hL}{k} = (0.037 Re_L^{4/5} - 871) Pr^{1/3}$$

$$Nu_L = [0.037(1.443 \times 10^7)^{4/5} - 871](0.706)^{1/3}$$

$$= 1.681 \times 10^4$$

so the average heat transfer coefficient is

$$h = \frac{Nu_L k}{L}$$

$$h = \frac{(1.681 \times 10^4)(0.0269 \text{ W/m·K})}{8.6 \text{ m}}$$

$$= 52.6 \text{ W/m}^2\text{·K}$$

Neglecting the stainless steel panels and interfaces, the total thermal resistance includes the conduction resistance of the insulation plus the resistance of the outside boundary layer,

$$R_{tot} = \frac{\Delta x}{k_p A} + \frac{1}{hA}$$

where $A = 3(2.3 \text{ m} \times 8.6 \text{ m}) = 59.34 \text{ m}^2$, the total surface area of the top and sides of the cargo compartment, and Δx is insulation thickness. Thus,

$$R_{tot} = \frac{0.025 \text{ m}}{(0.026 \text{ W/m·K})(59.34 \text{ m}^2)} + \frac{1}{(52.6 \text{ W/m}^2\text{·K})(59.34 \text{ m}^2)}$$

$$= 0.01652 \text{ K/W}$$

The heat gain to the refrigerated compartment is

$$q = \frac{T_\infty - T_{s,i}}{R_{tot}}$$

$$q = \frac{(36 - 4)°\text{C}}{0.01652 \text{ K/W}}$$

$$= 1937 \text{ W}$$

which is equivalent to the required cooling load for the refrigeration system. Hence,

$$q_{cool} = 1937 \text{ W} \times \frac{3.4121 \text{ Btu/h}}{1 \text{ W}} \times \frac{1 \text{ ton}}{12{,}000 \text{ Btu/h}}$$

$$= \underline{\underline{0.55 \text{ ton}}}$$

DISCUSSION

It is instructive to examine the effect of truck speed on cooling load. The graph below shows that cooling load changes significantly for low speeds only.

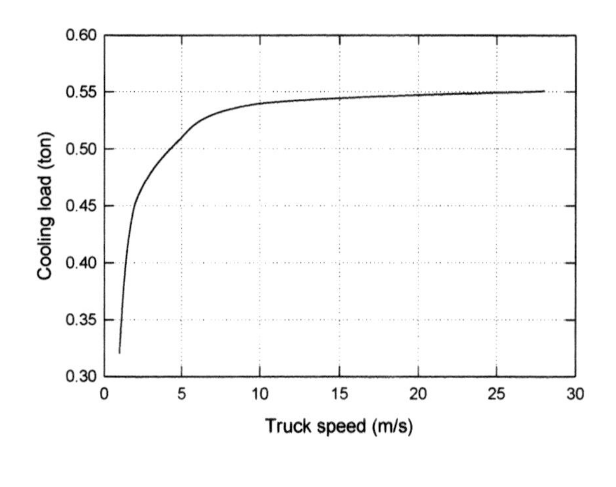

❐ ❐ ❐

PROBLEM 7.9 Temperatures of Cured Composite Plastic Parts on a Conveyor

A conveyor belt carries warm parts made of a composite plastic ($\rho = 1530$ kg/m^3, $c_p = 1260$ J/kg·K, $k = 0.85$ W/m·K) from a curing oven to a collection receptacle. The exit of the curing oven is 60 m away from the collection receptacle, and the conveyor belt travels at a constant speed of 0.3 m/s. The parts measure 6 cm × 6 cm × 8 mm thick and are spaced 2 cm apart as they lie flat on the conveyor belt such that their sides are parallel and perpendicular to their direction of travel. As the parts exit the curing oven, they have a uniform temperature of 75°C. If the temperature of the surrounding air is 20°C, estimate the surface temperatures of the parts as they enter the collection receptacle.

DIAGRAM

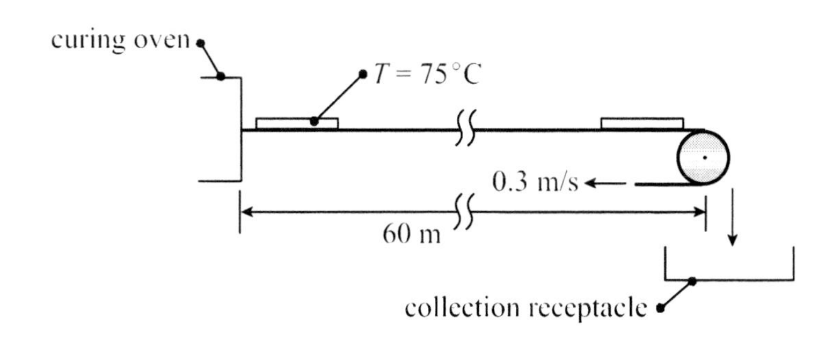

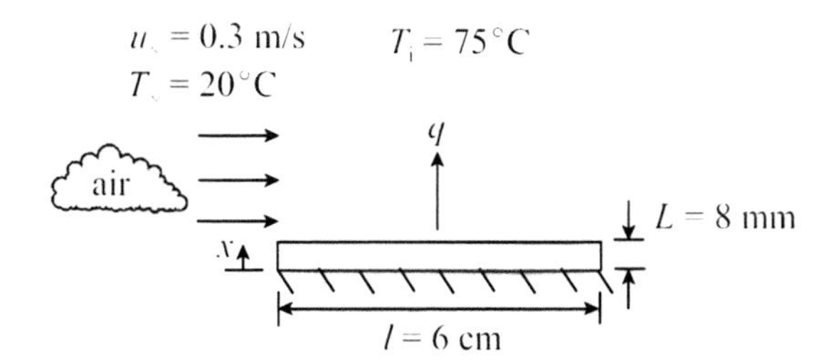

ASSUMPTIONS

1. The conveyor speed is constant.
2. Unsteady one-dimensional conduction.
3. The back side of the part is insulated.
4. Heat transfer from the edges of the part is neglected.
5. Radiation is neglected.
6. The top surface of each part is flat and smooth.
7. A new boundary layer begins at the leading edge of each part.
8. Constant properties.

PROPERTIES

composite plastic: $\rho = 1530\ \text{kg/m}^3$, $c_p = 1260\ \text{J/kg·K}$, $k = 0.45\ \text{W/m·K}$

Assume $T_s = 60°\text{C}$. $T_f = (T_s + T_\infty)/2 = (60°\text{C} + 20°\text{C})/2 = 40°\text{C} = 313\ \text{K}$

atmospheric air: $k = 0.0273\ \text{W/m·K}$, $\nu = 17.20 \times 10^{-6}\ \text{m}^2\text{/s}$, $Pr = 0.705$

ANALYSIS

The assumed surface temperature is an estimated average as the parts travel 60 m from the curing oven to the collection receptacle. The Reynolds number for each part is

$$Re_l = \frac{u_\infty l}{\nu}$$

$$Re_l = \frac{(0.3\ \text{m/s})(0.06\ \text{m})}{17.20 \times 10^{-6}\ \text{m}^2\text{/s}}$$

$$= 1047$$

The flow is laminar, so the average Nusselt number is

$$Nu_l = \frac{hl}{k} = 0.664 Re_l^{1/2} Pr^{1/3}$$

$$Nu_l = 0.664(1047)^{1/2}(0.705)^{1/3}$$

$$= 19.12$$

and the average heat transfer coefficient is

$$h = \frac{Nu_l k}{l}$$

$$h = \frac{(19.12)(0.0273\ \text{W/m·K})}{0.06\ \text{m}}$$

$$= 8.70\ \text{W/m}^2\text{·K}$$

Now that the heat transfer coefficient has been calculated, we can do an unsteady one-dimensional conduction analysis on the part. The thermal diffusivity of the plastic is

$$\alpha \equiv \frac{k}{\rho c_p}$$

$$\alpha = \frac{0.45\ \text{W/m·K}}{(1530\ \text{kg/m}^3)(1260\ \text{J/kg·K})}$$

$$= 2.334 \times 10^{-7}\ \text{m}^2\text{/s}$$

Treating each part as a large plate of half-thickness L, the Biot number is

$$Bi \equiv \frac{hL}{k}$$

$$Bi = \frac{(8.70\ \text{W/m}^2\text{·K})(0.008\ \text{m})}{0.45\ \text{W/m·K}}$$

$$= 0.155$$

For a large plate, with uniform initial temperature, that is suddenly exposed to a convective environment, the temperature may be found by a one-term approximation of an infinite series,

$$\frac{\theta}{\theta_i} = \frac{T(x,t) - T_\infty}{T_i - T_\infty} = C_1 \exp(-\zeta_1^2 Fo)\cos(\zeta_1 x/L)$$

where the parameters C_1 and ζ_1 are obtained from the Biot number Bi. These parameters have the values

$$C_1 = 1.0244 \quad , \quad \zeta_1 = 0.3834$$

The cooling time for each part is the length of the conveyor belt divided by its speed,

$$t = \frac{s}{u_\infty}$$

$$t = \frac{60\ \text{m}}{0.3\ \text{m/s}}$$

$$= 200\ \text{s}$$

Thus, the Fourier number is

$$Fo \equiv \frac{\alpha t}{L^2}$$

$$Fo = \frac{(2.334 \times 10^{-7}\ \text{m}^2\text{/s})(200\ \text{s})}{(0.008\ \text{m})^2}$$

$$= 0.7294$$

Substituting numerical values into the one-term approximation for $x = 0$, the back surface of the part,

$$T(0,t) = (75°C - 20°C)(1.0244)\exp[-(0.3834)^2(0.7294)]\cos(0) + 20°C$$

$$= \underline{\underline{70.6°C}}$$

and substituting numerical values into the one-term approximation for $x = L$, the top surface of the part,

$$T(L,t) = (75°C - 20°C)(1.0244)\exp[-(0.3834)^2(0.7294)]\cos(0.3834) + 20°C$$

$$= \underline{\underline{66.9°C}}$$

DISCUSSION

During 200 s of exposure to the 20°C air, the bottom surface of the part cools to 70.6°C, and the top surface of the part cools to 66.9°C. The part is moving so slowly that the convection could almost be characterized as natural convection rather than forced convection. This is why the heat transfer coefficient is low and why the part is still close to its cure temperature when it enters the collection receptacle.

The estimate of surface temperature for thermal properties was slightly low, so a correction would yield a modest gain in accuracy. Because convection is weak, the inclusion of radiation would yield significantly lower part temperatures. Also, depending upon the material from which the conveyor belt is made, conduction from the part into the belt could be significant.

❒ ❒ ❒

PROBLEM 7.10 Effect of Circuit Board Orientation on Device Temperatures in Forced Air Cooling

A 10 cm × 20 cm circuit board densely populated with devices on one side is to be cooled by forcing 40°C air over the board at a velocity of 3 m/s. The electronics packaging engineer has the option to orient the circuit board with either dimension parallel to the direction of air flow, as illustrated in the diagram. The total heat dissipated by the circuit board is 15 W. The maximum allowable case temperature, specified by the manufacturer of the devices, is 75°C. Find the average case temperature of the devices for both circuit board orientations, and discuss which orientation is preferable and why.

DIAGRAM

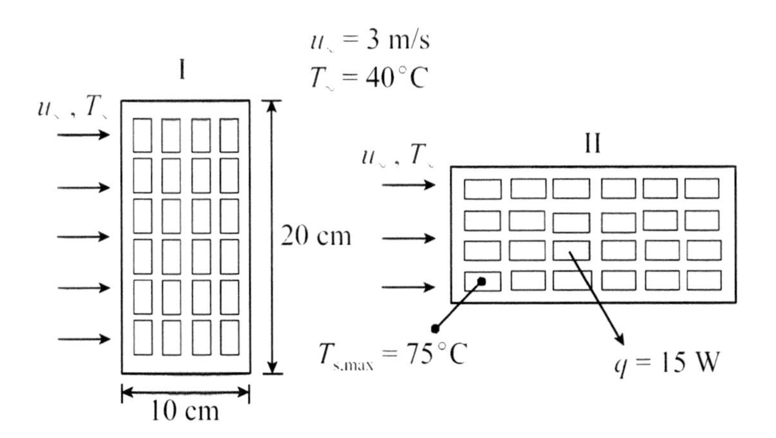

ASSUMPTIONS

1. Steady conditions.
2. Due to irregularities in the devices, turbulence exists over the entire circuit board.
3. The dissipated heat is spread uniformly over the circuit board.
4. The back side of the circuit board is insulated.
5. Heat transfer from the edges of the devices is neglected.
6. Radiation is neglected.
7. Constant properties.

PROPERTIES

Assume $T_s = 70°C$. $T_f = (T_s + T_\infty)/2 = (70°C + 40°C)/2 = 55°C = 328$ K

atmospheric air: $k = 0.0284$ W/m·K, $\nu = 18.71 \times 10^{-6}$ m^2/s, $Pr = 0.703$

ANALYSIS

Orientation I

The Reynolds number based on $L = 10$ cm is

$$Re_L = \frac{u_\infty L}{\nu}$$

$$Re_L = \frac{(3 \text{ m/s})(0.10 \text{ m})}{18.71 \times 10^{-6} \text{ m}^2\text{/s}}$$

$$= 1.603 \times 10^4$$

The average Nusselt number is

$$Nu_L = \frac{hL}{k} = 0.037\,Re_L^{4/5}\,Pr^{1/3}$$

$$Nu_L = 0.037(1.603 \times 10^4)^{4/5}(0.703)^{1/3}$$

$$= 76.06$$

so the average heat transfer coefficient is

$$h = \frac{Nu_L k}{L}$$

$$h = \frac{(76.06)(0.0284)}{0.10\text{ m}}$$

$$= 21.6\text{ W/m}^2\cdot\text{K}$$

From Newton's law of cooling,

$$q = hA(T_s - T_\infty)$$

the average surface temperature of the devices is

$$T_s = \frac{15\text{ W}}{(21.6\text{ W/m}^2\cdot\text{K})(0.1\text{ m} \times 0.2\text{ m})} + 40°\text{C}$$

$$= \underline{\underline{74.7°\text{C}}}$$

Orientation II

The Reynolds number based on $L = 20$ cm is

$$Re_L = \frac{u_\infty L}{\nu}$$

$$Re_L = \frac{(3\text{ m/s})(0.20\text{ m})}{18.71 \times 10^{-6}\text{ m}^2\text{/s}}$$

$$= 3.207 \times 10^4$$

The average Nusselt number is

$$Nu_L = \frac{hL}{k} = 0.037\,Re_L^{4/5}\,Pr^{1/3}$$

$$Nu_L = 0.037(3.207 \times 10^4)^{4/5}(0.703)^{1/3}$$

$$= 132.5$$

so the average heat transfer coefficient is

$$h = \frac{Nu_L k}{L}$$

$$h = \frac{(132.5)(0.0284)}{0.20 \text{ m}}$$

$$= 18.8 \text{ W/m}^2\cdot\text{K}$$

From Newton's law of cooling,

$$q = hA(T_s - T_\infty)$$

the average surface temperature of the devices is

$$T_s = \frac{15 \text{ W}}{(18.8 \text{ W/m}^2\cdot\text{K})(0.1 \text{ m} \times 0.2 \text{ m})} + 40°\text{C}$$

$$= \underline{\underline{79.9°\text{C}}}$$

DISCUSSION

The predicted case temperature for orientation I meets the manufacturer's specification, whereas the predicted case temperature for orientation II does not. Hence, orientation I is preferred. The reason that the surface temperature is lower for orientation I is because the average boundary layer thickness is smaller, thereby giving a higher average heat transfer coefficient. The difference between the two temperatures is only 5.2°C, but if the packaging engineer is faced with a critical thermal management problem in which a few degrees means a significant reduction in reliability, or even device failure, circuit board orientation could be important. Relaxing assumptions 4, 5 and 6 would most likely show that temperature specifications are indeed met for orientation II.

In this analysis, we found the average surface temperature of the devices, so our results do not represent a worst case. The devices with the highest case temperatures are those nearest the trailing edge of the circuit board. To find the case temperatures of these device, we would have to know individual heat dissipations and then calculate local heat transfer coefficients.

PROBLEM 7.11 Convective and Radiative Heat Transfer from a Long Cylinder in Cross Flow

A long cylinder experiences the cross flow of atmospheric air at 5°C. The velocity of the air upstream from the cylinder is 20 m/s, the diameter of the cylinder is 12 cm, and the emissivity of the cylinder's surface is 0.9. The cylinder is surrounded by a large enclosure at 10°C, and the surface temperature of the cylinder is 360°C. Find the convective and radiative heat transfer from the cylinder per unit length.

DIAGRAM

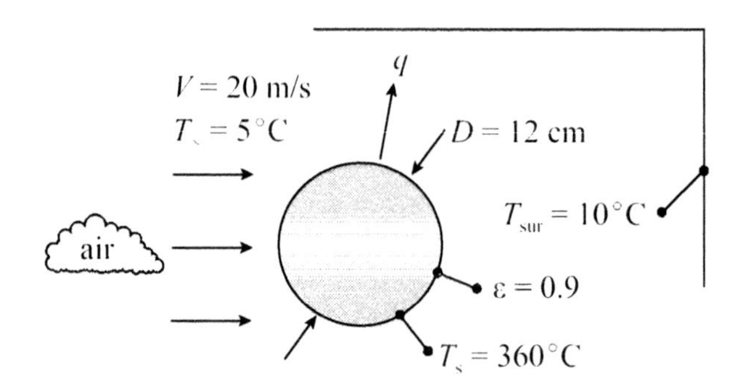

ASSUMPTIONS

1. Steady conditions.
2. Flow is normal to the axis of the cylinder.
3. The surface of the cylinder is smooth.
4. The surface of the cylinder is isothermal.
5. Heat transfer from the ends of the cylinder is neglected.
6. Constant properties.

PROPERTIES

$T_f = (T_s + T_\infty)/2 = (360°C + 5°C)/2 = 182.5°C = 456$ K

atmospheric air: $k = 0.0377$ W/m·K, $\nu = 33.16 \times 10^{-6}$ m^2/s, $Pr = 0.686$

ANALYSIS

The Reynolds number is

$$Re_D = \frac{VD}{\nu}$$

$$Re_D = \frac{(20 \text{ m/s})(0.12 \text{ m})}{33.16 \times 10^{-6} \text{ m}^2\text{/s}}$$

$$= 7.238 \times 10^4$$

and the average Nusselt number may be calculated using the empirical correlation

$$Nu_D = \frac{hD}{k} = CRe_D^m Pr^{1/3}$$

where the constants C and m depend on the Reynolds number Re_D. For $Re_D = 7.238 \times 10^4$, these constants have the values $C = 0.027$ and $m = 0.805$. Thus,

$$Nu_D = (0.027)(7.238 \times 10^4)^{0.805}(0.686)^{1/3}$$

$$= 194.4$$

The average heat transfer coefficient is

$$h = \frac{Nu_D k}{D}$$

$$h = \frac{(194.4)(0.0377 \text{ W/m}\cdot\text{K})}{0.12 \text{ m}}$$

$$= 61.1 \text{ W/m}^2\cdot\text{K}$$

The convective heat transfer per unit length is

$$q'_{conv} = q_{conv}/L = h\pi D(T_s - T_\infty)$$

$$q'_{conv} = (61.1 \text{ W/m}^2\cdot\text{K})\pi(0.12 \text{ m})(360 - 5)°\text{C}$$

$$= \underline{\underline{8177 \text{ W/m}}}$$

and the radiative heat transfer per unit length is

$$q'_{rad} = q_{rad}/L = \varepsilon\sigma\pi D(T_s^4 - T_{sur}^4)$$

$$q'_{rad} = (0.9)(5.669 \times 10^{-8} \text{ W/m}^2\cdot\text{K})\pi(0.12 \text{ m})[(633 \text{ K})^4 - (283 \text{ K})^4]$$

$$= \underline{\underline{2965 \text{ W/m}}}$$

Thus, the total heat transfer per unit length is

$$q' = q'_{conv} + q'_{rad}$$

$$q' = (8177 + 2965)\text{W/m}$$

$$= 1.114 \times 10^4 \text{ W/m}$$

DISCUSSION

Besides the one used here, there are other empirical correlations for finding the Nusselt number for a cylinder in cross flow. A correlation that satisfies $Re_D\, Pr > 0.2$ is

$$Nu_D = 0.3 + \frac{0.62 Re_D^{1/2} Pr^{1/3}}{[1+(0.4/Pr)^{2/3}]^{1/4}}\left[1+\left(\frac{Re_D}{282{,}000}\right)^{5/8}\right]^{4/5}$$

which yields $Nu_D = 171.6$ and $h = 53.9$ W/m^2·K. The resulting convective heat transfer per unit length is $q'_{conv} = 7214$ W/m. Any empirical correlation in convective heat transfer is subject to experimental uncertainties, typically around 15 to 20 percent, and should therefore not be regarded as highly accurate.

❐ ❐ ❐

PROBLEM 7.12 Temperature of a Current-Carrying Wire in Cross Flow

While carrying 10.5 A of electrical current, a long 30-gage aluminum wire ($D = 0.2546$ mm) is exposed to the cross flow of -40°C atmospheric nitrogen. The velocity of the nitrogen upstream from the wire is 16 m/s. If the electrical resistance of the wire is 0.555 Ω/m, what is the temperature of the wire?

DIAGRAM

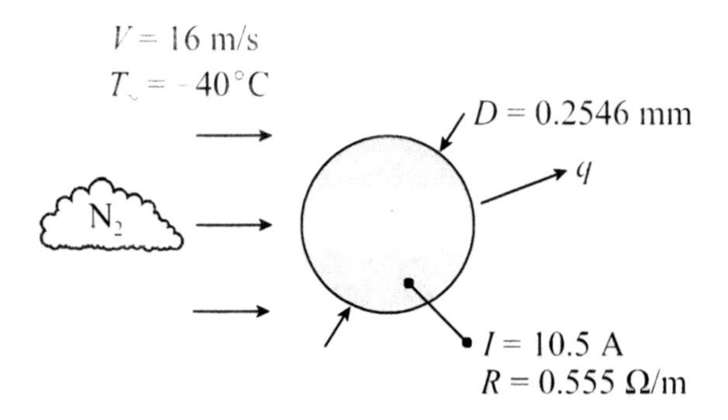

ASSUMPTIONS

1. Steady conditions.
2. Flow is normal to the axis of the wire.
3. The surface of the wire is smooth.
4. The wire is isothermal.
5. Radiation is neglected.
6. Constant properties.

PROPERTIES

30-gage aluminum wire: $R = 0.555\ \Omega/\text{m}$

Assume $T_s = 50°\text{C}$. $T_f = (T_s + T_\infty)/2 = (50°\text{C} - 40°\text{C})/2 = 5°\text{C} = 278\ \text{K}$

atmospheric nitrogen: $k = 0.0243\ \text{W/m·K}$, $\nu = 13.93 \times 10^{-6}\ \text{m}^2/\text{s}$, $Pr = 0.721$

ANALYSIS

The Reynolds number is

$$Re_D = \frac{VD}{\nu}$$

$$Re_D = \frac{(16\ \text{m/s})(0.2546 \times 10^{-3}\ \text{m})}{13.93 \times 10^{-6}\ \text{m}^2/\text{s}}$$

$$= 292.4$$

and the average Nusselt number may be calculated using the empirical correlation

$$Nu_D = \frac{hD}{k} = CRe_D^m Pr^{1/3}$$

where the constants C and m depend on the Reynolds number Re_D. For $Re_D = 292.4$, these constants have the values $C = 0.683$ and $m = 0.466$. Thus,

$$Nu_D = (0.683)(292.4)^{0.466}(0.721)^{1/3}$$

$$= 8.63$$

The average heat transfer coefficient is

$$h = \frac{Nu_D k}{D}$$

$$h = \frac{(8.63)(0.0243\ \mathrm{W/m\cdot K})}{0.2546 \times 10^{-3}\ \mathrm{m}}$$

$$= 824\ \mathrm{W/m^2\cdot K}$$

The rate of electrical energy generation within the wire equals the convective heat transfer from the wire's surface. Thus,

$$q'_{conv} = q_{conv}/L = I^2R = h\pi D(T_s - T_\infty)$$

Solving for T_s, the surface temperature of the wire,

$$T_s = \frac{(10.5\ \mathrm{A})^2(0.555\ \Omega/\mathrm{m})}{(824\ \mathrm{W/m^2\cdot K})\pi(0.2546 \times 10^{-3}\ \mathrm{m})} - 40°\mathrm{C}$$

$$= \underline{\underline{52.8°\mathrm{C}}}$$

Due to the small diameter of the wire, the wire is virtually isothermal.

DISCUSSION

The convective heat transfer per unit length is q'_{conv} = 61.2 W/m, which is equivalent to the rate of electrical energy generation. The predicted wire temperature is close to the assumed value, so a correction of thermal properties is not required. A wire's current-carrying capacity is ultimately limited by its melting point, although a wire is typically not operated near this condition. The graph below shows the temperature of the wire as a function of current. If the wire is constructed of pure aluminum, it melts at about 660°C, which corresponds to a current of about 29 A.

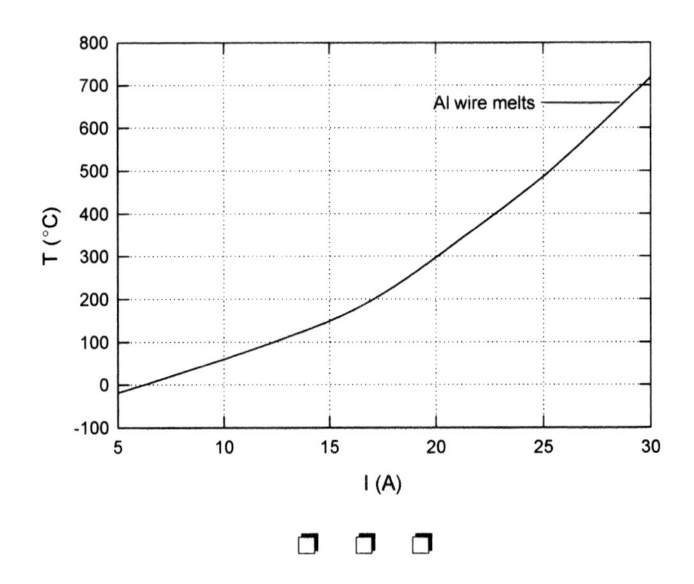

❒ ❒ ❒

PROBLEM 7.13 Heat Loss from an Insulated Steam Pipe in Cross Flow

An long steel pipe with magnesia insulation carries 130°C steam across a room where the air temperature is 20°C. The outside radius of the pipe is 6 cm, and the insulation thickness is 2.5 cm. Ventilation equipment in the room induces air to flow across the pipe at a velocity of 1.8 m/s. Find the heat loss per unit length of pipe. Assume that the heat transfer coefficient for the steam is large and that the pipe wall is thin.

DIAGRAM

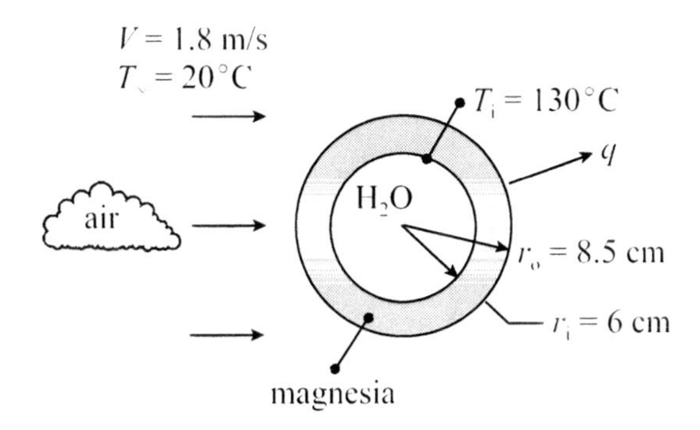

ASSUMPTIONS

1. Steady conditions.
2. Air flow is normal to the axis of the pipe.
3. The heat transfer coefficient for the steam is large.
4. Thermal resistance of the pipe wall is neglected.
5. The outside surface of the insulation is smooth and isothermal.
6. Radiation is neglected.
7. Thermal contact resistance is neglected.
8. Constant properties.

PROPERTIES

magnesia insulation: $k_m = 0.051$ W/m·K

Assume $T_o = 25$°C. $T_f = (T_o + T_\infty)/2 = (25°\text{C} + 20°\text{C})/2 = 22.5°\text{C} = 296$ K

atmospheric air: $k = 0.0260$ W/m·K, $\nu = 15.53 \times 10^{-6}$ m²/s, $Pr = 0.708$

ANALYSIS

The Reynolds number is

$$Re_D = \frac{VD}{\nu}$$

$$Re_D = \frac{(1.8 \text{ m/s})(0.17 \text{ m})}{15.53 \times 10^{-6} \text{ m}^2\text{/s}}$$

$$= 1.970 \times 10^4$$

and the average Nusselt number may be found using the empirical correlation

$$Nu_D = \frac{hD}{k} = CRe_D^m Pr^{1/3}$$

where the constants C and m depend on the Reynolds number Re_D. For $Re_D = 1.970 \times 10^4$, these constants have the values $C = 0.193$ and $m = 0.618$. Thus,

$$Nu_D = (0.193)(1.970 \times 10^4)^{0.618}(0.708)^{1/3}$$

$$= 77.54$$

The average heat transfer coefficient is

$$h = \frac{Nu_D k}{D}$$

$$h = \frac{(77.54)(0.0260 \text{ W/m·K})}{0.17 \text{ m}}$$

$$= 11.9 \text{ W/m}^2\text{·K}$$

Neglecting thermal resistance of the pipe wall and the interface, and assuming that the heat transfer coefficient for the steam is large, the total thermal resistance includes the conduction resistance of the insulation plus the resistance of the outside boundary layer. For a unit length of pipe,

$$R_{tot} = \frac{\ln(r_o/r_i)}{2\pi L k_m} + \frac{1}{2\pi r_o L h}$$

$$R_{tot} = \frac{\ln(0.085 \text{ m}/0.06 \text{ m})}{2\pi(1 \text{ m})(0.051 \text{ W/m·K})} + \frac{1}{2\pi(0.085 \text{ m})(1 \text{ m})(11.9 \text{ W/m}^2\text{·K})}$$

$$= 1.244 \text{ K/W}$$

The heat loss is

$$q = \frac{T_i - T_\infty}{R_{tot}}$$

$$q = \frac{(130 - 20)°C}{1.244 \text{ K/W}}$$

$$= \underline{\underline{88.4 \text{ W}}}$$

where a unit length of pipe, $L = 1$ m, is implied in the answer.

DISCUSSION

This result is the convective heat loss only and does not accurately reflect the outside surface temperature of the insulation. Because the heat transfer coefficient is low, radiation is probably significant and should not be neglected. The total heat loss is

$$q = q_{conv} + q_{rad} = 2\pi r_o Lh(T_o - T_\infty) + 2\pi r_o L\varepsilon\sigma(T_o^4 - T_{sur}^4)$$

and conduction through the insulation is

$$q_{cond} = \frac{T_i - T_o}{\left(\dfrac{\ln(r_o/r_i)}{2\pi L k_m}\right)}$$

For the radiation calculation, we assume $\varepsilon = 0.5$ and $T_{sur} = T_\infty$. Heat conducted through the insulation is equivalent to the heat loss from the surface by convection and radiation. Equating the two relations above, and numerically solving for the heat transfers and surface temperature,

$$q_{conv} = 72.4 \text{ W}, \; q_{rad} = 18.3 \text{ W}, \; T_o = 31.3°C.$$

Hence, the total heat loss per unit pipe length is 90.7 W, and radiation accounts for 20 percent, a significant amount. Most pipe insulation surfaces have a low emissivity to reduce radiative heat loss. Of course, a thicker layer of insulation can always be used, but there is an economic tradeoff between the energy cost of keeping the steam hot (or from condensing) and the initial cost of pipe insulation. It is left as an exercise to determine the total heat loss if the pipe is bare. The magnitude of the heat loss for this case will clearly show why steam pipes should be insulated.

The predicted surface temperature of the insulation is slightly higher than the assumed value, but little accuracy would be gained by making a correction of thermal properties.

❐ ❐ ❐

PROBLEM 7.14 Cooling Time for an Aluminum Extrusion in Cross Flow

In an aluminum manufacturing facility, 3-m long rods of pure aluminum with a 5 cm × 5 cm square cross section are extruded at a temperature of 550°C. The extrusions are immediately exposed to the cross flow of 30°C atmospheric air at a velocity of 6 m/s, where the flow is normal to one of the flat sides of the extrusion. Estimate the time required for the extrusion to cool to a safe handling temperature of 60°C.

DIAGRAM

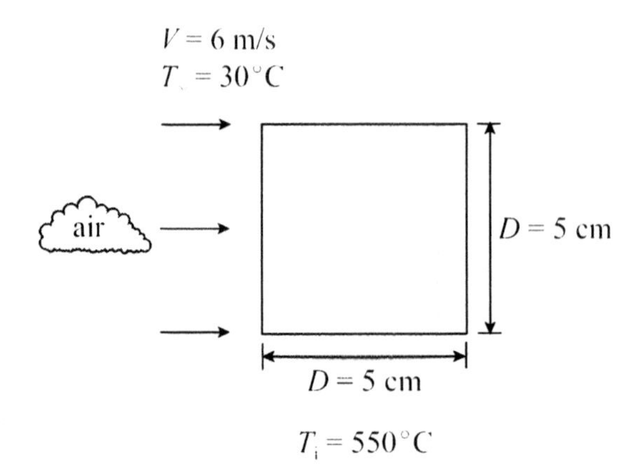

ASSUMPTIONS

1. Steady conditions.
2. The extrusion is surrounded by the cooling air.
3. The extrusion has a uniform initial temperature.
4. Air flow is normal to the axis of the extrusion and to one of the flat sides.
5. The surface of the extrusion is smooth and isothermal.
6. Heat transfer from the ends of the extrusion is neglected.
7. Radiation is neglected.
8. Constant properties.

PROPERTIES

pure aluminum: $\rho = 2702\ \text{kg/m}^3$, $c_p = 903\ \text{J/kg·K}$, $k_e = 237\ \text{W/m·K}$

Assume $T_s = 300°\text{C}$. $T_f = (T_s + T_\infty)/2 = (300°\text{C} + 30°\text{C})/2 = 165°\text{C} = 438\ \text{K}$

atmospheric air: $k = 0.0365\ \text{W/m·K}$, $\nu = 30.95 \times 10^{-6}\ \text{m}^2/\text{s}$, $Pr = 0.687$

ANALYSIS

To simplify the analysis, we assume a surface temperature that is roughly the average of the initial temperature of the extrusion and the handling temperature. The Reynolds number based on the diameter of the square cylinder is

$$Re_D = \frac{VD}{\nu}$$

$$Re_D = \frac{(6\text{ m/s})(0.05\text{ m})}{30.95 \times 10^{-6}\text{ m}^2\text{/s}}$$

$$= 9693$$

and the average Nusselt number may be found using the empirical correlation

$$Nu_D = \frac{hD}{k} = CRe_D^m Pr^{1/3}$$

where the constants C and m for a square cylinder are $C = 0.102$ and $m = 0.675$. Thus,

$$Nu_D = (0.102)(9693)^{0.675}(0.687)^{1/3}$$

$$= 44.17$$

The average heat transfer coefficient is

$$h = \frac{Nu_D k}{D}$$

$$h = \frac{(44.17)(0.0365\text{ W/m}\cdot\text{K})}{0.05\text{ m}}$$

$$= 32.2\text{ W/m}^2\cdot\text{K}$$

Next, we determine if the extrusion is lumped or if spatial temperature effects must be included. Because a long square cylinder is formed by the orthogonal intersection of two large plates of half-thickness L, the Biot number is

$$Bi \equiv \frac{hL}{k_e}$$

$$Bi = \frac{(32.2\text{ W/m}^2\cdot\text{K})(0.025\text{ m})}{237\text{ W/m}\cdot\text{K}}$$

$$= 3.40 \times 10^{-3} < 0.1$$

Thus, the extrusion is lumped. The temperature history of the extrusion is given by

$$\frac{\theta}{\theta_i} = \frac{T - T_\infty}{T_i - T_\infty} = \exp\left[-\left(\frac{hA}{\rho V c_p}\right)t\right]$$

where T is temperature at time t, A is surface area and V is volume. Noting that for a square cylinder, $A/V = 4/D$, and solving for t,

$$t = -\frac{\rho D c_p}{4h} \ln\left(\frac{T - T_\infty}{T_i - T_\infty}\right)$$

$$t = -\frac{(2702 \text{ kg/m}^3)(0.05 \text{ m})(903 \text{ J/kg·K})}{4(32.2 \text{ W/m}^2\text{·K})} \ln\left(\frac{60°\text{C} - 30°\text{C}}{550°\text{C} - 30°\text{C}}\right)$$

$= 2702 \text{ s} = \underline{\underline{45.0 \text{ min}}}$

DISCUSSION

The inclusion of radiation would decrease the cooling time. Furthermore, unless the extrusion is suspended in the cooling air by thin cords, conduction into a table, conveyor, or some other supporting structure, would also decrease the cooling time.

❐ ❐ ❐

PROBLEM 7.15 Wind Chill

A typical adult human loses 200 W/m^2 of heat from the surface of his body while engaged in brisk walking. Approximating an adult human as a cylinder with a length and diameter of 1.8 m and 45 cm, respectively, estimate the body surface temperature of a person exposed to a 4 m/s cross wind of 0°C air. Repeat the calculation for wind speeds up to 30 m/s, and plot the results. Explain the difference in the predicted surface temperatures in terms of the wind chill effect.

DIAGRAM

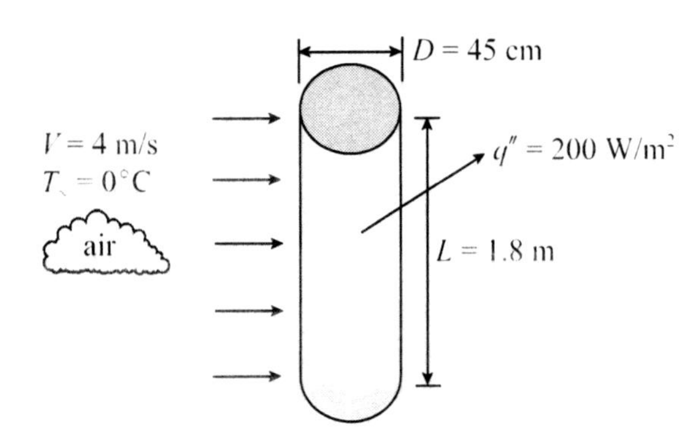

ASSUMPTIONS

1. Steady conditions.
2. An adult human is approximated as a cylinder.
3. Air flow is normal to the axis of the cylinder.
4. The surface of the cylinder is smooth and isothermal.
5. Heat transfer from the ends of the cylinder is neglected.
6. Radiation is neglected.
7. Constant properties.

PROPERTIES

Assume $T_s = 15°C$. $T_f = (T_s + T_\infty)/2 = (15°C + 0°C)/2 = 7.5°C = 281$ K

atmospheric air: $k = 0.0248$ W/m·K, $\nu = 14.20 \times 10^{-6}$ m²/s, $Pr = 0.712$

ANALYSIS

The Reynolds number is

$$Re_D = \frac{VD}{\nu}$$

$$Re_D = \frac{(4 \text{ m/s})(0.45 \text{ m})}{14.20 \times 10^{-6} \text{ m}^2/\text{s}}$$

$$= 1.268 \times 10^5$$

and the average Nusselt number may be found using the empirical correlation

$$Nu_D = 0.3 + \frac{0.62 Re_D^{1/2} Pr^{1/3}}{\left[1 + (0.4/Pr)^{2/3}\right]^{1/4}} \left[1 + \left(\frac{Re_D}{282{,}000}\right)^{5/8}\right]^{4/5}$$

$$Nu_D = 0.3 + \frac{0.62(1.268 \times 10^5)^{1/2}(0.712)^{1/3}}{[1 + (0.4/0.712)^{2/3}]^{1/4}} [1 + (1.268 \times 10^5/282{,}000)^{5/8}]^{4/5}$$

$$= 253.3$$

The average heat transfer coefficient is

$$h = \frac{Nu_D k}{D}$$

$$h = \frac{(253.3)(0.0248 \text{ W/m·K})}{0.45 \text{ m}}$$

$$= 14.0 \text{ W/m}^2\text{·K}$$

The convective heat transfer from the body is given by Newton's law of cooling,

$$q'' = h(T_s - T_\infty)$$

Solving for T_s, the surface temperature of the body,

$$T_s = \frac{200 \text{ W/m}^2}{14.0 \text{ W/m}^2\text{·K}} + 0°C$$

$= \underline{\underline{14.3°C}}$

Proceeding in the same manner for other wind speeds, we plot the surface temperature of the body as a function of wind speed in the graph below.

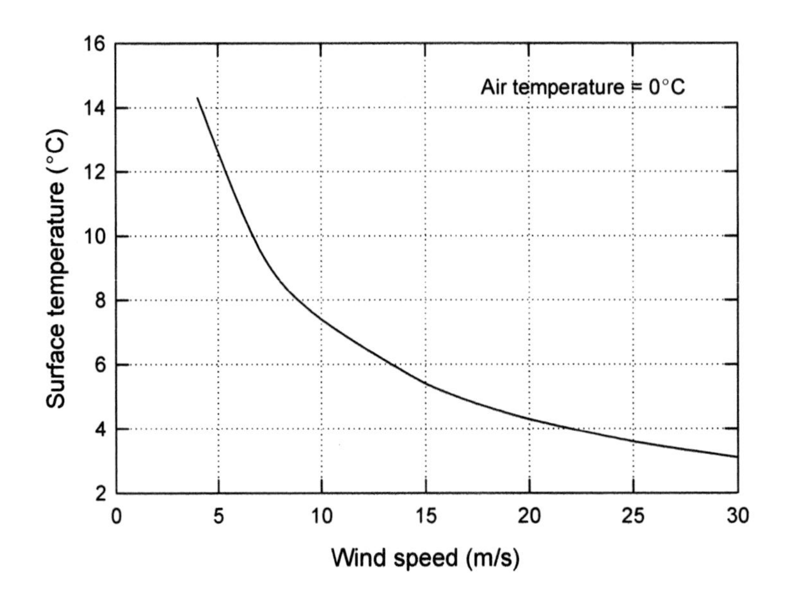

DISCUSSION

Wind chill is the suppression of the surface temperature of a human being due to the movement of cold air. As shown in the graph, the surface (skin) temperature decreases with increasing wind speed, consistent with experience. As wind speed increases, surface temperature asymptotically approaches the free stream air temperature because the heat transfer coefficient increases with velocity and the surface heat flux is fixed.

❒ ❒ ❒

PROBLEM 7.16 **Surface Temperature of an Electronic Device with a Pin Fin**

To minimize the temperature of a small electronic device, a copper pin fin is attached to its surface. The length and diameter of the pin fin are 12 mm and 3 mm, respectively. Atmospheric air at 25°C flows across the fin at a velocity of 10 m/s. If the devices dissipates 1.25 W, what is its surface temperature?

DIAGRAM

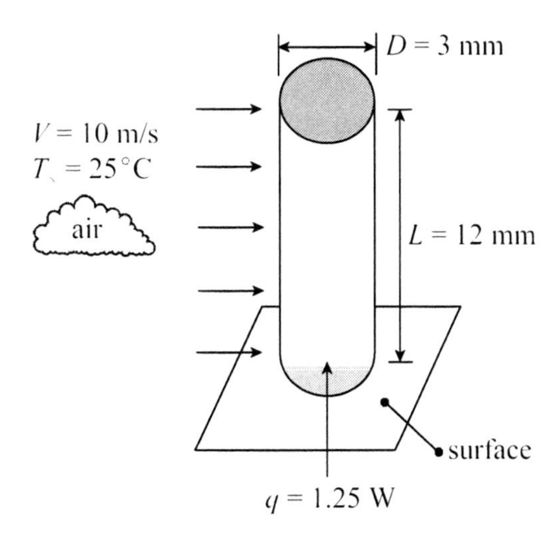

ASSUMPTIONS

1. Steady conditions.
2. All dissipated heat transfers into the fin.
3. Air flow is normal to the axis of the fin.
4. The tip of the fin is insulated.
5. The surface of the fin is smooth.
6. Radiation is neglected.
7. Thermal contact resistance at the attachment point is neglected.
8. Constant properties.

PROPERTIES

pure copper: $\rho = 8933\ \text{kg/m}^3$, $c_p = 385\ \text{J/kg·K}$, $k_f = 401\ \text{W/m·K}$

Assume $T_s = 40°\text{C}$. $T_f = (T_s + T_\infty)/2 = (40°\text{C} + 25°\text{C})/2 = 32.5°\text{C} = 306\ \text{K}$

atmospheric air: $k = 0.0267\ \text{W/m·K}$, $\nu = 16.49 \times 10^{-6}\ \text{m}^2\text{/s}$, $Pr = 0.706$

ANALYSIS

The Reynolds number is

$$Re_D = \frac{VD}{\nu}$$

$$Re_D = \frac{(10\ \text{m/s})(0.003\ \text{m})}{16.49 \times 10^{-6}\ \text{m}^2\text{/s}}$$

$= 1819$

and the average Nusselt number may be found using the empirical correlation

$$Nu_D = \frac{hD}{k} = CRe_D^m Pr^{1/3}$$

where the constants C and m depend on the Reynolds number. For $Re_D = 1819$, these constants have the values $C = 0.683$ and $m = 0.466$. Thus,

$Nu_D = (0.683)(1819)^{0.466}(0.706)^{1/3}$

$= 20.10$

The average heat transfer coefficient is

$$h = \frac{Nu_D k}{D}$$

$h = \frac{(20.10)(0.0267\ \text{W/m}\cdot\text{K})}{0.003\ \text{m}}$

$= 178.9\ \text{W/m}^2\cdot\text{K}$

Now that the heat transfer coefficient has been calculated, we can do a fin analysis. The heat transfer through the base of the fin is

$$q = (hPk_f A_c)^{1/2} \theta_b \tanh(mL)$$

where $P = \pi D$, $A_c = \pi D^2/4$, $\theta = T_b - T_\infty$ and

$$m = \left(\frac{hP}{k_f A_c}\right)^{1/2}$$

These parameters have the values

$P = \pi(0.003\ \text{m})$

$= 9.245 \times 10^{-3}\ \text{m}$

$A_c = \pi(0.003\ \text{m})^2/4$

$= 7.069 \times 10^{-6}\ \text{m}^2$

$$m = \left(\frac{(178.9 \text{ W/m}^2 \cdot \text{K})(9.245 \times 10^{-3} \text{ m})}{(401 \text{ W/m} \cdot \text{K})(7.069 \times 10^{-6} \text{ m}^2)} \right)^{1/2}$$

$$= 24.16 \text{ m}^{-1}$$

Solving for the base temperature T_b,

$$T_b = \frac{q}{(hPk_f A_c)^{1/2} \tanh(mL)} + T_\infty$$

$$T_b = \frac{1.25}{[(178.9)(9.245 \times 10^{-3})(401)(7.069 \times 10^{-6})]^{1/2} \tanh[(24.16)(0.012)]} + 25°\text{C}$$

$$= \underline{\underline{89.7°\text{C}}}$$

DISCUSSION

Because thermal contact resistance is neglected, the surface temperature of the device equals the base temperature of the fin. Assuming that the device has five times the surface area as the footprint of the fin and that the unfinned device has the same heat transfer coefficient, the surface temperature of the device without the fin is

$$T_s = \frac{q}{5hA_c} + T_\infty$$

$$T_s = \frac{1.25 \text{ W}}{5(178.9 \text{ W/m}^2 \cdot \text{K})(7.069 \times 10^{-6} \text{ m}^2)} + 25°\text{C}$$

$$= 223°\text{C}$$

which clearly illustrates the reason for extending the surface of electronic devices. The film temperature could be corrected to provide a more accurate heat transfer coefficient and hence a more accurate surface temperature.

❒ ❒ ❒

PROBLEM 7.17 Heat Loss from an Exhaust Stack

A thin-walled steel exhaust stack expels hot combustion gas into the atmosphere. The height and diameter of the stack are 2.4 m and 30 cm, respectively. The average surface temperature of the stack is 110°C, and its emissivity is 0.8. A steady wind induces a cross flow of 15°C air across the stack at a velocity of 7 m/s. The mass flow rate and specific heat of the combustion gas are 0.12 kg/s and 2350 J/kg·K, respectively. If the average temperature of the surrounding surfaces is 10°C, find the heat loss from the exhaust stack. Also find the temperature drop of the exhaust gas.

DIAGRAM

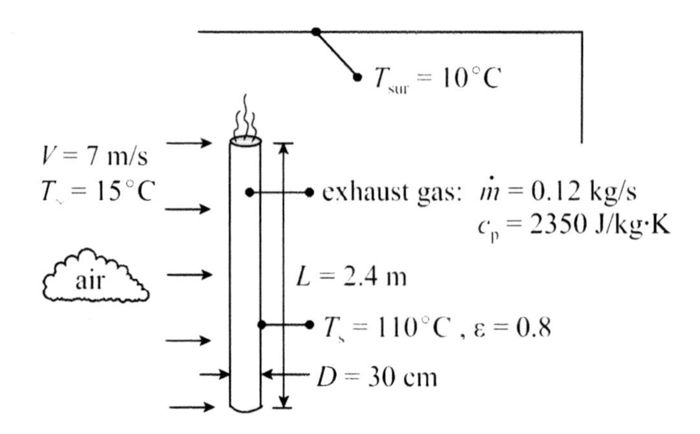

ASSUMPTIONS

1. Steady conditions.
2. The average temperature of the exhaust gas and stack are equal.
3. Air flow is normal to the axis of the stack.
4. The surface of the stack is smooth.
5. Constant properties.

PROPERTIES

exhaust gas: $c_p = 2350$ J/kg·K

$T_f = (T_s + T_\infty)/2 = (110°C + 15°C)/2 = 62.5°C = 336$ K

atmospheric air: $k = 0.0290$ W/m·K, $\nu = 19.51 \times 10^{-6}$ m²/s, $Pr = 0.702$

ANALYSIS

The Reynolds number is

$$Re_D = \frac{VD}{\nu}$$

$$Re_D = \frac{(7 \text{ m/s})(0.30 \text{ m})}{19.51 \times 10^{-6} \text{ m}^2/\text{s}}$$

$$= 1.076 \times 10^5$$

and the average Nusselt number may be found using the empirical correlation

$$Nu_D = \frac{hD}{k} = CRe_D^m Pr^{1/3}$$

where the constants C and m depend on the Reynolds number. For $Re_D = 1.076 \times 10^5$, these constants have the values $C = 0.027$ and $m = 0.805$. Thus,

$$Nu_D = (0.027)(1.076 \times 10^5)^{0.805}(0.702)^{1/3}$$

$$= 269.6$$

The average heat transfer coefficient is

$$h = \frac{Nu_D k}{D}$$

$$h = \frac{(269.6)(0.0290 \text{ W/m·K})}{0.30 \text{ m}}$$

$$= 26.1 \text{ W/m}^2\text{·K}$$

The heat loss from the stack is the sum of the convective and radiative heat transfer from the surface,

$$q = q_{conv} + q_{rad} = h\pi DL(T_s - T_\infty) + \varepsilon\sigma\pi DL(T_s^4 - T_{sur}^4)$$

$$q_{conv} = (26.1 \text{ W/m}^2\text{·K})\pi(0.30 \text{ m})(2.4 \text{ m})(110 - 15)°\text{C}$$

$$= 5608 \text{ W}$$

$$q_{rad} = (0.8)(5.669 \times 10^{-8} \text{ W/m}^2\text{·K})\pi(0.30 \text{ m})(2.4 \text{ m})[(383 \text{ K})^4 - (283 \text{ K})^4]$$

$$= 1549 \text{ W}$$

Hence,

$$q = (5608 + 1549)\text{W}$$

$$= \underline{\underline{7157 \text{ W}}}$$

A first law analysis of the exhaust gas yields

$$q = \dot{m}c_p\Delta T$$

where q is the heat transfer from the stack, and ΔT is the temperature drop of the exhaust gas,

$$\Delta T = \frac{7157 \text{ W}}{(0.12 \text{ kg/s})(2350 \text{ J/kg·K})}$$

$$= \underline{\underline{25.4°C}}$$

DISCUSSION

The heat loss from the exhaust stack is 7157 W, most of which is convective heat transfer, and the exhaust gas cools by 25.4°C as it flows through the 2.4 m long stack.

❒ ❒ ❒

PROBLEM 7.18 Air Temperature from Thermocouple Data in a Duct

A thermocouple is used to measure the temperature of hot air flowing in a duct. To protect the thermocouple from damage, a thermocouple well of low carbon steel (k_s = 51 W/m·K) is used. The well is a 16 cm long hollow tube with inside and outside diameters of 8 mm and 12 mm, respectively. One thermocouple (T_1) is soldered to the bottom of the well, and a second thermocouple (T_2) is attached to the duct wall adjacent to the thermocouple well. The thermocouples measure temperatures of T_1 = 152°C and T_2 = 117°C. If the mean velocity of the air in the duct is 2 m/s, what is the temperature of the air?

DIAGRAM

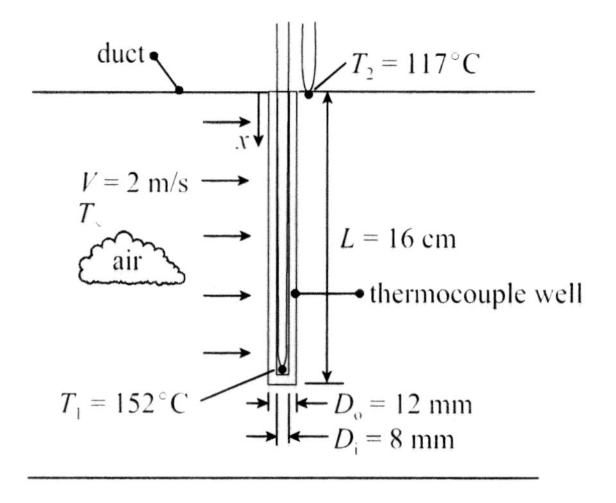

ASSUMPTIONS

1. Steady conditions.
2. Air flow is normal to the axis of the thermocouple well.
3. The surface of the thermocouple well is smooth.
4. The tip of the thermocouple well is insulated.

5. Thermal resistance of the tip of the thermocouple well is neglected.
6. Radiation is neglected.
7. Constant properties.

PROPERTIES

low carbon steel: $k_s = 51$ W/m·K

$T_s = (T_1 + T_2)/2 = (152°C + 117°C)/2 = 134.5°C = 408$ K

Assume $T_\infty = 160°C$. $T_f = (T_s + T_\infty)/2 = (134.5°C + 160°C)/2 = 147°C = 420$ K

atmospheric air: $k = 0.0352$ W/m·K, $\nu = 28.80 \times 10^{-6}$ m²/s, $Pr = 0.686$

ANALYSIS

Note that for the purpose of thermal property calculations, we assumed the surface temperature of the thermocouple well to be the average of the two thermocouple readings. The Reynolds number is

$$Re_D = \frac{VD}{\nu}$$

$$Re_D = \frac{(2 \text{ m/s})(0.012 \text{ m})}{28.80 \times 10^{-6} \text{ m}^2\text{/s}}$$

$$= 833.3$$

and the average Nusselt number may be found using the empirical correlation

$$Nu_D = \frac{hD}{k} = CRe_D^m Pr^{1/3}$$

where the constants C and m depend on the Reynolds number. For $Re_D = 833.3$, these constants have the values $C = 0.683$ and $m = 0.466$. Thus,

$$Nu_D = (0.683)(833.3)^{0.466}(0.686)^{1/3}$$

$$= 13.83$$

The average heat transfer coefficient is

$$h = \frac{Nu_D k}{D}$$

$$h = \frac{(13.83)(0.0352 \text{ W/m·K})}{0.012 \text{ m}}$$

$$= 40.6 \text{ W/m}^2\text{·K}$$

Now that the heat transfer coefficient is known, we can do an analysis of the thermocouple well, which acts as a pin fin. The temperature distribution in a fin with an insulated tip is given by

$$\frac{\theta}{\theta_b} = \frac{T(x) - T_\infty}{T_b - T_\infty} = \frac{\cosh[m(L-x)]}{\cosh(mL)}$$

where T_b and T_∞ are the fin base and fluid temperatures, respectively, and

$$m = \left(\frac{hP}{k_s A_c}\right)^{1/2}$$

where,

$$P = \pi(0.012 \text{ m})$$

$$= 0.0377 \text{ m}$$

$$A_c = \pi[(0.012 \text{ m})^2 - (0.008 \text{ m})^2]/4$$

$$= 6.283 \times 10^{-5} \text{ m}^2$$

Thus,

$$m = \left(\frac{(40.6 \text{ W/m}^2\text{·K})(0.0377 \text{ m})}{(51 \text{ W/m·K})(6.283 \times 10^{-5} \text{ m}^2)}\right)^{1/2}$$

$$= 21.86 \text{ m}^{-1}$$

Setting $x = L$, and introducing the thermocouple readings into the temperature distribution,

$$\frac{T_1 - T_\infty}{T_2 - T_\infty} = \frac{1}{\cosh(mL)} = C$$

Solving for T_∞, the temperature of the air in the duct,

$$T_\infty = \frac{CT_2 - T_1}{C - 1}$$

$$T_\infty = \frac{(0.06048)(117°\text{C}) - 152°\text{C}}{0.06048 - 1}$$

$= \underline{\underline{154.3°C}}$

DISCUSSION

Thus, the actual air temperature is 2.3°C higher than that indicated by T_1, which is due to conduction from the fin (thermocouple well) into the duct wall. Because the duct wall is much cooler than T_1, a radiation error is also present. The assumed value of T_∞ is close enough to the calculated value that a correction of thermal properties is not warranted.

❒ ❒ ❒

PROBLEM 7.19 Sheath Temperature of an Electric Heating Element

A process air heater, consisting of a duct with a single row of electric heating elements, is used to produce high-temperature air for a drying operation. The electric heating elements are cylinders with a length and diameter of 50 cm and 1.25 cm, respectively. Air upstream of the heating elements has an average velocity of 12 m/s and a temperature of 30°C. The thin sheath of each heating element is made of a nickel alloy with a melting point of 1370°C and an emissivity of 0.4. Each heating element draws 16.5 A of current at a line voltage of 220 V. If the duct surface temperature is equal to that of the air, find the temperature of the heating element sheath. Is the sheath at risk of melting?

DIAGRAM

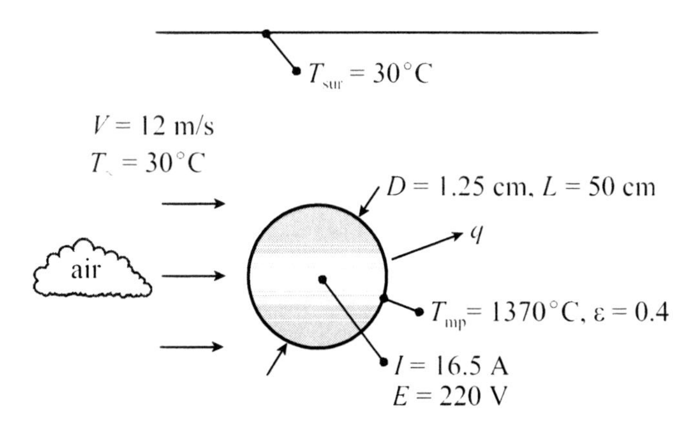

ASSUMPTIONS

1. Steady conditions.
2. Each heating element in the row behaves as an isolated cylinder in cross flow.
3. Air flow is normal to the axis of the cylinder.
4. The surface of the cylinder is smooth.
5. Conduction from the heating element to the duct wall is neglected.
6. The sheath is isothermal.
7. Constant properties.

PROPERTIES

nickel alloy: $T_{mp} = 1370°C$, $\varepsilon = 0.4$

Assume $T_s = 1100°C$. $T_f = (T_s + T_\infty)/2 = (1100°C + 30°C)/2 = 565°C = 838$ K

atmospheric air: $k = 0.0590$ W/m·K, $\nu = 91.67 \times 10^{-6}$ m²/s, $Pr = 0.714$

ANALYSIS

The Reynolds number is

$$Re_D = \frac{VD}{\nu}$$

$$Re_D = \frac{(12 \text{ m/s})(0.0125 \text{ m})}{91.67 \times 10^{-6} \text{ m}^2/\text{s}}$$

$$= 1636$$

and the average Nusselt number may be found using the empirical correlation

$$Nu_D = \frac{hD}{k} = CRe_D^m Pr^{1/3}$$

where the constants C and m depend on the Reynolds number. For $Re_D = 1636$, these constants have the values $C = 0.683$ and $m = 0.466$, so we have

$$Nu_D = (0.683)(1636)^{0.466}(0.714)^{1/3}$$

$$= 19.20$$

The average heat transfer coefficient is

$$h = \frac{Nu_D k}{D}$$

$$h = \frac{(19.20)(0.0590 \text{ W/m·K})}{0.0125 \text{ m}}$$

$$= 90.6 \text{ W/m}^2\text{·K}$$

The electrical energy generated by the heating element equals the convective and radiative heat transfer from the surface of the sheath,

$$q = IE = q_{conv} + q_{rad} = h\pi DL(T_s - T_\infty) + \varepsilon\sigma\pi DL(T_s^4 - T_{sur}^4)$$

where,

$$q = (16.5\text{A})(220 \text{ V})$$

$$= 3630 \text{ W}$$

Using temperature units of K in the radiation term, and numerically solving for T_s,

$$T_s = 1396 \text{ K} = \underline{\underline{1123\,°\text{C}}}$$

DISCUSSION

The predicted temperature is 247 °C below the melting point of the nickel alloy, so the sheath will not melt, but the mechanical properties of the sheath may be compromised at this temperature, inducing a structural failure. Because the surface of the heating element is the coolest region of the part, additional analyses are required to assess the thermal condition of the materials inside the sheath and the electrical conductor itself. Furthermore, this sheath temperature may not provide the proper air temperature required for the drying process. Accordingly, the electrical power could be reduced, or fins could be installed on the heating element.

❒ ❒ ❒

PROBLEM 7.20 Heat Transfer from a Rotating Shaft

A steel shaft rotating at 3200 rpm is heated by conduction from a motor to which it is attached, giving the shaft a surface temperature of 45 °C. The length and diameter of the shaft are 30 cm and 4.2 cm, respectively. If the temperature of the air surrounding the shaft is 15 °C, what is the heat transfer from the shaft? Hint: The average Nusselt number for a rotating cylinder in a still fluid may be calculated using the correlation

$$Nu_D = 0.133 Re_D^{2/3} Pr^{1/3}$$

where the Reynolds number is

$$Re_D = \frac{\Omega D^2}{\nu}$$

and Ω is angular velocity in rad/s.

DIAGRAM

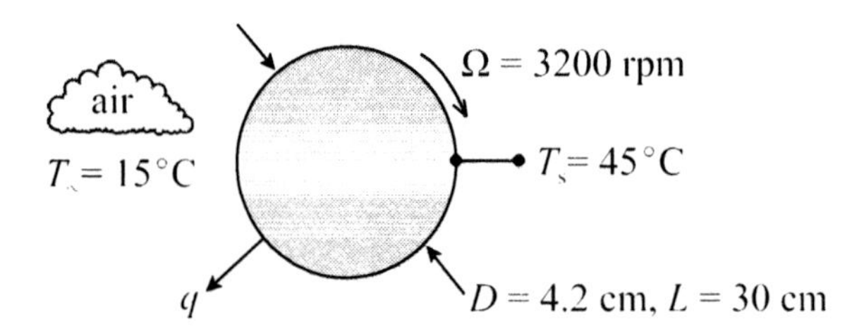

ASSUMPTIONS

1. Steady conditions.
2. The surrounding air is still.
3. The surface of the shaft is smooth and isothermal..
4. Radiation is neglected.
5. Constant properties.

PROPERTIES

$T_f = (T_s + T_\infty)/2 = (45°C + 15°C)/2 = 30°C = 303\ K$

atmospheric air: $k = 0.0265$ W/m·K, $\nu = 16.19 \times 10^{-6}$ m²/s, $Pr = 0.707$

ANALYSIS

The angular velocity of the shaft in rad/s is

$$\Omega = 3200 \text{ rev/min} \times 2\pi \text{ rad/rev} \times 1 \text{ min/60 s}$$

$$= 335.1 \text{ rad/s}$$

The Reynolds number is

$$Re_D = \frac{\Omega D^2}{\nu}$$

$$Re_D = \frac{(335.1 \text{ rad/s})(0.042 \text{ m})^2}{16.19 \times 10^{-6} \text{ m}^2/\text{s}}$$

$$= 3.651 \times 10^4$$

and the average Nusselt number is

$$Nu_D = 0.133 Re_D^{2/3} Pr^{1/3}$$

$Nu_D = 0.133(3.651 \times 10^4)^{2/3}(0.707)^{1/3}$

$= 130.4$

The average heat transfer coefficient is

$$h = \frac{Nu_D k}{D}$$

$h = \dfrac{(130.4)(0.0265 \text{ W/m·K})}{0.042 \text{ m}}$

$= 82.3 \text{ W/m}^2\text{·K}$

so the heat transfer from the shaft is

$$q = h\pi DL(T_s - T_\infty)$$

$q = (82.3 \text{ W/m}^2\text{·K})\pi(0.042 \text{ m})(0.30 \text{ m})(45 - 15)°\text{C}$

$= \underline{\underline{97.7 \text{ W}}}$

DISCUSSION

The effect of angular velocity on heat transfer from the shaft is shown in the graph below.

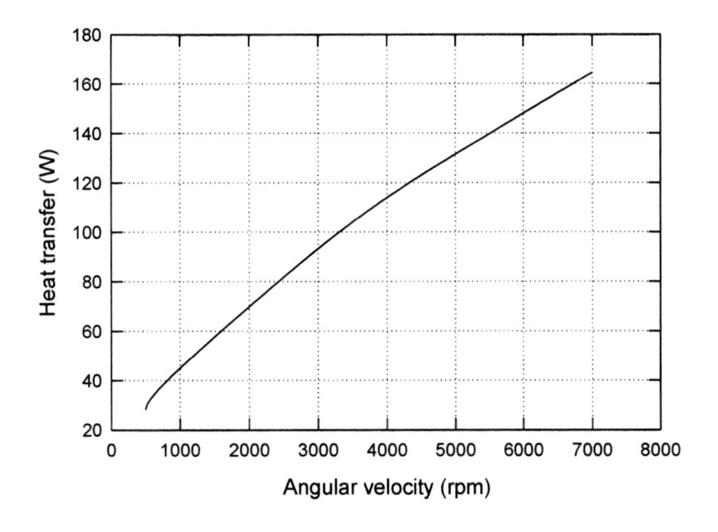

❒ ❒ ❒

PROBLEM 7.21 Heat Transfer from a Sphere in Cross Flow

A stream of 10 °C liquid water flows at a velocity of 2 m/s across a 1.4 cm diameter sphere. If the surface temperature of the sphere is 75 °C, find the heat transfer from the sphere.

DIAGRAM

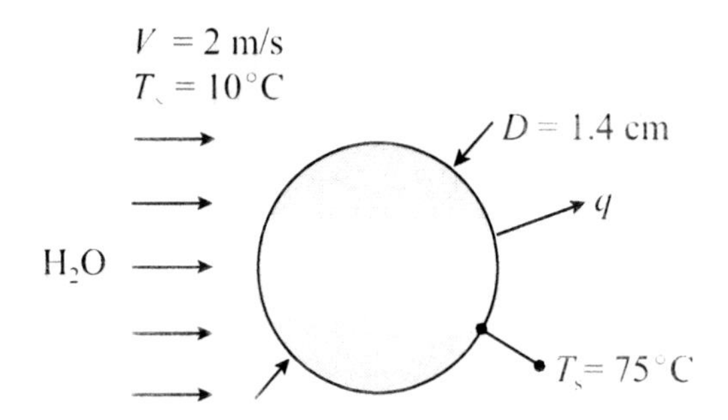

ASSUMPTIONS

1. Steady conditions.
2. The sphere is surrounded by water.
3. The surface of the sphere is smooth and isothermal.
4. Radiation is neglected.
5. Constant properties.

PROPERTIES

$T_f = (T_s + T_\infty)/2 = (75°C + 10°C)/2 = 42.5°C = 316$ K

liquid water: $\rho = 990.7$ kg/m^3, $k = 0.635$ W/m·K, $\mu = 620.2 \times 10^{-6}$ Pa·s, $Pr = 4.08$
$\nu = 6.260 \times 10^{-7}$ m^2/s, $\mu_s = 374.6 \times 10^{-6}$ Pa·s

ANALYSIS

Note that all thermal properties are evaluated at the film temperature, except μ_s, which is evaluated at the surface temperature. The Reynolds number is

$$Re_D = \frac{VD}{\nu}$$

$$Re_D = \frac{(2 \text{ m/s})(0.014 \text{ m})}{6.260 \times 10^{-7} \text{ m}^2\text{/s}}$$

$$= 4.473 \times 10^4$$

The average Nusselt number for a sphere in cross flow may be found using the empirical correlation

$$Nu_D = 2 + (0.4 Re_D^{1/2} + 0.06 Re_D^{2/3}) Pr^{0.4} \left(\frac{\mu}{\mu_s} \right)^{1/4}$$

$$Nu_D = 2 + [0.4(4.473 \times 10^4)^{1/2} + 0.06(4.473 \times 10^4)^{2/3}](4.08)^{0.4}(620.2 \times 10^{-6}/374.6 \times 10^{-6})^{1/4}$$

$$= 320.9$$

so the average heat transfer coefficient is

$$h = \frac{Nu_D k}{D}$$

$$h = \frac{(320.9)(0.635 \text{ W/m}\cdot\text{K})}{0.014 \text{ m}}$$

$$= 1.455 \times 10^4 \text{ W/m}^2\cdot\text{K}$$

The heat transfer from the sphere is given by Newton's law of cooling,

$$q = h\pi D^2 (T_s - T_\infty)$$

$$q = (1.455 \times 10^4 \text{ W/m}^2\cdot\text{K})\pi(0.014 \text{ m})^2(75 - 10)°\text{C}$$

$$= \underline{\underline{582 \text{ W}}}$$

DISCUSSION

Another commonly used Nusselt number correlation for a sphere in cross flow is

$$Nu_D = 2 + 0.6 Re_D^{1/2} Pr^{1/3}$$

In this correlation and the one used in our analysis, in the limit as $Re_D \rightarrow 0$, $Nu_D = 2$, which corresponds to heat transfer by conduction from a spherical surface to a stationary infinite medium surrounding the sphere. Either correlation may be used for a sphere in almost any type of cross flow application, but the first correlation is subject to the conditions

$$0.71 < Pr < 380,\ 3.5 < Re_D < 7.6 \times 10^4,\ 1.0 < (\mu/\mu_s) < 3.2$$

which are satisfied in the analysis.

❒ ❒ ❒

PROBLEM 7.22 Surface Temperature of an Ocean Mine

A spherical mine used to destroy surface ships is moored to the ocean bottom such that the mine is positioned about 3 m below the ocean surface. The mine is 1.2 m in diameter, and its electrical and chemical systems generate a total of 18 kW, which is convected from the skin of the device to the surrounding water. If a steady ocean current of 1.6 m/s exists in the region, what is the surface temperature of the mine? The ocean temperature is 15°C.

DIAGRAM

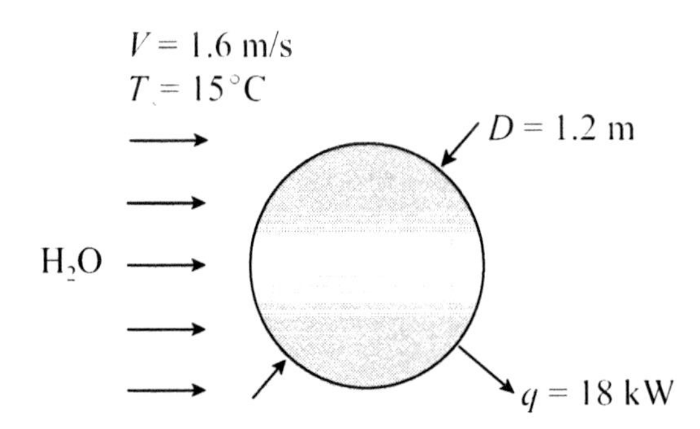

ASSUMPTIONS

1. Steady conditions.
2. The surface of the mine is smooth and isothermal.
3. Radiation is neglected.
4. Thermal properties of ocean water are equivalent to those of pure water.
5. Constant properties.

PROPERTIES

Assume $T_s = 19°C$. $T_f = (T_s + T_\infty)/2 = (19°C + 15°C)/2 = 17°C = 290\ K$

liquid water: $\rho = 999\ kg/m^3$, $k = 0.598\ W/m{\cdot}K$, $\mu = 1080 \times 10^{-6}\ Pa{\cdot}s$, $Pr = 7.56$
$\nu = 1.081 \times 10^{-6}\ m^2/s$

ANALYSIS

All thermal properties are evaluated at the film temperature. The Reynolds number is

$$Re_D = \frac{VD}{\nu}$$

$$Re_D = \frac{(1.6 \text{ m/s})(1.2 \text{ m})}{1.081 \times 10^{-6} \text{ m}^2/\text{s}}$$

$$= 1.776 \times 10^6$$

The average Nusselt number is found using the correlation

$$Nu_D = 2 + 0.6 Re_D^{1/2} Pr^{1/3}$$

$$Nu_D = 2 + 0.6(1.776 \times 10^6)^{1/2}(7.56)^{1/3}$$

$$= 1571$$

Thus, the average heat transfer coefficient is

$$h = \frac{Nu_D k}{D}$$

$$h = \frac{(1571)(0.598 \text{ W/m·K})}{1.2 \text{ m}}$$

$$= 783 \text{ W/m}^2\text{·K}$$

The total energy generated by the mine is convected from the its surface to the ocean water. From Newton's law of cooling,

$$q = h\pi D^2 (T_s - T_\infty)$$

Solving for surface temperature T_s,

$$T_s = \frac{18 \times 10^3 \text{ W}}{(783 \text{ W/m}^2\text{·K})\pi(1.2 \text{ m})^2} + 15°\text{C}$$

$$= \underline{\underline{20.1°\text{C}}}$$

DISCUSSION

The surface temperature of the mine is only 5.1 °C higher than the ocean temperature. If radiation was included in the analysis, the surface temperature would be slightly lower. Because the predicted surface temperature is very close to the assumed value, a thermal property correction is not required.

PROBLEM 7.23 Water Tank Height for Cooling Copper Spheres

Spheres of pure copper are quenched by being dropped into a large tank of water maintained at a temperature of 25°C. The spheres are 16 mm in diameter, and have a uniform initial temperature of 90°C. Find the height of the water tank required to cool the center of the spheres to 40°C. Assume that the terminal velocity of the spheres is achieved immediately after they enter the water.

DIAGRAM

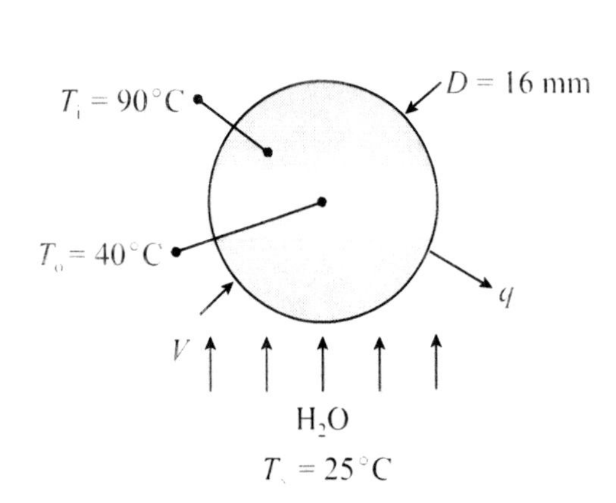

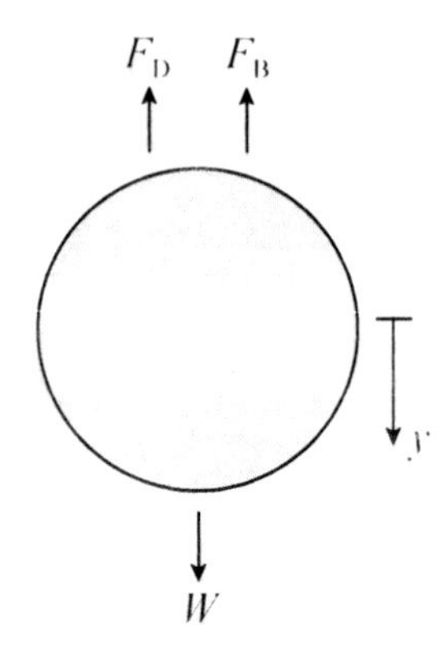

ASSUMPTIONS

1. The water tank is large.
2. Terminal velocity is achieved immediately after the sphere enters the water.
3. The sphere has a uniform initial temperature.
4. The surface of the sphere is smooth and isothermal.
5. Radiation is neglected.
6. Constant properties.

PROPERTIES

pure copper: $\rho_s = 8933$ kg/m^3, $c_p = 385$ J/kg·K, $k_s = 401$ W/m·K

Assume $T_s = (T_i + T_o)/2 = 65°C$. $T_f = (T_s + T_\infty)/2 = (65°C + 25°C)/2 = 45°C = 318$ K

liquid water: $\rho = 989.9$ kg/m^3, $k = 0.638$ W/m·K, $\mu = 599 \times 10^{-6}$ Pa·s, $Pr = 3.93$
$\nu = 6.051 \times 10^{-7}$ m^2/s, $\mu_s = 433 \times 10^{-6}$ Pa·s

ANALYSIS

Note that for the purpose of thermal property calculations, we assumed the surface temperature of the sphere to be the average of the initial and final center temperatures of the sphere. Because the sphere achieves terminal velocity immediately after entering the water, the sphere does not accelerate. Thus, a force balance in the vertical direction yields

$$\Sigma F_y = 0 = W - F_D - F_B$$

where W is weight, F_D is drag force and F_B is buoyancy force. These three forces are expressed as

$$W = mg = \rho_s V g = \rho_s \pi (D^3/6) g$$

$$F_D = C_D A \rho \frac{V^2}{2}$$

$$F_B = \rho V g = \rho \pi (D^3/6) g$$

where C_D is the drag coefficient for a sphere and $A = \pi(D^2/4)$ is frontal area. Substituting these expressions into the force balance and simplifying,

$$D(\rho - \rho_s)g - \tfrac{3}{4} C_D \rho V^2 = 0$$

The Reynolds number is

$$Re_D = \frac{VD}{\nu}$$

Using a graph of C_D as a function of Re_D for a sphere (refer to a fluid mechanics or heat transfer text), the force balance and Reynolds number expression are iteratively solved for C_D, Re_D and V, the terminal velocity of the sphere. Solving,

$C_D \approx 0.47$, $Re_D \approx 5.02 \times 10^4$, $V \approx 1.9$ m/s

The average Nusselt number for a sphere in cross flow may be found using the empirical correlation

$$Nu_D = 2 + (0.4 Re_D^{1/2} + 0.06 Re_D^{2/3}) Pr^{0.4} \left(\frac{\mu}{\mu_s} \right)^{1/4}$$

$$Nu_D = 2 + [0.4(5.02 \times 10^4)^{1/2} + 0.06(5.02 \times 10^4)^{2/3}](3.93)^{0.4}(599 \times 10^{-6}/433 \times 10^{-6})^{1/4}$$

$$= 323$$

so the average heat transfer coefficient is

$$h = \frac{Nu_D k}{D}$$

$$h = \frac{(323)(0.638 \text{ W/m·K})}{0.016 \text{ m}}$$

$$= 1.288 \times 10^4 \text{ W/m}^2\text{·K}$$

Now that the heat transfer coefficient is known, we determine if the sphere is lumped. The Biot number for a sphere is

$$Bi = \frac{h(V/A)}{k_s} = \frac{h(R/3)}{k_s}$$

$$Bi = \frac{(1.288 \times 10^4 \text{ W/m}^2\text{·K})(0.008 \text{ m/3})}{401 \text{ W/m·K}}$$

$$= 0.0857 < 0.1$$

Thus, the sphere is lumped, which means that the temperature is uniform at any given time. The time required for the sphere to cool from 90°C to 40°C is

$$t = -\frac{\rho_s V c_p}{hA} \ln\left(\frac{T_o - T_\infty}{T_i - T_\infty}\right)$$

$$t = -\frac{(8933 \text{ kg/m}^3)(0.008 \text{ m/3})(385 \text{ J/kg·K})}{1.288 \times 10^4 \text{ W/m}^2\text{·K}} \ln\left(\frac{40°\text{C} - 25°\text{C}}{90°\text{C} - 25°\text{C}}\right)$$

$$= 1.044 \text{ s}$$

Hence, the required tank height is

$$\Delta y = Vt$$

$$\Delta y = (1.9 \text{ m/s})(1.044 \text{ s})$$

$$= \underline{\underline{1.98 \text{ m}}}$$

DISCUSSION

A more accurate analysis could be performed by including radiation and time dependence of the heat transfer coefficient.

PROBLEM 7.24 Time Constant for a Spherical Thermocouple Sensor

A thermocouple, formed by soldering the ends of copper and constantan wires together, is used to measure the air temperature in a ventilation duct carrying air at an average velocity of 20 m/s. Upstream conditions cause the air temperature to undergo large variations with a characteristic frequency of 5 Hz, and we require the thermocouple to detect these temperature variations. The average temperature of the air in the duct is 140°C. If the thermocouple sensor is a spherical mass of solder with a diameter of 0.90 mm, is the thermocouple suitable for this application?

DIAGRAM

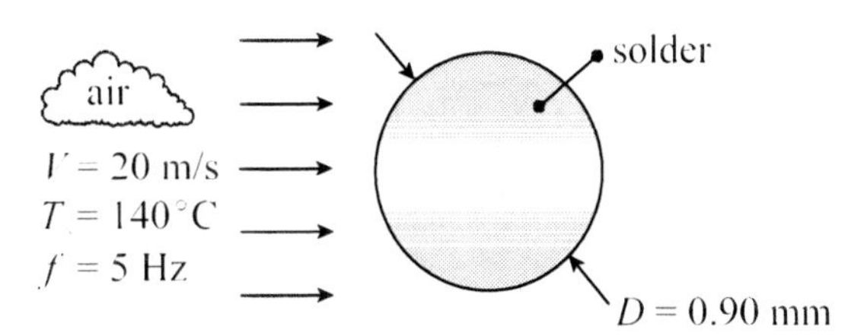

ASSUMPTIONS

1. The sensor is isothermal.
2. The surface of the sensor is smooth.
3. Radiation is neglected.
4. Effects of the wires connected to the sphere of solder are neglected.
5. Constant properties.

PROPERTIES

solder: $\rho_s = 8332$ kg/m^3, $c_p = 188$ J/kg·K, $k_s = 51$ W/m·K

$T_f = T_\infty = 140°\text{C} = 413$ K

atmospheric air: $k = 0.0347$ W/m·K, $\nu = 27.96 \times 10^{-6}$ m^2/s, $Pr = 0.689$

ANALYSIS

The Reynolds number is

$$Re_D = \frac{VD}{\nu}$$

$$Re_D = \frac{(20 \text{ m/s})(9.0 \times 10^{-4} \text{ m})}{27.96 \times 10^{-6} \text{ m}^2/\text{s}}$$

$$= 643.8$$

The average Nusselt number for a sphere in cross flow may be found using the empirical correlation

$$Nu_D = 2 + (0.4Re_D^{1/2} + 0.06Re_D^{2/3})Pr^{0.4}\left(\frac{\mu}{\mu_s}\right)^{1/4}$$

$$Nu_D = 2 + [0.4(643.8)^{1/2} + 0.06(643.8)^{2/3}](0.689)^{0.4}(1)^{1/4}$$

$$= 14.6$$

where we note that $\mu = \mu_s$. The the average heat transfer coefficient is

$$h = \frac{Nu_D k}{D}$$

$$h = \frac{(14.6)(0.0347\text{ W/m·K})}{9.0 \times 10^{-4}\text{ m}}$$

$$= 563\text{ W/m}^2\text{·K}$$

Now that the heat transfer coefficient is known, let's check the validity of the first assumption. The Biot number is

$$Bi = \frac{h(R/3)}{k_s}$$

$$Bi = \frac{(563\text{ W/m}^2\text{·K})(4.5 \times 10^{-4}\text{ m/3})}{51\text{ W/m·K}}$$

$$= 4.97 \times 10^{-3} < 0.1$$

so the sensor is isothermal (lumped). The time constant is

$$\tau_c = \frac{\rho_s c_p}{h}\frac{R}{3}$$

$$\tau_c = \frac{(8332\text{ kg/m}^3)(188\text{ J/kg·K})(4.5 \times 10^{-4}\text{ m})}{3(563\text{ W/m}^2\text{·K})}$$

$$= \underline{\underline{0.417\text{ s}}}$$

DISCUSSION

The frequency at which the air temperature changes is 5 Hz, which means that the period of the temperature cycles is $\tau = 1/f = 0.2$ s. Because the period is smaller than the time constant of the thermocouple sensor, the sensor cannot respond rapidly enough to the variations of air temperature. Hence, the thermocouple sensor cannot accurately track the temperature variations and should not be used in this application. A smaller sensor is needed.

❒ ❒ ❒

PROBLEM 7.25 Life Prediction for an Outdoor Lamp

A 60-W lamp designed for outdoor use is approximately spherical with a diameter of 9 cm. Under laboratory conditions when the lamp was undergoing testing, the illuminated lamp was continuously exposed to a steady 2 m/s cross flow of air until the filament burned out. To determine the effect of ambient air temperature on operating life, the air temperature was varied in the tests such that the surface temperature of the lamp, measured with a thermocouple, varied between 100°C and 175°C. After testing hundreds of lamps, technicians determined that the average life for surface temperatures of 100°C and 175°C is 4950 and 3600 hours, respectively. Assuming that lamp life varies linearly between these limits, predict the life for a lamp operating in a 2 m/s cross flow of 30°C air. The temperature of the surroundings is 20°C, and the emissivity of the lamp's surface is 0.8.

DIAGRAM

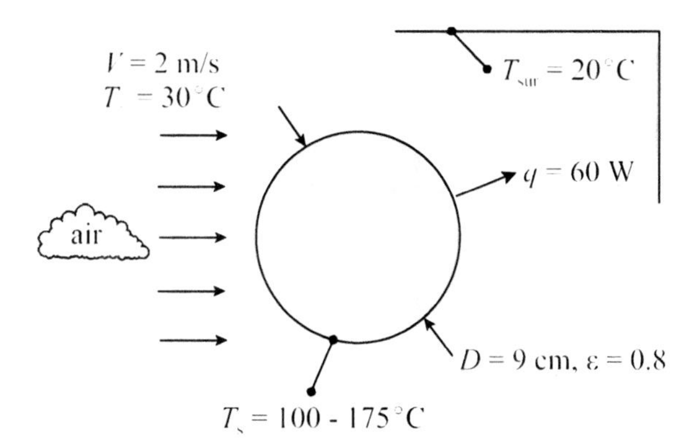

ASSUMPTIONS

1. Steady conditions.
2. The lamp is spherical.
3. All 60 W of dissipated power is convected and radiated from the lamp's surface.
4. The surface of the lamp is smooth and isothermal.
5. Constant properties.

PROPERTIES

lamp surface: $\varepsilon = 0.8$

Assume $T_s = 150°C$. $T_f = (T_s + T_\infty)/2 = (150°C + 30°C)/2 = 90°C = 363$ K

atmospheric air: $k = 0.0310$ W/m·K, $\nu = 22.35 \times 10^{-6}$ m^2/s, $Pr = 0.697$
$\mu = 213.9 \times 10^{-7}$ Pa·s, $\mu_s = 239.6 \times 10^{-7}$ Pa·s

ANALYSIS

The Reynolds number is

$$Re_D = \frac{VD}{\nu}$$

$$Re_D = \frac{(2 \text{ m/s})(0.09 \text{ m})}{22.35 \times 10^{-6} \text{ m}^2/\text{s}}$$

$$= 8054$$

The average Nusselt number for a sphere in cross flow may be found using the empirical correlation

$$Nu_D = 2 + (0.4 Re_D^{1/2} + 0.06 Re_D^{2/3}) Pr^{0.4} \left(\frac{\mu}{\mu_s} \right)^{1/4}$$

$$Nu_D = 2 + [0.4(8054)^{1/2} + 0.06(8054)^{2/3}](0.697)^{0.4}(213.9 \times 10^{-7}/239.6 \times 10^{-7})^{1/4}$$

$$= 52.5$$

The average heat transfer coefficient is

$$h = \frac{Nu_D k}{D}$$

$$h = \frac{(52.5)(0.0310 \text{ W/m·K})}{0.09 \text{ m}}$$

$$= 18.1 \text{ W/m}^2\text{·K}$$

The power dissipated by the lamp is the sum of the heat transfer convected and radiated from the surface of the lamp,

$$q = q_{conv} + q_{rad} = h\pi D^2 (T_s - T_\infty) + \varepsilon\sigma\pi D^2 (T_s^4 - T_{sur}^4)$$

Setting $q = 60$ W, and iteratively solving for T_s,

$T_s = 392 \text{ K} = 119°\text{C}$

Using linear interpolation on the lamp data, the predicted lamp life is

$$\text{life} = \underline{\underline{4608 \text{ h}}}$$

DISCUSSION

The Nusselt number correlation used in the analysis strictly applies for the ranges $0.71 < Pr < 380$, $3.5 < Re_D < 7.6 \times 10^4$ and $1.0 < \mu/\mu_s < 3.2$. In this problem, $Pr = 0.697$ and $\mu/\mu_s = 0.893$, so we have used the correlation somewhat outside its limits of applicability. Also, because our assumed and calculated surface temperatures are not close, a correction of thermal properties may be in order.

❒ ❒ ❒

PROBLEM 7.26 Surface Temperature and Air Conditioning Load for a Radome

A radome, a term composed from the words *radar* and *dome*, is a weatherproof enclosure that protects an antenna and prevents observers from deducing the direction of the antenna and therefore which satellite is being targeted. Radomes are typically constructed of fiberglass or other non-conductive material. Consider a spherical radome with a diameter of 2.8 m. The antenna and other electrical equipment within the enclosure dissipate a total power of 2.5 kW, and the system operates in an arid environment where the ambient air temperature is 45°C. The radome is subjected to an incident solar radiation of 650 W/m^2, and the solar absorptivity of the radome's surface is 0.1. If the radome is exposed to a steady 4 m/s wind, find the outside surface temperature of the radome, and determine the air conditioning load for the system.

DIAGRAM

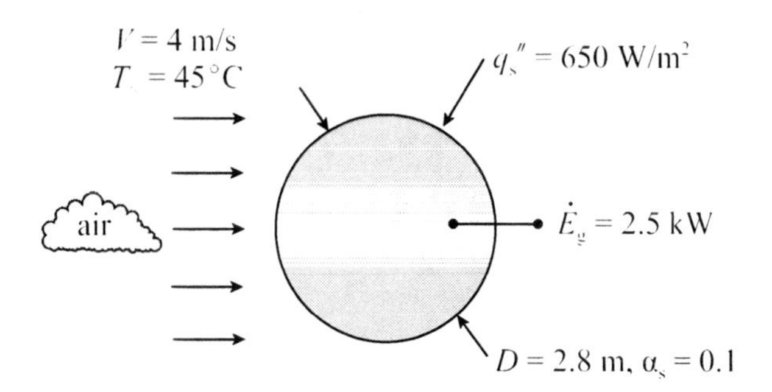

ASSUMPTIONS

1. Steady conditions.
2. The radome is surrounded by air.

3. The entire radome surface is subjected to solar radiation.
4. All 2.5 kW of dissipated power is transferred into the radome.
5. The surface of the radome is smooth and isothermal.
6. Radiation (except solar) is neglected.
7. Conduction into supporting structures is neglected.
8. Constant properties.

PROPERTIES

radome surface: $\alpha_s = 0.1$

Assume $T_s = 55°C$. $T_f = (T_s + T_\infty)/2 = (55°C + 45°C)/2 = 50°C = 323$ K

atmospheric air: $k = 0.0280$ W/m·K, $\nu = 18.20 \times 10^{-6}$ m²/s, $Pr = 0.704$
$\mu = 195.5 \times 10^{-7}$ Pa·s, $\mu_s = 197.8 \times 10^{-7}$ Pa·s

ANALYSIS

The Reynolds number is

$$Re_D = \frac{VD}{\nu}$$

$$Re_D = \frac{(4 \text{ m/s})(2.8 \text{ m})}{18.20 \times 10^{-6} \text{ m}^2/\text{s}}$$

$$= 6.154 \times 10^5$$

The average Nusselt number for a sphere in cross flow may be found using the empirical correlation

$$Nu_D = 2 + (0.4 Re_D^{1/2} + 0.06 Re_D^{2/3}) Pr^{0.4} \left(\frac{\mu}{\mu_s}\right)^{1/4}$$

$$Nu_D = 2 + [0.4(6.154 \times 10^5)^{½} + 0.06(6.154 \times 10^5)^{2/3}](0.704)^{0.4}(195.5 \times 10^{-7}/197.8 \times 10^{-7})^{1/4}$$

$$= 650$$

The average heat transfer coefficient is

$$h = \frac{Nu_D k}{D}$$

$$h = \frac{(650)(0.0280 \text{ W/m·K})}{2.8 \text{ m}}$$

$= 6.5\ \text{W/m}^2\cdot\text{K}$

An energy balance on the radome is written as

$$\dot{E}_{in} + \dot{E}_g - \dot{E}_{out} = \dot{E}_{st}$$

Under steady conditions the rate of energy storage is zero, so we have

$$\dot{E}_{in} + \dot{E}_g - \dot{E}_{out} = 0$$

where the first term is the absorbed solar radiation, the second term is the power dissipation of the equipment within the enclosure, and the third term is the convective heat transfer from the radome surface. Thus,

$$q_s''\alpha_s\pi D^2 + \dot{E}_g - h\pi D^2(T_s - T_\infty) = 0$$

Solving for surface temperature T_s,

$$T_s = \frac{(650\ \text{W/m}^2)(0.1)\pi(2.8\ \text{m})^2 + 2500\ \text{W}}{(650\ \text{W/m}^2)\pi(2.8\ \text{m})^2} + 45\,°\text{C}$$

$$= \underline{\underline{70.6\,°\text{C}}}$$

The air conditioning load is the rate of energy removal by a refrigeration system that offsets the total heat gain for the system,

$$q_{ac} = q_s''\alpha_s\pi D^2 + \dot{E}_g$$

$$q_{ac} = (650\ \text{W/m}^2)(0.1)\pi(2.8\ \text{m})^2 + 2500\ \text{W}$$

$$= \underline{\underline{4101\ \text{W}}}$$

DISCUSSION

In tons of refrigeration, the air conditioning load is

$$q_{ac} = 4101\ \text{W} \times \frac{3.4121\ \text{Btu/h}}{1\ \text{W}} \times \frac{1\ \text{ton}}{12{,}000\ \text{Btu/h}} = 1.2\ \text{ton}$$

which may be met by using a 2-ton unit, providing a factor of safety for potentially higher solar radiation and other contingencies.

The assumed surface temperature is lower than the calculated value, indicating that a correction of thermal properties could be made to improve accuracy of the surface temperature calculation. Also, including emitted radiation from the radome surface would improve accuracy, but the air conditioning load would not change unless the power dissipation, solar radiation, or solar absorptivity were different. As with Problem 7.25, we have used the Nusselt number correlation outside its ranges of applicability.

❐ ❐ ❐

PROBLEM 7.27 Heat Transfer for a Bank of Aligned Tubes

Atmospheric air at 20°C approaches a bank of aligned tubes at a velocity of 6 m/s. The thin-walled metal tubes carry 130°C steam, and have an outside diameter of 1.5 cm. The transverse and longitudinal pitches of the bank are 4.0 cm and 4.2 cm, respectively. There are five rows of tubes in the direction of the air flow and seven tubes per row. Find the heat transfer per unit length for the tube bank.

DIAGRAM

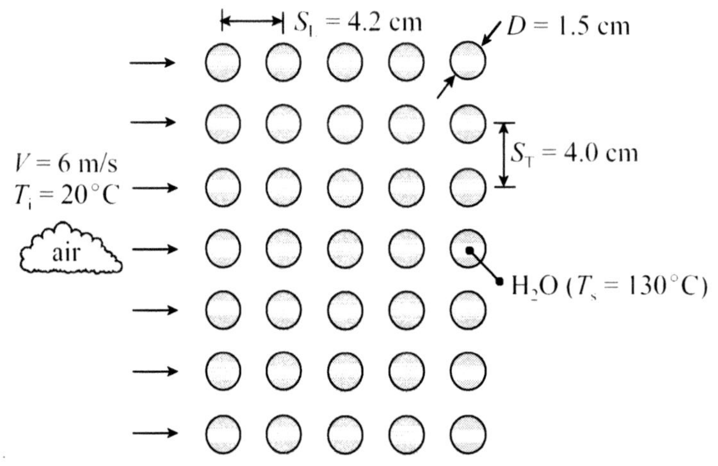

$S_l = 4.2$ cm
$D = 1.5$ cm
$S_T = 4.0$ cm
$V = 6$ m/s
$T_i = 20°C$
air
H_2O ($T_s = 130°C$)

ASSUMPTIONS

1. Steady conditions
2. The outside surface temperature of the tubes is equivalent to the steam temperature.
3. Radiation is neglected.
4. Constant properties.

PROPERTIES

Assume $T_o = 34°C$. $T_{av} = (T_o + T_i)/2 = (34°C + 20°C)/2 = 27°C = 300$ K

atmospheric air: $\rho = 1.1614$ kg/m^3, $c_p = 1007$ J/kg·K, $\nu = 15.89 \times 10^{-6}$ m^2/s, $k = 0.0263$ W/m·K, $Pr = 0.707$, $Pr_s = 0.690$

ANALYSIS

Note that all thermal properties, except Pr_s, are based on an average air temperature in the tube bank using an assumed outlet air temperature. For a bank of aligned tubes, the maximum velocity occurs at a transverse plane and is given by

$$V_{max} = \frac{S_T}{S_T - D} V$$

$$V_{max} = \frac{(0.04 \text{ m})(6 \text{ m/s})}{0.04 \text{ m} - 0.015 \text{ m}}$$

$$= 9.60 \text{ m/s}$$

Thus, the Reynolds number based on the maximum velocity is

$$Re_{D,max} = \frac{V_{max} D}{\nu}$$

$$Re_{D,max} = \frac{(9.60 \text{ m/s})(0.015 \text{ m})}{15.89 \times 10^{-6} \text{ m}^2/\text{s}}$$

$$= 9062$$

A widely used correlation for the average Nusselt number for a bank of tubes is

$$Nu_D = C_2 C Re_{D,max}^m Pr^{0.36} \left(\frac{Pr}{Pr_s} \right)^{1/4}$$

where the constants C and m depend on the type of tube bank (aligned or staggered) and the Reynolds number. For aligned tubes and $10^3 < Re_{D,max} < 2 \times 10^5$, $C = 0.27$ and $m = 0.63$. The constant C_2 is a correction factor to account for fewer than 20 rows of tubes. For five rows, $C_2 =$ 0.92. (Tables of these constants may be found in a heat transfer text). Hence,

$$Nu_D = (0.92)(0.27)(9062)^{0.63}(0.707)^{0.36}(0.707/0.690)^{1/4}$$

$$= 68.6$$

and the average heat transfer coefficient is

$$h = \frac{Nu_D k}{D}$$

$$h = \frac{(68.6)(0.0263 \text{ W/m·K})}{0.015 \text{ m}}$$

$= 120\ \text{W/m}^2\cdot\text{K}$

The outlet air temperature may be calculated from

$$\frac{T_s - T_o}{T_s - T_i} = \exp\left(-\frac{\pi D N h}{\rho V N_T S_T c_p}\right)$$

where N is the total number of tubes in the bank, and N_T is the number of tubes per row. Solving for the difference in surface and outlet air temperatures,

$$T_s - T_o = (T_s - T_i)\exp\left(\frac{-\pi(0.015\ \text{m})(35)(120\ \text{W/m}^2\cdot\text{K})}{(1.1614\ \text{kg/m}^3)(6\ \text{m/s})(7)(0.04\ \text{m})(1007\ \text{J/kg}\cdot\text{K})}\right)$$

$$= (130°\text{C} - 20°\text{C})(0.9042)$$

$$= 99.5°\text{C}$$

Thus, the outlet air temperature is

$$T_o = 130°\text{C} - 99.5°\text{C}$$

$$= 30.5°\text{C}$$

The appropriate form of ΔT for use in a heat transfer calculation is the log mean temperature difference,

$$\Delta T_{lm} = \frac{(T_s - T_i) - (T_s - T_o)}{\ln[(T_s - T_i)/(T_s - T_o)]}$$

$$\Delta T_{lm} = \frac{(130 - 20)°\text{C} - (130 - 30.5)°\text{C}}{\ln[(130 - 20)°\text{C}/(130 - 30.5)°\text{C}]}$$

$$= 104.7°\text{C}$$

The heat transfer per unit length for the tube bank is

$$q' = N h \pi D \Delta T_{lm}$$

$$q' = (35)(120\ \text{W/m}^2\cdot\text{K})\pi(0.015\ \text{m})(104.7°\text{C})$$

$$= 2.072 \times 10^4\ \text{W/m} = \underline{\underline{20.72\ \text{kW/m}}}$$

DISCUSSION

The predicted outlet air temperature is close enough to the assumed value that a thermal property correction is not needed. It is interesting to note that the number of rows, i.e., the number of tubes in the longitudinal direction, may be calculated as $N_L = N/N_T$, so the expression that was used to find the outlet air temperature may be written as

$$\frac{T_s - T_o}{T_s - T_i} = \exp\left(-\frac{\pi D N_L h}{\rho V S_T c_p}\right)$$

As the number of rows N_L increases, the outlet air temperature T_o asymptotically approaches the surface temperature T_s of the tubes. Hence, if the objective of the tube bank is to heat the air as much as possible, a point is reached where adding additional rows of tubes is of no advantage. Of course, that point was not even close to being reached in this problem.

❐ ❐ ❐

PROBLEM 7.28 Heat Transfer for a Bank of Staggered Tubes

Atmospheric air at 15°C approaches a bank of staggered tubes at a velocity of 10 m/s. The thin-walled metal tubes, which have an outside diameter of 4 cm, carry 75°C oil. The transverse and longitudinal pitches of the bank are 8 cm. There are ten rows of tubes in the direction of the air flow and six tubes per row. Find the heat transfer per unit length for the tube bank.

DIAGRAM

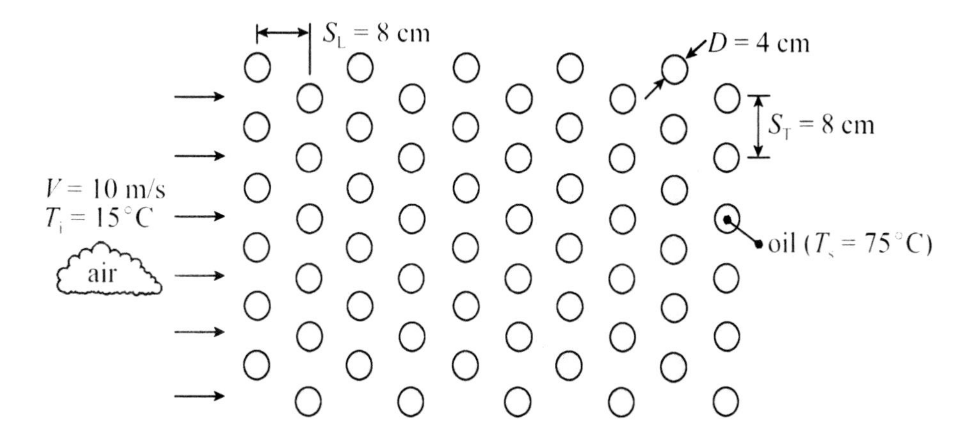

ASSUMPTIONS

1. Steady conditions
2. The outside surface temperature of the tubes is equivalent to the oil temperature.
3. Radiation is neglected.
4. Constant properties.

PROPERTIES

Assume $T_o = 30°C$. $T_{av} = (T_o + T_i)/2 = (30°C + 15°C)/2 = 22.5°C = 296$ K

atmospheric air: $\rho = 1.1801$ kg/m^3, $c_p = 1007$ J/kg·K, $\nu = 15.53 \times 10^{-6}$ m^2/s, $k = 0.0260$ W/m·K, $Pr = 0.708$, $Pr_s = 0.700$

ANALYSIS

Note that all thermal properties, except Pr_s, are based on an average air temperature in the tube bank using an assumed outlet air temperature. For a bank of staggered tubes, the maximum velocity can occur at either a transverse plane (a plane that passes through the centers of the tubes in the same row) or an oblique plane (a plane that passes through the centers of the diagonal tubes in adjacent rows). The maximum velocity occurs at an oblique plane if

$$S_D = \left[S_L^2 + \left(\frac{S_T}{2}\right)^2\right]^{1/2} < \frac{S_T + D}{2}$$

$S_D = [(0.08 \text{ m})^2 + (0.08 \text{ m}/2)^2]^{1/2}$

$= 0.0894$ m

$(S_T + D)/2 = (0.08 \text{ m} + 0.04 \text{ m})/2$

$= 0.060$ m

The condition above is not met, so the maximum velocity occurs at a transverse plane and is given by

$$V_{max} = \frac{S_T}{S_T - D} V$$

$$V_{max} = \frac{(0.08 \text{ m})(10 \text{ m/s})}{0.08 \text{ m} - 0.04 \text{ m}}$$

$= 20.0$ m/s

Thus, the Reynolds number based on the maximum velocity is

$$Re_{D,max} = \frac{V_{max} D}{\nu}$$

$$Re_{D,max} = \frac{(20.0 \text{ m/s})(0.04 \text{ m})}{15.53 \times 10^{-6} \text{ m}^2/\text{s}}$$

$= 5.151 \times 10^4$

A widely used correlation for the average Nusselt number for a bank of tubes is

$$Nu_D = C_2 C Re_{D,\max}^m Pr^{0.36}\left(\frac{Pr}{Pr_s}\right)^{1/4}$$

where the constants C and m depend on the type of tube bank (aligned or staggered) and the Reynolds number. For staggered tubes and $10^3 < Re_{D,\max} < 2 \times 10^5$, $C = 0.35(S_T/S_L)^{1/5} = 0.35$ and $m = 0.60$. The constant C_2 is a correction factor to account for fewer than 20 rows of tubes. For ten rows, $C_2 = 0.97$. (Tables of these constants may be found in a heat transfer text). Hence,

$$Nu_D = (0.97)(0.35)(5.151 \times 10^4)^{0.60}(0.708)^{0.36}(0.708/0.700)^{1/4}$$

$$= 201.9$$

The average heat transfer coefficient is

$$h = \frac{Nu_D k}{D}$$

$$h = \frac{(201.9)(0.0260\ \text{W/m·K})}{0.04\ \text{m}}$$

$$= 131\ \text{W/m}^2\text{·K}$$

The outlet air temperature may be calculated from

$$\frac{T_s - T_o}{T_s - T_i} = \exp\left(-\frac{\pi D N h}{\rho V N_T S_T c_p}\right)$$

where N is the total number of tubes in the bank, and N_T is the number of tubes per row. Solving for the difference in surface and outlet air temperatures,

$$T_s - T_o = (T_s - T_i)\exp\left(\frac{-\pi(0.04\ \text{m})(60)(131\ \text{W/m}^2\text{·K})}{(1.1801\ \text{kg/m}^3)(10\ \text{m/s})(6)(0.08\ \text{m})(1007\ \text{J/kg·K})}\right)$$

$$= (75°\text{C} - 15°\text{C})(0.8410)$$

$$= 50.5°\text{C}$$

Thus, the outlet air temperature is

$$T_o = 75°\text{C} - 50.5°\text{C}$$

$= 24.5°C$

The appropriate form of ΔT for use in a heat transfer calculation is the log mean temperature difference,

$$\Delta T_{lm} = \frac{(T_s - T_i) - (T_s - T_o)}{\ln[(T_s - T_i)/(T_s - T_o)]}$$

$$\Delta T_{lm} = \frac{(75 - 15)°C - (75 - 24.5)°C}{\ln[(75 - 15)°C/(75 - 24.5)°C]}$$

$= 55.1°C$

The heat transfer per unit length for the tube bank is

$$q' = Nh\pi D\Delta T_{lm}$$

$q' = (60)(131\ W/m^2{\cdot}K)\pi(0.04\ m)(55.1°C)$

$= 5.442 \times 10^4\ W/m = \underline{\underline{54.42\ kW/m}}$

DISCUSSION

The predicted outlet air temperature is close enough to the assumed value that a thermal property correction is not needed.

❒ ❒ ❒

PROBLEM 7.29 Air Cooled Steam Condenser

An air cooled steam condenser consists of a 20 × 20 square bank of aligned, thin-walled metal tubes with an outside diameter of 3 cm. The transverse and longitudinal pitches are 5 cm and 6 cm, respectively. Saturated steam at 300 kPa enters the tubes, and atmospheric air at 25°C approaches the tube bank at a velocity of 7 m/s. If the tubes are 1.75 m long, find the heat transfer and the rate of steam condensation for the tube bank.

DIAGRAM

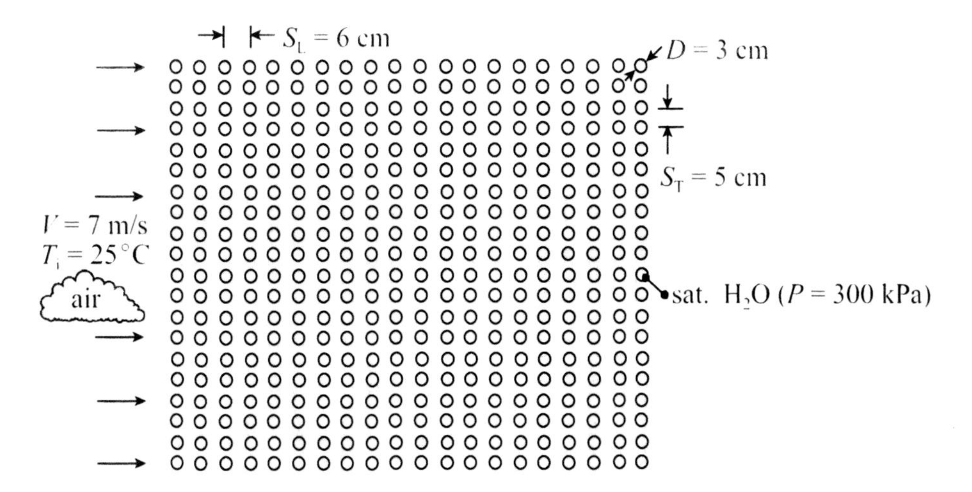

ASSUMPTIONS

1. Steady conditions
2. The outside surface temperature of the tubes is equivalent to the saturation temperature of the steam.
3. Radiation is neglected.
4. Constant properties.

PROPERTIES

saturated water: P_{sat} = 300 kPa, $T_{sat} = T_s$ = 133.5°C

Assume T_o = 35°C. $T_{av} = (T_o + T_i)/2 = (35°C + 25°C)/2 = 30°C = 303$ K

atmospheric air: $\rho = 1.1514$ kg/m^3, $c_p = 1007$ J/kg·K, $\nu = 16.19 \times 10^{-6}$ m^2/s,
$k = 0.0265$ W/m·K, $Pr = 0.707$, $Pr_s = 0.690$

ANALYSIS

Note that all thermal properties, except Pr_s, are based on an average air temperature in the tube bank using an assumed outlet air temperature. For a bank of aligned tubes, the maximum velocity occurs at a transverse plane and is given by

$$V_{max} = \frac{S_T}{S_T - D}V$$

$$V_{max} = \frac{(0.05 \text{ m})(7 \text{ m/s})}{0.05 \text{ m} - 0.03 \text{ m}}$$

$$= 17.5 \text{ m/s}$$

Thus, the Reynolds number based on the maximum velocity is

$$Re_{D,max} = \frac{V_{max} D}{\nu}$$

$$Re_{D,max} = \frac{(17.5 \text{ m/s})(0.03 \text{ m})}{16.19 \times 10^{-6} \text{ m}^2\text{/s}}$$

$$= 3.243 \times 10^4$$

A widely used correlation for the average Nusselt number for a bank of tubes is

$$Nu_D = C Re_{D,max}^m Pr^{0.36} \left(\frac{Pr}{Pr_s} \right)^{1/4}$$

where the constants C and m depend on the type of tube bank (aligned or staggered) and the Reynolds number. For aligned tubes and $10^3 < Re_{D,max} < 2 \times 10^5$, $C = 0.27$ and $m = 0.63$. (Tables of these constants may be found in a heat transfer text). Hence,

$$Nu_D = (0.27)(3.243 \times 10^4)^{0.63}(0.707)^{0.36}(0.707/0.690)^{1/4}$$

$$= 166.6$$

and the average heat transfer coefficient is

$$h = \frac{Nu_D k}{D}$$

$$h = \frac{(166.6)(0.0265 \text{ W/m·K})}{0.03 \text{ m}}$$

$$= 147 \text{ W/m}^2\text{·K}$$

The outlet air temperature may be calculated from

$$\frac{T_s - T_o}{T_s - T_i} = \exp\left(- \frac{\pi D N h}{\rho V N_T S_T c_p} \right)$$

where N is the total number of tubes in the bank, and N_T is the number of tubes per row. Solving for the difference in surface and outlet air temperatures,

$$T_s - T_o = (T_s - T_i)\exp\left(\frac{-\pi(0.03\text{ m})(400)(147\text{ W/m}^2\cdot\text{K})}{(1.1514\text{ kg/m}^3)(7\text{ m/s})(20)(0.05\text{ m})(1007\text{ J/kg}\cdot\text{K})}\right)$$

$$= (133.5°\text{C} - 25°\text{C})(0.5052)$$

$$= 54.8°\text{C}$$

Thus, the outlet air temperature is

$$T_o = 133.5°\text{C} - 54.8°\text{C}$$

$$= 78.7°\text{C}$$

The appropriate form of ΔT for use in a heat transfer calculation is the log mean temperature difference,

$$\Delta T_{lm} = \frac{(T_s - T_i) - (T_s - T_o)}{\ln[(T_s - T_i)/(T_s - T_o)]}$$

$$\Delta T_{lm} = \frac{(133.5 - 25)°\text{C} - (133.5 - 78.7)°\text{C}}{\ln[(133.5 - 25)°\text{C}/(133.5 - 78.7)°\text{C}]}$$

$$= 78.6°\text{C}$$

The tubes are 1.75 m long, so the heat transfer for the tube bank is

$$q = Nh\pi DL\Delta T_{lm}$$

$$q = (400)(147\text{ W/m}^2\cdot\text{K})\pi(0.03\text{ m})(1.75\text{ m})(78.6°\text{C})$$

$$= 7.623 \times 10^5\text{ W} = \underline{\underline{762\text{ kW}}}$$

The rate of steam condensation is the heat transfer divided by the latent heat associated with the phase change (refer to the steam tables of a thermodynamics text),

$$\dot{m}_{con} = \frac{q}{h_{fg}}$$

$$\dot{m}_{con} = \frac{7.623 \times 10^5\text{ W}}{2164 \times 10^3\text{ J/kg}}$$

$$= \underline{\underline{0.352\text{ kg/s}}}$$

DISCUSSION

The calculated temperature of the outlet air is significantly higher than the assumed value, so a correction of thermal properties should be made. The following results are obtained:

$Re_{D,max} = 2.852 \times 10^4$
$Nu_D = 153.2$
$h = 144$ W/m²·K
$T_o = 80.3$°C
$\Delta T_{lm} = 77.6$°C

$q = \underline{\underline{753 \text{ kW}}}$

$\dot{m}_{con} = \underline{\underline{0.348 \text{ kg/s}}}$

A correction of thermal properties changed the heat transfer and condensation rate by less than two percent.

❒ ❒ ❒

PROBLEM 7.30 Heat Transfer for an Impinging Jet

The top surface of a circular plate with a radius of 2 cm is cooled by an impinging jet of air exiting a 5-mm diameter round nozzle at a velocity and temperature of 25 m/s and 30°C, respectively. The centers of the nozzle and plate are aligned, and the plate surface is parallel with the exit plane of the nozzle. The plate has a uniform temperature of 110°C and is 4 cm from the nozzle. If the bottom surface of the plate is insulated, find the heat transfer from the plate.

DIAGRAM

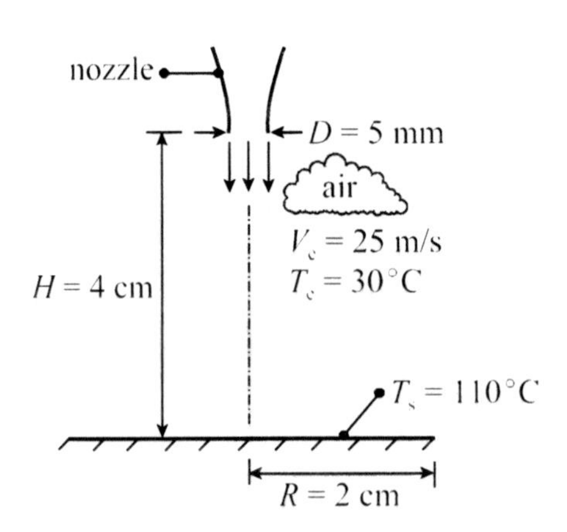

ASSUMPTIONS

1. Steady conditions.
2. The top surface of the plate is isothermal.
3. The bottom surface of the plate is insulated.
4. The impinging jet is directed normal to the plate.
5. Radiation is neglected.
6. Constant properties.

PROPERTIES

$T_f = (T_s + T_e)/2 = (110°C + 30°C)/2 = 70°C = 343$ K

atmospheric air: $k = 0.0295$ W/m·K, $\nu = 20.21 \times 10^{-6}$ m²/s, $Pr = 0.701$

ANALYSIS

The Reynolds number based on the exit velocity and diameter of the jet is

$$Re_D = \frac{V_e D}{\nu}$$

$$Re_D = \frac{(25 \text{ m/s})(0.005 \text{ m})}{20.21 \times 10^{-6} \text{ m}^2/\text{s}}$$

$$= 6185$$

A commonly used correlation for the average Nusselt number for an impinging jet from a single round nozzle is

$$Nu = \left(\frac{D}{R}\right)\frac{1 - 1.1(D/R)}{1 + 0.1(H/D - 6)\,D/R}2\,Re_D^{1/2}(1 + 0.005Re_D^{0.55})^{1/2}\,Pr^{0.42}$$

$$Nu = \frac{(0.005/0.02)[1 - 1.1(0.005/0.02)]2(6185)^{1/2}\,[1 + 0.005(6185)^{0.55}]^{1/2}\,(0.701)^{0.42}}{1 + 0.1[(0.04/0.005) - 6](0.005/0.02)}$$

$$= 25.5$$

The average heat transfer coefficient is

$$h = \frac{Nu_D k}{D}$$

$$h = \frac{(25.5)(0.0295 \text{ W/m·K})}{0.005 \text{ m}}$$

$= 150$ W/m²·K

The heat transfer from the plate is given by Newton's law of cooling,

$$q = h\pi R^2 (T_s - T_e)$$

$q = (150 \text{ W/m}^2\cdot\text{K})\pi(0.02 \text{ m})^2(110 - 30)°\text{C}$

$= \underline{\underline{15.1 \text{ W}}}$

DISCUSSION

The Nusselt number correlation used in the analysis is valid for the ranges $2000 < Re_D < 4 \times 10^5$, $2 < H/D < 12$, and $2.5 < R/D < 7.5$. Hence, we have used the correlation within its limits of applicability.

❐ ❐ ❐

Chapter 8

Internal Forced Convection

PROBLEM 8.1 **Pressure Drop and Power Requirement for Flow in a Cast Iron Pipe**

Water at a mean temperature of 27°C flows with a mean velocity of 2.4 m/s through a 750 m long cast iron pipe with an inside diameter of 12.5 cm. If the surface roughness of the pipe is 260 μm, find the pressure drop across the pipe and the power required to sustain the flow.

DIAGRAM

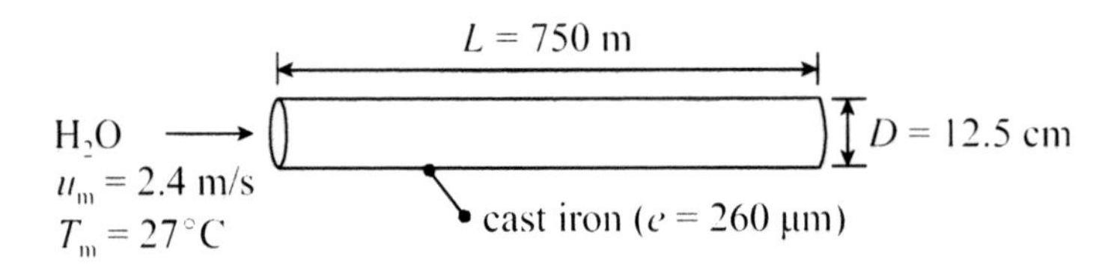

ASSUMPTIONS

1. Steady conditions.
2. Fully developed flow.
3. Constant properties.

PROPERTIES

cast iron surface roughness: $e = 260\ \mu m$

liquid water: $T_m = 27°C = 300\ K$ $\rho = 997\ kg/m^3$, $\mu = 855 \times 10^{-6}\ Pa{\cdot}s$

ANALYSIS

The Reynolds number is

$$Re_D = \frac{\rho u_m D}{\mu}$$

$$Re_D = \frac{(997\ kg/m^3)(2.4\ m/s)(0.125\ m)}{855 \times 10^{-6}\ Pa{\cdot}s}$$

$$= 3.498 \times 10^5$$

so the flow is turbulent. The relative roughness is

$$e/D = (260 \times 10^{-6}\ m)/(0.125\ m)$$

$$= 2.08 \times 10^{-3}$$

A useful correlation for friction factor for turbulent flow in a pipe is

$$f = \frac{0.25}{\left[\log\left(\frac{e/D}{3.7} + \frac{5.74}{Re_D^{0.9}}\right)\right]^2}$$

$$f = \frac{0.25}{\left[\log\left(\frac{2.08 \times 10^{-3}}{3.7} + \frac{5.74}{(3.498 \times 10^5)^{0.9}}\right)\right]^2}$$

$$= 0.0243$$

The pressure drop across the pipe is

$$\Delta p = f\frac{\rho u_m^2}{2D}L$$

$$\Delta p = \frac{(0.0243)(997\ \text{kg/m}^3)(2.4\ \text{m/s})^2(750\ \text{m})}{2(0.125\ \text{m})}$$

$$= 4.186 \times 10^5\ \text{Pa} = \underline{\underline{419\ \text{kPa}}}$$

and the power required to sustain the flow, i.e., the power needed to overcome resistance to flow due to friction, is pressure drop times volume flow rate,

$$P = \Delta p\dot{V} = \Delta p u_m(\pi D^2/4)$$

$$P = (4.186 \times 10^5\ \text{Pa})(2.4\ \text{m/s})\pi(0.125\ \text{m})^2/4$$

$$= 1.233 \times 10^4\ \text{W} = \underline{\underline{12.3\ \text{kW}}}$$

DISCUSSION

The pressure drop across the pipe is 419 kPa (60.8 psi), and the power required by a pump to sustain the flow is 12.3 kW (16.5 hp). As a first approximation, turbulent flow is fully developed if $L/D > 10$, and we have $L/D = (750\ \text{m})/(0.125\ \text{m}) = 6000$. Thus, the second assumption is justified.

PROBLEM 8.2 Mean Velocity and Temperature for Laminar Flow in a Pipe

At a given axial position in a pipe with an inside radius of $R = 2$ cm, the velocity and temperature profiles for laminar incompressible flow are given by the relations

$$u(r) = 0.2\left[1-(r/R)^2\right] \text{ m/s}$$

$$T(r) = 350 + 60(r/R)^2 - 20(r/R)^4 \text{ K}$$

Find the mean velocity and mean (bulk) temperature at this axial position.

DIAGRAM

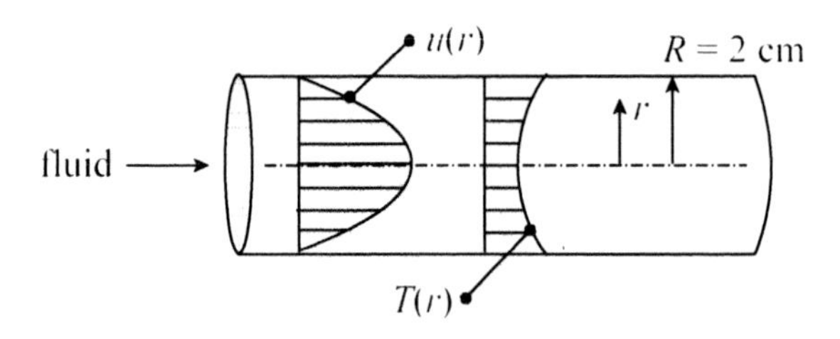

ASSUMPTIONS

1. Laminar incompressible flow.
2. Fully developed flow.

PROPERTIES

not applicable

ANALYSIS

The mean velocity for laminar incompressible flow in a circular pipe is given by

$$u_m = \frac{2}{R^2}\int_0^R u(r)rdr$$

so we have

$$u_m = \frac{2}{R^2}\int_0^R 0.2\left[1-(r/R)^2\right]rdr$$

Carrying out the integration, the mean velocity is

$$u_m = \frac{0.4}{R^2}\left[\frac{R^2}{2} - \frac{R^2}{4}\right]$$

u_m = 0.1 m/s

The mean temperature for laminar incompressible flow in a circular pipe is given by

$$T_m = \frac{2}{u_m R^2}\int_0^R u(r)T(r)rdr$$

so we have

$$T_m = \frac{2}{0.1R^2}\int_0^R 0.2\left[1-(r/R)^2\right]\left[350+60(r/R)^2-20(r/R)^4\right]rdr$$

Carrying out the integration, the mean temperature is

$$T_m = \frac{4}{R^2}\left[175R^2 + 15R^2 - 3.33R^2 - 87.5R^2 - 10R^2 + 2.5R^2\right]$$

T_m = 367 K

DISCUSSION

Using the velocity and temperature profiles, a quick calculation reveals that the velocity varies from 0 to 0.2 m/s and that the temperature varies from 350 to 390 K. Therefore, our results are reasonable. It is instructive to note that $u_{max} = 2u_m = 0.2$ m/s. It is also instructive to note that the mean velocity and temperature are independent of pipe radius *R*, which divides out in the calculations.

❒ ❒ ❒

PROBLEM 8.3 Temperature of Cooling Air in a Rack of Electronics at Elevation

A blower draws 25 °C atmospheric air through a large rack of electronic equipment. The volume flow rate is 0.3 m^3/s, and the total power dissipated by the modules in the rack is 6.4 kW. If the rack is located in a remote military installation at an elevation of 1700 m above sea level, find the temperature of the air as it exits the rack. Where within the rack should the modules with the highest power dissipations be located?

DIAGRAM

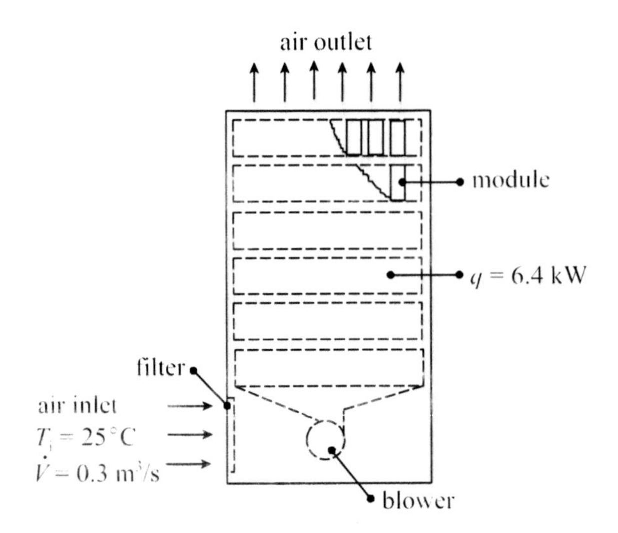

ASSUMPTIONS

1. Steady conditions.
2. All the dissipated heat transfers to the cooling air.
3. There are no air leaks in the rack.
4. Constant properties.

PROPERTIES

atmospheric air: elevation = 1700 m, T_i = 25°C = 298 K, c_p = 1007 J/kg·K

ANALYSIS

We first find the density of air at an elevation of 1700 m above sea level. From a table of atmospheric pressure at various elevations (refer to a fluid mechanics or thermodynamics text), the atmospheric pressure at 1700 m is p = 82.51 kPa. The ideal gas law is

$$p = \rho RT$$

where ρ is density, R is specific gas constant and T is absolute temperature. Thus, the density of the air is

$$\rho = \frac{82.51 \text{ kPa}}{(0.2870 \text{ kJ/kg·K})(25 + 273)\text{K}}$$

$$= 0.9646 \text{ kg/m}^3$$

Hence, the mass flow rate of air through the rack is

$$\dot{m} = \rho \dot{V}$$

$\dot{m}$ = (0.9646 kg/m^3)(0.3 m^3/s)

= 0.2894 kg/s

The first law of thermodynamics for the air flowing through the rack is written as

$$q = \dot{m} c_p (T_o - T_i)$$

Solving for the outlet air temperature T_o,

$$T_o = \frac{6400 \text{ W}}{(0.2894 \text{ kg/s})(1007 \text{ J/kg·K})} + 25\,°\text{C}$$

$$= \underline{\underline{47.0\,°\text{C}}}$$

DISCUSSION

The modules with the highest power dissipations should be located near the inlet of the rack where the air temperature is lowest. This assures that the temperatures of these modules will be minimized. Using Newton's law of cooling,

$$q = hA(T_s - T_\infty)$$

the surface temperature T_s of a module is

$$T_s = \frac{q}{hA} + T_\infty$$

which can be minimized, assuming a fixed heat transfer coefficient h and surface area A, if T_∞ is small. If all the modules have the same power dissipation, the temperature of the modules at the outlet of the rack are approximately 22 °C hotter than those near the inlet.

❒ ❒ ❒

PROBLEM 8.4 Power Requirement of a Resistance Heater

Water is to be heated from 15 °C to 55 °C as it flows through a thin-walled copper tube wrapped with an electric resistance heater. If a volume flow rate of 10 L/min is desired, what is the required power rating of the resistance heater? If the tube has a length and diameter of 6 m and 1.3 cm, respectively, what is the heat flux at the inside surface of the tube? The outside surface of the resistance heater is heavily insulated.

DIAGRAM

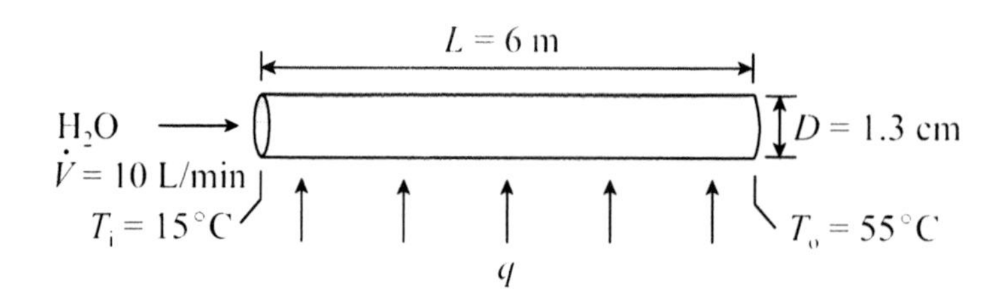

ASSUMPTIONS

1. Steady conditions.
2. All the heat dissipated by the resistance heater transfers to the water.
3. The heat flux induced by the resistance heater is uniform.
4. Constant properties.

PROPERTIES

$T_{av} = (T_i + T_o)/2 = (15° + 55°C)/2 = 35°C = 308$ K

liquid water: $\rho = 994$ kg/m^3, $c_p = 4178$ J/kg·K

ANALYSIS

The mass flow rate of the water is

$$\dot{m} = \rho \dot{V}$$

$$\dot{m} = (994 \text{ kg/m}^3)(10 \text{ L/min} \times 1 \text{ m}^3/1000 \text{ L} \times 1 \text{ min}/60 \text{ s})$$

$$= 0.1657 \text{ kg/s}$$

The first law of thermodynamics for the water flowing through the tube is written as

$$q = \dot{m}c_p(T_o - T_i)$$

where q is the heat dissipated by the resistance heater, which is equivalent to the power rating of the heater since all the electrical energy generated by the heater transfers to the water. Thus,

$$q = (0.1657 \text{ kg/s})(4178 \text{ J/kg·K})(55 - 15)°\text{C}$$

$$= 2.769 \times 10^4 \text{ W} = \underline{\underline{27.7 \text{ kW}}}$$

Heat flux is heat transfer divided by the surface area of the tube,

$$q'' = \frac{q}{\pi D L}$$

$$q'' = \frac{(2.769 \times 10^4 \text{ W})}{\pi(0.013 \text{ m})(6 \text{ m})}$$

$$= 1.130 \times 10^5 \text{ W/m}^2 = \underline{\underline{113 \text{ kW/m}^2}}$$

DISCUSSION

Because the tube has a thin wall, the heat flux at any point in the tube wall is approximately the same. The mean temperature of a fluid flowing in a tube, subject to a uniform heat flux, at any distance x from the inlet is given by

$$T(x) = \frac{\pi D q''}{\dot{m} c_p} x + T_i$$

which is simply the first law of thermodynamics for the fluid over a section of tube length x as measured from the inlet. This relation reveals that the fluid temperature is a linear function of heat flux. Substituting values, the temperature of the water is given by

$$T(x) = \frac{\pi(0.013 \text{ m})(1.130 \times 10^5 \text{ W/m}^2)}{(0.1657 \text{ kg/s})(4178 \text{ J/kg·K})} x + 15°\text{C}$$

$$= (6.67\, x + 15)°\text{C}$$

For example, at $x = 5$ m, $T(5) = (6.67°\text{C/m})(5 \text{ m}) + 15°\text{C} = 48.4°\text{C}$. It is straightforward to verify that $T(0) = 15°\text{C}$ and $T(6) = 55°\text{C}$, the inlet and outlet temperatures of the water.

If the temperature of the inside surface of the tube at distance x was known, we could calculate the local heat transfer coefficient using Newton's law of cooling,

$$h = \frac{q''}{T_s(x) - T(x)}$$

For example, assuming that $T_s = 75°\text{C}$ at the outlet of the tube, the local heat transfer coefficient is

$$h = \frac{1.130 \times 10^5 \text{ W/m}^2}{(75 - 55)°\text{C}}$$

$$= 5650 \text{ W/m}^2\text{·K}$$

❒ ❒ ❒

PROBLEM 8.5 Engine Oil Flowing in a Tube

Engine oil flows through a 3.6-cm diameter, 8-m long tube at a mass flow rate of 0.35 kg/s. The oil enters the tube at a temperature of 20°C. If the inside tube surface is maintained at 80°C, find the heat transfer and the outlet temperature of the oil.

DIAGRAM

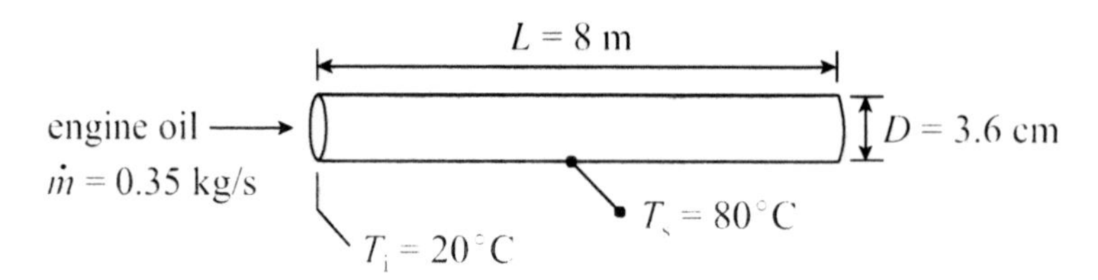

ASSUMPTIONS

1. Steady conditions.
2. The tube surface is isothermal.
3. Constant properties.

PROPERTIES

Assume $T_o = 30°C$. $T_{av} = (T_i + T_o)/2 = (20°C + 30°C)/2 = 25°C = 298$ K

engine oil: $c_p = 1901$ J/kg·K, $k = 0.145$ W/m·K, $\nu = 664 \times 10^{-6}$ m²/s, $Pr = 7700$
$\mu = 0.589$ Pa·s

ANALYSIS

The Reynolds number is

$$Re_D = \frac{4\dot{m}}{\pi D \mu}$$

$$Re_D = \frac{4(0.35 \text{ kg/s})}{\pi(0.036 \text{ m})(0.589 \text{ Pa·s})}$$

$$= 21.0$$

Hence, the flow is laminar. In order to use the appropriate Nusselt number correlation, we must determine if the flow is hydrodynamically or thermally fully developed. For laminar flow, the hydrodynamic entry length is approximated by

$$\left(\frac{x_{fd,h}}{D}\right)_{lam} \approx 0.05Re_D$$

Hence, the hydrodynamic entry length is

$$x_{fd,h} \approx 0.05(21.0)(0.036 \text{ m})$$

$$= 0.0378 \text{ m}$$

which is less than the length of the tube, so the flow is hydrodynamically fully developed. The thermal entry length is approximated by

$$\left(\frac{x_{fd,t}}{D}\right)_{lam} \approx 0.05Re_D Pr$$

Hence, the thermal entry length is

$$x_{fd,t} \approx 0.05(21.0)(7700)(0.036 \text{ m})$$

$$= 291 \text{ m}$$

which is greater than the length of the tube, so the flow is not thermally fully developed. A widely used correlation for the average Nusselt number for thermally developing laminar flow is

$$Nu_D = 3.66 + \frac{0.0668(D/L)Re_D Pr}{1 + 0.04[(D/L)Re_D Pr]^{2/3}}$$

$$Nu_D = 3.66 + \frac{0.0668(0.036 \text{ m}/8 \text{ m})(21.0)(7700)}{1 + 0.04[(0.036 \text{ m}/8 \text{ m})(21.0)(7700)]^{2/3}}$$

$$= 15.13$$

The average heat transfer coefficient is

$$h = \frac{Nu_D k}{D}$$

$$h = \frac{(15.13)(.0145 \text{ W/m·K})}{0.036 \text{ m}}$$

$$= 60.9 \text{ W/m}^2\text{·K}$$

For the case of constant surface temperature, an energy balance yields

$$\frac{T_s - T_o}{T_s - T_i} = \exp\left(-\frac{\pi DL}{\dot{m} c_p} h \right)$$

$$T_s - T_o = (T_s - T_i) \exp \left(-\frac{\pi(0.036\ \text{m})(8\ \text{m})(60.9\ \text{W/m}^2\cdot\text{K})}{(0.35\ \text{kg/s})(1901\ \text{J/kg}\cdot\text{K})} \right)$$

$$= (80°\text{C} - 20°\text{C})(0.9205)$$

$$= 55.2°\text{C}$$

Thus, the outlet oil temperature is

$$T_o = 80°\text{C} - 55.2°\text{C}$$

$$= \underline{\underline{24.8°\text{C}}}$$

and the heat transfer is

$$q = \dot{m} c_p (T_o - T_i)$$

$$q = (0.35\ \text{kg/s})(1901\ \text{J/kg}\cdot\text{K})(24.8 - 20)°\text{C}$$

$$= \underline{\underline{3194\ \text{W}}}$$

DISCUSSION

If we had assumed thermally fully developed flow, the Nusselt number would have been

$$Nu_D = 3.66$$

which can be obtained by solving the energy equation for fully developed laminar flow. Also note that this result is obtained from the Nusselt number correlation above for the case $D/L \rightarrow 0$. The resulting heat transfer coefficient, outlet oil temperature and heat transfer are

$h = 14.7\ \text{W/m}^2\cdot\text{K}$
$T_o = 21.2°\text{C}$
$q = 798\ \text{W}$

Clearly, assuming thermally fully developed flow leads to significant errors.

❒ ❒ ❒

PROBLEM 8.6 Water Flowing in a Plastic Pipe

A plastic pipe with a diameter of 5 cm carries water at a mean velocity of 3 m/s. The water enters the pipe at 10°C, and the inside surface of the pipe is maintained at 75°C. If the pipe is 16 m long, find the heat transfer and the outlet temperature of the water.

DIAGRAM

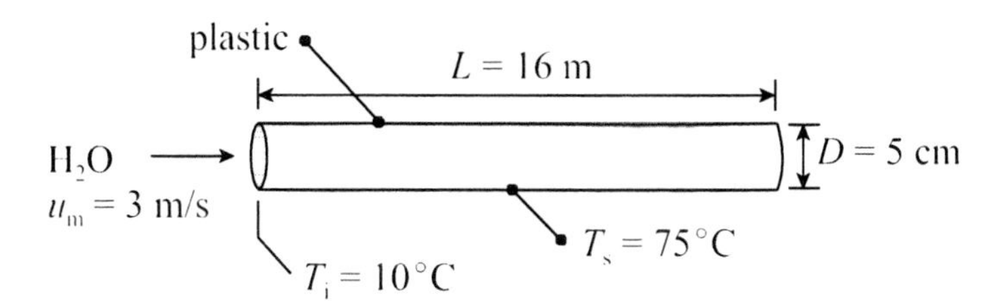

ASSUMPTIONS

1. Steady conditions.
2. The tube surface is isothermal and smooth.
3. Constant properties.

PROPERTIES

Assume $T_o = 44°C$. $T_{av} = (T_i + T_o)/2 = (10°C + 44°C)/2 = 27°C = 300$ K

liquid water: $\rho = 997$ kg/m^3, $c_p = 4179$ J/kg·K, $k = 0.613$ W/m·K, $Pr = 5.83$
$\nu = 8.576 \times 10^{-7}$ m^2/s

ANALYSIS

The Reynolds number is

$$Re_D = \frac{u_m D}{\nu}$$

$$Re_D = \frac{(3 \text{ m/s})(0.05 \text{ m})}{8.576 \times 10^{-7} \text{ m}^2\text{/s}}$$

$$= 1.749 \times 10^5$$

so the flow is turbulent. Fully developed turbulent flow (hydrodynamic and thermal) is generally achieved at a distance of ten pipe diameters from the inlet. The pipe is 16 m long, and ten diameters is only 0.5 m, so we can assume fully developed flow in the entire pipe. A commonly used correlation for the average Nusselt number for fully developed turbulent flow in a smooth circular pipe is

$$Nu_D = 0.023 Re_D^{4/5} Pr^n$$

where $n = 0.4$ when the fluid is heated, and $n = 0.3$ when the fluid is cooled. We have

$$Nu_D = 0.023(1.749 \times 10^5)^{4/5}(5.83)^{0.4}$$

$$= 728.2$$

and the average heat transfer coefficient is

$$h = \frac{Nu_D k}{D}$$

$$h = \frac{(728.2)(0.613\ \text{W/m·K})}{0.05\ \text{m}}$$

$$= 8928\ \text{W/m}^2\text{·K}$$

Mass flow rate is

$$\dot{m} = \rho u_m \pi D^2/4$$

$$\dot{m} = (997\ \text{kg/m}^3)(3\ \text{m/s})\pi(0.05\ \text{m})^2/4$$

$$= 5.873\ \text{kg/s}$$

For the case of constant surface temperature, an energy balance yields

$$\frac{T_s - T_o}{T_s - T_i} = \exp\left(-\frac{\pi DL}{\dot{m}c_p}h\right)$$

$$T_s - T_o = (T_s - T_i)\exp\left(-\frac{\pi(0.05\ \text{m})(16\ \text{m})(8928\ \text{W/m}^2\text{·K})}{(5.873\ \text{kg/s})(4179\ \text{J/kg·K})}\right)$$

$$= (75°\text{C} - 10°\text{C})(0.4008)$$

$$= 26.1°\text{C}$$

Thus, the outlet water temperature is

$$T_o = 75°\text{C} - 26.1°\text{C}$$

$$= \underline{\underline{48.9°\text{C}}}$$

and the heat transfer is

$$q = \dot{m}c_p(T_o - T_i)$$

q = (5.873 kg/s)(4179 J/kg·K)(48.9 - 10)°C

= 9.547 × 10^5 W = 955 kW

DISCUSSION

A more accurate Nusselt number correlation is

$$Nu_D = \frac{(f/8)(Re_D - 1000)Pr}{1 + 12.7(f/8)^{1/2}(Pr^{2/3} - 1)}$$

where f is friction factor obtained from the Moody diagram, or from the relation

$$f = \frac{0.25}{\left[\log\left(\frac{e/D}{3.7} + \frac{5.74}{Re_D^{0.9}}\right)\right]^2}$$

Plastic is assumed to be smooth, so $e = 0$, and the relation above reduces to

$$f = \frac{0.25}{\left[\log\left(\frac{5.74}{Re_D^{0.9}}\right)\right]^2}$$

$$f = \frac{0.25}{\{\log[5.74/(1.749 \times 10^5)^{0.9}]\}^2}$$

= 0.0159

and the Nusselt number is

$$Nu_D = \frac{(0.0159/8)(1.749 \times 10^5 - 1000)(5.83)}{1 + 12.7(0.0159/8)^{1/2}(5.83^{2/3} - 1)}$$

= 888.5

which yields a heat transfer coefficient of h = 1.089 × 10^4 W/m²·K. The outlet water temperature and heat transfer, respectively, are T_o = 53.7°C and q = 1.32 MW.

PROBLEM 8.7 Water Flowing in a Steel Pipe Subjected to a Cross Flow of Air

A bare, 2½ inch schedule 40 steel pipe (D_i = 62.7 mm, D_o = 73.0 mm) carries water at a mean velocity of 1.2 m/s while subjected to the cross flow of atmospheric air at 4 m/s and 15°C. The water enters the pipe at 80°C. If the pipe is 750 m long, find the heat transfer from the pipe and the outlet temperature of the water. The thermal conductivity and inside surface roughness of the steel pipe are k_p = 60 W/m·K and e = 46 μm, respectively.

DIAGRAM

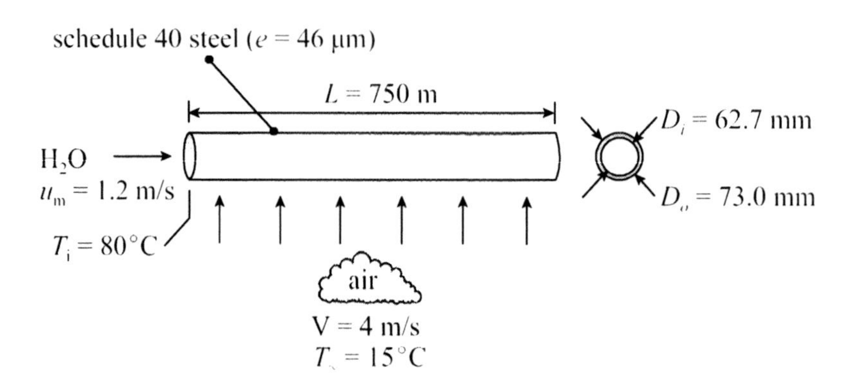

ASSUMPTIONS

1. Steady conditions.
2. Air flow is normal to the axis of the pipe.
3. The outside surface of the pipe is smooth.
4. Fully developed flow.
5. Radiation is neglected.
6. Constant properties.

PROPERTIES

steel pipe: k_p = 60 W/m·K, e = 46 μm

Assume T_o = 65°C. $T_{av} = (T_i + T_o)/2$ = (80°C + 65°C)/2 = 72.5°C = 346 K

liquid water: ρ = 976 kg/m^3, c_p = 4192 J/kg·K, k_1 = 0.668 W/m·K, Pr_1 = 2.42
$\nu_1 = 3.937 \times 10^{-7}$ m^2/s

Assume T_s = 70°C. $T_f = (T_s + T_\infty)/2$ = (70°C + 15°C)/2 = 42.5°C = 316 K

atmospheric air: k_2 = 0.0275 W/m·K, $\nu_2 = 17.50 \times 10^{-6}$ m^2/s, Pr_2 = 0.705

ANALYSIS

We must calculate two heat transfer coefficients, one for the internal water flow and one for the external air flow. For the internal flow, the Reynolds number is

$$Re_D = \frac{u_m D_i}{\nu_1}$$

$$Re_D = \frac{(1.2 \text{ m/s})(0.0627 \text{ m})}{3.937 \times 10^{-7} \text{ m}^2/\text{s}}$$

$$= 1.911 \times 10^5$$

so the flow is turbulent. The relative roughness is

$$e/D_i = (46 \times 10^{-6} \text{ m})/(0.0627 \text{ m})$$

$$= 7.34 \times 10^{-4}$$

so the friction factor is

$$f = \frac{0.25}{\left[\log\left(\frac{e/D_i}{3.7} + \frac{5.74}{Re_D^{0.9}}\right)\right]^2}$$

$$f = \frac{0.25}{\left[\log\left(\frac{7.34 \times 10^{-4}}{3.7} + \frac{5.74}{(1.911 \times 10^5)^{0.9}}\right)\right]^2}$$

$$= 0.0201$$

The Nusselt number is

$$Nu_D = \frac{(f/8)(Re_D - 1000)Pr_1}{1 + 12.7(f/8)^{1/2}(Pr_1^{2/3} - 1)}$$

$$Nu_D = \frac{(0.0201/8)(1.911 \times 10^5 - 1000)(2.42)}{1 + 12.7(0.0201/8)^{1/2}(2.42^{2/3} - 1)}$$

$$= 765.0$$

Thus, the heat transfer coefficient is

$$h_1 = \frac{Nu_D k_1}{D_i}$$

$$h_1 = \frac{(765.0)(0.668 \text{ W/m·K})}{0.0627 \text{ m}}$$

$$= 8150 \text{ W/m}^2\text{·K}$$

Mass flow rate is

$$\dot{m} = \rho u_m \pi D_i^2 / 4$$

$$\dot{m} = (976 \text{ kg/m}^3)(1.2 \text{ m/s})\pi(0.0627 \text{ m})^2/4$$

$$= 3.616 \text{ kg/s}$$

For the external flow, the Reynolds number is

$$Re_D = \frac{VD_o}{\nu_2}$$

$$Re_D = \frac{(4 \text{ m/s})(0.073 \text{ m})}{17.50 \times 10^{-6} \text{ m}^2\text{/s}}$$

$$= 1.669 \times 10^4$$

and the average Nusselt number is

$$Nu_D = C Re_D^m Pr_2^{1/3}$$

where, for $Re_D = 1.669 \times 10^4$, $C = 0.193$ and $m = 0.618$. Thus,

$$Nu_D = (0.193)(1.669 \times 10^4)^{0.618}(0.705)^{1/3}$$

$$= 69.9$$

and the average heat transfer coefficient is

$$h_2 = \frac{Nu_D k_2}{D_o}$$

$$h_2 = \frac{(69.9)(0.0275 \text{ W/m·K})}{0.073 \text{ m}}$$

$= 26.3\ \text{W/m}^2\cdot\text{K}$

Now that both heat transfer coefficients have been found, we calculate the total thermal resistance,

$$R_{tot} = \frac{1}{h_1 \pi D_i L} + \frac{\ln(D_o/D_i)}{2\pi k_p L} + \frac{1}{h_2 \pi D_o L}$$

$$R_{tot} = \frac{1}{(8150\ \text{W/m}^2\cdot\text{K})\pi(0.0627\ \text{m})(750\ \text{m})} + \frac{\ln(0.073\ \text{m}/0.0627\ \text{m})}{2\pi(60\ \text{W/m}\cdot\text{K})(750\ \text{m})}$$

$$+ \frac{1}{(26.3\ \text{W/m}^2\cdot\text{K})\pi(0.073\ \text{m})(750\ \text{m})}$$

$$= (8.306 \times 10^{-7} + 5.379 \times 10^{-7} + 2.211 \times 10^{-4})\ \text{K/W}$$

$$= 2.224 \times 10^{-4}\ \text{K/W}$$

Finally, to find the outlet temperature of the water, we use the relation

$$\frac{T_\infty - T_o}{T_\infty - T_i} = \exp\left(-\frac{1}{\dot{m} c_p R_{tot}}\right)$$

$$T_\infty - T_o = (T_\infty - T_i)\exp\left(\frac{-1}{(3.616\ \text{kg/s})(4192\ \text{J/kg}\cdot\text{K})(2.224 \times 10^{-4}\ \text{K/W})}\right)$$

$$= (15°\text{C} - 80°\text{C})(0.7433)$$

$$= -48.3°\text{C}$$

Thus, the outlet water temperature is

$$T_o = 15°\text{C} + 48.3°\text{C}$$

$$= \underline{\underline{63.3°\text{C}}}$$

and the heat transfer is

$$q = \dot{m} c_p (T_i - T_o)$$

$$q = (3.616\ \text{kg/s})(4192\ \text{J/kg}\cdot\text{K})(80 - 63.3)°\text{C}$$

$$= 2.531 \times 10^5\ \text{W} = \underline{\underline{253\ \text{kW}}}$$

DISCUSSION

We note that the thermal resistance of the outside boundary layer is roughly three orders of magnitude larger than that of the inside boundary layer and the pipe wall. Hence, to a first approximation, we could have ignored the thermal resistance of the inside boundary layer and pipe wall. Our initial estimate of the outlet water temperature is close to the calculated value, so no thermal property corrections are required for the water. To check our estimate of the outside surface temperature of the pipe, we apply Newton's law of cooling,

$$q = h_2 \pi D_o L (T_s - T_\infty)$$

Solving for surface temperature T_s,

$$T_s = \frac{2.531 \times 10^5 \text{ W}}{(26.3 \text{ W/m}^2\cdot\text{K})\pi(0.073 \text{ m})(750 \text{ m})} + 15°\text{C}$$

$$= 71.0°\text{C}$$

Since the surface temperature of the pipe varies along the pipe's length, this value is to be regarded as an average. Again, the calculated value is very close to our initial estimate.

❒ ❒ ❒

PROBLEM 8.8 Heat Loss from a Square Sheet Metal Duct

A 30 m long duct with a 40 cm × 40 cm cross section carries warm air across a warehouse where the operating space is maintained at 20°C. Air flows through the duct, which is constructed of thin sheet metal, at a mean velocity of 2 m/s. Natural air currents inside the warehouse induce a steady 0.8 m/s flow of air across the duct. If the warm air enters the duct at 45°C, find the heat loss from the duct and the temperature of the air at the duct outlet.

DIAGRAM

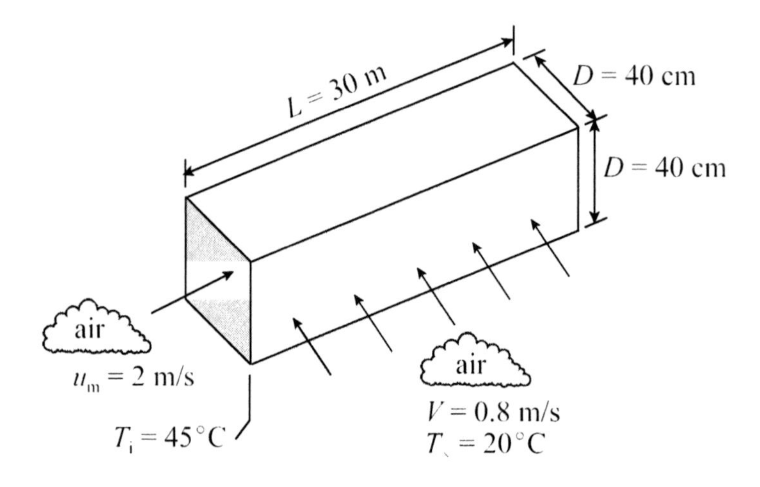

ASSUMPTIONS

1. Steady conditions.
2. Air flow is normal to the axis and one side of the duct.
3. The inside and outside surfaces of the duct are smooth.
4. Fully developed flow.
5. Radiation is neglected.
6. Constant properties.

PROPERTIES

Assume $T_o = 30°C$. $T_{av} = T_f = (T_i + T_o)/2 = (45°C + 30°C)/2 = 37.5°C = 311$ K

atmospheric air: $\rho = 1.1248$ kg/m³, $c_p = 1007$ J/kg·K, $k = 0.0271$ W/m·K, $\nu = 17.00 \times 10^{-6}$ m²/s, $Pr = 0.706$

ANALYSIS

To simplify the analysis, we have based thermal properties of duct air and warehouse air on the same temperature. Two heat transfer coefficients are required, one for the internal air flow and one for the external air flow. For the internal flow, Reynolds number is based on the hydraulic diameter of the duct,

$$D_h = \frac{4A}{P}$$

where cross sectional area is $A = D^2$ and wetted perimeter is $P = 4D$. Thus,

$$D_h = \frac{4(0.40 \text{ m})^2}{4(0.40 \text{ m})}$$

$$= 0.40 \text{ m}$$

The Reynolds number is

$$Re_{D_h} = \frac{u_m D_h}{\nu}$$

$$Re_{Dh} = \frac{(2 \text{ m/s})(0.40 \text{ m})}{17.00 \times 10^{-6} \text{ m}^2/\text{s}}$$

$$= 4.706 \times 10^4$$

so the flow is turbulent. A correlation for the Nusselt number is

$$Nu_D = \frac{(f/8)(Re_D - 1000)\,Pr}{1 + 12.7(f/8)^{1/2}(Pr^{2/3} - 1)}$$

where f is friction factor obtained from the relation

$$f = \frac{0.25}{\left[\log\left(\frac{e/D}{3.7} + \frac{5.74}{Re_D^{0.9}}\right)\right]^2}$$

Sheet metal is assumed to be smooth, so $e = 0$, and the relation above reduces to

$$f = \frac{0.25}{\left[\log\left(\frac{5.74}{Re_D^{0.9}}\right)\right]^2}$$

$$f = \frac{0.25}{\{\log[5.74/(4.706 \times 10^4)^{0.9}]\}^2}$$

$$= 0.0210$$

The Nusselt number is

$$Nu_D = \frac{(0.0210/8)(4.706 \times 10^4 - 1000)(0.706)}{1 + 12.7(0.0210/8)^{1/2}(0.706^{2/3} - 1)}$$

$$= 98.7$$

so the heat transfer coefficient is

$$h_1 = \frac{Nu_D k}{D_h}$$

$$h_1 = \frac{(98.7)(0.0271\ \text{W/m·K})}{0.40\ \text{m}}$$

$$= 6.69\ \text{W/m}^2\text{·K}$$

Mass flow rate is

$$\dot{m} = \rho u_m D^2$$

$$\dot{m} = (1.1248\ \text{kg/m}^3)(2\ \text{m/s})(0.40\ \text{m})^2$$

$$= 0.3599\ \text{kg/s}$$

For the external flow, the Reynolds number is

$$Re_D = \frac{VD}{\nu}$$

$$Re_D = \frac{(0.8 \text{ m/s})(0.40 \text{ m})}{17.00 \times 10^{-6} \text{ m}^2/\text{s}}$$

$$= 1.882 \times 10^4$$

and the average Nusselt number is

$$Nu_D = C Re_D^m Pr_2^{1/3}$$

where, for $Re_D = 1.882 \times 10^4$, $C = 0.102$ and $m = 0.675$. Thus,

$$Nu_D = (0.102)(1.882 \times 10^4)^{0.675}(0.706)^{1/3}$$

$$= 69.8$$

and the average heat transfer coefficient is

$$h_2 = \frac{Nu_D k}{D}$$

$$h_2 = \frac{(69.8)(0.0271 \text{ W/m·K})}{0.40 \text{ m}}$$

$$= 4.73 \text{ W/m}^2\text{·K}$$

Now that both heat transfer coefficients have been found, we calculate the total thermal resistance. Neglecting the thermal resistance of the duct wall,

$$R_{tot} = \frac{1}{PL}\left(\frac{1}{h_1} + \frac{1}{h_2}\right)$$

$$R_{tot} = \frac{1}{4(0.40 \text{ m})(30 \text{ m})}\left(\frac{1}{6.69 \text{ W/m}^2\text{·K}} + \frac{1}{4.73 \text{ W/m}^2\text{·K}}\right)$$

$$= 7.519 \times 10^{-3} \text{ K/W}$$

To find the outlet temperature of the warm air, we use the relation

$$\frac{T_\infty - T_o}{T_\infty - T_i} = \exp\left(-\frac{1}{\dot{m} c_p R_{tot}}\right)$$

$$T_\infty - T_o = (T_\infty - T_i)\exp\left(\frac{-1}{(0.3599\text{ kg/s})(1007\text{ J/kg·K})(7.519\times10^{-3}\text{ K/W})}\right)$$

$$= (20°\text{C} - 45°\text{C})(0.6928)$$

$$= -17.3°\text{C}$$

Thus, the outlet air temperature is

$$T_o = 20°\text{C} + 17.3°\text{C}$$

$$= \underline{\underline{37.3°\text{C}}}$$

and the heat loss from the duct is

$$q = \dot{m}c_p(T_i - T_o)$$

$$q = (0.3599\text{ kg/s})(1007\text{ J/kg·K})(45 - 37.3)°\text{C}$$

$$= \underline{\underline{2791\text{ W}}}$$

DISCUSSION

The thermal resistance of the duct wall was neglected because the wall is thin and made of a material with a high thermal conductivity. The thermal resistance of neither boundary layer could be neglected, however, because the heat transfer coefficients are comparable. If radiation had been included, the heat loss from the duct would be larger. Our calculated outlet air temperature is close enough to the initial estimate that a correction of thermal properties is not warranted.

❐ ❐ ❐

PROBLEM 8.9 Design of a Geothermal Air Conditioning System

A geothermal air conditioning system for a house consists of several closed loops of polybutylene tubing (D_i = 20.9 mm, D_o = 26.7 mm, k_p = 0.18 W/m·K), buried deep in the ground, that circulate a water/ethylene glycol coolant through a liquid-to-air heat exchanger inside the house. The average deep ground temperature during the summer is 9°C, and the mean velocity of the coolant is 1.5 m/s. If the coolant is 18°C at the inlet of the tube, how long must each loop be for the coolant temperature to be 10°C at the outlet of the tube?

DIAGRAM

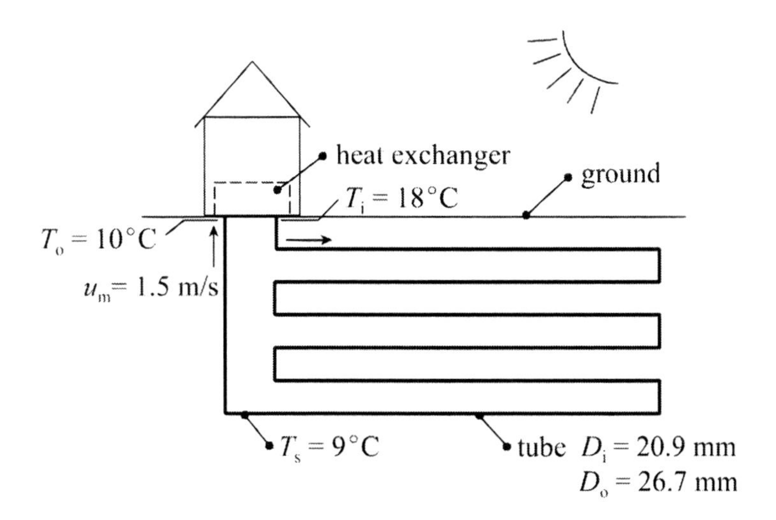

ASSUMPTIONS

1. Steady conditions.
2. The inside surface of the tube is smooth.
3. The outside surface of the tube is constant at the deep ground temperature.
4. Fully developed flow.
5. Constant properties.

PROPERTIES

polybutylene: k_p = 0.18 W/m·K

$T_{av} = (T_i + T_o)/2 = (18°C + 10°C)/2 = 14°C = 287$ K

H_2O/ethylene glycol: ρ = 1076 kg/m^3, c_p = 3258 J/kg·K, k = 0.376 W/m·K,
ν = 4.53 × 10^{-6} m^2/s, Pr = 42.3

ANALYSIS

The Reynolds number is

$$Re_D = \frac{u_m D_i}{\nu}$$

$$Re_D = \frac{(1.5 \text{ m/s})(0.0209 \text{ m})}{4.53 \times 10^{-6} \text{ m}^2\text{/s}} = 6921$$

The flow is turbulent. Friction factor f is obtained from the relation

$$f = \frac{0.25}{\left[\log\left(\frac{e/D_i}{3.7} + \frac{5.74}{Re_D^{0.9}}\right)\right]^2}$$

The polybutylene surface is assumed to be smooth, so $e = 0$, and the relation above reduces to

$$f = \frac{0.25}{\left[\log\left(\frac{5.74}{Re_D^{0.9}}\right)\right]^2}$$

$$f = \frac{0.25}{\{\log[5.74/(6921)^{0.9}]\}^2}$$

$$= 0.0344$$

The Nusselt number is

$$Nu_D = \frac{(f/8)(Re_D - 1000)Pr_1}{1 + 12.7(f/8)^{1/2}(Pr_1^{2/3} - 1)}$$

$$Nu_D = \frac{(0.0344/8)(6921 - 1000)(42.3)}{1 + 12.7(0.0344/8)^{1/2}(42.3^{2/3} - 1)}$$

$$= 104.8$$

so the heat transfer coefficient is

$$h = \frac{Nu_D k}{D_i}$$

$$h = \frac{(104.8)(0.376 \text{ W/m·K})}{0.0209 \text{ m}}$$

$$= 1885 \text{ W/m}^2\text{·K}$$

Mass flow rate is

$$\dot{m} = \rho u_m \pi D_i^2/4$$

$$\dot{m} = (1076 \text{ kg/m}^3)(1.5 \text{ m/s})\pi(0.0267 \text{ m})^2/4$$

$$= 0.9037 \text{ kg/s}$$

The total thermal resistance is

$$R_{tot} = \frac{1}{L}\left(\frac{1}{\pi D_i h} + \frac{\ln(D_o/D_i)}{2\pi k_p}\right)$$

$$R_{tot} = \frac{1}{L}\left(\frac{1}{\pi(0.0209\text{ m})(1885\text{ W/m}^2\cdot\text{K})} + \frac{\ln(0.0267\text{ m}/0.0209\text{ m})}{2\pi(0.18\text{ W/m}\cdot\text{K})}\right)$$

$$= 0.2246/L \text{ K/W}$$

To find the required length of tube, we use the relation

$$\frac{T_s - T_o}{T_s - T_i} = \exp\left(-\frac{1}{\dot{m}c_p R_{tot}}\right)$$

where T_s is the outside surface temperature of the tube. Solving for L, the length of one tube,

$$L = -0.2246\dot{m}c_p \ln\left(\frac{T_s - T_o}{T_s - T_i}\right)$$

$$L = -(0.2246)(0.9037\text{ kg/s})(3258\text{ J/kg}\cdot\text{K}) \ln\left(\frac{9°\text{C} - 10°\text{C}}{9°\text{C} - 18°\text{C}}\right)$$

$$= \underline{\underline{1453\text{ m}}}$$

DISCUSSION

Each loop of tubing must be at least 1453 m long. Several tubing loops would be required because one loop does not have enough surface area for the heat exchanger to handle the cooling load of the house. The heat exchanger would consist of several tubes, formed into coils to increase surface area even more, across which the air in the house is forced by a blower. As the air flows across the cold tubes, heat from the air transfers to the coolant, thereby providing air conditioning for the house. That heat is transported by the coolant and is subsequently transferred to the earth, which acts as a large heat sink. If the tubes are spaced too closely in the earth, local ground temperatures could be affected, thereby reducing the effectiveness of the air conditioning system.

❒ ❒ ❒

PROBLEM 8.10 A Liquid Cooled Microchannel Heat Sink

A technique for cooling a high performance electronic device is to attach it to a metal heat sink with microchannels through which a liquid coolant flows. For a 2.5 cm × 2.5 cm aluminum heat sink with ten equally spaced 0.8 mm diameter microchannels, what is the total mass flow rate of water required for a maximum heat sink surface temperature of 85°C if the heat dissipation of the device is 110 W? The inlet temperature of the water is 25°C. What is the pressure drop?

DIAGRAM

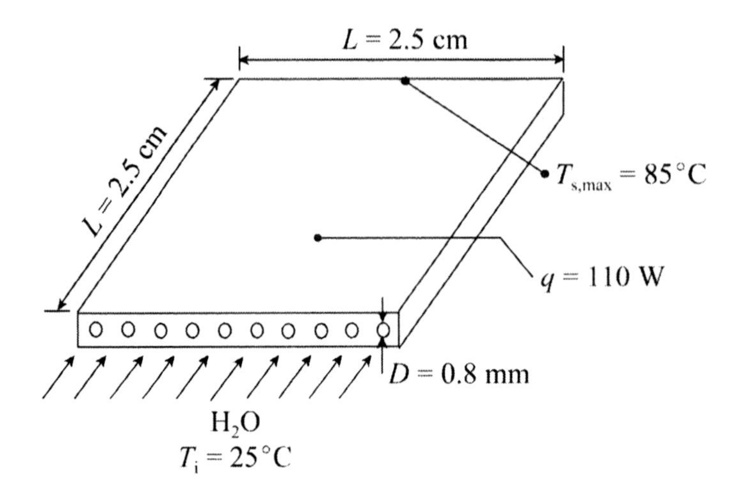

ASSUMPTIONS

1. Steady conditions.
2. Fully developed flow.
3. The surface of each channel is smooth.
4. Heat dissipated by the device is spread uniformly over the heat sink surface and channels.
5. All dissipated heat is transferred to the water.
6. The mass flow rate is equal in all channels.
7. Constant properties.

PROPERTIES

Assume $T_o = 29°\text{C}$. $T_{av} = (T_i + T_o)/2 = (25°\text{C} + 29°\text{C})/2 = 27°\text{C} = 300\ \text{K}$

liquid water: $\rho = 997\ \text{kg/m}^3$, $c_p = 4179\ \text{J/kg·K}$, $k = 0.613\ \text{W/m·K}$,
$\mu = 855 \times 10^{-6}\ \text{Pa·s}$, $\nu = 8.576 \times 10^{-7}\ \text{m}^2\text{/s}$

ANALYSIS

Conservation of energy and the fourth and fifth assumptions demand that the heat dissipated by the device transfers uniformly to the water. Thus,

$$q = Nq_s''\pi DL$$

where N is the number of channels and q''_s is the heat flux at the surface of each channel. So,

$$q_s'' = \frac{q}{N\pi DL}$$

$$q''_s = \frac{110 \text{ W}}{10\pi(0.8 \times 10^{-3} \text{ m})(0.025 \text{ m})}$$

$$= 1.751 \times 10^5 \text{ W/m}^2$$

Conservation of energy applied to the water in one channel is written as

$$q_s''\pi DL = \dot{m}_1 c_p (T_o - T_i)$$

where $\dot{m}_1 = \dot{m}/N$ is the mass flow rate in one channel. The maximum heat sink surface temperature occurs at the outlet of the channels. Writing Newton's law of cooling for this location,

$$q_s'' = h(T_{s,\max} - T_o)$$

If we assume fully developed laminar flow for a constant heat flux surface,

$$Nu_D = \frac{hD}{k} = 4.36$$

so the heat transfer coefficient is

$$h = \frac{4.36(0.613 \text{ W/m·K})}{0.8 \times 10^{-3} \text{ m}}$$

$$= 3341 \text{ W/m}^2\text{·K}$$

Substituting this quantity into Newton's law of cooling, and solving for T_o,

$$T_o = 85°\text{C} - (1.751 \times 10^5 \text{ W/m}^2)/(3341 \text{ W/m}^2\text{·K})$$

$$= 32.6°\text{C}$$

Hence, the mass flow rate for one channel is

$$\dot{m}_1 = \frac{q_s''\pi DL}{c_p (T_o - T_i)}$$

$$\dot{m}_1 = \frac{(1.751 \times 10^5 \text{ W/m}^2)\pi(0.8 \times 10^{-3} \text{ m})(0.025 \text{ m})}{4179 \text{ J/kg·K}(32.6 - 25)°\text{C}}$$

$$= 3.464 \times 10^{-4} \text{ kg/s}$$

There are ten channels, so the total mass flow rate is

$$\dot{m} = 10(3.464 \times 10^{-4}\ \text{kg/s})$$

$$= \underline{\underline{3.464 \times 10^{-3}\ \text{kg/s}}}$$

To find the pressure drop across the heat sink, we first need the Reynolds number,

$$Re_D = \frac{4\dot{m}_1}{\pi D \mu}$$

$$Re_D = \frac{4(3.464 \times 10^{-4}\ \text{kg/s})}{\pi(0.8 \times 10^{-3}\ \text{m})(855 \times 10^{-6}\ \text{Pa·s})}$$

$$= 645$$

The flow is laminar, so the friction factor f is

$$f = \frac{64}{Re_D}$$

$$f = 64/645$$

$$= 0.0992$$

The mean velocity is

$$u_m = \frac{4\dot{m}}{\rho \pi D^2}$$

$$u_m = \frac{4(3.464 \times 10^{-4}\ \text{kg/s})}{(997\ \text{kg/m}^3)\pi(0.8 \times 10^{-3}\ \text{m})^2}$$

$$= 0.691\ \text{m/s}$$

Finally, the pressure drop is

$$\Delta p = f \frac{\rho u_m^2}{2D} L$$

$$\Delta p = \frac{(0.0992)(997\ \text{kg/m}^3)(0.691\ \text{m/s})^2(0.025\ \text{m})}{2(0.8 \times 10^{-3}\ \text{m})}$$

$$= \underline{\underline{738\ \text{Pa}}}$$

DISCUSSION

Based on the calculated mean velocity, the residence time for the water in the heat sink channels is

$$t = L/u_m$$

$t = (0.025 \text{ m})/(0.691 \text{ m/s})$

$= 0.0362 \text{ s} = 36.2 \text{ ms}$

No thermal property correction is needed because the calculated and assumed values for the outlet temperature of the water are close.

❐ ❐ ❐

PROBLEM 8.11 An Automobile Exhaust Pipe

A typical automobile exhaust pipe carries exhaust gas at a mean velocity of 6 m/s from the manifold to the muffler. If the gas leaves the manifold at 450°C, and the muffler is 2.6 m away, find the temperature of the exhaust gas as it enters the muffler. Also, find the heat loss from the exhaust pipe. The ambient air temperature is 35°C, with a heat transfer coefficient of 12 W/m²·K for the outside surface of the exhaust pipe. The exhaust pipe is thin walled with a diameter of 10 cm and has a surface roughness of 20 μm.

DIAGRAM

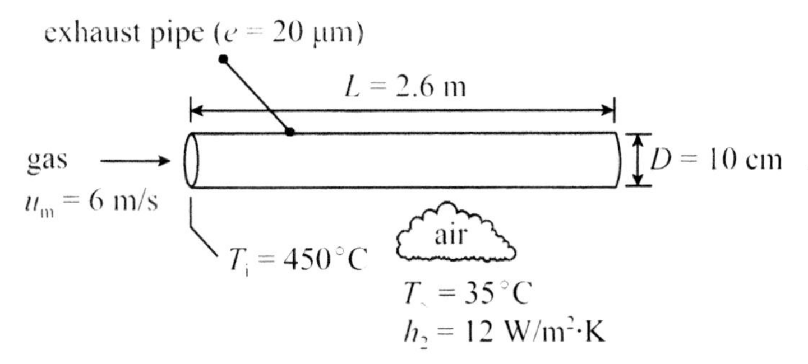

ASSUMPTIONS

1. Steady conditions.
2. Thermal resistance of the pipe wall is neglected.
3. Fully developed flow.
4. Exhaust gas is approximated as atmospheric air.
5. Radiation is neglected.
6. Constant properties.

PROPERTIES

Assume $T_o = 375°C$. $T_{av} = (T_i + T_o)/2 = (450°C + 375°C)/2 = 412.5°C = 686\ K$

atmospheric air: $\rho = 0.5082\ kg/m^3$, $c_p = 1072\ J/kg{\cdot}K$, $k = 0.0516\ W/m{\cdot}K$, $\nu = 65.89 \times 10^{-6}\ m^2/s$, $Pr = 0.694$

ANALYSIS

The Reynolds number is

$$Re_D = \frac{u_m D}{\nu}$$

$$Re_D = \frac{(6\ m/s)(0.10\ m)}{65.89 \times 10^{-6}\ m^2/s}$$

$$= 9106$$

so the flow is turbulent. Friction factor f is obtained from the relation

$$f = \frac{0.25}{\left[\log\left(\frac{e/D}{3.7} + \frac{5.74}{Re_D^{0.9}}\right)\right]^2}$$

The exhaust pipe has a surface roughness of $e = 20\ \mu m$, giving a relative roughness of

$$e/D = (20 \times 10^{-6}\ m)/(0.10\ m)$$

$$= 2.00 \times 10^{-4}\ m$$

so the friction factor is

$$f = \frac{0.25}{\left[\log\left(\frac{2.00 \times 10^{-4}}{3.7} + \frac{5.74}{(9106)^{0.9}}\right)\right]^2}$$

$$= 0.0321$$

The Nusselt number is

$$Nu_D = \frac{(f/8)(Re_D - 1000)Pr}{1 + 12.7(f/8)^{1/2}(Pr^{2/3} - 1)}$$

$$Nu_D = \frac{(0.0321/8)(9106 - 1000)(0.694)}{1 + 12.7(0.0321/8)^{1/2}(0.694^{2/3} - 1)}$$

$= 27.3$

The heat transfer coefficient is

$$h_1 = \frac{Nu_D k}{D}$$

$$h_1 = \frac{(27.3)(0.0516 \text{ W/m·K})}{0.10 \text{ m}}$$

$= 14.1 \text{ W/m}^2\text{·K}$

Neglecting thermal resistance of the pipe wall, total thermal resistance is

$$R_{tot} = \frac{1}{\pi DL}\left(\frac{1}{h_1} + \frac{1}{h_2}\right)$$

$$R_{tot} = \frac{1}{\pi(0.10 \text{ m})(2.6 \text{ m})}\left(\frac{1}{14.1 \text{ W/m}^2\text{·K}} + \frac{1}{12 \text{ W/m}^2\text{·K}}\right)$$

$= 0.1888 \text{ K/W}$

Mass flow rate is

$$\dot{m} = \rho u_m \pi D^2/4$$

$\dot{m} = (0.5082 \text{ kg/m}^3)(6 \text{ m/s})\pi(0.10 \text{ m})^2/4$

$= 0.0239 \text{ kg/s}$

To find the outlet temperature of the exhaust gas, we use the relation

$$\frac{T_\infty - T_o}{T_\infty - T_i} = \exp\left(-\frac{1}{\dot{m}c_p R_{tot}}\right)$$

$$T_\infty - T_o = (T_\infty - T_i)\exp\left(\frac{-1}{(0.0239 \text{ kg/s})(1072 \text{ J/kg·K})(0.1888 \text{ K/W})}\right)$$

$= (35°\text{C} - 450°\text{C})(0.8132)$

$= -337°\text{C}$

Thus, the outlet gas temperature is

$$T_o = 35°C + 337°C$$

$$= \underline{\underline{372°C}}$$

and the heat loss from the exhaust pipe is

$$q = \dot{m}c_p(T_i - T_o)$$

$$q = (0.0239 \text{ kg/s})(1072 \text{ J/kg·K})(450 - 372)°C$$

$$= \underline{\underline{1998 \text{ W}}}$$

DISCUSSION

Automobile exhaust gases are a mixture of carbon monoxide, nitrous oxides and other gases. The bulk thermal properties of this mixture were approximated using atmospheric air at the estimated average temperature between the inlet and outlet of the exhaust pipe. The calculated value of the outlet temperature was very close to our initial estimate, making a property correction unnecessary.

❒ ❒ ❒

PROBLEM 8.12 Design of a Driveway Deicing System

A system for melting ice and snow on a concrete driveway consists of a parallel array of thin-walled plastic tubes buried within the concrete slab. A heat exchanger, located indoors, regulates the temperature of a water/ethylene glycol mixture at 22°C as the mixture enters the tubes. To prevent the buildup of ice and snow on the driveway, its surface must be maintained at 15°C. Each tube has a length and diameter of 20 m and 2.4 cm, respectively, and the tubes are spaced 12 cm apart at a depth of 7 cm. If the desired outlet temperature of the mixture is 20°C, find the required mass flow rate in each tube.

DIAGRAM

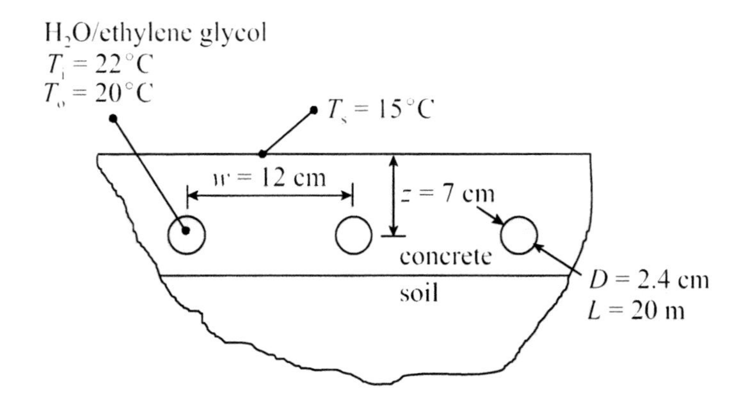

ASSUMPTIONS

1. Steady conditions.
2. Thermal resistance of the tube wall is neglected.
3. Contact resistance of the tube/concrete interface is neglected.
4. Fully developed flow.
5. The inside surface of each tube is smooth.
6. Thermal conductivity of the underlying soil is equivalent to that of concrete.
7. The driveway deicing system is modeled as a two-dimensional semi-infinite region.
8. Convection and radiation at the driveway surface are neglected.
9. Constant properties.

PROPERTIES

concrete: $k_c = 1.4$ W/m·K

$T_{av} = (T_i + T_o)/2 = (22 + 20)/2 = 21\,°C = 294$ K

H_2O/ethylene glycol: $\rho = 1073$ kg/m^3, $c_p = 3285$ J/kg·K, $k = 0.381$ W/m·K,
$\mu = 3.84 \times 10^{-3}$ Pa·s, $\nu = 3.58 \times 10^{-6}$ m^2/s, $Pr = 33.1$

ANALYSIS

Under the stated assumptions, there are two thermal resistances in the deicing system: the boundary layer and the concrete/soil, which is assumed to be a semi-infinite region with the thermal conductivity of concrete. The first thermal resistance is

$$R_{t,1} = \frac{1}{\pi DLh}$$

where h is the heat transfer coefficient, which is unknown at this point. The second thermal resistance is

$$R_{t,2} = \frac{1}{Sk_c}$$

where S is the conduction shape factor given by

$$S = \frac{2\pi L}{\ln\left[\frac{2w}{\pi D}\sinh(2\pi z/w)\right]}$$

The total thermal resistance is

$$R_{tot} = R_{t,1} + R_{t,2}$$

Under the assumption of fully developed flow, we can further assume that the flow is laminar. If this assumption is correct, the Nusselt number for a circular pipe with an isothermal surface is

$$Nu_D = \frac{hD}{k} = 3.66$$

Hence, the heat transfer coefficient is

$$h = \frac{(3.66)(0.381\ \text{W/m}\cdot\text{K})}{0.024\ \text{m}}$$

$$= 58.1\ \text{W/m}^2\cdot\text{K}$$

The conduction shape factor is

$$S = \frac{2\pi(20\ \text{m})}{\ln\left\{\frac{2(0.12\ \text{m})}{\pi(0.024\ \text{m})}\sinh\left[2\pi(0.07\ \text{m})/0.12\ \text{m}\right]\right\}}$$

$$S = 71.25\ \text{m}$$

The total thermal resistance is

$$R_{tot} = \frac{1}{\pi(0.024\ \text{m})(20\ \text{m})(58.1\ \text{W/m}^2\cdot\text{K})} + \frac{1}{(71.25\ \text{m})(1.4\ \text{W/m}\cdot\text{K})}$$

$$= 0.0214\ \text{K/W}$$

The water/ethylene glycol mixture changes only 2°C from the inlet to the outlet, so we can approximate the temperature of the mixture as $T_{av} = 21\,°\text{C}$. Thus, the heat transfer is

$$q = \frac{T_{av} - T_s}{R_{tot}} = \dot{m}c_p(T_i - T_o)$$

Solving for mass flow rate,

$$\dot{m} = \frac{T_{av} - T_s}{R_{tot}c_p(T_i - T_o)}$$

$$\dot{m} = \frac{(21 - 15)°\text{C}}{(0.0214\ \text{K/W})(3285\ \text{J/kg·K})(22 - 20)°\text{C}}$$

$$= \underline{\underline{0.0427\ \text{kg/s}}}$$

The assumption of laminar flow must be verified. The Reynolds number is

$$Re_D = \frac{4\dot{m}}{\pi D\mu}$$

$$Re_D = \frac{4(0.0427\ \text{kg/s})}{\pi(0.024\ \text{m})(3.84 \times 10^{-3}\ \text{Pa·s})}$$

$$= 590$$

As assumed, the flow is laminar.

DISCUSSION

In the above relation for mass flow rate, we can see that mass flow rate is a strong function of T_s, the surface temperature of the concrete driveway. For example, to maintain an ice free surface at $T_s = 10°\text{C}$, the required mass flow rate is $\dot{m} = 0.0783$ kg/s, if all other parameters are held constant. In a complete deicing design, the variation of all parameters–tube diameter, tube spacing, tube depth, flow rate, inlet and outlet temperatures, and driveway surface temperature–should be considered.

The effects of varying tube spacing and depth are considered in a similar deicing system in Problem 4.6.

PROBLEM 8.13 Design of a Parabolic Solar Collector

The hot water requirements of a house are met by heating water from 14 °C to 70 °C in a parabolic solar collector. Occupants of the house consume 0.2 m^3 (52.8 gal) of hot water per day for washing clothes and dishes and bathing. Water flows through a 3.0-cm diameter thin-walled aluminum tube whose outside surface is black anodized to increase solar absorption. The tube is positioned at the focal line of the parabolic collector where it receives solar energy at a rate of 110 W per meter of length. Find the required length of the collector to meet the hot water requirements. Also, find the surface temperature of the tube at the outlet. Assume that hot water is used for only six hours during a typical day.

DIAGRAM

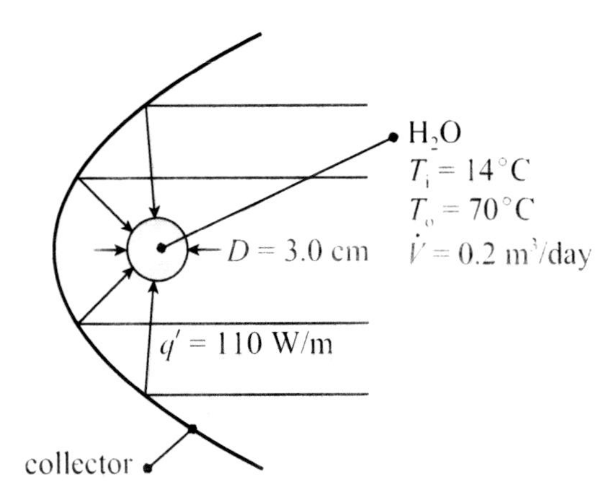

ASSUMPTIONS

1. Steady conditions.
2. The solar heat flux is uniform.
3. Fully developed flow.
4. The inside surface of the aluminum tube is smooth.
5. All solar energy incident on the tube is absorbed.
6. Convection and radiation from the outside surface of the tube are neglected.
7. Constant properties.

PROPERTIES

$T_{av} = (T_i + T_o)/2 = (14°C + 70°C)/2 = 42°C = 315$ K

liquid water: $\rho = 991$ kg/m^3, $c_p = 4179$ J/kg·K, $k = 0.634$ W/m·K, $\mu = 631 \times 10^{-6}$ Pa·s, $\nu = 6.367 \times 10^{-7}$ m^2/s, $Pr = 4.16$

ANALYSIS

Converting the daily volume flow rate to useful units, and assuming hot water usage for six hours per day, we have

$$0.2 \text{ m}^3/\text{day} \times 24/6 \times 1 \text{ day}/24 \text{ h} \times 1 \text{ h}/3600 \text{ s} = 9.259 \times 10^{-6} \text{ m}^3/\text{s}$$

Mass flow rate is

$$\dot{m} = \rho \dot{V}$$

$$\dot{m} = (991 \text{ kg/m}^3)(9.259 \times 10^{-6} \text{ m}^3/\text{s})$$

$$= 9.176 \times 10^{-3} \text{ kg/s}$$

and the heat transfer is

$$q = q' L = \dot{m} c_p (T_o - T_i)$$

so the required length of tube (solar collector) is

$$L = \frac{(9.176 \times 10^{-3} \text{ kg/s})(4179 \text{ J/kg·K})(70 - 14)°\text{C}}{110 \text{ W/m}}$$

$$= \underline{\underline{19.5 \text{ m}}}$$

The Reynolds number is

$$Re_D = \frac{4\dot{V}}{\pi D \nu}$$

$$Re_D = \frac{4(9.259 \times 10^{-6} \text{ m}^3/\text{s})}{\pi(0.030 \text{ m})(6.367 \times 10^{-7} \text{ m}^2/\text{s})}$$

$$= 617$$

so the flow is laminar. For fully developed laminar flow in a circular pipe with a uniform heat flux at the surface, the Nusselt number is

$$Nu_D = \frac{hD}{k} = 4.36$$

Thus, the heat transfer coefficient is

$$h = \frac{(4.36)(0.634 \text{ W/m·K})}{0.030 \text{ m}}$$

$= 92.1\ \mathrm{W/m^2{\cdot}K}$

Now that the heat transfer coefficient is known, the surface temperature of the tube at the outlet may be found using Newton's law of cooling

$$q'' = \frac{q}{\pi D L} = \frac{q'}{\pi D} = h(T_s - T_o)$$

Solving for the surface temperature,

$$T_s = \frac{(110\ \mathrm{W/m})}{\pi(0.03\ \mathrm{m})(92.1\ \mathrm{W/m^2{\cdot}K})} + 70°\mathrm{C}$$

$= \underline{\underline{82.7°\mathrm{C}}}$

DISCUSSION

It is instructive to examine how solar energy affects collector length. Assuming the same hot water demand and the same inlet and outlet temperatures, the graph below shows the relationship between solar energy per unit length and collector length. Clearly, the amount of solar radiation significantly affects the collector length required to meet the hot water demands of the house. For prolonged periods of cloudy conditions, a supplementary heating system may be required.

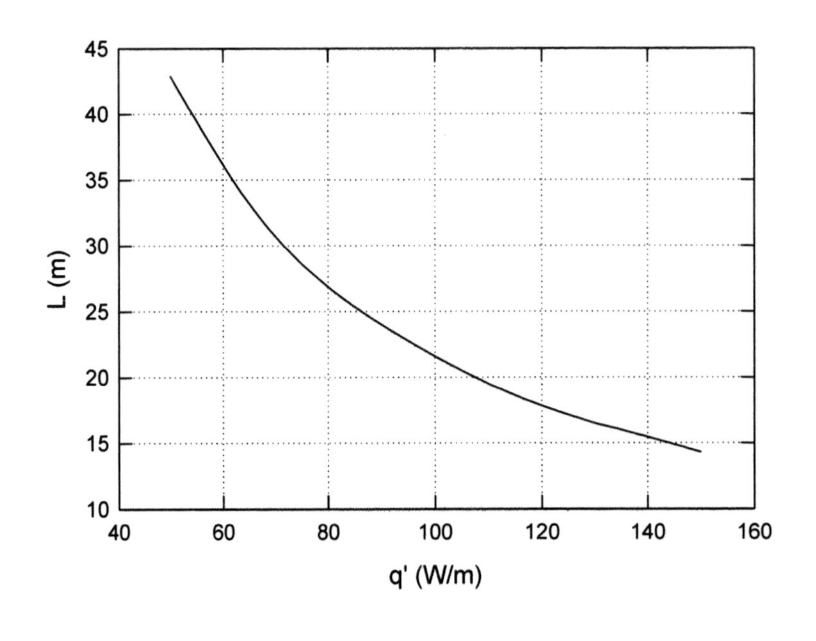

❐ ❐ ❐

PROBLEM 8.14 Liquid Mercury Flow in a Nuclear Reactor

A 4-cm diameter thin-walled stainless steel tube in a nuclear reactor transports liquid mercury at a mean velocity of 1.3 m/s. The outside surface of the tube is exposed to a uniform radiant heat flux of 5 MW/m^2. The mercury enters the 1-m long tube at 25°C. Find the heat transfer coefficient and outlet temperature of the mercury.

DIAGRAM

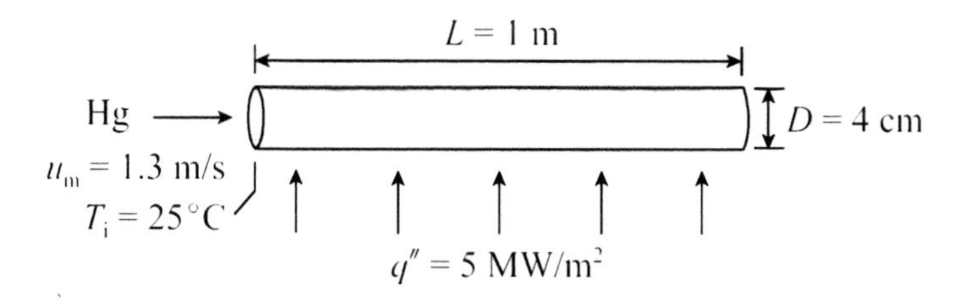

ASSUMPTIONS

1. Steady conditions.
2. The radiant heat flux is uniform.
3. Fully developed flow.
4. The inside surface of the stainless steel tube is smooth.
5. All radiant energy incident on the tube is absorbed.
6. Convection and radiation from the outside surface of the tube are neglected.
7. Constant properties.

PROPERTIES

Assume $T_o = 250°\text{C}$. $T_{av} = (T_i + T_o)/2 = (25°\text{C} + 250°\text{C})/2 = 137.5°\text{C} = 411\text{ K}$

liquid mercury: $\rho = 13{,}261\text{ kg/m}^3$, $c_p = 136.3\text{ J/kg·K}$, $k = 9.932\text{ W/m·K}$,
$\nu = 0.0867 \times 10^{-6}\text{ m}^2\text{/s}$, $Pr = 0.0158$

ANALYSIS

The Reynolds number is

$$Re_D = \frac{u_m D}{\nu}$$

$$Re_D = \frac{(1.3\text{ m/s})(0.04\text{ m})}{0.0867 \times 10^{-6}\text{ m}^2\text{/s}}$$

$$= 5.998 \times 10^5$$

so the flow is turbulent. The Peclet number is defined as

$$Pe_D \equiv Re_D Pr$$

$Pe_D = (5.998 \times 10^5)(0.0158)$

$= 9.477 \times 10^4$

A commonly used Nusselt number correlation for the fully developed turbulent flow of liquid metals in a smooth circular pipe with a uniform heat flux is

$$Nu_D = 4.82 + 0.0185 Pe_D^{0.827}$$

$Nu_D = 4.82 + 0.0185(9.477 \times 10^4)^{0.827}$

$= 246.3$

Hence, the heat transfer coefficient is

$$h = \frac{Nu_D k}{D}$$

$$h = \frac{(246.3)(9.932 \text{ W/m·K})}{0.04 \text{ m}}$$

$= \underline{\underline{6.116 \times 10^4 \text{ W/m}^2\text{·K}}}$

Mass flow rate is

$$\dot{m} = \rho u_m \pi D^2/4$$

$\dot{m} = (13{,}261 \text{ kg/m}^3)(1.3 \text{ m/s})\pi(0.04 \text{ m})^2/4$

$= 21.66 \text{ kg/s}$

and the heat transfer is

$$q = q''\pi DL = \dot{m}c_p(T_o - T_i)$$

Solving for the outlet temperature of the mercury,

$$T_o = \frac{(5 \times 10^6 \text{ W/m}^2)\pi(0.04 \text{ m})(1 \text{ m})}{(21.66 \text{ kg/s})(136.3 \text{ J/kg·K})} + 25°\text{C}$$

$= \underline{\underline{238°\text{C}}}$

DISCUSSION

The Nusselt number correlation used here is valid for the ranges $3.6 \times 10^3 < Re_D < 9.05 \times 10^5$ and $10^2 < Pe_D < 10^4$. Thus, we have used the correlation within its range of applicability for Reynolds number but not for Peclet number. We should therefore regard our results as approximate. The maximum surface temperature of the tube occurs at the outlet and may be found using Newton's law of cooling

$$q'' = h(T_s - T_o)$$

Solving for surface temperature,

$$T_s = \frac{5 \times 10^6 \text{ W/m}^2}{6.116 \times 10^4 \text{ W/m}^2\cdot\text{K}} + 238°\text{C}$$

$$= 320°\text{C}$$

No correction of thermal properties is necessary.

❒ ❒ ❒

PROBLEM 8.15 Liquid Sodium Flow in a Manufacturing Operation

In a chemical manufacturing operation, a 2.5-cm diameter thin-walled stainless steel tube carries liquid sodium at a mass flow rate of 18 kg/s. The tube is 4 m long, and its surface is maintained at a uniform temperature of 60°C. If the liquid sodium enters the tube at 400°C, find the heat transfer and the outlet temperature of the liquid sodium.

DIAGRAM

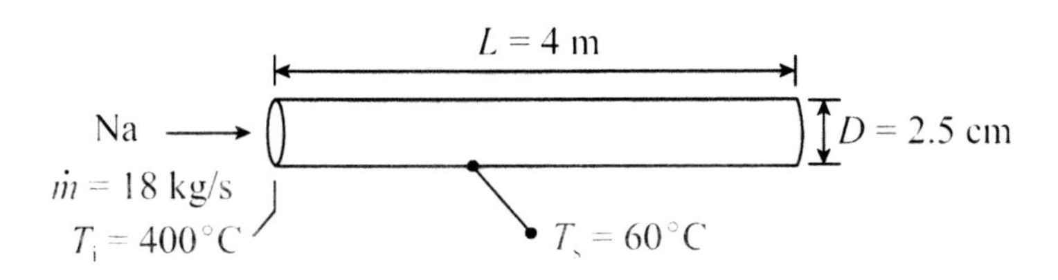

ASSUMPTIONS

1. Steady conditions.
2. Uniform surface temperature.
3. Fully developed flow.
4. The inside surface of the stainless steel tube is smooth.
5. Constant properties.

PROPERTIES

Assume $T_o = 100°C$. $T_{av} = (T_i + T_o)/2 = (100°C + 400°C)/2 = 250°C = 523$ K

liquid sodium: $\rho = 890$ kg/m^3, $c_p = 1335$ J/kg·K, $k = 78.4$ W/m·K,
$\mu = 4.555 \times 10^{-4}$ Pa·s, $\nu = 5.118 \times 10^{-7}$ m^2/s, $Pr = 0.0077$

ANALYSIS

The Reynolds number is

$$Re_D = \frac{4\dot{m}}{\pi D \mu}$$

$$Re_D = \frac{4(18 \text{ kg/s})}{\pi(0.025 \text{ m})(4.555 \times 10^{-4} \text{ Pa·s})}$$

$$= 2.013 \times 10^6$$

so the flow is turbulent. The Peclet number is defined as

$$Pe_D \equiv Re_D Pr$$

$$Pe_D = (2.013 \times 10^6)(0.0077)$$

$$= 1.550 \times 10^4$$

A commonly used Nusselt number correlation for the fully developed turbulent flow of liquid metals in a smooth circular pipe with a constant surface temperature is

$$Nu_D = 5.0 + 0.025 Pe_D^{0.8}$$

$$Nu_D = 5.0 + 0.025(1.550 \times 10^4)^{0.8}$$

$$= 61.3$$

Hence, the heat transfer coefficient is

$$h = \frac{Nu_D k}{D}$$

$$h = \frac{(61.3)(78.4 \text{ W/m·K})}{0.025 \text{ m}}$$

$$= 1.922 \times 10^5 \text{ W/m}^2\text{·K}$$

The outlet temperature of the liquid sodium may be found using the relation

$$\frac{T_s - T_o}{T_s - T_i} = \exp\left(-\frac{\pi DL}{\dot{m}c_p}h\right)$$

$$T_s - T_o = (T_s - T_i)\exp\left(\frac{-\pi(0.025\text{ m})(4\text{ m})(1.922\times10^5\text{ W/m}^2\cdot\text{K}}{(18\text{ kg/s})(1335\text{ J/kg}\cdot\text{K})}\right)$$

$$= (60°\text{C} - 400°\text{C})(0.0810)$$

$$= -27.5°\text{C}$$

Thus, the outlet temperature is

$$T_o = 60°\text{C} + 27.5°\text{C}$$

$$= \underline{\underline{87.5°\text{C}}}$$

Now that the outlet temperature is known, the heat transfer may be found.

$$q = \dot{m}c_p(T_i - T_o)$$

$$q = (18\text{ kg/s})(1335\text{ J/kg}\cdot\text{K})(400 - 87.5)°\text{C}$$

$$= 7.51\times10^6\text{ W} = \underline{\underline{7.51\text{ MW}}}$$

DISCUSSION

The Nusselt number correlation used here is valid for $Pe_D > 100$, so we have used the correlation within its range of applicability. Liquid metal flows typically have very high heat transfer coefficients, which is evident in this problem. The calculated outlet temperature of the liquid sodium is close enough to our initial estimate that a correction of thermal properties is not required.

❐ ❐ ❐

PROBLEM 8.16 Flow of Glycerin Between Two Parallel Plates

Glycerin flows between two 1 m × 1 m parallel plates spaced 1.25 cm apart. The mean velocity and inlet temperature of the glycerin are 38 cm/s and 40°C, respectively. If both plates are maintained at 5°C, find the heat transfer and the power required to sustain the flow. Assume that the sides of the channel are blocked such that the glycerin is confined by the plates. Also, assume that the back sides of the plates are insulated.

DIAGRAM

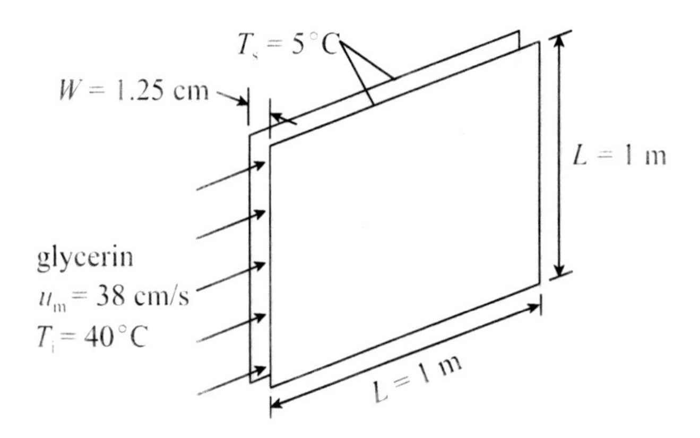

ASSUMPTIONS

1. Steady conditions.
2. Both plates have the same uniform surface temperature.
3. Fully developed flow.
4. The sides of the channel are blocked, confining the glycerin between the plates.
5. The back sides of the plates are insulated.
6. Constant properties.

PROPERTIES

Assume $T_o = 30°C$. $T_{av} = (T_i + T_o)/2 = (40°C + 30°C)/2 = 35°C = 308$ K

glycerin: $\rho = 1255$ kg/m^3, $c_p = 2477$ J/kg·K, $k = 0.286$ W/m·K, $\nu = 352 \times 10^{-6}$ m^2/s

ANALYSIS

The hydraulic diameter of the narrow rectangular channel is

$$D_h = \frac{4A}{P} \approx \frac{4WL}{2L} = 2W$$

$$D_h = 2(0.0125 \text{ m})$$

$$= 0.025 \text{ m}$$

so the Reynolds number is

$$Re_D = \frac{u_m D_h}{\nu}$$

$$Re_D = \frac{(0.38 \text{ m/s})(0.025 \text{ m})}{352 \times 10^{-6} \text{ m}^2\text{/s}}$$

$$= 27.0$$

Hence, the flow is laminar. Because the plates are large compared to the space between them, we treat the aspect ratio of the channel as infinite. Thus, for fully developed laminar flow with constant surface temperature, the Nusselt number is

$$Nu_D = \frac{hD_h}{k} = 7.54$$

so the heat transfer coefficient is

$$h = \frac{Nu_D k}{D_h}$$

$$h = \frac{(7.54)(0.286 \text{ W/m·K})}{0.025 \text{ m}}$$

$$= 86.3 \text{ W/m}^2\text{·K}$$

Mass flow rate is

$$\dot{m} = \rho u_m A$$

$$\dot{m} = (1255 \text{ kg/m}^3)(0.38 \text{ m/s})(0.0125 \text{ m} \times 1 \text{ m})$$

$$= 5.961 \text{ kg/s}$$

The outlet temperature of the glycerin may be found using the relation

$$\frac{T_s - T_o}{T_s - T_i} = \exp\left(-\frac{\pi D_h L}{\dot{m} c_p} h \right)$$

$$T_s - T_o = (T_s - T_i) \exp\left(-\frac{\pi(0.025 \text{ m})(1 \text{ m})(86.3 \text{W/m}^2\text{·K})}{(5.961 \text{ kg/s})(2477 \text{ J/kg·K})} \right)$$

$$= (5°\text{C} - 40°\text{C})(0.9995)$$

$$= -34.98°\text{C}$$

Thus, the outlet temperature is

$$T_o = 5°C + 34.98°C$$

$$= 39.98°C \approx \underline{\underline{40.0°C}}$$

Now that the outlet temperature is known, the heat transfer is

$$q = \dot{m}c_p(T_i - T_o)$$

$$q = (5.961\ \text{kg/s})(2477\ \text{J/kg·K})(40 - 39.98)°C$$

$$= \underline{\underline{295\ \text{W}}}$$

For fully developed laminar flow in a channel with an infinite aspect ratio and constant surface temperature, the friction factor is

$$f = \frac{96}{Re_D}$$

$$f = 96/27.0$$

$$= 3.56$$

and the pressure drop is

$$\Delta p = f\frac{\rho u_m^2}{2D_h}L$$

$$\Delta p = \frac{(3.56)(1255\ \text{kg/m}^3)(0.38\ \text{m/s})^2(1\ \text{m})}{2(0.025\ \text{m})}$$

$$= 1.290 \times 10^4\ \text{Pa} = 12.9\ \text{kPa}$$

The power required to sustain the flow is

$$P = \Delta p u_m A$$

$$P = (1.290 \times 10^4\ \text{Pa})(0.38\ \text{m/s})(0.0125\ \text{m} \times 1\ \text{m})$$

$$= \underline{\underline{61.3\ \text{W}}}$$

DISCUSSION

The outlet temperature of the glycerin is only a fraction of a degree lower than the inlet temperature. The power required to sustain the flow of glycerin is the output power of a pump to overcome the resistance due to friction in the channel. Other sources of loss would also have to be accounted for in the selection or design of a pump for this application. A slight improvement of the results could be made by correcting thermal properties.

❒ ❒ ❒

PROBLEM 8.17 Air Flow Between Two Circuit Boards

A fan forces air at a mean velocity of 1.8 m/s through the channel formed by two facing circuit boards. Due to different heights of the devices mounted on the boards, the effective channel width is 3.0 cm, as shown in the diagram. Each circuit board, which measures 16 cm × 24 cm, dissipates 35 W of power uniformly over its surface. The temperature of the air entering the channel is 25°C. Find the outlet temperature of the air, and estimate the maximum surface temperature of the devices. Two pieces of sheet metal block the sides of the channel, confining the air flow between the circuit boards. Assume a friction factor of 0.1.

DIAGRAM

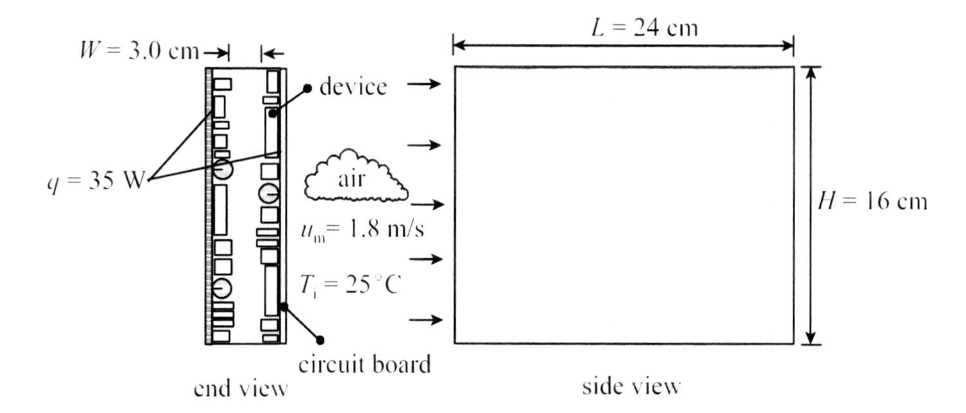

ASSUMPTIONS

1. Steady conditions.
2. Each circuit board produces a uniform heat flux at its surface.
3. The friction factor is 0.1.
4. The sides of the channel are blocked, confining the air between the circuit boards.

5. The back sides of the circuit boards are insulated.
6. Radiation is neglected.
7. Constant properties.

PROPERTIES

Assume $T_o = 33°C$. $T_{av} = (T_i + T_o)/2 = (25°C + 33°C)/2 = 29°C = 302$ K

atmospheric air: $\rho = 1.1547$ kg/m³, $c_p = 1007$ J/kg·K, $k = 0.0264$ W/m·K,
$\nu = 16.09 \times 10^{-6}$ m²/s, $Pr = 0.707$

ANALYSIS

Thermal properties of air are based on the average of the inlet and outlet temperatures. The outlet air temperature is calculated from the energy balance

$$q_t = \dot{m}c_p(T_o - T_i)$$

where $q_t = 2q = 70$ W, the total heat dissipated by the circuit boards. Mass flow rate is

$$\dot{m} = \rho u_m WH$$

$$\dot{m} = (1.1547 \text{ kg/m}^3)(1.8 \text{ m/s})(0.03 \text{ m})(0.16 \text{ m})$$

$$= 9.977 \times 10^{-3} \text{ kg/s}$$

Hence, the outlet temperature of the air is

$$T_o = \frac{(70 \text{ W})}{(9.977 \times 10^{-3} \text{ kg/s})(1007 \text{ J/kg·K})} + 25°C$$

$$= \underline{\underline{32.0°C}}$$

which is close to our initial estimate. The hydraulic diameter of the channel is

$$D_h = \frac{4WH}{P}$$

$$D_h = \frac{4(0.03 \text{ m})(0.16 \text{ m})}{2(0.03 \text{ m} + 0.16 \text{ m})}$$

$$= 0.0505 \text{ m}$$

so the Reynolds number is

$$Re_D = \frac{u_m D_h}{\nu}$$

$$Re_D = \frac{(1.8 \text{ m/s})(0.0505 \text{ m})}{16.09 \times 10^{-6} \text{ m}^2/\text{s}}$$

$$= 5652$$

Thus, the flow is turbulent. A length of approximately ten diameters are required for turbulent flow to become fully developed, which means that the circuit board channel would have to be at least 50 cm long. A Nusselt number correlation for developing turbulent flow is

$$Nu_D = \frac{(f/8)(Re_D - 1000)Pr}{1 + 12.7(f/8)^{1/2}(Pr^{2/3} - 1)}\left[1 + (D_h/L)^{2/3}\right]$$

$$Nu_D = \frac{(0.1/8)(5652 - 1000)(0.707)}{1 + 12.7(0.1/8)^{1/2}(0.707^{2/3} - 1)} [1 + (0.0505 \text{ m}/0.24 \text{ m})^{2/3}]$$

$$= 78.7$$

Note that as $D_h/L \rightarrow 0$, the correlation reduces to that for fully developed turbulent flow. The heat transfer coefficient is

$$h = \frac{Nu_D k}{D_h}$$

$$h = \frac{(78.7)(0.0264 \text{ W/m·K})}{0.0505 \text{ m}}$$

$$= 41.1 \text{ W/m}^2\text{·K}$$

The heat flux at the surface of each circuit board is the dissipated heat divided by the surface area of the circuit board.

$$q'' = \frac{q}{LH}$$

$$q'' = \frac{35 \text{ W}}{(0.24 \text{ m})(0.16 \text{ m})}$$

$$= 911.5 \text{ W/m}^2$$

The maximum surface temperature of the devices occurs at the outlet of the channel, which may be found using Newton's law of cooling,

$$q'' = h(T_s - T_o)$$

Solving for surface temperature T_s,

$$T_s = \frac{911.5\ \text{W/m}^2}{41.1\ \text{W/m}^2\cdot\text{K}} + 32.0°\text{C}$$

$$= \underline{\underline{54.2°\text{C}}}$$

DISCUSSION

Let's see if our assumed value for the friction factor is reasonable by calculating the "roughness" of the channel due to the height differences of the devices. Using the relation

$$f = \frac{0.25}{\left[\log\left(\dfrac{\varepsilon}{3.7D_h} + \dfrac{5.74}{Re_D^{0.9}}\right)\right]^2}$$

and solving for ε,

$$\varepsilon = 4.445 \times 10^{-3}\ \text{m} = 4.445\ \text{mm}$$

This is a reasonable value for the mean roughness of the channel.

❐ ❐ ❐

PROBLEM 8.18 Liquid Cooled Power Transformer

An electrical transformer in a power substation dissipates 1.2 kW. The length and diameter of the transformer are 60 cm and 40 cm, respectively. To maintain the surface of the transformer at 45 °C, a 20-mm thin-walled diameter tube, which carries glycerin coolant, is welded to the outer surface of the transformer, as shown in the diagram. The inlet temperature of the coolant is 20 °C. If the maximum temperature rise of the coolant is 7 °C, find the required coolant mass flow rate, the total length of tubing, and the spacing S between turns of the tubing.

DIAGRAM

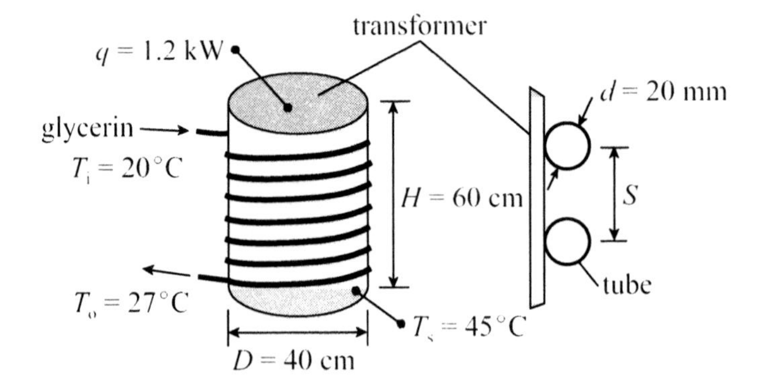

ASSUMPTIONS

1. Steady conditions.
2. Fully developed flow.
3. All dissipated heat transfers to the glycerin.
4. The inside surface of the tube is smooth.
5. Uniform surface temperature.
6. Constant properties.

PROPERTIES

$T_{av} = (T_i + T_o)/2 = (20°C + 27°C)/2 = 23.5°C = 297$ K

glycerin: $\rho = 1262$ kg/m^3, $c_p = 2409$ J/kg·K, $k = 0.286$ W/m·K, $\mu = 1.11$ Pa·s, $\nu = 882 \times 10^{-6}$ m^2/s

ANALYSIS

The mass flow rate of glycerin is obtained from the energy balance

$$q = \dot{m} c_p (T_o - T_i)$$

Solving for mass flow rate,

$$\dot{m} = \frac{1200 \text{ W}}{(2409 \text{ J/kg·K})(27 - 20)°\text{C}}$$

$$= \underline{\underline{0.0712 \text{ kg/s}}}$$

The Reynolds number is

$$Re_d = \frac{4\dot{m}}{\pi d \mu}$$

$$Re_d = \frac{4(0.0712 \text{ kg/s})}{\pi(0.020 \text{ m})(1.11 \text{ Pa·s})}$$

$$= 4.08$$

so the flow is laminar. For fully developed laminar flow in a circular pipe with a constant surface temperature, the Nusselt number is

$$Nu_d = \frac{hd}{k} = 3.66$$

Thus, the heat transfer coefficient is

$$h = \frac{Nu_d k}{d}$$

$$h = \frac{(3.66)(0.286\ \text{W/m·K})}{0.020\ \text{m}}$$

$$= 52.3\ \text{W/m}^2\text{·K}$$

The required length of tubing is found using the relation

$$\frac{T_s - T_o}{T_s - T_i} = \exp\left(-\frac{\pi d L}{\dot{m} c_p} h\right)$$

Solving for length L,

$$L = -\frac{\dot{m} c_p}{\pi d h} \ln\left(\frac{T_s - T_o}{T_s - T_i}\right)$$

$$L = -\frac{(0.0712\ \text{kg/s})(2409\ \text{J/kg·K})}{\pi(0.020\ \text{m})(52.3\ \text{W/m}^2\text{·K})} \ln\left(\frac{45°\text{C} - 27°\text{C}}{45°\text{C} - 20°\text{C}}\right)$$

$$= \underline{\underline{17.1\ \text{m}}}$$

The required number of turns of tubing is

$$N = \frac{L}{\pi D}$$

$$N = \frac{17.1\ \text{m}}{\pi(0.40\ \text{m})}$$

$$= 13.6$$

so the spacing between turns of tubing is

$$S = H/N$$

$$S = (0.60\ \text{m})/(13.6)$$

$$= 0.0441\ \text{m} = \underline{\underline{4.41\ \text{cm}}}$$

DISCUSSION

Let's examine the effect of power dissipation on tube spacing. The graph below shows how tube spacing decreases with increasing power dissipation. The graph shows that the tubes are in direct contact ($S = 2.0$ cm) for a power dissipation of approximately 2.6 kW. Hence, if no other design parameters are changed in the cooling system, this is the maximum allowable power dissipation for the transformer.

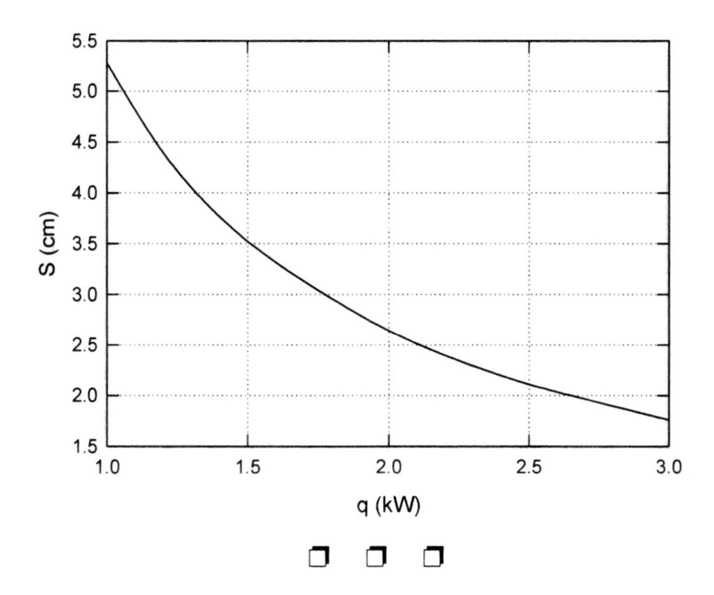

❒ ❒ ❒

PROBLEM 8.19 Steam Condensation in an Insulated Pipe

A 9-cm diameter thin-walled steel pipe carries superheated steam at atmospheric pressure across a room where the air temperature is 20°C. The pipe is covered with a 2-cm thick layer of magnesia insulation. The steam enters the pipe at 130°C with a mean velocity of 8 m/s. If the heat transfer coefficient on the outside surface of the insulation is 11 W/m²·K, find the point along the pipe at which the steam begins condensing.

DIAGRAM

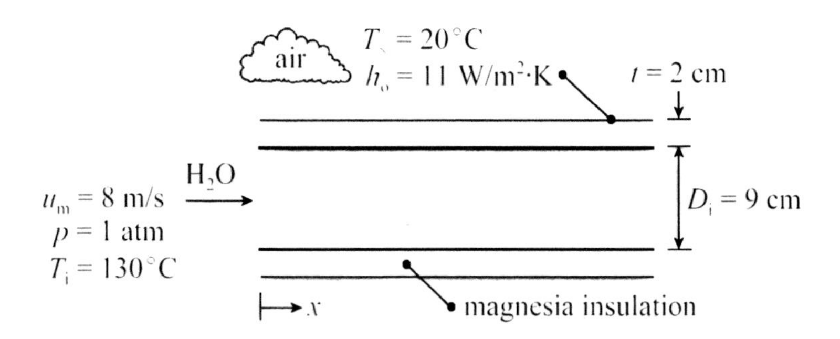

ASSUMPTIONS

1. Steady conditions.
2. Fully developed flow.
3. The inside surface of the pipe is smooth.
4. Thermal properties of steam are equivalent to those of saturated water vapor.
5. Radiation is neglected.
6. Thermal contact resistance is neglected.
7. Constant properties.

PROPERTIES

magnesia insulation: $k_m = 0.055$ W/m·K

$T_{av} = (T_i + T_x)/2 = (130°C + 100°C)/2 = 115°C = 388$ K

saturated water vapor: $\rho = 0.957$ kg/m³, $c_p = 2094$ J/kg·K, $k = 0.0261$ W/m·K, $\nu = 1.318 \times 10^{-5}$ m²/s, $Pr = 1.009$

ANALYSIS

The Reynolds number is

$$Re_D = \frac{u_m D_i}{\nu}$$

$$Re_D = \frac{(8 \text{ m/s})(0.09 \text{ m})}{1.318 \times 10^{-5} \text{ m}^2/\text{s}}$$

$$= 5.463 \times 10^4$$

so the flow is turbulent. For a smooth pipe, the roughness is zero, so the friction factor is

$$f = \frac{0.25}{\left[\log\left(\frac{5.74}{Re_D^{0.9}}\right)\right]^2}$$

$$f = \frac{0.25}{\{\log[5.74/(5.463 \times 10^4)^{0.9}]\}^2}$$

$$= 0.0204$$

The Nusselt number is

$$Nu_D = \frac{(f/8)(Re_D - 1000)Pr}{1 + 12.7(f/8)^{1/2}(Pr^{2/3} - 1)}$$

$$Nu_D = \frac{(0.0204/8)(5.463 \times 10^4 - 1000)(1.009)}{1 + 12.7(0.0204/8)^{1/2}(1.009^{2/3} - 1)}$$

$$= 137.5$$

so the heat transfer coefficient is

$$h_i = \frac{Nu_D k}{D_i}$$

$$h_i = \frac{(137.5)(0.0261\ \text{W/m·K})}{0.09\ \text{m}}$$

$$= 39.9\ \text{W/m}^2\text{·K}$$

The outside diameter of the insulation is

$$D_o = D_i + 2t$$

$$D_o = 0.09\ \text{m} + 2(0.02\ \text{m})$$

$$= 0.13\ \text{m}$$

The overall heat transfer coefficient, based on the inside surface of the thin-walled pipe, is

$$U_i = \left[\frac{1}{h_i} + \frac{D_i \ln(D_o/D_i)}{2k_m} + \frac{D_i}{D_o}\frac{1}{h_o}\right]^{-1}$$

$$U_i = \left[\frac{1}{39.9\ \text{W/m}^2\text{·K}} + \frac{(0.09\ \text{m})\ln(0.13\ \text{m}/0.09\ \text{m})}{2(0.055\ \text{W/m·K})} + \frac{0.09\ \text{m}}{0.13\ \text{m}}\ \frac{1}{11\ \text{W/m}^2\text{·K}}\right]^{-1}$$

$$= 2.57\ \text{W/m}^2\text{·K}$$

The mass flow rate of steam is

$$\dot{m} = \rho u_m \pi D_i^2/4$$

$$\dot{m} = (0.957\ \text{kg/m}^3)(8\ \text{m/s})\pi(0.09\ \text{m})^2/4$$

$$= 0.0487\ \text{kg/s}$$

Condensation is a surface phenomena, so condensation will occur where the pipe temperature reaches 100°C. Heat transfer from the steam to the surrounding air is equivalent to the heat transfer from the steam to the surface of the pipe, so we can write

$$U_i \pi D_i x (T_x - T_\infty) = h_i \pi D_i x (T_x - T_s)$$

where x denotes the location along the pipe where condensation begins, T_x is the steam temperature at x, and T_s is the surface temperature of the pipe. Solving for T_x,

$$T_x = \frac{T_\infty - (h_i/U_i)T_s}{1-(h_i/U_i)}$$

$$T_x = \frac{20°\text{C} - (39.9\ \text{W/m}^2\cdot\text{K}/2.57\ \text{W/m}^2\cdot\text{K})(100°\text{C})}{1 - (39.9\ \text{W/m}^2\cdot\text{K}/2.57\ \text{W/m}^2\cdot\text{K})}$$

$$= 105.5°\text{C}$$

An energy balance on the pipe yields

$$\frac{T_\infty - T_x}{T_\infty - T_i} = \exp\left(-\frac{\pi D_i x}{\dot{m} c_p} U_i\right)$$

Solving for x,

$$x = -\frac{\dot{m} c_p}{\pi D_i U_i} \ln\left(\frac{T_\infty - T_x}{T_\infty - T_i}\right)$$

$$x = -\frac{(0.0487\ \text{kg/s})(2094\ \text{J/kg}\cdot\text{K})}{\pi(0.09\ \text{m})(2.57\ \text{W/m}^2\cdot\text{K})} \ln\left(\frac{20°\text{C} - 105.5°\text{C}}{20°\text{C} - 130°\text{C}}\right)$$

$$= \underline{\underline{35.4\ \text{m}}}$$

DISCUSSION

A value of T_x = 100°C was assumed for the determination of thermal properties of steam. A correction of thermal properties would yield only a modest improvement in accuracy. Besides, the entire analysis is based on the assumption that superheated steam has the thermal properties of saturated vapor.

PROBLEM 8.20 Dishwasher Water Supply Line

In a residential plumbing system, the water heater heats water to a temperature of 60°C. The supply line for the dishwasher is a bare copper pipe (e = 1.5 μm) with a length and diameter of 9 m and 2 cm, respectively. When the dishwasher is operating, the flow rate of hot water in the supply line is 1.2 m^3/h. Find the temperature of the water as it enters the dishwasher if the temperature of the air surrounding the pipe is 18°C. The heat transfer coefficient for the outside surface of the pipe is 10 W/m^2·K.

DIAGRAM

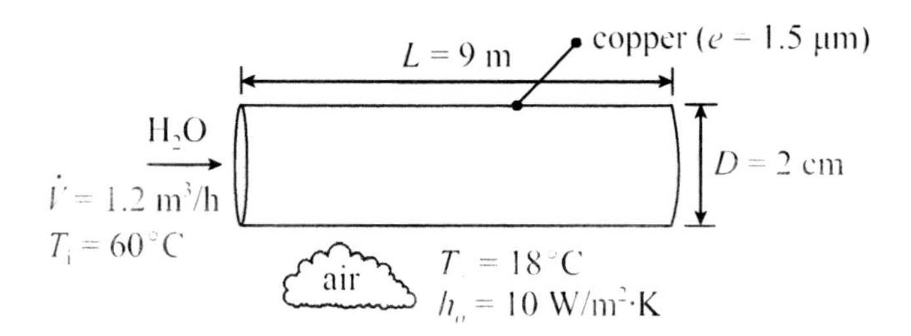

ASSUMPTIONS

1. Steady conditions.
2. Fully developed flow.
3. Thermal resistance of the pipe wall is neglected.
4. Radiation is neglected.
5. Constant properties.

PROPERTIES

Assume T_o = 55°C. $T_{av} = (T_i + T_o)/2$ = (60°C + 55°C)/2 = 57.5°C = 331 K

liquid water: ρ = 984 kg/m^3, c_p = 4184 J/kg·K, k = 0.651 W/m·K, Pr = 3.10
$\nu = 4.896 \times 10^{-7}$ m^2/s

ANALYSIS

Converting volume flow rate,

$$1.2 \text{ m}^3/\text{h} \times 1 \text{ h}/3600 \text{ s} = 3.333 \times 10^{-4} \text{ m}^3/\text{s}$$

The Reynolds number is

$$Re_D = \frac{4\dot{V}}{\pi D \nu}$$

$$Re_D = \frac{4(3.333 \times 10^{-4}\ \text{m}^3/\text{s})}{\pi(0.02\ \text{m})(4.896 \times 10^{-7}\ \text{m}^2/\text{s})}$$

$$= 4.334 \times 10^4$$

Thus, the flow is turbulent. The relative roughness is

$$e/D = (1.5 \times 10^{-6}\ \text{m})/(0.02\ \text{m})$$

$$= 7.5 \times 10^{-5}$$

so the friction factor is

$$f = \frac{0.25}{\left[\log\left(\frac{e/D}{3.7} + \frac{5.74}{Re_D^{0.9}}\right)\right]^2}$$

$$f = \frac{0.25}{\left[\log\left(\frac{7.5 \times 10^{-5}}{3.7} + \frac{5.74}{(4.334 \times 10^4)^{0.9}}\right)\right]^2}$$

$$= 0.0217$$

The Nusselt number is

$$Nu_D = \frac{(f/8)(Re_D - 1000)\,Pr}{1 + 12.7(f/8)^{1/2}(Pr^{2/3} - 1)}$$

$$Nu_D = \frac{(0.0217/8)(4.334 \times 10^4 - 1000)(3.10)}{1 + 12.7(0.0217/8)^{1/2}(3.10^{2/3} - 1)}$$

$$= 204.0$$

Thus, the heat transfer coefficient is

$$h_i = \frac{Nu_D k}{D}$$

$$h_i = \frac{(204.0)(0.651\ \text{W/m}\cdot\text{K})}{0.02\ \text{m}}$$

$$= 6640\ \text{W/m}^2\cdot\text{K}$$

Mass flow rate is

$$\dot{m} = \rho \dot{V}$$

$\dot{m}$ = (984 kg/m^3)(3.333 × 10^{-4} m^3/s)

= 0.328 kg/s

Neglecting the thermal resistance of the pipe wall, the total thermal resistance is

$$R_{tot} = \frac{1}{\pi D L}\left(\frac{1}{h_i} + \frac{1}{h_o}\right)$$

$$R_{tot} = \frac{1}{\pi(0.02\text{ m})(9\text{ m})}\left(\frac{1}{6640\text{ W/m}^2\cdot\text{K}} + \frac{1}{10\text{ W/m}^2\cdot\text{K}}\right)$$

= 0.1771 K/W

To find the temperature of the water as it enters the dishwasher (the outlet temperature of the pipe), we use the relation

$$\frac{T_\infty - T_o}{T_\infty - T_i} = \exp\left(-\frac{1}{\dot{m} c_p R_{tot}}\right)$$

$$T_\infty - T_o = (T_\infty - T_i)\exp\left(\frac{-1}{(0.328\text{ kg/s})(4184\text{ J/kg}\cdot\text{K})(0.1771\text{ K/W})}\right)$$

= (18°C − 60°C)(0.9959)

= −41.8°C

Thus, the outlet water temperature is

T_o = 18°C + 41.8°C

= 59.8°C

DISCUSSION

It is instructive to examine the effect of mass flow rate on the temperature of the water as it enters the dishwasher. The graph below shows this relationship. As indicated in the graph, the water temperature varies significantly at low flow rates but asymptotically approaches the temperature of the water in the water heater at high flow rates, as expected.

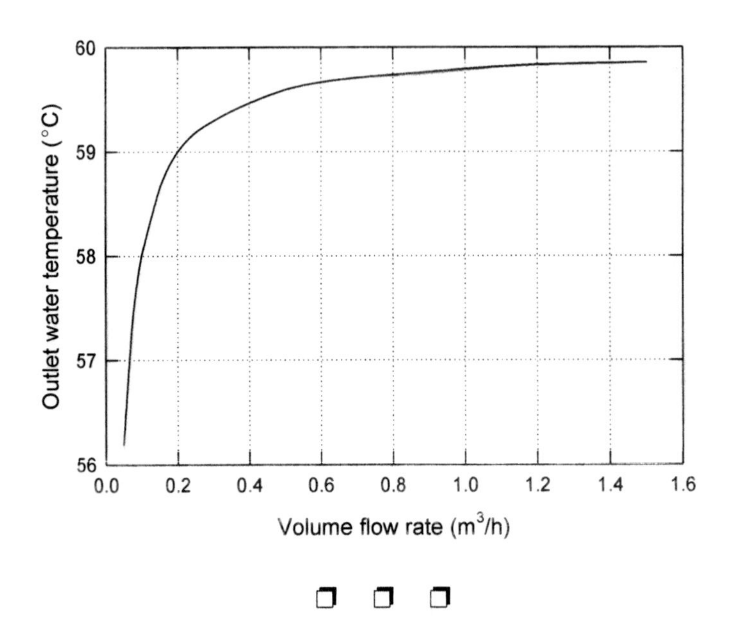

❒ ❒ ❒

PROBLEM 8.21 **Geothermal Piping System**

A geothermal district heating system involves the transport of 110°C water from a geothermal well to a small city located 13 km away. The water flows at 1.4 m^3/s in a well insulated steel pipe (e = 45 μm) with a diameter of 50 cm. Find the pressure drop from the wellhead to the distribution point at the city, the electric power consumption for the pump motor, and the daily cost of power if the cost of electricity is $0.06/kWh. Assume that the wellhead and distribution point have the same elevation and that the efficiency of the pump motor is 60 percent. Neglect minor flow losses.

DIAGRAM

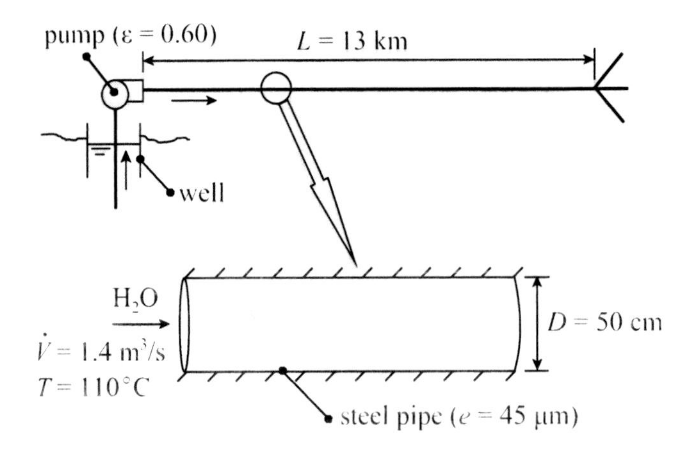

ASSUMPTIONS

1. Steady conditions.
2. Fully developed flow.
3. No heat transfer from the pipe.
4. Minor flow losses are neglected.
5. Thermal properties are those of saturated liquid water.
6. Constant properties.

PROPERTIES

$T = 110°C = 383$ K

saturated liquid water: $\rho = 951$ kg/m^3, $\nu = 2.658 \times 10^{-7}$ m^2/s

ANALYSIS

The Reynolds number is

$$Re_D = \frac{4\dot{V}}{\pi D \nu}$$

$$Re_D = \frac{4(1.4 \text{ m}^3/\text{s})}{\pi(0.50 \text{ m})(2.658 \times 10^{-7} \text{ m}^2/\text{s})}$$

$$= 1.341 \times 10^7$$

so the flow is turbulent. The relative roughness of the pipe is

$$e/D = (45 \times 10^{-6} \text{ m})/(0.50 \text{ m})$$

$$= 9.0 \times 10^{-4}$$

so the friction factor is

$$f = \frac{0.25}{\left[\log\left(\frac{e/D}{3.7} + \frac{5.74}{Re_D^{0.9}}\right)\right]^2}$$

$$f = \frac{0.25}{\left[\log\left(\frac{9.0 \times 10^{-4}}{3.7} + \frac{5.74}{(1.341 \times 10^7)^{0.9}}\right)\right]^2}$$

$$= 0.0192$$

The mean velocity is

$$u_m = \frac{\dot{V}}{\pi D^2/4}$$

$$u_m = \frac{1.4\ \text{m}^3/\text{s}}{\pi(0.50\ \text{m})^2/4}$$

$$= 7.13\ \text{m/s}$$

and the pressure drop across the pipe is

$$\Delta p = f\frac{\rho u_m^2}{2D}L$$

$$\Delta p = \frac{(0.0192)(951\ \text{kg/m}^3)(7.13\ \text{m/s})^2(13\times10^3\ \text{m})}{2(0.50\ \text{m})}$$

$$= 1.207\times10^7\ \text{Pa} = \underline{\underline{12.1\ \text{MPa}}}$$

so the power required to overcome the resistance due to friction is

$$P = \Delta p\dot{V}$$

$$P = (1.207\times10^7\ \text{Pa})(1.4\ \text{m}^3/\text{s})$$

$$= 1.690\times10^7\ \text{W}$$

This is the power added to the flow by the pump. The electric power consumption for the pump motor, i.e., the power input to the pump motor, is the power added to the flow divided by efficiency,

$$P_i = P/\varepsilon$$

$$P_i = (1.690\times10^7\ \text{W})/0.60$$

$$= 2.817\times10^7\ \text{W} = \underline{\underline{28.2\ \text{MW}}}$$

The cost of power to run the pump is

$$\text{cost} = 2.817\times10^7\ \text{W}\times(1\ \text{kWh}/1000\ \text{Wh})\times(\$0.06/\text{kWh})\times(24\ \text{h/day})$$

$$= \underline{\underline{\$40{,}565/\text{day}}}$$

DISCUSSION

From the relation for pressure drop we see that the daily cost of power to run the pump is directly proportional to the distance (pipe length L) between the wellhead and the distribution point at the city. The graph below shows the daily cost as a function of this distance, leaving all other parameters unchanged.

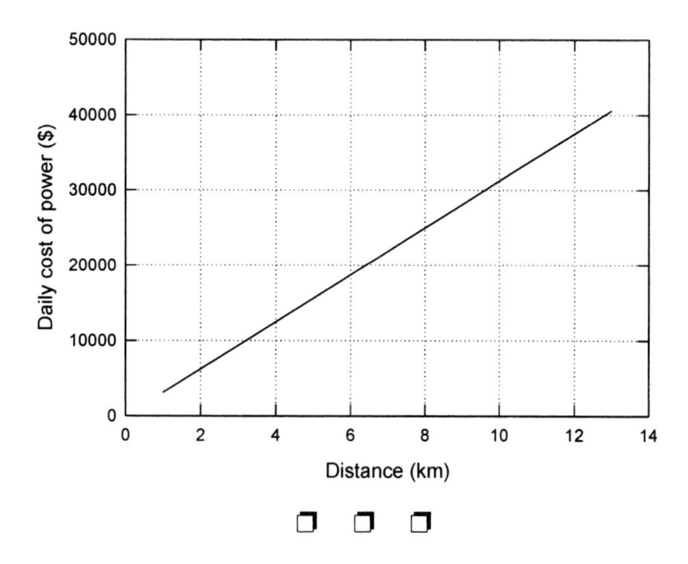

❒ ❒ ❒

PROBLEM 8.22 Heat Transfer in a Concentric Tube Annulus

A concentric tube annulus carries glycerin at a mass flow rate of 60 kg/s. The surface of the inner tube is maintained at a constant temperature of −40°C by a condensing refrigerant vapor that flows through it, and the outside surface of the outer tube is insulated. The outside diameter of the inner tube is 15 cm, and the inside diameter of the outer tube is 30 cm. If the inlet temperature of the glycerin is 45°C, how long must the concentric tube be for an outlet glycerin temperature of 30°C? What is the heat transfer?

DIAGRAM

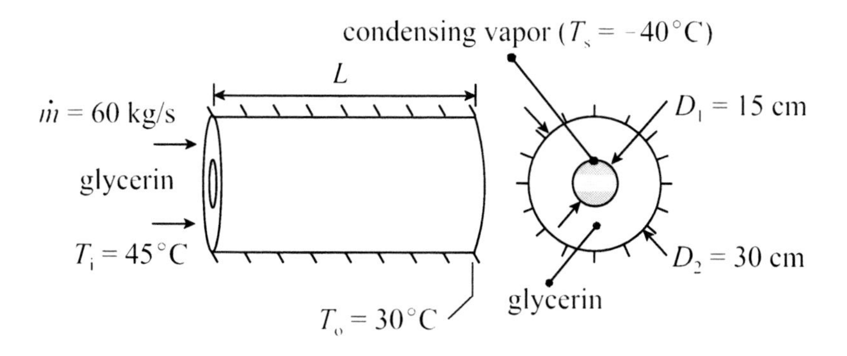

ASSUMPTIONS

1. Steady conditions.
2. Fully developed flow.
3. The surface of the inner pipe is isothermal.
4. The outside surface of the outer pipe is insulated.
5. Constant properties.

PROPERTIES

$T_{av} = (T_i + T_o)/2 = (45°C + 30°C)/2 = 37.5°C = 311$ K

glycerin: $\rho = 1253.2$ kg/m^3, $c_p = 2497$ J/kg·K, $k = 0.286$ W/m·K,
$\nu = 270 \times 10^{-6}$ m^2/s, $Pr = 2941$

ANALYSIS

The hydraulic diameter of the annulus is

$$D_h = \frac{4A}{P} = \frac{4\pi(D_2^2 - D_1^2)/4}{\pi(D_2 + D_1)} = D_2 - D_1$$

$D_h = 0.30 \text{ m} - 0.15 \text{ m}$

$= 0.15$ m

and the mean velocity of the glycerin is

$$u_m = \frac{\dot{m}}{\rho A}$$

$$u_m = \frac{60 \text{ kg/s}}{(1253.2 \text{ kg/m}^3)(0.0530 \text{ m}^2)}$$

$= 0.903$ m/s

The Reynolds number is

$$Re_D = \frac{u_m D_h}{\nu}$$

$$Re_D = \frac{(0.903 \text{ m/s})(0.15 \text{ m})}{270 \times 10^{-6} \text{ m}^2\text{/s}}$$

$= 502$

Hence, the flow is laminar. For fully developed laminar flow in a concentric tube annulus with one surface insulated and the other surface isothermal, the Nusselt number is constant. For a diameter ratio of 0.50, the Nusselt number is

$$Nu_D = \frac{hD_h}{k} = 5.74$$

so the heat transfer coefficient for the outside surface of the inner tube is

$$h = \frac{Nu_D k}{D_h}$$

$$h = \frac{(5.74)(0.286\ \text{W/m·K})}{0.15\ \text{m}}$$

$$= 10.9\ \text{W/m}^2\text{·K}$$

An energy balance on the glycerin yields

$$\frac{T_s - T_o}{T_s - T_i} = \exp\left(-\frac{\pi D_i L}{\dot{m} c_p} h\right)$$

Solving for tube length L,

$$L = -\frac{\dot{m} c_p}{\pi D_i h} \ln\left(\frac{T_s - T_o}{T_s - T_i}\right)$$

$$L = -\frac{(60\ \text{kg/s})(2497\ \text{J/kg·K})}{\pi(0.15\ \text{m})(10.9\ \text{W/m}^2\text{·K})} \ln\left(\frac{-40°\text{C} - 30°\text{C}}{-40°\text{C} - 45°\text{C}}\right)$$

$$= \underline{\underline{5663\ \text{m}}}$$

The outer tube is insulated, so the heat transfer is from the glycerin to the outside surface of the inner tube.

$$q = \dot{m} c_p (T_i - T_o)$$

$$q = (60\ \text{kg/s})(2497\ \text{J/kg·K})(45 - 30)°\text{C}$$

$$= 2.25 \times 10^6\ \text{W} = \underline{\underline{2.25\ \text{MW}}}$$

DISCUSSION

As shown in the expression above, tube length is directly proportional to mass flow rate. The graph below shows this variation for a range of mass flow rates, leaving all other parameters unchanged.

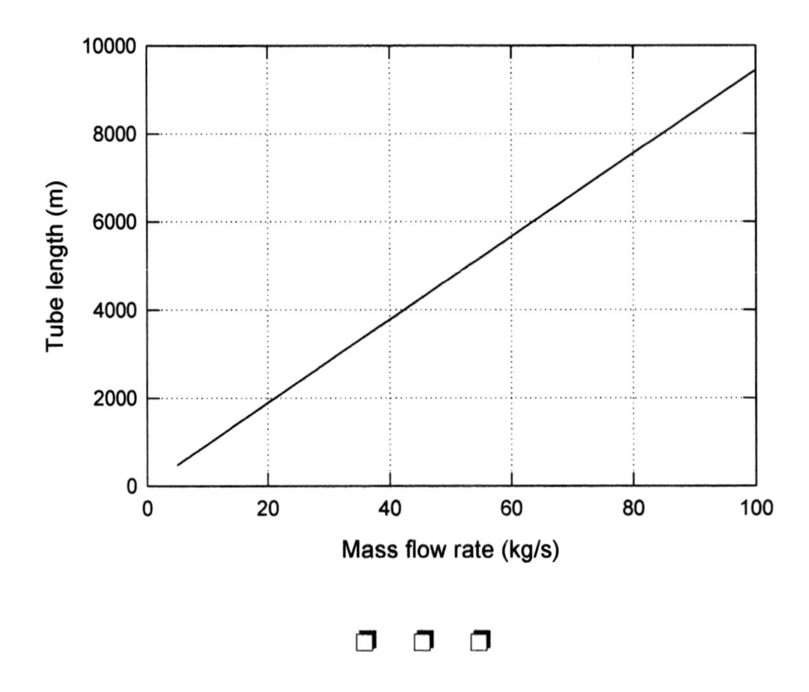

❒ ❒ ❒

Chapter 9

Natural Convection

PROBLEM 9.1 **Natural Convection from a Vertical Plate**

A 1 m × 1 m vertical plate is surrounded by an infinite expanse of 20 °C quiescent air. The temperature of the plate is 85 °C. If the plate convects from both sides, find the total heat transfer.

DIAGRAM

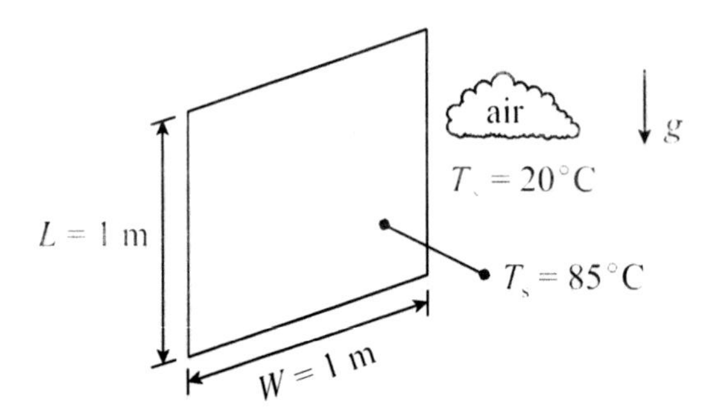

ASSUMPTIONS

1. Steady conditions.
2. Infinite expanse of stagnant atmospheric air.
3. Plate is isothermal.
4. Radiation is neglected.
5. Constant properties.

PROPERTIES

$T_f = (T_s + T_\infty)/2 = (85\,°C + 20\,°C)/2 = 52.5\,°C = 326$ K

atmospheric air: $\nu = 18.51 \times 10^{-6}$ m²/s, $k = 0.0282$ W/m·K, $\alpha = 26.3 \times 10^{-6}$ m²/s, $Pr = 0.703$, $\beta = 1/T_f = 3.067 \times 10^{-3}\ \text{K}^{-1}$

ANALYSIS

The Rayleigh number is

$$Ra_L = \frac{g\beta(T_s - T_\infty)L^3}{\nu\alpha}$$

$$Ra_L = \frac{(9.81\ \text{m/s}^2)(3.067 \times 10^{-3}\ \text{K}^{-1})(85 - 20)\,°\text{C}(1\ \text{m})^3}{(18.51 \times 10^{-6}\ \text{m}^2/\text{s})(26.3 \times 10^{-6}\ \text{m}^2/\text{s})}$$

$$= 4.017 \times 10^9$$

Because $Ra_L > 10^9$, the flow is turbulent. The average Nusselt number may be calculated from the relation

$$Nu_L = \frac{hL}{k} = \left\{ 0.825 + \frac{0.387 Ra_L^{1/6}}{\left[1 + (0.492/Pr)^{9/16}\right]^{8/27}} \right\}^2$$

$$Nu_L = \left\{ 0.825 + \frac{0.387(4.017 \times 10^9)^{1/6}}{[1 + (0.492/0.703)^{9/16}]^{8/27}} \right\}^2$$

$$= 189.1$$

Hence, the heat transfer coefficient is

$$h = \frac{Nu_L k}{L}$$

$$h = \frac{(189.1)(0.0282 \text{ W/m·K})}{1\text{m}}$$

$$= 5.33 \text{ W/m}^2\text{·K}$$

Neglecting radiation, the total heat transfer from the plate is

$$q = 2LWh(T_s - T_\infty)$$

$$q = 2(1 \text{ m})(1 \text{ m})(5.33 \text{ W/m}^2\text{·K})(85 - 20)°\text{C}$$

$$= \underline{\underline{693 \text{ W}}}$$

DISCUSSION

The Rayleigh number is the product of the Grashof and Prandtl numbers,

$$Ra_L = Gr_L Pr$$

The Grashof number plays the same role in natural convection as the Reynolds number plays in forced convection. The Reynolds number is the ratio of the inertial force to the viscous force acting on the fluid, whereas the Grashof number is the ratio of the buoyancy force to the viscous force acting on the fluid.

PROBLEM 9.2 Natural Convection from a Vertical Plate in Two Orientations

A 40 cm × 80 cm vertical plate is surrounded by 25°C quiescent air, and the temperature of the plate is 50°C. Compare the convective heat transfer when the plate is oriented in the two positions shown in the diagram.

DIAGRAM

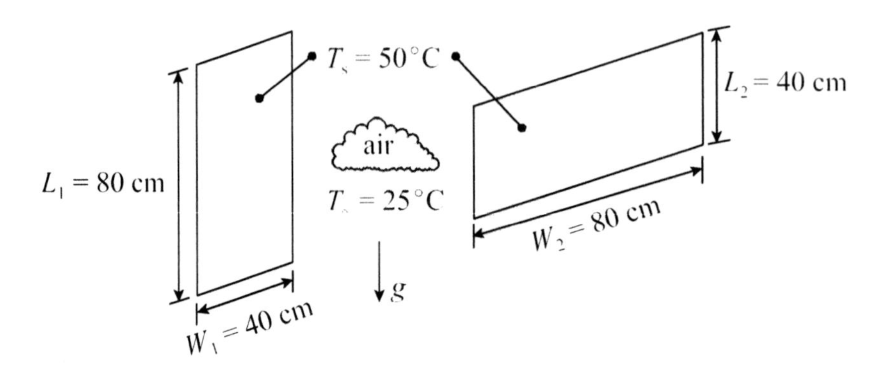

ASSUMPTIONS

1. Steady conditions.
2. Infinite expanse of stagnant atmospheric air.
3. Plate is isothermal.
4. Radiation is neglected.
5. Constant properties.

PROPERTIES

$T_f = (T_s + T_\infty)/2 = (50°C + 25°C)/2 = 37.5°C = 311$ K

atmospheric air: $\nu = 17.00 \times 10^{-6}$ m²/s, $k = 0.0271$ W/m·K, $\alpha = 24.1 \times 10^{-6}$ m²/s, $Pr = 0.706$, $\beta = 1/T_f = 3.215 \times 10^{-3}$ K^{-1}

ANALYSIS

For the first orientation, the Rayleigh number is

$$Ra_{L1} = \frac{g\beta(T_s - T_\infty)L_1^3}{\nu\alpha}$$

$$Ra_{L1} = \frac{(9.81 \text{ m/s}^2)(3.215 \times 10^{-3} \text{ K}^{-1})(50 - 25)°\text{C}(0.80 \text{ m})^3}{(17.00 \times 10^{-6} \text{ m}^2/\text{s})(24.1 \times 10^{-6} \text{ m}^2/\text{s})}$$

$= 9.854 \times 10^8$

Because $Ra_{L1} < 10^9$, the flow is laminar. For laminar flow, the average Nusselt number may be calculated from the relation

$$Nu_{L1} = \frac{h_1 L_1}{k} = 0.68 + \frac{0.670 Ra_{L1}^{1/4}}{\left[1 + (0.492/Pr)^{9/16}\right]^{4/9}}$$

$$Nu_{L1} = 0.68 + \frac{0.670(9.854 \times 10^8)^{1/4}}{[1 + (0.492/0.706)^{9/16}]^{4/9}}$$

$= 91.7$

so the heat transfer coefficient is

$$h_1 = \frac{Nu_{L1} k}{L_1}$$

$$h_1 = \frac{(91.7)(0.0271 \text{ W/m·K})}{0.80 \text{ m}}$$

$= 3.11 \text{ W/m}^2\text{·K}$

and the heat transfer is

$$q_1 = 2 L_1 W_1 h_1 (T_s - T_\infty)$$

$q_1 = 2(0.80 \text{ m})(0.40 \text{ m})(3.11 \text{ W/m}^2\text{·K})(50 - 25)°\text{C}$

$= \underline{\underline{49.8 \text{ W}}}$

For the second orientation, the Rayleigh number is

$$Ra_{L2} = \frac{g\beta(T_s - T_\infty)L_2^3}{\nu\alpha}$$

$$Ra_{L2} = \frac{(9.81 \text{ m/s}^2)(3.215 \times 10^{-3} \text{ K}^{-1})(50 - 25)°\text{C}(0.40 \text{ m})^3}{(17.00 \times 10^{-6} \text{ m}^2\text{/s})(24.1 \times 10^{-6} \text{ m}^2\text{/s})}$$

$= 1.232 \times 10^8$

As with the first orientation, the flow is laminar. For laminar flow, the average Nusselt number may be calculated from the relation

$$Nu_{L2} = \frac{h_2 L_2}{k} = 0.68 + \frac{0.670 Ra_{L2}^{1/4}}{\left[1 + (0.492/Pr)^{9/16}\right]^{4/9}}$$

$$Nu_{L2} = 0.68 + \frac{0.670(1.232 \times 10^8)^{1/4}}{[1 + (0.492/0.706)^{9/16}]^{4/9}}$$

$$= 54.8$$

so the heat transfer coefficient is

$$h_2 = \frac{Nu_{L2} k}{L_2}$$

$$h_2 = \frac{(54.8)(0.0271 \text{ W/m·K})}{0.40 \text{ m}}$$

$$= 3.71 \text{ W/m}^2\text{·K}$$

and the heat transfer is

$$q_2 = 2 L_2 W_2 h_2 (T_s - T_\infty)$$

$$q_2 = 2(0.40 \text{ m})(0.80 \text{ m})(3.71 \text{ W/m}^2\text{·K})(50 - 25)°\text{C}$$

$$= \underline{\underline{59.4 \text{ W}}}$$

DISCUSSION

The ratio of the heat transfer for the two plate orientations is

$$q_1/q_2 = (49.8 \text{ W})/(59.4 \text{ W})$$

$$= 0.838$$

Why is the heat transfer higher for the second orientation? For both orientations, the boundary layer is laminar, but for the second orientation the boundary layer is not as thick, i.e., not as developed, as for the first orientation. Hence, the heat transfer coefficient is higher for the second orientation.

PROBLEM 9.3 Maximum Temperature of Devices on a Vertical Circuit Board

A circuit board measuring 18 cm × 18 cm is oriented vertically in 20°C quiescent air. The board is populated with closely spaced identical devices. The total heat dissipation for the circuit board is 9 W. Assuming a uniform heat flux at the device surfaces, what is the maximum surface temperature of the devices? The back side of the circuit board is insulated.

DIAGRAM

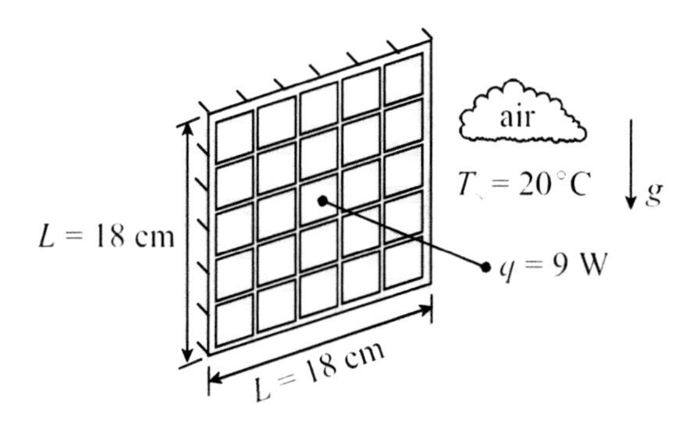

ASSUMPTIONS

1. Steady conditions.
2. Infinite expanse of stagnant atmospheric air.
3. Uniform heat flux at surface.
4. Back side of circuit board is insulated.
5. Radiation is neglected.
6. Constant properties.

PROPERTIES

Assume $T_s(L/2) = 70°\text{C}$. $T_f = [T_s(L/2) + T_\infty]/2 = (70°\text{C} + 20°\text{C})/2 = 45°\text{C} = 318\ \text{K}$

atmospheric air: $\nu = 17.70 \times 10^{-6}\ \text{m}^2/\text{s}$, $k = 0.0276\ \text{W/m·K}$, $\alpha = 25.2 \times 10^{-6}\ \text{m}^2/\text{s}$, $Pr = 0.705$, $\beta = 1/T_f = 3.145 \times 10^{-3}\ \text{K}^{-1}$

ANALYSIS

For a uniform heat flux surface condition, the Rayleigh number is based on the temperature difference of the surface's vertical midpoint and the ambient fluid. Thus,

$$Ra_L = \frac{g\beta\,[T_s(L/2) - T_\infty]L^3}{\nu\alpha}$$

$$Ra_L = \frac{(9.81\ \text{m/s}^2)(3.145 \times 10^{-3}\ \text{K}^{-1})(70 - 20)°\text{C}(0.18\ \text{m})^3}{(17.70 \times 10^{-6}\ \text{m}^2/\text{s})(25.2 \times 10^{-6}\ \text{m}^2/\text{s})}$$

$$= 2.017 \times 10^7$$

which indicates the flow is laminar. For laminar flow, the average Nusselt number may be calculated from the relation

$$Nu_L = \frac{hL}{k} = 0.68 + \frac{0.670 Ra_L^{1/4}}{[1 + (0.492/Pr)^{9/16}]^{4/9}}$$

$$Nu_L = 0.68 + \frac{0.670(2.017 \times 10^7)^{1/4}}{[1 + (0.492/0.705)^{9/16}]^{4/9}}$$

$$= 35.1$$

so the heat transfer coefficient is

$$h = \frac{Nu_L k}{L}$$

$$h = \frac{(35.1)(0.0276\ \text{W/m·K})}{0.18\ \text{m}}$$

$$= 5.38\ \text{W/m}^2\text{·K}$$

Using our initial estimate for the midpoint surface temperature and calculated heat transfer coefficient, the heat flux at the surface of the devices is

$$q'' = h[T_s(L/2) - T_\infty]$$

$$q'' = (5.38\ \text{W/m}^2\text{·K})(70 - 20)°\text{C}$$

$$= 269\ \text{W/m}^2$$

but the actual heat flux is the total heat dissipation divided by surface area,

$$q'' = q/L^2$$

$$q'' = (9\ \text{W})/(0.18\ \text{m})^2$$

$$= 278\ \text{W/m}^2$$

The predicted heat flux is 3 percent lower than the actual value, which is acceptable given the experimental errors inherent in convection heat transfer correlations. Had our predicted heat flux been significantly different than the actual value, additional iterations on $T_s(L/2)$ would have been required, eventually leading to convergence with the actual heat flux of 278 W/m^2. The maximum surface temperature occurs at the top of the circuit board. An approximate relation for the surface temperature variation is

$$T_s(x) - T_\infty \approx 1.15\left(\frac{x}{L}\right)^{1/5} [T_s(L/2) - T_\infty]$$

where x is a vertical coordinate measured from the bottom of the vertical surface. Thus, the surface temperature of the devices at $x = L$, the top of the board, is

$$T_s(L) = T_{\max} = T_\infty + 1.15\,[T_s(L/2) - T_\infty]$$

$$T_{\max} = 20°\text{C} + 1.15(70 - 20)°\text{C}$$

$$= \underline{\underline{77.5°\text{C}}}$$

DISCUSSION

Many electronic solid state devices, particularly those used in military applications, have a maximum surface (case) temperature specification of 85°C.

❒ ❒ ❒

PROBLEM 9.4 Heat Loss from a Window

A single pane window in a building has a height and width of 2 m and 1.5 m, respectively. The outside surface of the window experiences a steady 8 m/s wind, while the inside surface is exposed to quiescent air. The ambient air temperature outside the building is −6°C, and the glass is 4 mm thick. If the temperature of the air inside the building is 21°C, what is the heat loss from the window?

DIAGRAM

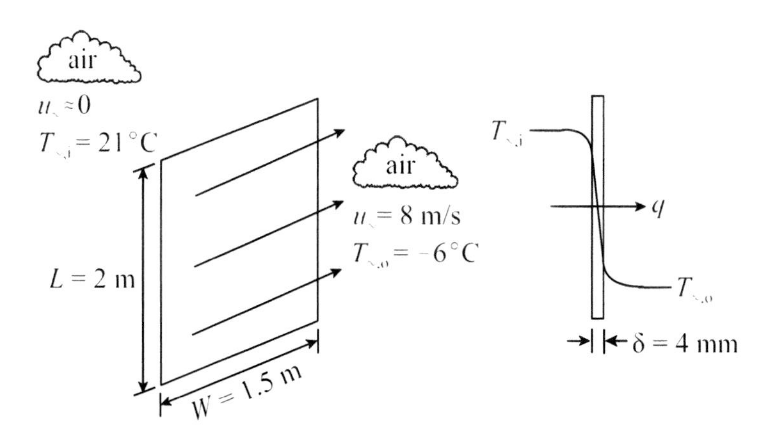

ASSUMPTIONS

1. Steady conditions.
2. Infinite expanse of stagnant atmospheric air at the inside surface of the window.
3. Steady horizontal flow of air across the outside surface of the window.
4. Radiation is neglected.
5. Constant properties.

PROPERTIES

window glass: $k_g = 1.4$ W/m·K

atmospheric air:

inside surface
assume $T_{s,i} = 12°C$. $T_{f,i} = (T_{s,i} + T_{\infty,i})/2 = (12°C + 21°C)/2 = 16.5°C = 290$ K

$\nu_i = 15.00 \times 10^{-6}$ m²/s, $k_i = 0.0293$ W/m·K, $\alpha = 21.2 \times 10^{-6}$ m²/s,
$Pr_i = 0.710$, $\beta = 1/T_f = 3.448 \times 10^{-3}$ K^{-1}

outside surface
assume $T_{s,o} = -3°C$. $T_{f,o} = (T_{s,o} + T_{\infty,o})/2 = (-3°C - 6°C)/2 = -4.5°C = 269$ K

$\nu_o = 13.13 \times 10^{-6}$ m²/s, $k_o = 0.0238$ W/m·K, $Pr_o = 0.715$

ANALYSIS

Two heat transfer coefficients are needed, one for each surface of the window glass. The outside surface is exposed to the parallel flow of ambient air. The Reynolds number for this flow is

$$Re_W = \frac{u_\infty W}{\nu_o}$$

$$Re_W = \frac{(8\text{ m/s})(1.5\text{ m})}{13.13 \times 10^{-6}\text{ m}^2/\text{s}}$$

$$= 9.139 \times 10^5$$

Because $Re_W > 5 \times 10^5$, the flow is mixed, i.e., the window experiences both laminar and turbulent flow conditions. The average Nusselt number may be found using the relation

$$Nu_o = \frac{h_o W}{k_o} = (0.037 Re_W^{4/5} - 871) Pr_o^{1/3}$$

$$Nu_o = [0.037(9.139 \times 10^5)^{4/5} - 871](0.715)^{1/3}$$

$= 1164$

and the heat transfer coefficient is

$$h_o = \frac{Nu_o k_o}{W}$$

$$h_o = \frac{(1164)(0.0238 \text{ W/m·K})}{1.5 \text{ m}}$$

$= 18.47 \text{ W/m}^2\text{·K}$

The inside surface of the window is exposed to natural convection. The Rayleigh number is

$$Ra_L = \frac{g\beta(T_{\infty,i} - T_{s,i})L^3}{\nu_i \alpha}$$

$$Ra_L = \frac{(9.81 \text{ m/s}^2)(3.448 \times 10^{-3} \text{ K}^{-1})(21 - 12)°\text{C}(2 \text{ m})^3}{(15.00 \times 10^{-6} \text{ m}^2\text{/s})(21.2 \times 10^{-6} \text{ m}^2\text{/s})}$$

$= 7.658 \times 10^9$

which indicates turbulent flow. The average Nusselt number may be calculated from the relation

$$Nu_L = \frac{h_i L}{k_i} = \left\{0.825 + \frac{0.387 Ra_L^{1/6}}{\left[1 + (0.492/Pr_i)^{9/16}\right]^{8/27}}\right\}^2$$

$$Nu_L = \left\{0.825 + \frac{0.387(7.658 \times 10^9)^{1/6}}{[1 + (0.492/0.710)^{9/16}]^{8/27}}\right\}^2$$

$= 232$

Hence, the heat transfer coefficient is

$$h_i = \frac{Nu_L k_i}{L}$$

$$h_i = \frac{(232)(0.0293 \text{ W/m·K})}{2 \text{ m}}$$

$= 3.399 \text{ W/m}^2\text{·K}$

Now that both heat transfer coefficients have been calculated, we find the total thermal resistance,

$$R_{tot} = \frac{1}{A}\left(\frac{1}{h_i} + \frac{\delta}{k_g} + \frac{1}{h_o}\right)$$

$$R_{tot} = \frac{1}{(2\text{ m})(1.5\text{ m})}\left(\frac{1}{3.399\text{ W/m}^2\cdot\text{K}} + \frac{0.004\text{ m}}{1.4\text{ W/m}\cdot\text{K}} + \frac{1}{18.47\text{ W/m}^2\cdot\text{K}}\right)$$

$$= 0.1171\text{ K/W}$$

The heat transfer is

$$q = \frac{T_{\infty,i} - T_{\infty,o}}{R_{tot}}$$

$$q = \frac{[21 - (-6)]°\text{C}}{0.1171\text{ K/W}}$$

$$= \underline{\underline{230.6\text{ W}}}$$

Our initial estimates of the inside and outside window surfaces should be checked. The heat transfer may be written as

$$q = h_i A(T_{\infty,i} - T_{s,i})$$

Solving for $T_{s,i}$, the inside surface temperature,

$$T_{s,i} = 21°\text{C} - \frac{230.6\text{ W}}{(3.399\text{ W/m}^2\cdot\text{K})(2\text{ m})(1.5\text{ m})}$$

$$= -1.61°\text{C}$$

The heat transfer may also be written as

$$q = h_o A(T_{s,o} - T_{\infty,o})$$

Solving for $T_{s,o}$, the outside surface temperature,

$$T_{s,o} = -6°\text{C} + \frac{(230.6\text{ W})}{(18.47\text{ W/m}^2\cdot\text{K})(2\text{ m})(1.5\text{ m})}$$

$$= -1.84°\text{C}$$

The calculated outside surface temperature is close to the initial estimate, but the calculated inside surface temperature is off. A correction of thermal properties for the heat transfer coefficient on the inside surface of the window would yield slightly more accurate results. This is left as an exercise for the reader.

DISCUSSION

The conduction through the glass is given by Fourier's law,

$$q = kA\frac{T_{s,i} - T_{s,o}}{\delta}$$

$$q = \frac{(1.4\ \text{W/m·K})(2\ \text{m})(1.5\ \text{m})[-1.61 - (-1.84)]°\text{C}}{0.004\ \text{m}}$$

$$= 241\ \text{W}$$

Under steady conditions, the heat conducted through the glass is equal to the heat convected to the glass from the inside air and the heat convected from the glass to the outside air. The difference in heat transfer answers is due to numerical roundoff.

❒ ❒ ❒

PROBLEM 9.5　　Convection and Radiation from a Vertical Plate

A thin metal plate measuring 40 cm × 40 cm is surrounded by quiescent atmospheric nitrogen at 10°C. The temperature and emissivity of the plate are 550°C and 0.8, respectively. Find the total heat transfer from the plate.

DIAGRAM

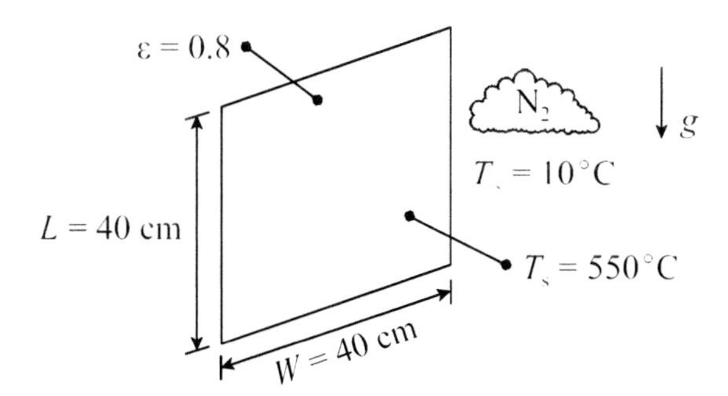

ASSUMPTIONS

1. Steady conditions.
2. Infinite expanse of stagnant atmospheric nitrogen.
3. Temperature of surroundings for radiation is equivalent to the nitrogen temperature.
4. Constant properties.

PROPERTIES

metal plate: $\varepsilon = 0.8$

$T_f = (T_s + T_\infty)/2 = (550°C + 10°C)/2 = 280°C = 553$ K

atmospheric nitrogen: $\nu = 45.28 \times 10^{-6}$ m²/s, $k = 0.0419$ W/m·K, $\alpha = 64.5 \times 10^{-6}$ m²/s, $Pr = 0.702$, $\beta = 1/T_f = 1.808 \times 10^{-3}$ K⁻¹

ANALYSIS

The Rayleigh number is

$$Ra_L = \frac{g\beta(T_s - T_\infty)L^3}{\nu\alpha}$$

$$Ra_L = \frac{(9.81 \text{ m/s}^2)(1.808 \times 10^{-3} \text{ K}^{-1})(550 - 10)°\text{C}(0.40 \text{ m})^3}{(45.28 \times 10^{-6} \text{ m}^2/\text{s})(64.5 \times 10^{-6} \text{ m}^2/\text{s})}$$

$$= 2.099 \times 10^8$$

which indicates laminar flow. For laminar flow, the average Nusselt number may be calculated from the relation

$$Nu_L = \frac{hL}{k} = 0.68 + \frac{0.670 Ra_L^{1/4}}{\left[1 + (0.492/Pr)^{9/16}\right]^{4/9}}$$

$$Nu_L = 0.68 + \frac{0.670(2.099 \times 10^8)^{1/4}}{[1 + (0.492/0.702)^{9/16}]^{4/9}}$$

$$= 62.5$$

so the heat transfer coefficient is

$$h = \frac{Nu_L k}{L}$$

$$h = \frac{(62.5)(0.0419 \text{ W/m·K})}{0.40 \text{ m}}$$

$$= 6.55 \text{ W/m}^2\text{·K}$$

The convective heat transfer is

$$q_{conv} = 2hA(T_s - T_\infty)$$

$q_{\text{conv}} = 2(6.55\ \text{W/m}^2\cdot\text{K})(0.40\ \text{m})^2(550 - 10)°\text{C}$

$= 1132\ \text{W}$

and the radiative heat transfer is

$$q_{rad} = 2\varepsilon\sigma A(T_s^4 - T_{sur}^4)$$

$q_{\text{rad}} = 2(0.8)(5.669 \times 10^{-8}\ \text{W/m}^2\cdot\text{K}^4)(0.40\ \text{m})^2\ (823^4 - 283^4)\text{K}^4$

$= 6565\ \text{W}$

The total heat transfer is

$$q = q_{conv} + q_{rad}$$

$q = (1132 + 6565)\text{W}$

$= \underline{\underline{7697\ \text{W}}}$

DISCUSSION

The radiative heat transfer is over 85 percent of the total heat transfer. This is not uncommon in natural convection systems in which surface temperatures are high. The graph below shows how the ratio q_{rad}/q varies with surface temperature T_s, holding all other quantities fixed.

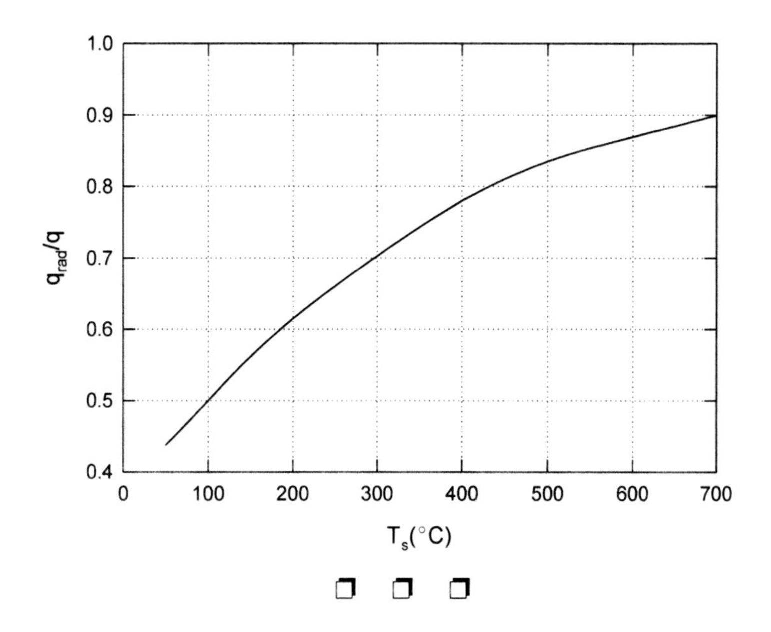

❐ ❐ ❐

PROBLEM 9.6 Convection from a Horizontal Plate with Upper Surface Heated

A metal plate measuring 50 cm × 50 cm is oriented horizontally in an infinite expanse of quiescent air. The upper surface of the plate is isothermal, having a temperature of 180°C, and the lower surface is insulated. If the temperature of the air is 30°C, find the heat transfer from the plate.

DIAGRAM

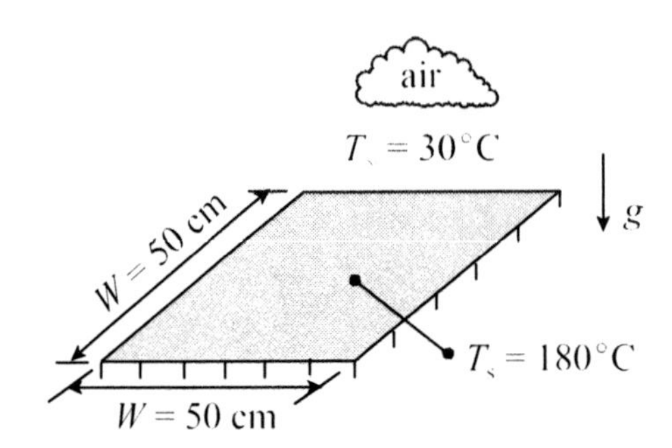

ASSUMPTIONS

1. Steady conditions.
2. Infinite expanse of stagnant atmospheric air.
3. Lower surface of plate is insulated.
4. Radiation is neglected.
5. Constant properties.

PROPERTIES

$T_f = (T_s + T_\infty)/2 = (180°C + 30°C)/2 = 105°C = 378$ K

atmospheric air: $\nu = 23.99 \times 10^{-6}$ m²/s, $k = 0.0321$ W/m·K, $\alpha = 34.6 \times 10^{-6}$ m²/s, $Pr = 0.694$, $\beta = 1/T_f = 2.646 \times 10^{-3}$ K^{-1}

ANALYSIS

For a horizontal plate, the characteristic length for use in the Rayleigh number is given by the relation

$$L = \frac{A}{P}$$

where $A = W^2$, the surface area of the plate, and $P = 4W$, the perimeter. Thus,

$L = (0.50 \text{ m})/4$

$= 0.125 \text{ m}$

The Rayleigh number is

$$Ra_L = \frac{g\beta(T_s - T_\infty)L^3}{\nu\alpha}$$

$$Ra_L = \frac{(9.81 \text{ m/s}^2)(2.646 \times 10^{-3} \text{ K}^{-1})(180 - 30)°\text{C}(0.125 \text{ m})^3}{(23.99 \times 10^{-6} \text{ m}^2/\text{s})(34.6 \times 10^{-6} \text{ m}^2/\text{s})}$$

$= 9.162 \times 10^6$

The average Nusselt number may be found using the relation

$$Nu_L = \frac{hL}{k} = 0.54 Ra_L^{1/4}$$

$Nu_L = (0.54)(9.162 \times 10^6)^{1/4}$

$= 29.7$

so the heat transfer coefficient is

$$h = \frac{Nu_L k}{L}$$

$$h = \frac{(29.7)(0.0321 \text{ W/m·K})}{0.125 \text{ m}}$$

$= 7.63 \text{ W/m}^2\text{·K}$

The heat transfer from the plate is

$$q = hA(T_s - T_\infty)$$

$q = (7.63 \text{ W/m}^2\text{·K})(0.50 \text{ m})^2(180 - 30)°\text{C}$

$= \underline{\underline{286 \text{ W}}}$

DISCUSSION

The relation for average Nusselt number used here also applies to the lower surface of a cooled horizontal plate for the range of Rayleigh numbers, $10^4 < Ra_L < 10^7$. For the Rayleigh number range, $10^7 < Ra_L < 10^{11}$, a recommended relation is

$$Nu_L = 0.15Ra_L^{1/3}$$

Heat transfer is more effective for heated plates facing upward and cooled plates facing downward than for heated plates facing downward and cooled plates facing upward because in these configurations the tendency of the fluid to ascend and descend, respectively, is impeded by the plate. This is illustrated in Problem 9.7.

❒ ❒ ❒

PROBLEM 9.7 Convection from a Horizontal Plate with Lower Surface Heated

The plate in Problem 9.6 is oriented horizontally in an infinite expanse of quiescent air. The lower surface of the plate is isothermal, having a temperature of 180°C, and the upper surface is insulated. If the temperature of the air is 30°C, find the heat transfer from the plate.

DIAGRAM

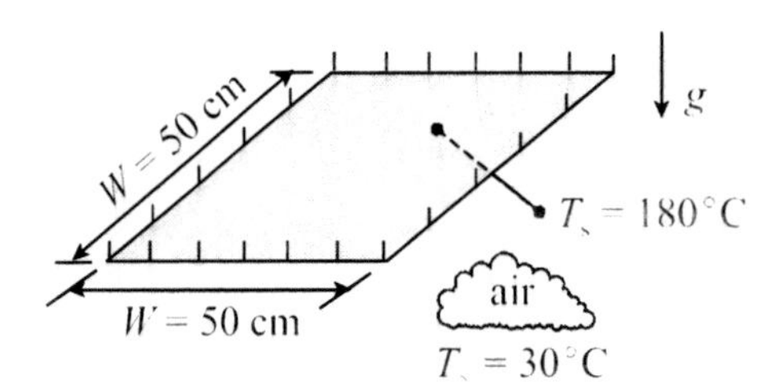

ASSUMPTIONS

1. Steady conditions.
2. Infinite expanse of stagnant atmospheric air.
3. Lower surface of plate is insulated.
4. Radiation is neglected.
5. Constant properties.

PROPERTIES

$T_f = (T_s + T_\infty)/2 = (180°\text{C} + 30°\text{C})/2 = 105°\text{C} = 378\ \text{K}$

atmospheric air: $\nu = 23.99 \times 10^{-6}\ \text{m}^2/\text{s}$, $k = 0.0321\ \text{W/m·K}$, $\alpha = 34.6 \times 10^{-6}\ \text{m}^2/\text{s}$, $Pr = 0.694$, $\beta = 1/T_f = 2.646 \times 10^{-3}\ \text{K}^{-1}$

ANALYSIS

The Rayleigh number is the same as in Problem 9.6, but the relation for average Nusselt number is

$$Nu_L = \frac{hL}{k} = 0.27\,Ra_L^{1/4}$$

$Nu_L = (0.27)(9.162 \times 10^6)^{1/4}$

$= 14.9$

half the previous value. The heat transfer coefficient is

$$h = \frac{Nu_L k}{L}$$

$$h = \frac{(14.9)(0.0321\ \text{W/m·K})}{0.125\ \text{m}}$$

$= 3.83\ \text{W/m}^2\text{·K}$

Thus, the heat transfer from the plate is

$$q = hA(T_s - T_\infty)$$

$q = (3.83\ \text{W/m}^2\text{·K})(0.50\ \text{m})^2(180 - 30)°\text{C}$

$= \underline{\underline{143\ \text{W}}}$

DISCUSSION

The heat transfer is only half the value for the plate in Problem 9.6 because the plate impedes the air as it ascends, thereby making natural convection less effective. The relation for average Nusselt number used here applies for the range of Rayleigh numbers, $10^5 < Ra_L < 10^{10}$. Radiation would represent a major fraction of the total heat transfer in Problems 9.6 and 9.7, particularly if the plate has a high emissivity.

❒ ❒ ❒

PROBLEM 9.8 Convection from a Long Horizontal Cylinder

A long 25-cm diameter cylinder is horizontally oriented in 20°C quiescent air. If the temperature of the surface is 240°C, find the heat transfer per unit length of cylinder.

DIAGRAM

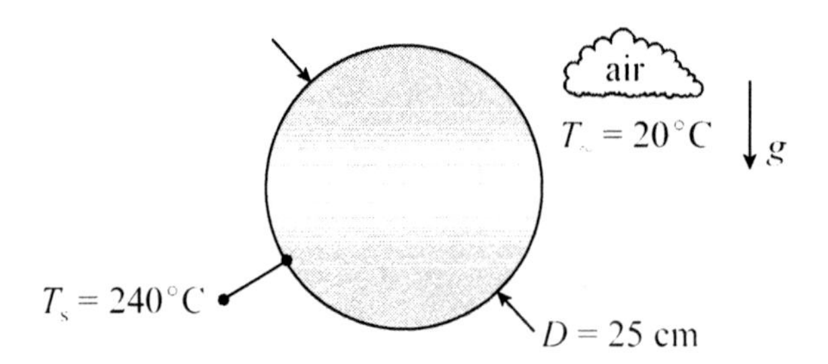

ASSUMPTIONS

1. Steady conditions.
2. Infinite expanse of stagnant atmospheric air.
3. Infinitely long cylinder.
4. Radiation is neglected.
5. Constant properties.

PROPERTIES

$T_f = (T_s + T_\infty)/2 = (240°C + 20°C)/2 = 130°C = 403$ K

atmospheric air: $\nu = 26.77 \times 10^{-6}$ m²/s, $k = 0.0340$ W/m·K, $\alpha = 38.8 \times 10^{-6}$ m²/s, $Pr = 0.690$, $\beta = 1/T_f = 2.481 \times 10^{-3}$ K^{-1}

ANALYSIS

The Rayleigh number for a horizontal cylinder is

$$Ra_D = \frac{g\beta(T_s - T_\infty)D^3}{\nu\alpha}$$

$$Ra_D = \frac{(9.81 \text{ m/s}^2)(2.481 \times 10^{-3} \text{ K}^{-1})(240 - 20)°\text{C}(0.25 \text{ m})^3}{(26.77 \times 10^{-6} \text{ m}^2/\text{s})(38.8 \times 10^{-6} \text{ m}^2/\text{s})}$$

$$= 8.055 \times 10^7$$

The average Nusselt number may be calculated using the relation

$$Nu_D = \frac{hD}{k} = CRa_D^n$$

where the quantities C and n depend on the Rayleigh number. For the range $10^7 < Ra_D < 10^{12}$, C

= 0.125 and n = 0.333. Thus, the average Nusselt number is

$$Nu_D = (0.125)(8.055 \times 10^7)^{0.333}$$

$$= 53.7$$

so the heat transfer coefficient is

$$h = \frac{Nu_D k}{D}$$

$$h = \frac{(53.7)(0.0340 \text{ W/m·K})}{0.25 \text{ m}}$$

$$= 7.30 \text{ W/m}^2\text{·K}$$

The heat transfer per meter of length is

$$q' = q/L = h\pi D(T_s - T_\infty)$$

$$q' = (7.30 \text{ W/m}^2\text{·K})\pi(0.25 \text{ m})(240 - 20)°\text{C}$$

$$= \underline{\underline{1261 \text{ W/m}}}$$

DISCUSSION

A single Nusselt number correlation for the Rayleigh number range, $Ra_D < 10^{12}$, is

$$Nu_D = \left\{0.60 + \frac{0.387 Ra_D^{1/6}}{\left[1 + (0.559/Pr)^{9/16}\right]^{8/27}}\right\}^2$$

which yields an average Nusselt number and heat transfer coefficient, respectively, of

$$Nu_D = 52.7 \ , \ h = 7.17 \text{ W/m}^2\text{·K}$$

The heat transfer per meter of length is

$$q' = 1239 \text{ W/m}$$

❒ ❒ ❒

PROBLEM 9.9 Convection from a Current-Carrying Wire

A long 14-gage copper wire carries 34 A of electrical current. The wire is bare and runs horizontally across a room where the air temperature is 30°C. Find the temperature of the wire. A 14-gage copper wire has a diameter of 1.628 mm and a resistance of 8.285 Ω/km.

DIAGRAM

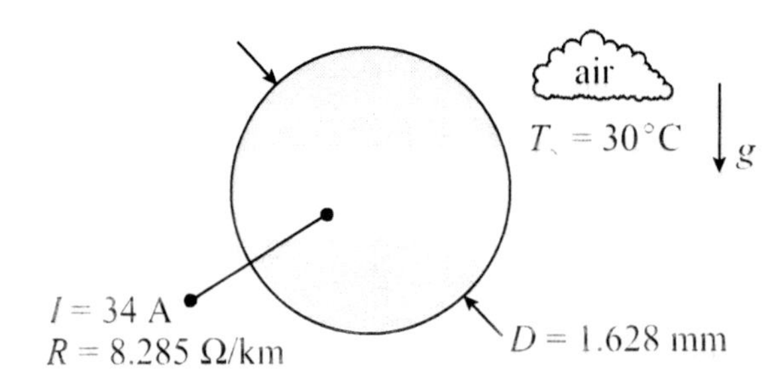

ASSUMPTIONS

1. Steady conditions.
2. Infinite expanse of stagnant atmospheric air.
3. Infinitely long wire.
4. Wire is isothermal.
5. Radiation is neglected.
6. Constant properties.

PROPERTIES

Assume $T_s = 100°C$. $T_f = (T_s + T_\infty)/2 = (100°C + 30°C)/2 = 65°C = 338$ K

atmospheric air: $\nu = 19.71 \times 10^{-6}$ m²/s, $k = 0.0291$ W/m·K, $\alpha = 28.1 \times 10^{-6}$ m²/s, $Pr = 0.702$, $\beta = 1/T_f = 2.959 \times 10^{-3}$ K^{-1}

ANALYSIS

The Rayleigh number for a horizontal cylinder is

$$Ra_D = \frac{g\beta(T_s - T_\infty)D^3}{\nu\alpha}$$

$$Ra_D = \frac{(9.81 \text{ m/s}^2)(2.959 \times 10^{-3} \text{ K}^{-1})(100 - 30)°\text{C}(1.628 \times 10^{-3} \text{ m})^3}{(19.71 \times 10^{-6} \text{ m}^2/\text{s})(28.1 \times 10^{-6} \text{ m}^2/\text{s})}$$

$= 15.83$

The average Nusselt number may be calculated using the relation

$$Nu_D = \frac{hD}{k} = CRa_D^n$$

where the quantities C and n depend on the Rayleigh number. For the range $10^{-2} < Ra_D < 10^2$, C = 1.02 and n = 0.148. Thus, the average Nusselt number is

$$Nu_D = (1.02)(15.83)^{0.148}$$

$$= 1.535$$

so the heat transfer coefficient is

$$h = \frac{Nu_D k}{D}$$

$$h = \frac{(1.535)(0.0291 \text{ W/m·K})}{1.628 \times 10^{-3} \text{ m}}$$

$$= 27.4 \text{ W/m}^2\text{·K}$$

The electrical power dissipated by the wire is equivalent to the heat transfer from the wire. Hence,

$$q = I^2 R = h\pi DL(T_s - T_\infty)$$

For a unit length of wire,

$$q = (34 \text{ A})^2(8.285 \times 10^{-3}\ \Omega)$$

$$= 9.578 \text{ W}$$

Solving for wire temperature,

$$T_s = \frac{q}{h\pi DL} + T_\infty$$

$$T_s = \frac{(7.457 \text{ W})}{(27.4 \text{ W/m}^2\text{·K})\pi(1.628 \times 10^{-3} \text{ m})(1 \text{ m})} + 30°\text{C}$$

$$= \underline{\underline{98.3°\text{C}}}$$

DISCUSSION

The assumption that the wire is isothermal can be checked by calculating the Biot number. Assuming that the wire is made of pure copper (k = 401 W/m·K),

$$Bi = \frac{hR}{k}$$

$$Bi = \frac{(27.4\ \text{W/m}^2\cdot\text{K})(8.140 \times 10^{-4}\ \text{m})}{401\ \text{W/m}\cdot\text{K}}$$

$$= 5.56 \times 10^{-5} < 0.1$$

Thus, the wire is lumped, i.e., it is isothermal. Also, the assumed wire temperature is sufficiently close to the calculated value that a correction of air properties is not justified.

Clearly, wire temperature increases with electrical current. The graph below shows this relationship.

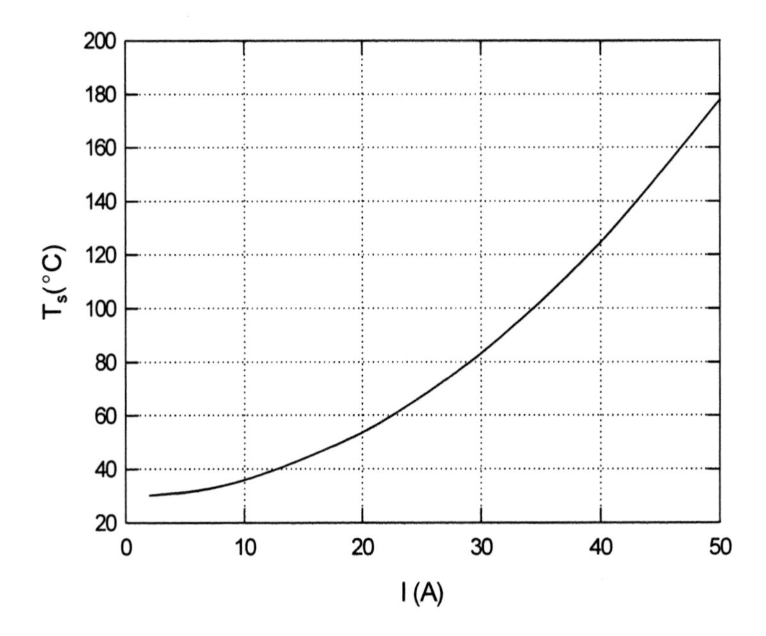

The melting point of pure copper is 1085 °C. Substituting this value into the relation for heat transfer and solving for electrical current, we obtain $I = 134$ A.

❒ ❒ ❒

PROBLEM 9.10 Heat Loss from an Uninsulated Steam Pipe

A horizontal uninsulated steam pipe passes through a large room whose walls and air are maintained at 25 °C. The pipe has an outside diameter of 11.4 cm and an emissivity of 0.8. Condensing steam in the pipe results in a surface temperature of 150 °C. Find the heat loss per unit length of pipe.

DIAGRAM

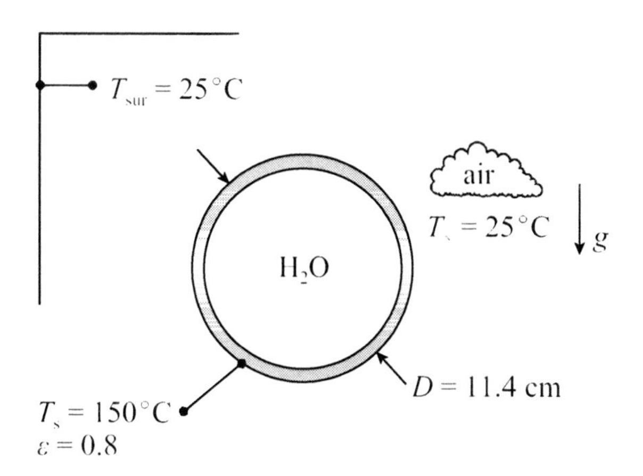

ASSUMPTIONS

1. Steady conditions.
2. Infinite expanse of stagnant atmospheric air.
3. Large room (enclosure).
4. Infinitely long pipe.
5. Constant properties.

PROPERTIES

$T_f = (T_s + T_\infty)/2 = (150°C + 25°C)/2 = 87.5°C = 361$ K

atmospheric air: $\nu = 22.13 \times 10^{-6}$ m²/s, $k = 0.0308$ W/m·K, $\alpha = 31.7 \times 10^{-6}$ m²/s, $Pr = 0.698$, $\beta = 1/T_f = 2.770 \times 10^{-3}$ K^{-1}

pipe: $\varepsilon = 0.8$

ANALYSIS

The Rayleigh number for a horizontal cylinder is

$$Ra_D = \frac{g\beta(T_s - T_\infty)D^3}{\nu\alpha}$$

$$Ra_D = \frac{(9.81 \text{ m/s}^2)(2.770 \times 10^{-3} \text{ K}^{-1})(150 - 25)°\text{C}(0.114 \text{ m})^3}{(22.13 \times 10^{-6} \text{ m}^2/\text{s})(31.7 \times 10^{-6} \text{ m}^2/\text{s})}$$

$$= 7.174 \times 10^6$$

The average Nusselt number may be calculated using the relation

$$Nu_D = \frac{hD}{k} = CRa_D^n$$

where the quantities C and n are functions of Rayleigh number. For the range $10^4 < Ra_D < 10^7$, $C = 0.480$ and $n = 0.250$. Hence, the average Nusselt number is

$$Nu_D = (0.480)(7.174 \times 10^6)^{0.250}$$

$$= 24.84$$

and the heat transfer coefficient is

$$h = \frac{Nu_D k}{D}$$

$$h = \frac{(24.84)(0.0308 \text{ W/m·K})}{0.114 \text{ m}}$$

$$= 6.71 \text{ W/m}^2\text{·K}$$

The total heat loss per unit length of pipe is

$$q' = q/L = h\pi D(T_s - T_\infty) + \varepsilon\sigma\pi D(T_s^4 - T_{sur}^4)$$

$$q' = (6.71 \text{ W/m}^2\text{·K})\pi(0.114 \text{ m})(150 - 25)°\text{C}$$

$$+ (0.8)(5.669 \times 10^{-8} \text{ W/m}^2\text{·K}^4)\pi(0.114 \text{ m})(423^4 - 298^4)\text{K}^4$$

$$= 300.4 \text{ W/m} + 391.9 \text{ W/m}$$

$$= \underline{\underline{692.3 \text{ W/m}}}$$

DISCUSSION

Note that the radiative heat loss, 391.9 W/m, is slightly higher than the convective heat loss. This is not unusual in natural convection systems in which heat transfer coefficients are low.

❐ ❐ ❐

PROBLEM 9.11 Design of an Electric Immersion Heater

An electric immersion heater, rated at 7250 W, is to heat water in a large tank. The heater consists of a an array of horizontally oriented tubes with a diameter of 6.5 mm. If the surface temperature of the tubes is 120°C, what is the required total length of the tubes if the water is to be heated from 15°C?

DIAGRAM

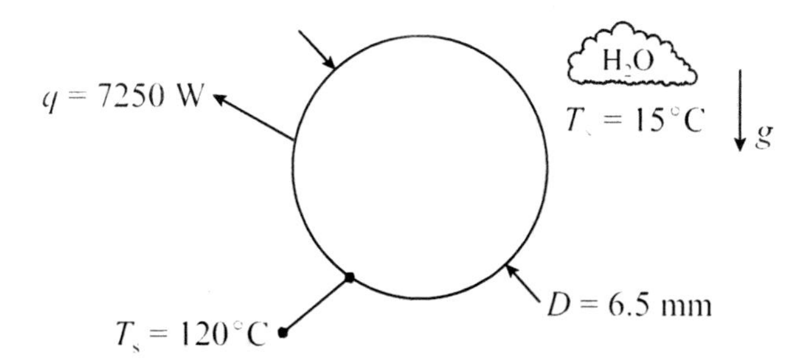

ASSUMPTIONS

1. Steady conditions.
2. Infinite expanse of stagnant water.
3. Each tube in array behaves as an isolated tube.
4. Radiation is neglected.
5. Constant properties.

PROPERTIES

$T_f = (T_s + T_\infty)/2 = (120°C + 15°C)/2 = 67.5°C = 341$ K

liquid water: $\nu = 4.225 \times 10^{-7}$ m²/s, $k = 0.662$ W/m·K, $\alpha = 2.89 \times 10^{-7}$ m²/s, $Pr = 2.62$, $\beta = 571.9 \times 10^{-6}$ K^{-1}

ANALYSIS

The Rayleigh number for a horizontal cylinder is

$$Ra_D = \frac{g\beta(T_s - T_\infty)D^3}{\nu\alpha}$$

$$Ra_D = \frac{(9.81\ \text{m/s}^2)(571.9 \times 10^{-6}\ \text{K}^{-1})(120 - 15)°\text{C}(6.5 \times 10^{-3}\ \text{m})^3}{(4.225 \times 10^{-7}\ \text{m}^2/\text{s})(2.89 \times 10^{-7}\ \text{m}^2/\text{s})}$$

$$= 1.325 \times 10^6$$

The average Nusselt number may be calculated using the relation

$$Nu_D = \frac{hD}{k} = CRa_D^n$$

where the quantities C and n are functions of Rayleigh number. For the range $10^4 < Ra_D < 10^7$,

$C = 0.480$ and $n = 0.250$. Hence, the average Nusselt number is

$$Nu_D = (0.480)(1.325 \times 10^6)^{0.250}$$

$$= 16.29$$

so the heat transfer coefficient is

$$h = \frac{Nu_D k}{D}$$

$$h = \frac{(16.29)(0.662 \text{ W/m·K})}{6.5 \times 10^{-3} \text{ m}}$$

$$= 1659 \text{ W/m}^2\text{·K}$$

The heat transfer from the heater is given by

$$q = h\pi DL(T_s - T_\infty)$$

Hence, the total length of tube required is

$$L = \frac{q}{h\pi D(T_s - T_\infty)}$$

$$L = \frac{(7250 \text{ W})}{(1659 \text{ W/m}^2\text{·K})\pi(6.5 \times 10^{-3} \text{ m})(120 - 15)°\text{C}}$$

$$= \underline{\underline{2.04 \text{ m}}}$$

DISCUSSION

A surface temperature of 120°C would cause the water to boil if the pressure was lower than about 200 kPa, the corresponding saturation pressure. Boiling is a phase change that results in heat transfer coefficients of roughly 2500 to 100,000 W/m²·K, which are much higher than those in natural convection systems. Electric immersion heaters typically consist of a bank of hairpin tubular elements brazed or welded into a screw plug and provided with wiring boxes where electrical connections are made.

❒ ❒ ❒

PROBLEM 9.12 Cooling Time for a Heat Treated Steel Rod

In a heat treating operation, a long rod of low carbon steel ($\rho = 7800$ kg/m³, $c_p = 434$ J/kg·K, $k =$ 64 W/m·K) emerges from an oven at a uniform temperature of 325°C into a room where the air temperature is 25°C. If the diameter of the rod is 1.75 cm, find the time required for the rod to achieve a temperature of 60°C.

DIAGRAM

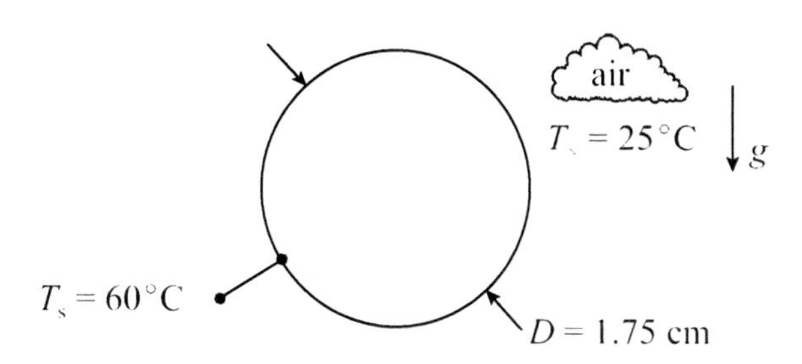

ASSUMPTIONS

1. Rod is surrounded by infinite expanse of stagnant atmospheric air.
2. Rod is isothermal at any given time.
3. Radiation is neglected.
4. Constant properties.

PROPERTIES

carbon steel: $\rho = 7800$ kg/m^3, $c_p = 434$ J/kg·K, $k = 64$ W/m·K

Let $T_s = (325 + 60)/2 = 192.5°C$ (average temperature during cooling time)

$T_f = (T_s + T_\infty)/2 = (192.5°C + 25°C)/2 = 108.8°C = 382$ K

atmospheric air: $\nu = 24.43 \times 10^{-6}$ m^2/s, $k = 0.0324$ W/m·K, $\alpha = 35.3 \times 10^{-6}$ m^2/s, $Pr = 0.694$, $\beta = 1/T_f = 2.618 \times 10^{-3}$ K^{-1}

ANALYSIS

The Rayleigh number for a horizontal cylinder is

$$Ra_D = \frac{g\beta(T_s - T_\infty)D^3}{\nu\alpha}$$

$$Ra_D = \frac{(9.81 \text{ m/s}^2)(2.618 \times 10^{-3} \text{ K}^{-1})(192.5 - 25)°\text{C}(0.0175 \text{ m})^3}{(24.43 \times 10^{-6} \text{ m}^2/\text{s})(35.3 \times 10^{-6} \text{ m}^2/\text{s})}$$

$$= 2.673 \times 10^4$$

The average Nusselt number may be calculated using the relation

$$Nu_D = \frac{hD}{k} = CRa_D^n$$

where the quantities C and n depend on the Rayleigh number. For the range $10^4 < Ra_D < 10^7$, $C = 0.480$ and $n = 0.250$. Thus, the average Nusselt number is

$$Nu_D = (0.480)(2.673 \times 10^4)^{0.250}$$

$$= 6.138$$

so the heat transfer coefficient is

$$h = \frac{Nu_D k}{D}$$

$$h = \frac{(6.138)(0.0324 \text{ W/m·K})}{0.0175 \text{ m}}$$

$$= 11.4 \text{ W/m}^2\text{·K}$$

We now check the validity of assumption 2. For a long cylinder the Biot number is

$$Bi \equiv \frac{h(R/2)}{k}$$

$$Bi = \frac{(11.4 \text{ W/m}^2\text{·K})(0.0175 \text{ m/4})}{64 \text{ W/m·K}}$$

$$= 7.79 \times 10^{-4} < 0.1$$

Thus, assumption 2 is valid, i.e., the rod is lumped. The temperature history of the rod is given by the relation

$$\frac{\theta}{\theta_i} = \frac{T - T_\infty}{T_i - T_\infty} = \exp\left[-\left(\frac{hA}{\rho V c_p}\right)t\right]$$

where T is temperature at time t, A is surface area and V is volume. Noting that $A/V = 2/R$, and solving for t,

$$t = -\frac{\rho R c_p}{2h}\ln\left(\frac{T - T_\infty}{T_i - T_\infty}\right)$$

$$t = -\frac{(7800 \text{ kg/m}^3)(0.0175 \text{ m/2})(434 \text{ J/kg·K})}{2(11.4 \text{ W/m}^2\text{·K})}\ln\left(\frac{60 - 25}{325 - 25}\right)$$

$$= 5582 \text{ s} = \underline{\underline{1.55 \text{ h}}}$$

DISCUSSION

The calculated cooling time is an estimate because the air properties were based on an average rod temperature during the cooling time. Higher accuracy could be achieved if we "stepped" in time through the cooling period, using a heat transfer coefficient that reflects the instantaneous temperature of the rod. Furthermore, the inclusion of radiation in the analysis would yield a more realistic cooling time since this mode of heat transfer is significant in most natural convection applications.

PROBLEM 9.13 Convection from an Electrically Heated Sphere

A 10-cm diameter sphere contains an embedded electric heater. Find the power required to maintain the surface of the sphere at 450°C if the sphere is surrounded by quiescent atmospheric air at 20°C.

DIAGRAM

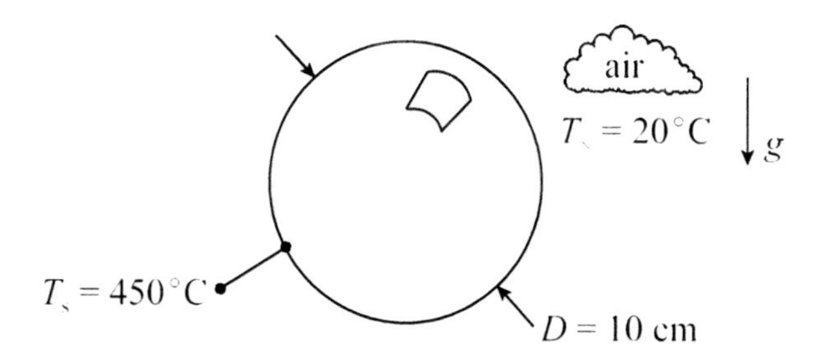

ASSUMPTIONS

1. Steady conditions.
2. Sphere is surrounded by infinite expanse of stagnant atmospheric air.
3. Radiation is neglected.
4. Constant properties.

PROPERTIES

$T_f = (T_s + T_\infty)/2 = (450°C + 20°C)/2 = 235°C = 508$ K

atmospheric air: $\nu = 39.87 \times 10^{-6}$ m^2/s, $k = 0.0412$ W/m·K, $\alpha = 58.3 \times 10^{-6}$ m^2/s, $Pr = 0.684$, $\beta = 1/T_f = 1.969 \times 10^{-3}$ K^{-1}

ANALYSIS

The Rayleigh number for a sphere is

$$Ra_D = \frac{g\beta(T_s - T_\infty)D^3}{\nu\alpha}$$

$$Ra_D = \frac{(9.81\ m/s^2)(1.969 \times 10^{-3}\ K^{-1})(450 - 20)°C(0.10\ m)^3}{(39.87 \times 10^{-6}\ m^2/s)(58.3 \times 10^{-6}\ m^2/s)}$$

$$= 3.573 \times 10^6$$

The average Nusselt number may be calculated using the relation

$$Nu_D = \frac{hD}{k} = 2 + \frac{0.589 Ra_D^{1/4}}{\left[1 + (0.469/Pr)^{9/16}\right]^{4/9}}$$

$$Nu_D = 2 + \frac{0.589(3.573 \times 10^6)^{1/4}}{[1 + (0.469/0.684)^{9/16}]^{4/9}}$$

$$= 21.68$$

so the heat transfer coefficient is

$$h = \frac{Nu_D k}{D}$$

$$h = \frac{(21.68)(0.0412\ W/m \cdot K)}{0.10\ m}$$

$$= 8.93\ W/m^2 \cdot K$$

The heat transfer from the sphere, and thus the power of the electric heater, is

$$q = h\pi D^2 (T_s - T_\infty)$$

$$q = (8.93\ W/m^2 \cdot K)\pi(0.10\ m)^2(450 - 20)°C$$

$$= \underline{\underline{121\ W}}$$

DISCUSSION

The result is independent of the sphere's thermal properties. All the power dissipated by the electric heater is convected from the surface of the sphere, regardless of the material from which the sphere is made. The heat flux at the surface of the sphere is

$$q'' = \frac{q}{A} = \frac{q}{\pi D^2}$$

$$q'' = \frac{120.6 \text{ W}}{\pi(0.10 \text{ m})^2}$$

$$= 3839 \text{ W/m}^2$$

❐ ❐ ❐

PROBLEM 9.14 Temperature of a 40-Watt Incandescent Light Bulb

A 40-watt incandescent light bulb illuminates a room where the temperature of the air, ceiling, floor and walls is 20°C. Approximating the light bulb as a 7.0-cm diameter sphere, find the surface temperature of the light bulb if the glass has an emissivity of 0.85.

DIAGRAM

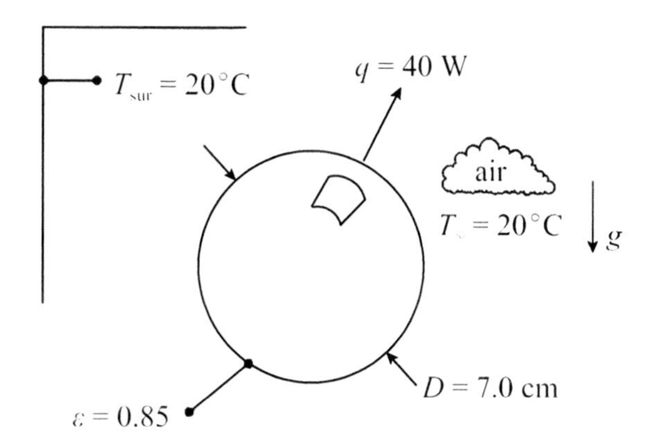

ASSUMPTIONS

1. Steady conditions.
2. Light bulb is a sphere.
3. Sphere is surrounded by infinite expanse of stagnant atmospheric air.
4. Conduction into light bulb base is neglected.
5. Constant properties.

PROPERTIES

Assume $T_s = 150°C$. $T_f = (T_s + T_\infty)/2 = (150°C + 20°C)/2 = 85°C = 358$ K

atmospheric air: $\nu = 21.80 \times 10^{-6}$ m²/s, $k = 0.0306$ W/m·K, $\alpha = 31.2 \times 10^{-6}$ m²/s, $Pr = 0.698$, $\beta = 1/T_f = 2.793 \times 10^{-3}$ K^{-1}

glass: $\varepsilon = 0.85$

ANALYSIS

The Rayleigh number for a sphere is

$$Ra_D = \frac{g\beta(T_s - T_\infty)D^3}{\nu\alpha}$$

$$Ra_D = \frac{(9.81 \text{ m/s}^2)(2.793 \times 10^{-3} \text{ K}^{-1})(150 - 20)°\text{C}(0.070 \text{ m})^3}{(21.80 \times 10^{-6} \text{ m}^2/\text{s})(31.2 \times 10^{-6} \text{ m}^2/\text{s})}$$

$$= 1.796 \times 10^6$$

The average Nusselt number may be calculated using the relation

$$Nu_D = \frac{hD}{k} = 2 + \frac{0.589 Ra_D^{1/4}}{\left[1 + (0.469/Pr)^{9/16}\right]^{4/9}}$$

$$Nu_D = 2 + \frac{0.589(1.796 \times 10^6)^{1/4}}{[1 + (0.469/0.698)^{9/16}]^{4/9}}$$

$$= 18.61$$

so the heat transfer coefficient is

$$h = \frac{Nu_D k}{D}$$

$$h = \frac{(18.61)(0.0306 \text{ W/m·K})}{0.070 \text{ m}}$$

$$= 8.14 \text{ W/m}^2\text{·K}$$

The heat transfer from the sphere is the sum of the convective and radiative components,

$$q = h\pi D^2(T_s - T_{sur}) + \varepsilon\sigma\pi D^2(T_s^4 - T_{sur}^4)$$

Setting $q = 40$ W and numerically solving for T_s, the surface temperature of the glass,

$$T_s = 438 \text{ K} = \underline{\underline{165°\text{C}}}$$

DISCUSSION

The convective and radiative heat transfer components are comparable.

$$q_{\text{conv}} = 18.2 \text{ W} \quad , \quad q_{\text{rad}} = 21.8 \text{ W}$$

Because the calculated surface temperature is only 15°C higher than the estimated value, a correction of thermal properties would yield only a modest improvement in accuracy.

PROBLEM 9.15 Design of a Burner for a Hot-Air Balloon

A hot-air balloon measuring 14 m in diameter is neutrally buoyant in 18°C air at an altitude of 1500 m above sea level. The temperature of the fabric, which reflects the average temperature of the air inside the balloon, must be 50°C for the balloon to maintain its vertical position. Find the rate at which heat must be supplied to the air in the balloon by its propane burner system.

DIAGRAM

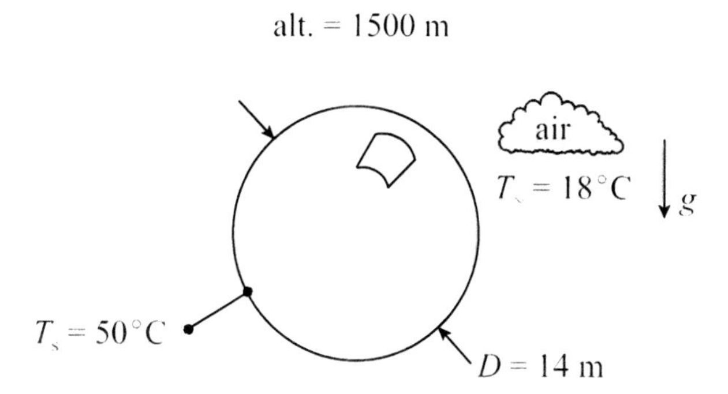

ASSUMPTIONS

1. Steady conditions.
2. Balloon is a sphere.
3. Sphere is surrounded by infinite expanse of stagnant atmospheric air at 1500 m altitude.
4. Radiation is neglected.
5. Constant properties.

PROPERTIES

$T_f = (T_s + T_\infty)/2 = (50°C + 18°C)/2 = 34°C = 307$ K

atmospheric air: $\mu = 187.9 \times 10^{-7}$ m²/s, $k = 0.0268$ W/m·K, $\beta = 1/T_f = 3.257 \times 10^{-3}$ K^{-1}

air at altitude: 1500 m altitude: $\rho = 1.059$ kg/m³
$\nu = \mu/\rho = 1.774 \times 10^{-5}$ m²/s
$\alpha = k/\rho c_p = (0.0268 \text{ W/m·K})/[(1.059 \text{ kg/m}^3)(1007 \text{ J/kg·K})]$
$= 2.513 \times 10^{-5}$ m²/s
$Pr = \nu/\alpha = (1.774 \times 10^{-5} \text{ m}^2/\text{s})/(2.513 \times 10^{-5} \text{ m}^2/\text{s}) = 0.706$

ANALYSIS

The Rayleigh number for a sphere is

$$Ra_D = \frac{g\beta(T_s - T_\infty)D^3}{\nu\alpha}$$

$$Ra_D = \frac{(9.81 \text{ m/s}^2)(3.257 \times 10^{-3} \text{ K}^{-1})(50 - 18)°\text{C}(14 \text{ m})^3}{(1.774 \times 10^{-5} \text{ m}^2/\text{s})(2.513 \times 10^{-5} \text{ m}^2/\text{s})}$$

$$= 6.293 \times 10^{12}$$

The average Nusselt number may be calculated using the relation

$$Nu_D = \frac{hD}{k} = 2 + \frac{0.589 Ra_D^{1/4}}{\left[1 + \left(0.469/Pr\right)^{9/16}\right]^{4/9}}$$

$$Nu_D = 2 + \frac{0.589(6.293 \times 10^{12})^{1/4}}{[1 + (0.469/0.706)^{9/16}]^{4/9}}$$

$$= 721$$

so the heat transfer coefficient is

$$h = \frac{Nu_D k}{D}$$

$$h = \frac{(721)(0.0268 \text{ W/m·K})}{14 \text{ m}}$$

$$= 1.38 \text{ W/m}^2\text{·K}$$

The heat transfer from the balloon's surface, and hence the rate at which the propane burner must supply heat to the air in the balloon, is

$$q = h\pi D^2 (T_s - T_\infty)$$

$q = (1.38\ \text{W/m}^2\cdot\text{K})\pi(14\ \text{m})^2\ (50 - 18)°\text{C}$

$= 2.719 \times 10^4\ \text{W} = \underline{\underline{27.2\ \text{kW}}}$

DISCUSSION

The correlation used to find the average Nusselt number is strictly valid for $Ra_D < 10^{11}$, so we have used it outside of its range of applicability. Also, radiation would be a significant fraction of the total heat transfer from the balloon because the heat transfer coefficient is very low. Assuming that $\varepsilon = 1$ and $T_{sur} = T_\infty$, the radiative component is

$$q_{rad} = \varepsilon\sigma\pi D^2 (T_s^4 - T_{sur}^4)$$

$q_{rad} = (1.0)(5.669 \times 10^{-8}\ \text{W/m}^2\cdot\text{K}^4)\pi(14\ \text{m})^2(323^4 - 291^4)\text{K}^4$

$= 1.296 \times 10^5\ \text{W} = 130\ \text{kW}$

which means that a more realistic heating rate for the burner is

$q = (27.2 + 130)\ \text{kW} = 157.2\ \text{kW}$

Neglecting radiation in this analysis results in a gross under design of the burner system. Furthermore, if the balloon's motion was included in the analysis, a heat transfer coefficient that takes both natural and forced convection into account would have to be calculated, in which case the convective heat transfer would be somewhat higher.

❐ ❐ ❐

PROBLEM 9.16 Natural Convection in an Isothermal Vertical Channel

A vertical channel consists of two isothermal, parallel plates spaced 10 cm apart. The channel is 50 cm high, and each plate is maintained at 200°C. Find the heat flux if the channel draws in atmospheric nitrogen at 30°C.

DIAGRAM

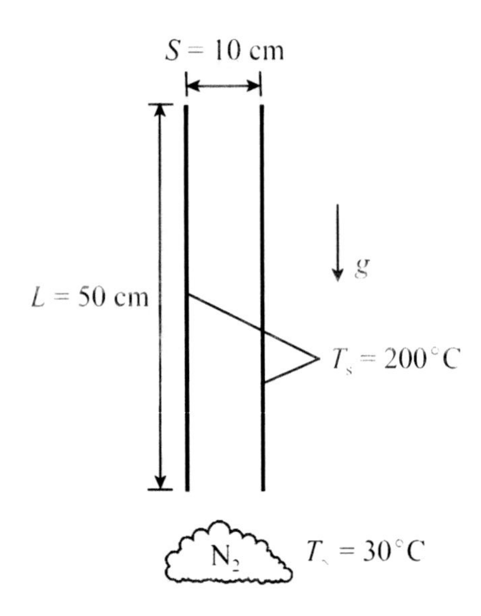

ASSUMPTIONS

1. Steady conditions.
2. Back sides of plates are insulated.
3. Plates are identical.
4. Constant properties.

PROPERTIES

$\overline{T} = (T_s + T_\infty)/2 = (200°\text{C} + 30°\text{C})/2 = 115°\text{C} = 388\ \text{K}$

atmospheric nitrogen: $\nu = 24.89 \times 10^{-6}\ \text{m}^2/\text{s}$, $k = 0.0319\ \text{W/m·K}$, $\alpha = 35.2 \times 10^{-6}\ \text{m}^2/\text{s}$, $Pr = 0.706$, $\beta = 1/T_f = 2.577 \times 10^{-3}\ \text{K}^{-1}$

ANALYSIS

The Rayleigh number for a vertical channel is

$$Ra_S = \frac{g\beta(T_s - T_\infty)S^3}{\nu\alpha}$$

$$Ra_S = \frac{(9.81\ \text{m/s}^2)(2.577 \times 10^{-3}\ \text{K}^{-1})(200 - 30)°\text{C}(0.50\ \text{m})^3}{(24.89 \times 10^{-6}\ \text{m}^2/\text{s})(35.2 \times 10^{-6}\ \text{m}^2/\text{s})}$$

$= 6.132 \times 10^8$

and the average Nusselt number may be found using the relation

$$Nu_S = \left[\frac{C_1}{(Ra_S\, S/L)^2} + \frac{C_2}{(Ra_S\, S/L)^{1/2}} \right]^{-1/2}$$

where, for isothermal plates, $C_1 = 576$ and $C_2 = 2.87$. Thus, the average Nusselt number is

$$Nu_S = \left[\frac{576}{[(6.132 \times 10^8)(0.10\text{ m})/(0.50\text{ m})]^2} + \frac{2.87}{[(6.132 \times 10^8)(0.10\text{ m})/(0.50\text{ m})]^{1/2}} \right]^{-1/2}$$

$= 62.1$

The heat flux is

$$q'' = Nu_S \frac{k}{S}(T_s - T_\infty)$$

$$q'' = (62.1)\frac{(0.0319\text{ W/m}\cdot\text{K})}{0.10\text{ m}}(200 - 30)°\text{C}$$

$= \underline{\underline{3368\text{ W/m}^2}}$

DISCUSSION

The equation for average Nusselt number is a semi-empirical relation applicable to the complete range of *S/L*. For small values of *S/L*, the boundary layers developing on opposing surfaces eventually merge to form a fully developed condition. For large values of *S/L*, the boundary layers on each surface develop independently, yielding a condition that corresponds to an isolated plate in an infinite quiescent fluid.

To find the heat transfer, *q*, the width of the channel would have to be known.

❒ ❒ ❒

PROBLEM 9.17　　Cooling an Array of Vertical Circuit Boards

An array of vertical circuit boards is cooled by natural convection in atmospheric air at 35°C. Both sides of each circuit board are densely populated with identical electrical devices such that each board dissipates the same amount of power uniformly over the surface. The boards measure 16 cm × 16 cm with an average spacing between the tops of the devices of 2.5 cm. If the maximum allowable surface temperature is 85°C, what is the maximum allowable power dissipation for each circuit board?

DIAGRAM

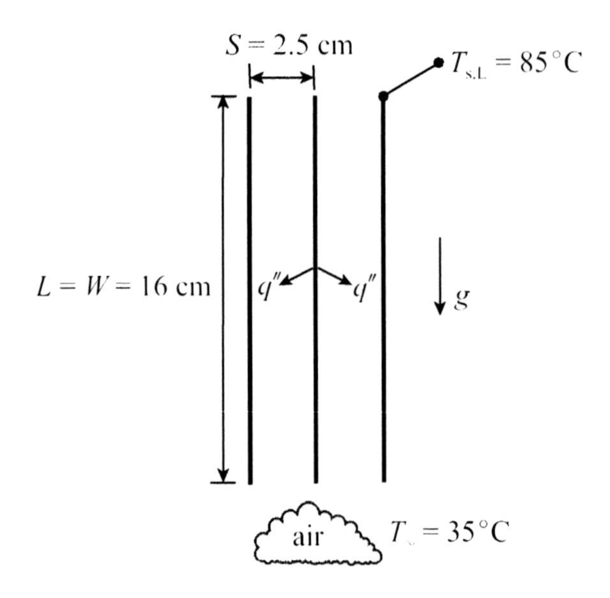

ASSUMPTIONS

1. Steady conditions.
2. Array is surrounded by quiescent atmospheric air.
3. Circuit boards are identical and modeled as flat plates.
4. Constant properties.

PROPERTIES

$\overline{T} = (T_{s,L} + T_\infty)/2 = (85°C + 35°C)/2 = 60°C = 333$ K

atmospheric air: $\nu = 19.21 \times 10^{-6}$ m²/s, $k = 0.0287$ W/m·K, $\alpha = 27.4 \times 10^{-6}$ m²/s, $Pr = 0.702$, $\beta = 1/T_f = 3.003 \times 10^{-3}$ K^{-1}

ANALYSIS

For isoflux surfaces, a modified Rayleigh number for a vertical channel is

$$Ra_S^* = \frac{g\beta q'' S^4}{k\nu\alpha}$$

$$Ra_S^* = \frac{(9.81 \text{ m/s}^2)(3.003 \times 10^{-3} \text{ K}^{-1})(0.025 \text{ m})^4}{(0.0287 \text{ W/m·K})(19.21 \times 10^{-6} \text{ m}^2\text{/s})(27.4 \times 10^{-6} \text{ m}^2\text{/s})} q''$$

$= 761.8\, q''$

where q'' is heat flux. The local Nusselt number may be found using the semi-empirical relation

$$Nu_{S,L} = \left[\frac{C_1}{Ra_S^* S/L} + \frac{C_2}{\left(Ra_S^* S/L\right)^{2/5}} \right]^{-1/2}$$

where, for symmetric isoflux surfaces, $C_1 = 48$ and $C_2 = 2.51$. The local Nusselt number is defined in terms of the surface temperature at the top of the channel,

$$Nu_{S,L} = \left(\frac{q''}{T_{S,L} - T_\infty} \right) \frac{S}{k}$$

Upon substituting $Ra_S^* = 761.8\, q''$ into the semi-empirical relation above and equating the two expressions for $Nu_{S,L}$, the heat flux q'' may be found numerically,

$q'' = 251 \text{ W/m}^2$

Hence, the maximum allowable power dissipation for one circuit board is

$$q = 2LWq''$$

$q = 2(0.16 \text{ m})(0.16 \text{ m})(251 \text{ W/m}^2)$

$= \underline{\underline{12.9 \text{ W}}}$

DISCUSSION

The maximum surface temperature for symmetric isoflux surfaces occurs at the top of the channel. Thus, the devices at the top of the channel are the most critical from a reliability standpoint.

❐ ❐ ❐

PROBLEM 9.18 Convection in a Horizontal Rectangular Cavity

A horizontal rectangular cavity containing atmospheric air has a height and length of 3.0 cm and 15 cm, respectively. The bottom and top surfaces of the cavity are maintained at temperatures of 180°C and 10°C, respectively. Find the heat transfer.

DIAGRAM

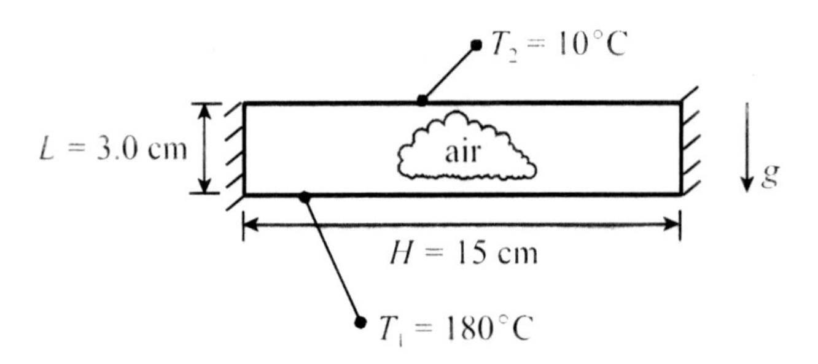

ASSUMPTIONS

1. Steady conditions.
2. Bottom surface is isothermal, and top surface is isothermal.
3. Radiation is neglected.
4. Constant properties.

PROPERTIES

$\bar{T} = (T_1 + T_2)/2 = (180°C + 10°C)/2 = 95°C = 368$ K

atmospheric air: $\nu = 22.90 \times 10^{-6}$ m²/s, $k = 0.0314$ W/m·K, $\alpha = 32.9 \times 10^{-6}$ m²/s, $Pr = 0.696$, $\beta = 1/T_f = 2.717 \times 10^{-3}$ K^{-1}

The Rayleigh number for a horizontal rectangular cavity is

$$Ra_L = \frac{g\beta(T_1 - T_2)L^3}{\nu\alpha}$$

$$Ra_L = \frac{(9.81 \text{ m/s}^2)(2.717 \times 10^{-3} \text{ K}^{-1})(180 - 10)°\text{C}(0.030 \text{ m})^3}{(22.90 \times 10^{-6} \text{ m}^2/\text{s})(32.9 \times 10^{-6} \text{ m}^2/\text{s})}$$

$$= 1.624 \times 10^5$$

and the average Nusselt number may be obtained from the relation

$$Nu_L = 0.069 Ra_L^{1/3} Pr^{0.074}$$

$$Nu_L = (0.069)(1.624 \times 10^5)^{1/3} (0.696)^{0.074}$$

$$= 3.665$$

so the heat transfer coefficient is

$$h = \frac{Nu_L k}{L}$$

$$h = \frac{(3.665)(0.0314\ \text{W/m·K})}{0.030\ \text{m}}$$

$$= 3.836\ \text{W/m}^2\text{·K}$$

The heat transfer per unit depth of cavity is

$$q' = hH(T_1 - T_2)$$

$$q' = (3.836\ \text{W/m}^2\text{·K})(0.15\ \text{m})(180 - 10)°\text{C}$$

$$= \underline{\underline{97.8\ \text{W/m}}}$$

DISCUSSION

The relation used for the average Nusselt number is strictly valid for $3 \times 10 < Ra_L < 7 \times 10^9$, so our result may be slightly inaccurate.

For Rayleigh numbers less than the critical value of 1708, buoyancy forces cannot overcome viscous forces, so there are no circulation currents in the cavity. Under such conditions, heat transfer across the cavity occurs by pure conduction through a stagnant fluid layer. Setting the Rayleigh number to its critical value, $Ra_L = 1708$, and solving for L, the height of the cavity, keeping all other quantities the same, we obtain $L = 6.58 \times 10^{-3}$ m = 6.58 mm. The heat transfer for this case may be calculated using Fourier's law

$$q' = k\frac{H}{L}(T_1 - T_2)$$

$$q' = (0.0314\ \text{W/m·K})(0.15\ \text{m})/(6.58 \times 10^{-3}\ \text{m})(180 - 10)°\text{C}$$

$$= 122\ \text{W/m}$$

which is 24.7 percent higher than our previous result.

❒ ❒ ❒

PROBLEM 9.19 Convection in a Vertical Rectangular Cavity

A vertical rectangular cavity containing atmospheric carbon dioxide has a height and length of 60 cm and 12 cm, respectively. The bottom and top surfaces of the cavity are insulated. If the vertical surfaces are maintained at temperatures of 275°C and 40°C, respectively, find the heat transfer.

DIAGRAM

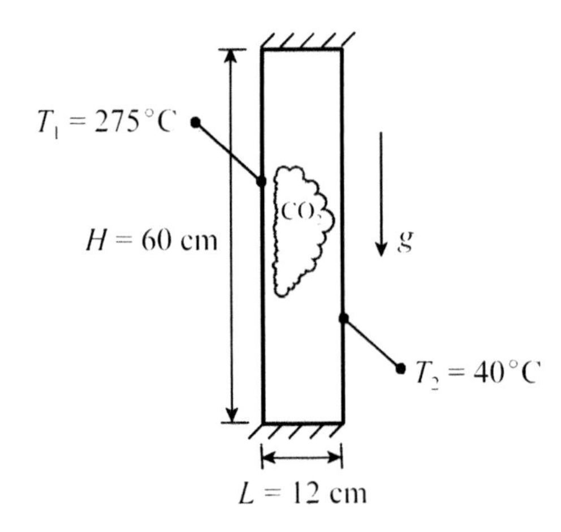

ASSUMPTIONS

1. Steady conditions.
2. Left surface is isothermal, and right surface is isothermal.
3. Radiation is neglected.
4. Constant properties.

PROPERTIES

$\overline{T} = (T_1 + T_2)/2 = (275°\text{C} + 40°\text{C})/2 = 158°\text{C} = 431\text{ K}$

atmospheric carbon dioxide: $\nu = 16.47 \times 10^{-6}\text{ m}^2/\text{s}$, $k = 0.0268\text{ W/m·K}$, $\alpha = 22.6 \times 10^{-6}\text{ m}^2/\text{s}$, $Pr = 0.731$, $\beta = 1/T_f = 2.320 \times 10^{-3}\text{ K}^{-1}$

The Rayleigh number for a vertical rectangular cavity is

$$Ra_L = \frac{g\beta(T_1 - T_2)L^3}{\nu\alpha}$$

$$Ra_L = \frac{(9.81\text{ m/s}^2)(2.320 \times 10^{-3}\text{ K}^{-1})(275 - 40)°\text{C}(0.12\text{ m})^3}{(16.47 \times 10^{-6}\text{ m}^2/\text{s})(22.6 \times 10^{-6}\text{ m}^2/\text{s})}$$

$$= 2.483 \times 10^7$$

The aspect ratio for the cavity is $H/L = (0.60\text{ m})/(0.12\text{ m}) = 5$. For vertical rectangular cavities with aspect ratios in the range, $2 < H/L < 10$, the average Nusselt number may be obtained from the relation

$$Nu_L = \frac{hL}{k} = 0.22\left(\frac{Pr}{0.2 + Pr} Ra_L\right)^{0.28} \left(\frac{H}{L}\right)^{-1/4}$$

$$Nu_L = 0.22 \left(\frac{0.731}{0.2 + 0.731} (2.483 \times 10^7)\right)^{0.28} [(0.60 \text{ m})/(0.12 \text{ m})]^{-1/4}$$

$$= 73.53$$

so the heat transfer coefficient is

$$h = \frac{Nu_L k}{L}$$

$$h = \frac{(73.53)(0.0218 \text{ W/m·K})}{0.12 \text{ m}}$$

$$= 13.4 \text{ W/m}^2\text{·K}$$

The heat transfer per unit depth of cavity is

$$q' = hH(T_1 - T_2)$$

$$q' = (13.4 \text{ W/m}^2\text{·K})(0.60 \text{ m})(275 - 40)°\text{C}$$

$$= \underline{\underline{1889 \text{ W/m}}}$$

DISCUSSION

Note that the relation for Nusselt number is not a function of the cavity depth W. For large values of the aspect ratio W/L, the Nusselt number is a weak function of W and is therefore neglected in the analysis.

❒ ❒ ❒

PROBLEM 9.20 Optimum Air Space for a Double-Pane Window

Derive an expression for the average Nusselt number that minimizes the convective heat transfer across the air space of a double-pane window. Verify that the Nusselt number obtained yields the optimum condition by graphing the average heat transfer coefficient as a function of air space width for a window height of 1.20 m. Let the inside surface temperatures of the warm and cool glass panes be 10°C and −6°C, respectively.

DIAGRAM

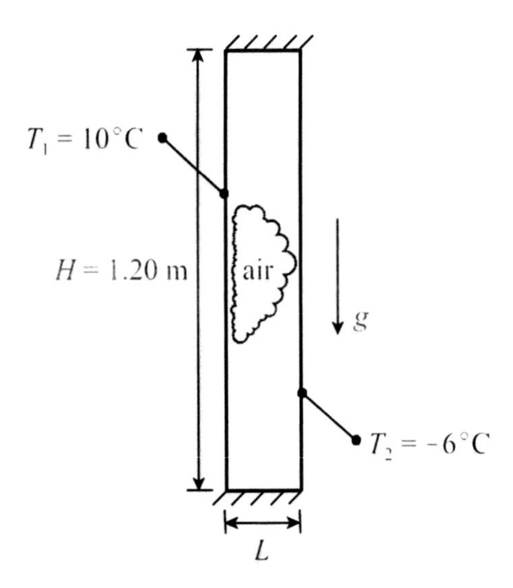

ASSUMPTIONS

1. Steady conditions.
2. Left surface is isothermal, and right surface is isothermal.
3. Bottom and top surfaces of cavity are insulated.
4. Radiation is neglected.
5. Constant properties.

PROPERTIES

$\overline{T} = (T_1 + T_2)/2 = (10°C - 6°C)/2 = 2°C = 275$ K

atmospheric air: $\nu = 13.67 \times 10^{-6}$ m²/s, $k = 0.0243$ W/m·K, $\alpha = 19.2 \times 10^{-6}$ m²/s, $Pr = 0.714$, $\beta = 1/T_f = 3.636 \times 10^{-3}$ K^{-1}

For vertical rectangular cavities with high aspect ratios ($H/L > 40$) the average Nusselt number may be found using the relation

$$Nu_L = \frac{hL}{k} = 1 + \frac{Ra_L}{720}\left(\frac{L}{H}\right)$$

Solving for h,

$$h = \frac{k}{L}\left[1 + \frac{Ra_L}{720}\left(\frac{L}{H}\right)\right] = k\left(\frac{1}{L} + \frac{CL^3}{720H}\right)$$

where

$$C = \frac{Ra_L}{L^3}$$

The air space width, L, that yields the minimum heat transfer coefficient is obtained by differentiating the expression for h with respect to L and setting the result to zero.

$$\frac{dh}{dL} = k\left(-\frac{1}{L} + \frac{3CL^2}{720H}\right) = 0$$

Reintroducing the Rayleigh number, Ra_L, through the parameter C,

$$\frac{L}{H} = \frac{240}{Ra_L}$$

Substituting this relation into the relation for Nusselt number,

$$Nu_L = \frac{4}{3}$$

Hence, the average heat transfer coefficient that minimizes the heat transfer through a double-pane window is

$$h = \frac{4}{3}\frac{k}{L}$$

To check the validity of our result for a specific case, we consider a double-pane window of height $H = 1.20$ m whose inside and outside panes are 10°C and -6°C, respectively. The graph below shows the variation of Nu_L and h with air space width, L. Note that $Nu_L = 1$ for small air spaces, indicating heat transfer by pure conduction. The Nusselt number increases with L, but the heat transfer coefficient has a minimum value at about $L = 1.9$ cm. This minimum occurs at the point where $Nu_L = 4/3$, as predicted in the analysis. Recalling that the convective thermal resistance for a plane layer is

$$R_{th} = \frac{1}{hA}$$

it is clear that, for a given surface area A, a minimum value of h yields the maximum value of R_{th} and thus a minimum heat transfer.

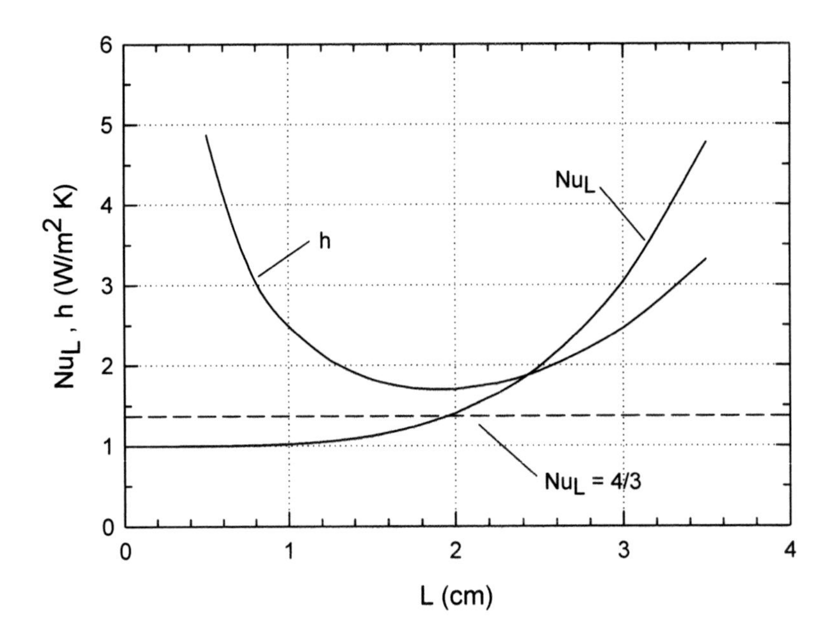

DISCUSSION

Some window manufacturers fill and seal the space between the glass panes with argon gas, which has a lower thermal conductivity than air. At the average temperature in this problem, argon has a thermal conductivity of 0.0165 W/m·K, which is 32 percent lower than the value for air. A lower spacing between glass panes can therefore be used without compromising the insulating ability of the window. Over a long period of time, however, the argon may leak out, being displaced by air.

Another method for increasing the insulating ability of windows is to coat the surface with an invisible reflective film. This type of glass is referred to as "low-e," which means low-emissivity. This type of glass reduces heat transfer by reflecting more solar radiation than standard glass.

❒ ❒ ❒

PROBLEM 9.21 Surface Temperature of a Wood-Burning Stove

A freestanding wood-burning stove with a width, depth and height of 70 cm, 82 cm and 60 cm, respectively, burns wood at a steady rate of 6 kg/h. Thirty percent of the stove's heat is lost to the outdoors in the combustion gases. If the heat of combustion of wood is 18 MJ/kg, estimate the surface temperature of the stove. The stove is located in a room where the temperature of the air and surroundings is 20°C, and the emissivity of the stove's surfaces is 0.95.

DIAGRAM

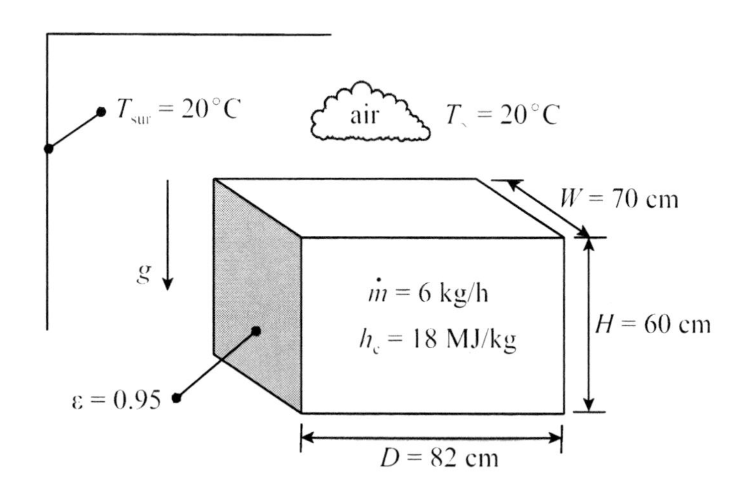

ASSUMPTIONS

1. Steady conditions.
2. Stove is surrounded by infinite expanse of stagnant atmospheric air.
3. Entire stove surface is isothermal.
4. Constant properties.

PROPERTIES

Assume $T_s = 275°C$. $T_f = (T_s + T_\infty)/2 = (275°C + 20°C)/2 = 148°C = 421$ K

atmospheric air: $\nu = 28.92 \times 10^{-6}$ m²/s, $k = 0.0353$ W/m·K, $\alpha = 42.0 \times 10^{-6}$ m²/s, $Pr = 0.688$, $\beta = 1/T_f = 2.375 \times 10^{-3}$ K^{-1}

stove surfaces: $\varepsilon = 0.95$

ANALYSIS

Natural convection and radiation occur at all six sides of the stove. As far as natural convection is concerned, three separate surfaces must be analyzed: vertical surface (the four sides), upper surface of a heated horizontal plate (the top), and lower surface of a heated horizontal plate (the bottom). The area of the vertical surface is

$$A_1 = 2(W + D)H$$

$A_1 = 2(0.70 \text{ m} + 0.82 \text{ m})(0.60 \text{ m})$

$= 1.824 \text{ m}^2$

The area of the top and bottom surfaces is

$$A_2 = WD$$

$A_2 = (0.70\text{ m})(0.82\text{ m})$

$= 0.574\text{ m}^2$

for which the characteristic length is

$$L = \frac{A_2}{P} = \frac{A_2}{2(D+W)}$$

$$L = \frac{0.574\text{ m}^2}{2(0.82\text{ m} + 0.70\text{ m})}$$

$= 0.1888\text{ m}$

Based on our assumed surface temperature, the Rayleigh number for the vertical surface is

$$Ra_H = \frac{g\beta(T_s - T_\infty)H^3}{\nu\alpha}$$

$$Ra_H = \frac{(9.81\text{ m/s}^2)(2.375 \times 10^{-3}\text{ K}^{-1})(275 - 20)^\circ\text{C}(0.60\text{ m})^3}{(28.92 \times 10^{-6}\text{ m}^2/\text{s})(42.0 \times 10^{-6}\text{ m}^2/\text{s})}$$

$= 1.778 \times 10^8$

which indicates the flow is laminar. The Rayleigh number for the top and bottom surfaces is

$$Ra_L = \frac{g\beta(T_s - T_\infty)L^3}{\nu\alpha}$$

$$Ra_L = \frac{(9.81\text{ m/s}^2)(2.375 \times 10^{-3}\text{ K}^{-1})(275 - 20)^\circ\text{C}(0.1888\text{ m})^3}{(28.92 \times 10^{-6}\text{ m}^2/\text{s})(42.0 \times 10^{-6}\text{ m}^2/\text{s})}$$

$= 3.292 \times 10^7$

The average Nusselt number for the vertical surface may be calculated using the relation

$$Nu_1 = \frac{h_1 H}{k} = 0.68 + \frac{0.670 Ra_H^{1/4}}{[1 + (0.492/Pr)^{9/16}]^{4/9}}$$

$$Nu_1 = 0.68 + \frac{0.670(1.778 \times 10^8)^{1/4}}{[1 + (0.492/0.688)^{9/16}]^{4/9}}$$

$= 59.9$

so the heat transfer coefficient is

$$h_1 = \frac{Nu_1 k}{H}$$

$$h_1 = \frac{(59.9)(0.0353 \text{ W/m·K})}{0.60 \text{ m}}$$

$= 3.52 \text{ W/m}^2\text{·K}$

The average Nusselt number for the top surface may be found using the relation

$$Nu_2 = \frac{h_2 L}{k} = 0.15 Ra_L^{1/3}$$

$Nu_2 = (0.15)(1.778 \times 10^8)^{1/3}$

$= 84.3$

so the heat transfer coefficient is

$$h_2 = \frac{Nu_2 k}{L}$$

$$h_2 = \frac{(84.3)(0.0353 \text{ W/m·K})}{0.1888 \text{ m}}$$

$= 15.8 \text{ W/m}^2\text{·K}$

The average Nusselt number for the bottom surface may be found using the relation

$$Nu_3 = \frac{h_3 L}{k} = 0.27 Ra_L^{1/4}$$

$Nu_3 = (0.27)(1.778 \times 10^8)^{1/4}$

$= 31.2$

so the heat transfer coefficient is

$$h_3 = \frac{Nu_3 k}{L}$$

$$h_3 = \frac{(31.2)(0.0353 \text{ W/m·K})}{0.1888 \text{ m}}$$

$= 5.83 \text{ W/m}^2\text{·K}$

The total heat transfer from the stove is the sum of the convective and radiative components

$$q = (h_1A_1 + h_2A_2 + h_3A_2)(T_s - T_\infty) + \varepsilon\sigma A(T_s^4 - T_{sur}^4)$$

where A is the total surface area of the stove and

$$q = 0.6\dot{m}h_c$$

$$q = (0.7)(6 \text{ kg/h} \times 1 \text{ h/3600 s})(18 \times 10^6 \text{ J/kg})$$

$$= 21.0 \times 10^3 \text{ W}$$

The factor 0.7 accounts for thirty percent of the stove's heat lost in the combustion gases. Numerically solving the relation above for T_s, the surface temperature of the stove,

$$T_s = 570 \text{ K} = \underline{\underline{297°\text{C}}}$$

DISCUSSION

Our initial estimate of the surface temperature is sufficiently close to the predicted value, so a property correction is not justified.

❒ ❒ ❒

PROBLEM 9.22 Surface Temperature of a Transistor

A transistor packaged in a "TO-5 can" is mounted on a horizontal circuit board maintained at 30°C. The height and diameter of the package are 6.4 mm and 8.1 mm, respectively. The transistor has three kovar ($k_l = 14$ W/m·K) leads with a length and diameter of 3.5 mm and 0.511 mm, respectively. Quiescent atmospheric air at 20°C surrounds the transistor. If the transistor dissipates 200 mW and the emissivity of its surface is 0.8, find its surface temperature.

DIAGRAM

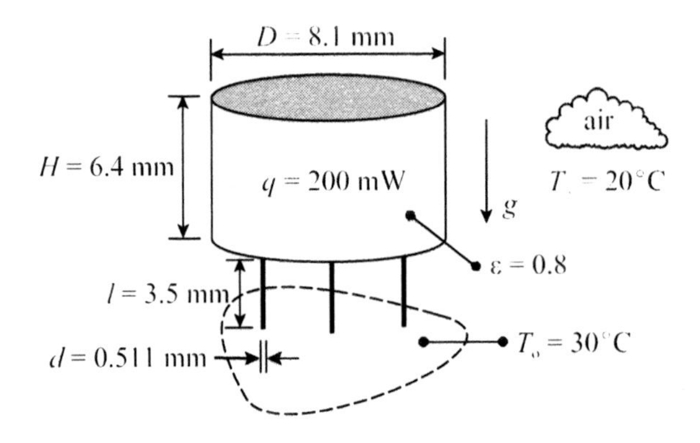

ASSUMPTIONS

1. Steady conditions.
2. Transistor is surrounded by infinite expanse of quiescent atmospheric air.
3. Transistor surface is isothermal.
4. Temperature of surroundings is equivalent to the air temperature.
5. Convection and radiation from leads are neglected.
6. Heat conducted in each lead is equal.
7. Constant properties.

PROPERTIES

Assume $T_s = 45°C$. $T_f = (T_s + T_\infty)/2 = (45°C + 20°C)/2 = 32.5°C = 306$ K

atmospheric air: $\nu = 16.49 \times 10^{-6}$ m^2/s, $k = 0.0267$ W/m·K, $\alpha = 23.4 \times 10^{-6}$ m^2/s, $Pr = 0.706$, $\beta = 1/T_f = 3.268 \times 10^{-3}$ K^{-1}

kovar leads: $k_l = 14$ W/m·K
package surface: $\varepsilon = 0.8$

ANALYSIS

Natural convection and radiation occur at all surfaces of the transistor package. As far as natural convection is concerned, three separate surfaces must be analyzed: vertical surface (the curved surface), upper surface of a heated horizontal plate (the top), and lower surface of a heated horizontal plate (the bottom). Conduction occurs in the three leads. The area of the vertical surface of the transistor package is

$$A_1 = \pi DH$$

$$A_1 = \pi(8.1 \times 10^{-3}\text{ m})(6.4 \times 10^{-3}\text{ m})$$

$$= 1.629 \times 10^{-4}\text{ m}^2$$

The surface area of the top and bottom of the package is

$$A_2 = \pi D^2/4$$

$$A_2 = \pi(8.1 \times 10^{-3}\text{ m})^2/4$$

$$= 5.153 \times 10^{-5}\text{ m}^2$$

where the footprints of the leads have been neglected for the bottom. The characteristic length for the top and bottom surfaces is

$$L = \frac{A_2}{P} = D/4$$

$L = (8.1 \times 10^{-3}\ \text{m})/4$

$= 2.025 \times 10^{-3}\ \text{m}$

Based on our assumed surface temperature, the Rayleigh number for the vertical surface is

$$Ra_H = \frac{g\beta(T_s - T_\infty)H^3}{\nu\alpha}$$

$$Ra_H = \frac{(9.81\ \text{m/s}^2)(3.268 \times 10^{-3}\ \text{K}^{-1})(45 - 20)°\text{C}(6.4 \times 10^{-3}\ \text{m})^3}{(16.49 \times 10^{-6}\ \text{m}^2/\text{s})(23.4 \times 10^{-6}\ \text{m}^2/\text{s})}$$

$= 545$

and the Rayleigh number for the top and bottom surfaces is

$$Ra_L = \frac{g\beta(T_s - T_\infty)L^3}{\nu\alpha}$$

$$Ra_L = \frac{(9.81\ \text{m/s}^2)(3.268 \times 10^{-3}\ \text{K}^{-1})(45 - 20)°\text{C}(2.025 \times 10^{-3}\ \text{m})^3}{(16.49 \times 10^{-6}\ \text{m}^2/\text{s})(23.4 \times 10^{-6}\ \text{m}^2/\text{s})}$$

$= 17.2$

Treating the vertical surface of the transistor package as a flat plate, the average Nusselt number may be calculated using the relation

$$Nu_1 = \frac{h_1 H}{k} = 0.68 + \frac{0.670 Ra_H^{1/4}}{[1 + (0.492/Pr)^{9/16}]^{4/9}}$$

$$Nu_1 = 0.68 + \frac{0.670(545)^{1/4}}{[1 + (0.492/0.706)^{9/16}]^{4/9}}$$

$= 3.16$

so the heat transfer coefficient is

$$h_1 = \frac{Nu_1 k}{H}$$

$$h_1 = \frac{(3.16)(0.0267\ \text{W/m·K})}{6.4 \times 10^{-3}\ \text{m}}$$

$= 13.2\ \text{W/m}^2\text{·K}$

The average Nusselt number for the top surface may be found using the relation

$$Nu_2 = \frac{h_2 L}{k} = 0.54\, Ra_L^{1/4}$$

$$Nu_2 = (0.54)(17.2)^{1/4}$$

$$= 1.10$$

so the heat transfer coefficient is

$$h_2 = \frac{Nu_2 k}{L}$$

$$h_2 = \frac{(1.10)(0.0267 \text{ W/m·K})}{2.025 \times 10^{-3} \text{ m}}$$

$$= 14.5 \text{ W/m}^2\text{·K}$$

The average Nusselt number for the bottom surface may be found using the relation

$$Nu_3 = \frac{h_3 L}{k} = 0.27\, Ra_L^{1/4}$$

$$Nu_3 = (0.27)(17.2)^{1/4}$$

$$= 0.550$$

so the heat transfer coefficient is

$$h_3 = \frac{Nu_3 k}{L}$$

$$h_3 = \frac{(0.550)(0.0267 \text{ W/m·K})}{2.025 \times 10^{-3} \text{ m}}$$

$$= 7.25 \text{ W/m}^2\text{·K}$$

The total heat transfer from the transistor is the sum of the conductive, convective and radiative components,

$$q = \frac{3k_l A_l}{l_1}(T_s - T_o) + (h_1 A_1 + h_2 A_2 + h_3 A_2)(T_s - T_\infty) + \varepsilon\sigma A(T_s^4 - T_{sur}^4)$$

where

$$A_l = \frac{\pi d^2}{4}$$

$$A_1 = \frac{\pi\,(0.511 \times 10^{-3}\ \text{m})^2}{4}$$

$$= 2.051 \times 10^{-7}\ \text{m}^2$$

and

$$A = A_1 + 2A_2$$

$$A = 1.629 \times 10^{-4}\ \text{m}^2 + 2(5.153 \times 10^{-5}\ \text{m}^2)$$

$$= 2.660 \times 10^{-4}\ \text{m}^2$$

Upon substituting $q = 0.200$ W and numerically solving the relation for total heat transfer for T_s, the surface temperature of the transistor,

$$T_s = 324\ \text{K} = \underline{\underline{51.3\,°\text{C}}}$$

DISCUSSION

It is instructive to examine the conductive, convective and radiative components separately to determine which mode is dominant. The heat transfers are

$$q_{\text{cond}} = 52.4\ \text{mW}$$

$$q_{\text{conv}} = 103\ \text{mW}$$

$$q_{\text{rad}} = 44.7\ \text{mW}$$

Hence, convection is largest, followed by conduction and radiation, which are comparable. Note that radiation from the package to the 30°C circuit board was not modeled, but all radiation was assumed to take place between the package and 20°C surroundings.

It should be noted that the values of the Rayleigh numbers in this problem are outside the ranges of applicability for the Nusselt number relations, so the heat transfer coefficients are questionable.

A correction of thermal properties is not required since the predicted surface temperature is close to the initial estimate.

❐ ❐ ❐

PROBLEM 9.23 **Combined Natural and Forced Convection for a Horizontal Cylinder**

A long, horizontal 5-cm diameter cylinder is subjected to a 0.6 m/s cross flow of 20°C atmospheric air. If the surface of the cylinder is maintained at 300°C, what is the convective heat transfer from the cylinder?

DIAGRAM

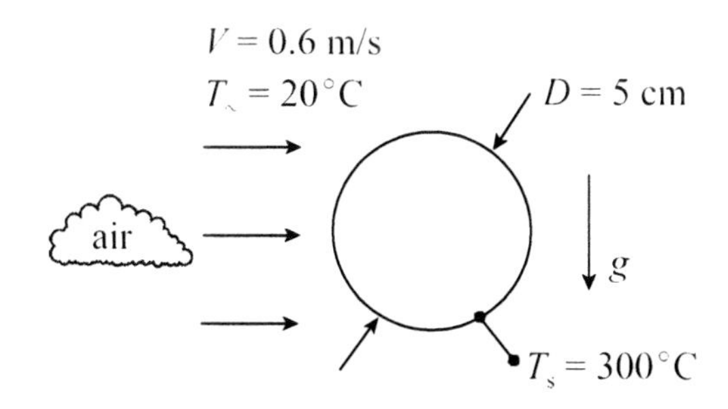

ASSUMPTIONS

1. Steady conditions.
2. Cylinder surface is isothermal.
3. Radiation is neglected.
4. Constant properties.

PROPERTIES

$T_f = (T_s + T_\infty)/2 = (300°C + 20°C)/2 = 160°C = 433$ K

atmospheric air: $\nu = 30.36 \times 10^{-6}$ m²/s, $k = 0.0361$ W/m·K, $\alpha = 44.2 \times 10^{-6}$ m²/s, $Pr = 0.687$, $\beta = 1/T_f = 2.309 \times 10^{-3}$ K^{-1}

ANALYSIS

The cylinder experiences natural and forced convection. First, we find the Reynolds number,

$$Re_D = \frac{VD}{\nu}$$

$$Re_D = \frac{(0.6 \text{ m/s})(0.05 \text{ m})}{30.36 \times 10^{-6} \text{ m}^2\text{/s}}$$

$$= 988$$

The average Nusselt number for a cylinder in cross flow may be found using the relation

$$Nu_{D,F} = \frac{hD}{k} = CRe_D^n Pr^{1/3}$$

where the subscript F refers to forced convection. The constants C and n depend on the Reynolds number. For the range $40 < Re_D < 4000$, $C = 0.683$ and $n = 0.466$. Thus,

$Nu_{D,F} = (0.683)(988)^{0.466}(0.687)^{1/3}$

$= 15.0$

The Rayleigh number for a horizontal cylinder is

$$Ra_D = \frac{g\beta(T_s - T_\infty)D^3}{\nu\alpha}$$

$$Ra_D = \frac{(9.81\ \text{m/s}^2)(2.309 \times 10^{-3}\ \text{K}^{-1})(300 - 20)^\circ\text{C}(0.05\ \text{m})^3}{(30.36 \times 10^{-6}\ \text{m}^2/\text{s})(44.2 \times 10^{-6}\ \text{m}^2/\text{s})}$$

$= 5.908 \times 10^5$

The average Nusselt number for natural convection may be calculated using the relation

$$Nu_{D,N} = \frac{hD}{k} = CRa_D^n$$

where the subscript N refers to natural convection. The quantities C and n depend on the Rayleigh number. For the range $10^4 < Ra_D < 10^7$, $C = 0.480$ and $n = 0.250$. Hence,

$Nu_{D,N} = (0.480)(5.908 \times 10^5)^{0.250}$

$= 13.3$

We see that the Nusselt numbers for forced and natural convection are comparable. A Nusselt number correlation for combined forced and natural convection is

$$Nu_D^n = \left| Nu_{D,F}^n \pm Nu_{D,N}^n \right|$$

where the exponent n depends on the geometry and the relative flow directions. For a horizontal cylinder with forced transverse flow, $n = 4$, and we use a positive sign. The absolute value must be used to preclude the possibility of a negative Nusselt number for opposing flows. We have

$Nu_D^4 = (15.0^4 + 13.3^4)$

$= 8.192 \times 10^4$

Hence,

$Nu_D = (8.192 \times 10^4)^{1/4}$

$= 16.9$

so the average heat transfer coefficient for combined forced and natural convection is

$$h = \frac{Nu_D k}{D}$$

$$h = \frac{(16.9)(0.0361\ \text{W/m}\cdot\text{K})}{0.05\ \text{m}}$$

$$= 12.2\ \text{W/m}^2\cdot\text{K}$$

The convective heat transfer per unit length of cylinder is

$$q' = h\pi D(T_s - T_\infty)$$

$$q' = (12.2\ \text{W/m}^2\cdot\text{K})\pi(0.05\ \text{m})(300 - 20)^\circ\text{C}$$

$$= \underline{\underline{537\ \text{W/m}}}$$

DISCUSSION

If radiation was included in the analysis, and assuming $\varepsilon = 0.9$ and $T_{sur} = T_\infty$,

$$q' = \pi D[h(T_s - T_\infty) + \varepsilon\sigma(T_s^4 - T_{sur}^4)]$$

$$q' = \pi(0.05\ \text{m})[(12.2\ \text{W/m}^2\cdot\text{K})(300 - 20)^\circ\text{C} + (0.9)(5.669 \times 10^{-8}\ \text{W/m}^2\cdot\text{K}^4)(573^4 - 293^4)\text{K}^4]$$

$$= 1341\ \text{W/m}$$

Radiation is clearly the dominant heat transfer mode in this problem, which demonstrates that in combined natural and forced convection applications, radiation should be considered.

Chapter 10

Boiling and Condensation

PROBLEM 10.1 **Boiling Water in a Copper Pan on the Stove**

A polished copper pan containing water at 1 atm is placed on the stove. The diameter of the pan is 20 cm, and the depth of the water is initially 12 cm. The burner maintains the bottom of the pan at 115°C. Find the power required to boil the water, the heat transfer coefficient, the vaporization rate, and the time required to vaporize all the water in the pan.

DIAGRAM

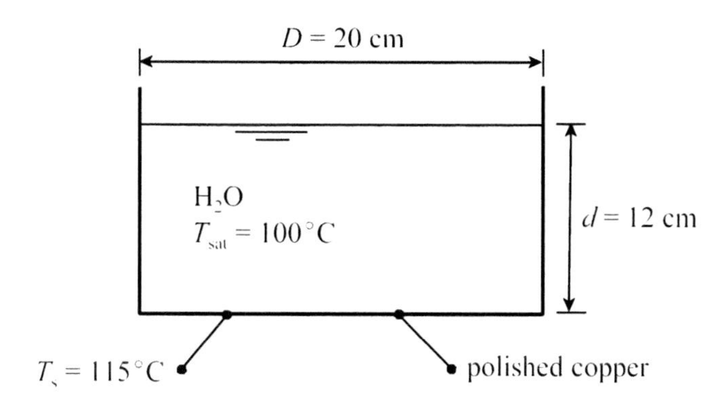

ASSUMPTIONS

1. Steady conditions.
2. Water exposed to standard atmospheric pressure.
3. Water is isothermal at saturation temperature.
4. Heat losses to surroundings are neglected.

PROPERTIES

saturated water, liquid, 100°C: $\rho_l = 957.9$ kg/m^3, $c_{p,l} = 4217$ J/kg·K, $\mu_l = 279 \times 10^{-6}$ Pa·s, $Pr_l = 1.76$, $h_{fg} = 2257$ kJ/kg, $\sigma = 58.9 \times 10^{-3}$ N/m

saturated water, vapor, 100°C: $\rho_v = 0.5955$ kg/m^3

ANALYSIS

The excess temperature is

$$\Delta T_e = T_s - T_{sat}$$

$$\Delta T_e = (115 - 100)°C$$

$$= 15°C = 15 \text{ K}$$

According to the boiling curve, nucleate pool boiling occurs in the range $5°C < \Delta T_e < 30°C$. The recommended correlation for calculating the surface heat flux in this range is

$$q_s'' = \mu_l h_{fg}\left[\frac{g(\rho_l - \rho_v)}{\sigma}\right]^{1/2}\left(\frac{c_{p,l}\Delta T_e}{C_{s,f} h_{fg} Pr_l^n}\right)^3$$

where the values of the quantities $C_{s,f}$ and n depend on the surface-fluid combination. For a polished copper surface and water, $C_{s,f} = 0.0130$ and $n = 1.0$. The surface heat flux is

$$q''_s = (279 \times 10^{-6})(2257 \times 10^3)\left[\frac{(9.81)(957.9 - 0.5955)}{58.9 \times 10^{-3}}\right]^{1/2}\left(\frac{(4217)(15)}{(0.0130)(2257 \times 10^3)(1.76)}\right)^3$$

$$= 4.62 \times 10^5 \text{ W/m}^2$$

The power required to boil the water is the product of the surface heat flux and the surface area of the bottom of the pan.

$$q = q_s'' \pi D^2/4$$

$$q = (4.62 \times 10^5 \text{ W/m}^2)\, \pi\, (0.20 \text{ m})^2/4$$

$$= 1.45 \times 10^4 \text{ W} = \underline{\underline{14.5 \text{ kW}}}$$

The heat transfer coefficient is the surface heat flux divided by the excess temperature.

$$h = \frac{q_s''}{\Delta T_e}$$

$$h = \frac{4.62 \times 10^5 \text{ W/m}^2}{15 \text{ K}}$$

$$= \underline{\underline{3.08 \times 10^4 \text{ W/m}^2\cdot\text{K}}}$$

The vaporization rate is a mass flow rate which may be calculated by recognizing that all the liquid that is converted into vapor in the boiling process leaves the pan across the free surface of the water. We divide the heat transfer by the heat of vaporization,

$$\dot{m} = \frac{q}{h_{fg}}$$

$$\dot{m} = \frac{1.45 \times 10^4 \text{ W}}{2257 \times 10^3 \text{ J/kg}}$$

$$= 6.42 \times 10^{-3} \text{ kg/s} = \underline{\underline{0.385 \text{ kg/min}}}$$

The time required to vaporize all the water in the pan is the original mass of water divided by the vaporization rate.

$$t = \frac{m}{\dot{m}} = \frac{\rho_l(\pi D^2/4)d}{\dot{m}}$$

$$t = \frac{(957.9\ \text{kg/m}^3)[\pi(0.20\ \text{m})^2/4](0.12\ \text{m})}{6.42 \times 10^{-3}\ \text{kg/s}}$$

$$= 562\ \text{s} = \underline{\underline{9.37\ \text{min}}}$$

DISCUSSION

As shown in this problem, heat transfer coefficients for boiling are extremely large compared with those for processes in which phase changes do not occur. Heat transfer coefficients for condensation are also very large.

❒ ❒ ❒

PROBLEM 10.2 Nickel Cylinder Submerged in Pressurized Water

A long cylinder of nickel is immersed in saturated water maintained at 250 kPa in a pressure vessel. The diameter of the cylinder is 1.8 cm. If the surface of the cylinder is maintained at 145°C, what is the heat transfer from the cylinder?

DIAGRAM

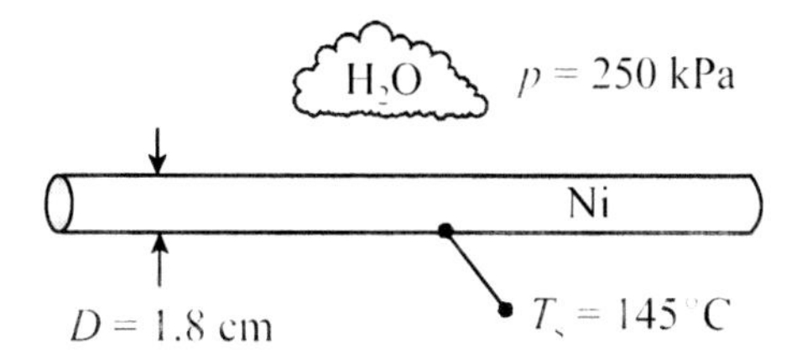

ASSUMPTIONS

1. Steady conditions.
2. Surface of cylinder is isothermal.
3. Water is isothermal at saturation temperature.
4. Radiation is neglected.
5. Constant properties.

PROPERTIES

$p = p_{sat} = 250$ kPa: $\quad T = T_{sat} \approx 127°C$

saturated water, liquid, 127°C: $\rho_l = 937.2$ kg/m^3, $c_{p,l} = 4256$ J/kg·K, $\mu_l = 217 \times 10^{-6}$ Pa·s, $Pr_l = 1.34$, $h_{fg} = 2183$ kJ/kg, $\sigma = 53.6 \times 10^{-3}$ N/m

saturated water, vapor, 127°C: $\rho_v = 1.368$ kg/m^3

ANALYSIS

The excess temperature is

$$\Delta T_e = T_s - T_{sat}$$

$$\Delta T_e = (145 - 127)°C$$

$$= 18°C = 18 \text{ K}$$

The boiling curve shows that nucleate pool boiling occurs in the range $5°C < \Delta T_e < 30°C$. The recommended correlation for calculating the surface heat flux in this range is

$$q''_s = \mu_l h_{fg} \left[\frac{g(\rho_l - \rho_v)}{\sigma} \right]^{1/2} \left(\frac{c_{p,l} \Delta T_e}{C_{s,f} h_{fg} Pr_l^n} \right)^3$$

where the values of the quantities $C_{s,f}$ and n depend on the surface-fluid combination. For a nickel surface and water, $C_{s,f} = 0.006$ and $n = 1.0$. The surface heat flux is

$$q''_s = (217 \times 10^{-6})(2183 \times 10^3) \left[\frac{(9.81)(937.2 - 1.368)}{53.6 \times 10^{-3}} \right]^{1/2} \left(\frac{(4256)(18)}{(0.006)(2183 \times 10^3)(1.34)} \right)^3$$

$$= 1.63 \times 10^7 \text{ W/m}^2$$

The heat transfer per unit length of cylinder is

$$q' = q''_s \pi D$$

$$q' = (1.63 \times 10^7 \text{ W/m}^2)\pi(0.018 \text{ m})$$

$$= \underline{\underline{9.22 \times 10^5 \text{ W/m}}}$$

DISCUSSION

The heat transfer coefficient is the surface heat flux divided by the excess temperature.

$$h = \frac{q''_s}{\Delta T_e}$$

$$h = \frac{1.63 \times 10^7 \text{ W/m}^2}{18 \text{ K}}$$

$$= 9.06 \times 10^5 \text{ W/m}^2\text{·K}$$

❒ ❒ ❒

PROBLEM 10.3 Burnout Current for a Platinum Wire

Find the electrical current at which a 28-gage platinum wire will burn out when submerged in water at one atmosphere pressure. The resistivity of platinum is 10.6 μΩ-cm.

DIAGRAM

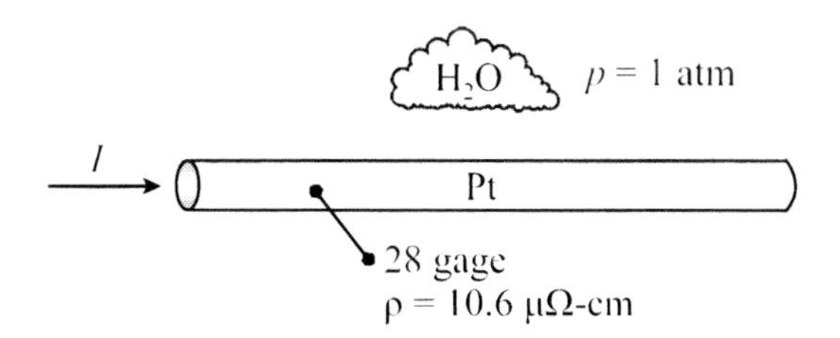

ASSUMPTIONS

1. Steady conditions.
2. Wire is isothermal.
3. Water is isothermal at saturation temperature.
4. Radiation is neglected.
5. Constant properties

PROPERTIES

saturated water, liquid, 100°C: $\rho_l = 957.9$ kg/m^3, $c_{p,l} = 4217$ J/kg·K, $\mu_l = 279 \times 10^{-6}$ Pa·s, $Pr_l = 1.76$, $h_{fg} = 2257$ kJ/kg, $\sigma = 58.9 \times 10^{-3}$ N/m

saturated water, vapor, 100°C: $\rho_v = 0.5955$ kg/m^3

ANALYSIS

The diameter of a 28-gage wire is D = 0.3211 mm. A burnout condition will occur if the electrical power dissipation creates a surface heat flux that exceeds the critical heat flux, q''_{max}. The critical heat flux is

$$q''_{max} = 0.149 h_{fg} \rho_v \left[\frac{\sigma g(\rho_l - \rho_v)}{\rho_v^2} \right]^{1/4}$$

$$q''_{max} = (0.149)(2257 \times 10^3)(0.5955) \left[\frac{(58.9 \times 10^{-3})(9.81)(957.9 - 0.5955)}{(0.5955)^2} \right]^{1/4}$$

$$= 1.26 \times 10^6 \text{ W/m}^2$$

The electrical resistance of a unit length of 28-gage platinum wire is

$$R = \frac{\rho L}{A} = \frac{4\rho L}{\pi D^2}$$

$$R = \frac{4(10.6 \times 10^{-6}\ \Omega\text{-cm})(100 \text{ cm})}{\pi(0.03211 \text{ cm})^2}$$

$$= 1.309\ \Omega$$

Equating the heat transfer to the electrical power dissipation,

$$q''_{max} \pi D = I^2 R$$

and solving for the current I,

$$I = \left(\frac{q''_{max} \pi D}{R} \right)^{1/2}$$

$$I = \left(\frac{(1.26 \times 10^6 \text{ W/m}^2)\pi(0.3211 \times 10^{-3} \text{ m})}{1.309\ \Omega} \right)^{1/2}$$

$$= \underline{\underline{31.2 \text{ A}}}$$

DISCUSSION

If the electrical current is higher than 31.2 A, the platinum wire will experience a burnout condition, which means that the excess temperature will be sufficiently high such that the wire's temperature will exceed the melting point of the wire material. The melting point of platinum is 1772°C.

❒ ❒ ❒

PROBLEM 10.4 Direct Immersion Cooled Electronic Devices

A high density array of identical electronic devices are cooled by direct immersion in a saturated dielectric fluid maintained at 50°C. The heat flux at the surface of each device is 80 kW/m^2, and the nucleate boiling constants for this application are $C_{s,f} = 0.007$ and $n = 1.7$. What is the surface temperature of the devices? Thermal properties of the dielectric fluid are given below.

DIAGRAM

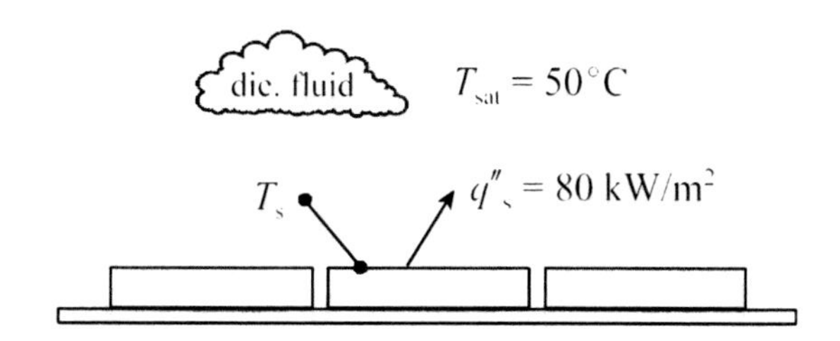

ASSUMPTIONS

1. Steady conditions.
2. Devices are isothermal.
3. Dielectric fluid is isothermal at saturation temperature.
4. Radiation is neglected.
5. Constant properties

PROPERTIES

saturated dielectric fluid, liquid, 50°C: $\rho_l = 1620$ kg/m^3, $c_{p,l} = 1150$ J/kg·K, $\mu_l = 430 \times 10^{-6}$ Pa·s, $Pr_l = 8.72$, $h_{fg} = 82.3$ kJ/kg, $\sigma = 7.9 \times 10^{-3}$ N/m

saturated dielectric fluid, vapor, 50°C: $\rho_v = 13.8$ kg/m^3

ANALYSIS

The recommended correlation for calculating the surface heat flux in the nucleate pool boiling range is

$$q_s'' = \mu_l h_{fg}\left[\frac{g(\rho_l - \rho_v)}{\sigma}\right]^{1/2}\left(\frac{c_{p,l}\Delta T_e}{C_{s,f} h_{fg} Pr_l^n}\right)^3$$

where the excess temperature is

$$\Delta T_e = T_s - T_{sat}$$

Substituting the relation for excess temperature into the relation for surface heat flux, and solving for T_s, the surface temperature,

$$T_s = \left(\frac{q_s'' h_{fg}^2}{\mu_l}\right)^{1/3}\left(\frac{C_{s,f} Pr_l^n}{c_{p,l}}\right)\left[\frac{g(\rho_l - \rho_v)}{\sigma}\right]^{-1/6} + T_{sat}$$

$$T_s = \left(\frac{(80 \times 10^3)(82.3 \times 10^3)^2}{430 \times 10^{-6}}\right)^{1/3}\left(\frac{(0.007)(8.72)^{1.7}}{1150}\right)\left[\frac{(9.81)(1620 - 13.8)}{7.9 \times 10^{-3}}\right]^{-1/6} + 50$$

$$= \underline{\underline{73.3°C}}$$

DISCUSSION

The graph below shows the variation of surface temperature with surface heat flux. The maximum allowable surface temperature for many solid state devices used in commercial applications is 85°C, which corresponds to a surface heat flux of about 275 kW/m^2. But the critical heat flux is only 152 kW/m^2, which may found using the relation

$$q''_{\max} = 0.149 h_{fg}\rho_v\left[\frac{\sigma g(\rho_l - \rho_v)}{\rho_v^2}\right]^{1/4}$$

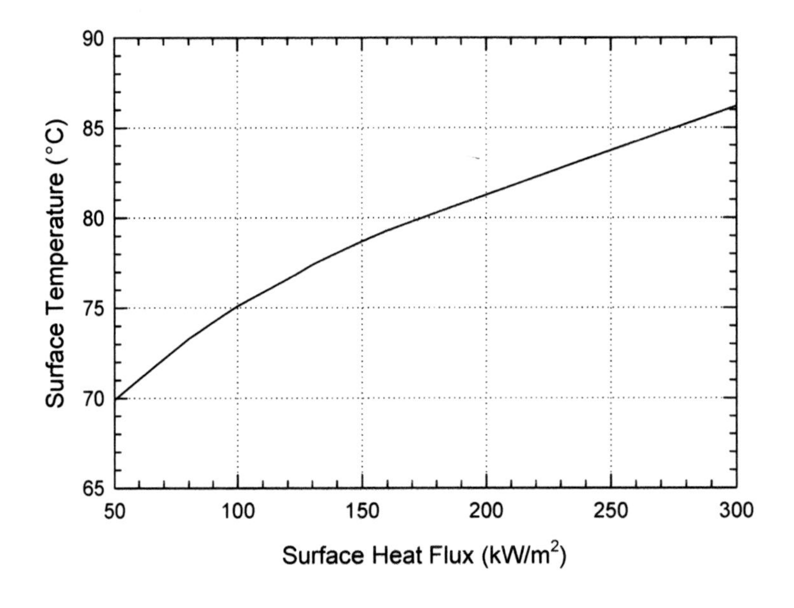

❒ ❒ ❒

PROBLEM 10.5 Nucleate Pool Boiling Experiment

A device for conducting boiling experiments consists of a stainless steel (k = 14.9 W/m·K) bar that is exposed to boiling water at one end and encapsulates an electrical heater at the other end. The surface exposed to the boiling water has been coated with a special film. The other surfaces of the bar are well insulated. Thermocouples measure the temperature of the bar at distances of x_1 = 5 cm and x_2 = 12 cm from the surface in contact with the water. It is observed that nucleate pool boiling is maintained in saturated water at atmospheric pressure when the temperatures are T_1 =145.3°C and T_2 = 197.8°C. If n = 1, what value of $C_{s,f}$ is associated with the nucleate pool boiling correlation?

DIAGRAM

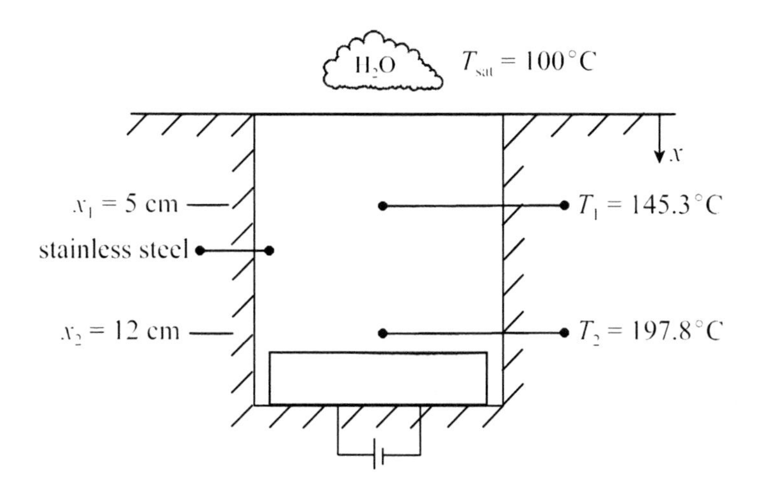

ASSUMPTIONS

1. Steady conditions.
2. One-dimensional steady conduction in bar.
3. Water is isothermal at saturation temperature.
4. Constant properties

PROPERTIES

saturated water, liquid, 100°C: ρ_l = 957.9 kg/m^3, $c_{p,l}$ = 4217 J/kg·K, μ_l = 279 × 10^{-6} Pa·s, Pr_l = 1.76, h_{fg} = 2257 kJ/kg, σ = 58.9 × 10^{-3} N/m

saturated water, vapor, 100°C: ρ_v = 0.5955 kg/m^3

stainless steel: k = 14.9 W/m·K

ANALYSIS

From the thermocouple data, the heat flux, which is the same throughout the bar, can be calculated using Fourier's law of conduction,

$$q'' = q_s'' = k\frac{(T_2 - T_1)}{(x_2 - x_1)}$$

$$q''_s = \frac{(14.9\ \text{W/m·K})(197.8 - 145.3)°\text{C}}{(0.12 - 0.05)\ \text{m}}$$

$$= 1.118 \times 10^4\ \text{W/m}^2$$

The recommended correlation for calculating the surface heat flux in the nucleate pool boiling range is

$$q_s'' = \mu_l h_{fg}\left[\frac{g(\rho_l - \rho_v)}{\sigma}\right]^{1/2}\left(\frac{c_{p,l}\Delta T_e}{C_{s,f} h_{fg} Pr_l^n}\right)^3$$

where the excess temperature is

$$\Delta T_e = T_s - T_{sat}$$

The quantity T_s is the surface temperature of the stainless steel bar, which can also be found using Fourier's law of conduction,

$$q'' = k\frac{(T_1 - T_s)}{x_1}$$

Solving for T_s,

$$T_s = T_1 - \frac{q'' x_1}{k}$$

$$T_s = 145.3°\text{C} - \frac{(1.118 \times 10^4\ \text{W/m}^2)(0.05\ \text{m})}{14.9\ \text{W/m·K}}$$

$$= 107.8°\text{C}$$

Thus, $\Delta T_e = 7.8°\text{C}$. Solving the correlation above for the boiling constant $C_{s,f}$,

$$C_{s,f} = \frac{c_{p,l}\Delta T_e}{Pr^n}\left(\frac{\mu_l}{q_s'' h_{fg}^2}\right)^{1/3}\left[\frac{g(\rho_l - \rho_v)}{\sigma}\right]^{1/6}$$

$$C_{s,f} = \frac{(4217)(7.8)}{1.76}\left(\frac{(279 \times 10^{-6})}{(1.118 \times 10^{4})(2257 \times 10^{3})^2}\right)^{1/3}\left[\frac{(9.81)(957.9 - 0.5955)}{58.9 \times 10^{-3}}\right]^{1/6}$$

$$= \underline{\underline{0.0234}}$$

DISCUSSION

The critical heat flux for water at atmospheric pressure is

$$q''_{\max} = 0.149 h_{fg}\rho_v\left[\frac{\sigma g(\rho_l - \rho_v)}{\rho_v^2}\right]^{1/4}$$

$$q''_{\max} = (0.149)(2257 \times 10^3)(0.5955)\left[\frac{(58.9 \times 10^{-3})(9.81)(957.9 - 0.5955)}{(0.5955)^2}\right]^{1/4}$$

$$= 1.26 \times 10^6 \text{ W/m}^2$$

For this heat flux, the temperature difference between the first thermocouple and the surface would be

$$T_1 - T_s = \frac{q''x_1}{k}$$

$$T_1 - T_s = \frac{(1.26 \times 10^6 \text{ W/m}^2)(0.05 \text{ m})}{14.9 \text{ W/m·K}}$$

$$= 4228°\text{C}$$

which shows that the critical heat flux cannot be achieved in this experiment because the melting point of the stainless steel bar is only about 1400°C.

❐ ❐ ❐

PROBLEM 10.6 Film Boiling for a Cylinder Submerged in Water

A long cylinder of diameter 4.5 cm is horizontally submerged in water at atmospheric pressure. If the surface of the cylinder is maintained at 260°C, what is the heat transfer?

DIAGRAM

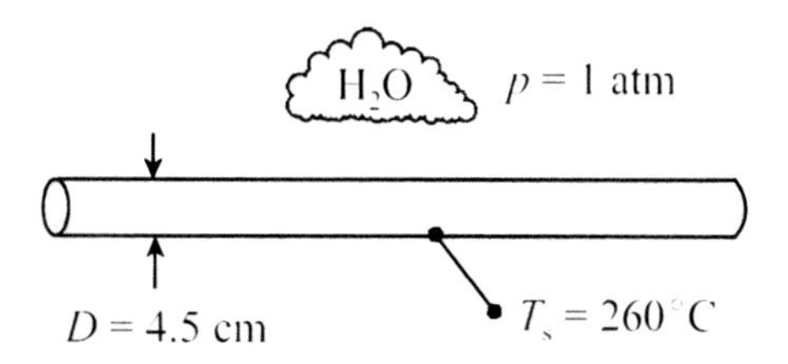

ASSUMPTIONS

1. Steady conditions.
2. Surface of cylinder is isothermal.
3. Water is isothermal at saturation temperature.
4. Radiation is neglected.
5. Constant properties.

PROPERTIES

saturated water, liquid, 100°C: $\rho_l = 957.9$ kg/m^3, $h_{fg} = 2257$ kJ/kg

$T_f = (T_s + T_{sat})/2 = (160°C + 100°C)/2 = 130°C = 403$ K

saturated water, vapor, 130°C: $\rho_v = 1.476$ kg/m^3, $c_{p,v} = 2177$ J/kg·K, $\nu_v = 8.917 \times 10^{-6}$ m^2/s, $k_v = 0.0275$ W/m·K

ANALYSIS

The excess temperature is

$$\Delta T_e = T_s - T_{sat}$$

$$\Delta T_e = (260 - 100)°C$$

$$= 160°C = 160 \text{ K}$$

The region on the boiling curve between the critical heat flux and the Leidenfrost point is termed transition boiling. For excess temperatures beyond the Leidenfrost point ($\Delta T_e > 120°C$), film boiling exists. A correlation for the average Nusselt number for film boiling that applies to spheres and horizontal cylinders of diameter D is

$$Nu_D = \frac{hD}{k_v} = C\left[\frac{g(\rho_l - \rho_v)h'_{fg}D^3}{\nu_v k_v (T_s - T_{sat})}\right]^{1/4}$$

where $C = 0.62$ for horizontal cylinders, and h'_{fg} is a corrected heat of vaporization,

$$h'_{fg} = h_{fg} + 0.80c_{p,v}(T_s - T_{sat})$$

$$h'_{fg} = 2257 \times 10^3 \text{ J/kg} + 0.80(2177 \text{ J/kg·K})(260 - 100)\text{K}$$

$$= 2536 \times 10^3 \text{ J/kg}$$

Hence, the average Nusselt number is

$$Nu_D = 0.62 \left[\frac{(9.81)(957.9 - 1.476)(2536 \times 10^3)(0.045)^3}{(8.917 \times 10^{-6})(0.0275)(260 - 100)} \right]^{1/4}$$

$$= 300.6$$

so the average heat transfer coefficient is

$$h = \frac{Nu_D k_v}{D}$$

$$h = \frac{(300.6)(0.0275 \text{ W/m·K})}{0.045 \text{ m}}$$

$$= 184 \text{ W/m}^2\text{·K}$$

The heat transfer per unit length of cylinder is

$$q' = h\pi D(T_s - T_{sat})$$

$$q' = (184 \text{ W/m}^2\text{·K})\pi(0.045 \text{ m})(260 - 100)°\text{C}$$

$$= \underline{\underline{4162 \text{ W/m}}}$$

DISCUSSION

If radiation was included in the analysis, we would use the following transcendental equation for the average combined heat transfer coefficient h,

$$h^{4/3} = h_{conv}^{4/3} + h_{rad} h^{1/3}$$

where the radiation coefficient h_{rad} is defined as

$$h_{rad} = \frac{\varepsilon\sigma(T_s^4 - T_{sat}^4)}{T_s - T_{sat}}$$

Letting $\varepsilon = 0.9$, and using our previous result, $h_{conv} = 184$ W/m²·K for the convective heat transfer coefficient, we numerically solve the transcendental equation for the combined heat transfer coefficient. We obtain $h = 199$ W/m²·K and $h_{rad} = 19.6$ W/m²·K. Therefore, the heat transfer is

$$q' = (199 \text{ W/m}^2\text{·K})\pi(0.045 \text{ m})(260 - 100) \text{ K}$$

$$= 4501 \text{ W/m}$$

❒ ❒ ❒

PROBLEM 10.7 Film Boiling for Heat-Treated Ball Bearings

A batch of 50 9-mm diameter ball bearings have a uniform temperature of 440 °C when suddenly removed from a heat treating oven. The bearings are immediately submerged in water at atmospheric pressure. If the surface of the bearings has an emissivity of 0.9, what is the heat transfer from the batch of bearings at the moment the batch is submerged?

DIAGRAM

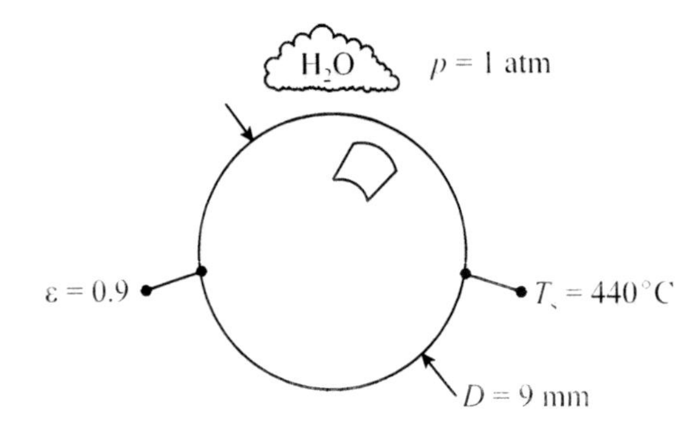

ASSUMPTIONS

1. Steady conditions.
2. Surface of sphere is isothermal.
3. Water is isothermal at saturation temperature.
4. No heat transfer in transit between oven and water.
5. Boiling phenomena at each bearing is not affected by nearby bearings.
6. Constant properties.

PROPERTIES

saturated water, liquid, 100°C: $\rho_l = 957.9\ \text{kg/m}^3$, $h_{fg} = 2257\ \text{kJ/kg}$

$T_f = (T_s + T_{sat})/2 = (440°\text{C} + 100°\text{C})/2 = 270°\text{C} = 543\ \text{K}$

saturated water, vapor, 270°C: $\rho_v = 27.96\ \text{kg/m}^3$, $c_{p,v} = 4381\ \text{J/kg·K}$, $\nu_v = 6.527 \times 10^{-7}\ \text{m}^2/\text{s}$, $k_v = 0.0553\ \text{W/m·K}$

ANALYSIS

The excess temperature is

$$\Delta T_e = T_s - T_{sat}$$

$$\Delta T_e = (440 - 100)°\text{C}$$

$$= 340°\text{C} = 340\ \text{K}$$

A correlation for the average Nusselt number for film boiling that applies to spheres and horizontal cylinders of diameter D is

$$Nu_D = \frac{h_{\text{conv}} D}{k_v} = C\left[\frac{g(\rho_l - \rho_v)h'_{fg} D^3}{\nu_v k_v (T_s - T_{sat})}\right]^{1/4}$$

where $C = 0.67$ for spheres, and h'_{fg} is a corrected heat of vaporization,

$$h'_{fg} = h_{fg} + 0.80 c_{p,v}(T_s - T_{sat})$$

$h'_{\text{fg}} = 2257 \times 10^3$ J/kg + 0.80(4381 J/kg·K)(440 − 100)K

= 3449×10^3 J/kg

Hence, the average Nusselt number is

$$Nu_{\text{D}} = 0.62 \left[\frac{(9.81)(957.9 - 27.96)(3449 \times 10^3)(0.009)^3}{(6.527 \times 10^{-7})(0.0553)(440 - 100)}\right]^{1/4}$$

= 128.9

so the average heat transfer coefficient is

$$h_{\text{conv}} = \frac{Nu_D k_v}{D}$$

$$h_{\text{conv}} = \frac{(128.9)(0.0553 \text{ W/m·K})}{0.009 \text{ m}}$$

= 792 W/m^2·K

Because radiation is included, we use the following transcendental equation for the average combined heat transfer coefficient h,

$$h^{4/3} = h_{conv}^{4/3} + h_{rad} h^{1/3}$$

where the radiation coefficient h_{rad} is

$$h_{rad} = \frac{\varepsilon\sigma(T_s^4 - T_{sat}^4)}{T_s - T_{sat}}$$

$$h_{\text{rad}} = \frac{(0.9)(5.669 \times 10^{-8} \text{ W/m}^2\text{·K}^4)(713^4 - 373^4)\text{K}^4}{(440 - 100)\text{ K}}$$

= 35.9 W/m^2·K

Numerically solving the transcendental equation for the combined heat transfer coefficient,

$$h = 819 \text{ W/m}^2\cdot\text{K}$$

Hence, the heat transfer for the batch of 50 bearings is

$$q = Nh\pi D^2 (T_s - T_{sat})$$

$$q = (50)\,(819 \text{ W/m}^2\cdot\text{K})\pi(0.009 \text{ m})^2(440 - 100)°\text{C}$$

$$= \underline{\underline{3543 \text{ W}}}$$

DISCUSSION

It should be emphasized that our result is the heat transfer at the instant the bearings are plunged into the water because the surface temperature of the bearings decreases with time, thereby decreasing the heat transfer. Of course, after a long period of time, the bearings achieve the same temperature as the surrounding water.

❒ ❒ ❒

PROBLEM 10.8 Film Boiling Time for a Nickel Rod

A long, 1.5-cm diameter rod of nickel has a uniform temperature of 600°C when it is suddenly submerged in a large pool of water at atmospheric pressure. If the rod is maintained in a horizontal position in the water and its emissivity is 0.1, find the time required for film boiling to cease.

DIAGRAM

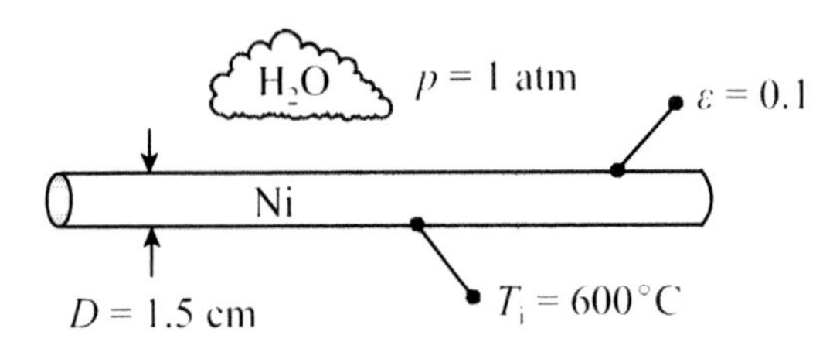

ASSUMPTIONS

1. Rod is isothermal at any given time, i.e., rod is lumped.
2. Water is isothermal at saturation temperature.

PROPERTIES

saturated water, liquid, 100°C: ρ_l = 957.9 kg/m^3, h_{fg} = 2257 kJ/kg

$T_f = (T_s + T_{sat})/2 = (600°C + 100°C)/2 = 350°C = 623$ K

saturated water, vapor, 350°C: ρ_v = 112.9 kg/m^3, $c_{p,v}$ = 17,140 J/kg·K, $\nu_v = 2.353 \times 10^{-7}$ m^2/s, k_v = 0.118 W/m·K

nickel, 600°C: ρ = 8900 kg/m^3, c_p = 542 J/kg·K

ANALYSIS

The time required for film boiling to cease will be determined by analyzing the rod as a lumped system, marching in time until the excess temperature is 120°C, the value at which the minimum heat flux occurs on the boiling curve. This is the Leidenfrost point. As the temperature of the rod changes with time, the properties of the water vapor and nickel are updated.

The relation for the temperature history of a lumped system is

$$\frac{T - T_{sat}}{T_i - T_{sat}} = \exp\left[-\left(\frac{hA}{\rho V c_p}\right)t\right]$$

where A and V are the surface area and volume of the rod, respectively, and t is time. The quantity T_i is the initial temperature of the rod. A relation for the average Nusselt number for film boiling that applies to spheres and horizontal cylinders of diameter D is

$$Nu_D = \frac{h_{conv}D}{k_v} = C\left[\frac{g(\rho_l - \rho_v)h'_{fg}D^3}{\nu_v k_v (T_s - T_{sat})}\right]^{1/4}$$

where $C = 0.62$ for horizontal cylinders, and h'_{fg} is a corrected heat of vaporization,

$$h'_{fg} = h_{fg} + 0.80c_{p,v}(T_s - T_{sat})$$

$$h'_{fg} = 2257 \times 10^3 \text{ J/kg} + 0.80(17{,}140 \text{ J/kg·K})(600 - 100)\text{K}$$

$$= 9113 \times 10^3 \text{ J/kg}$$

Radiation may be important in film boiling. We use the following transcendental equation for the average combined heat transfer coefficient h,

$$h^{4/3} = h_{conv}^{4/3} + h_{rad}h^{1/3}$$

where the radiation coefficient h_{rad} is

$$h_{rad} = \frac{\varepsilon\sigma(T_s^4 - T_{sat}^4)}{T_s - T_{sat}}$$

Using the properties listed above and the foregoing relations, we obtain the following results:

$$Nu_D = 228\,,\ h_{conv} = 1795\ \mathrm{W/m^2{\cdot}K}\,,\ h_{rad} = 6.37\ \mathrm{W/m^2{\cdot}K}\,,\ h = 1800\ \mathrm{W/m^2{\cdot}K}$$

Substituting $t = 1$ s, $T_i = 600°$C and $h = 1800$ W/m²·K into the relation for a lumped system,

$$T = T_s = 562°\mathrm{C}$$

which is the temperature of the rod after 1 s of boiling. This temperature becomes the new value of T_i for the next calculation, which requires that we update the properties of the water vapor and nickel. Choosing $t = 1$ s, the same time step as before, we obtain:

$$Nu_D = 162\,,\ h_{conv} = 1419\ \mathrm{W/m^2{\cdot}K}\,,\ h_{rad} = 6.37\ \mathrm{W/m^2{\cdot}K}\,,\ h = 1424\ \mathrm{W/m^2{\cdot}K}$$

As before, substituting $t = 1$ s, $T_i = 562°$C and $h = 1424$ W/m²·K into the relation for a lumped system,

$$T = T_s = 533°\mathrm{C}$$

which is the temperature of the rod after 2 s of film boiling. We continue this process until the excess temperature is $\Delta T_e = 120°$C. To facilitate the repetitive calculations, a computer is used. The approximate time required for film boiling to cease is

$$t \approx \underline{\underline{49\ \mathrm{s}}}$$

The calculations are summarized in the table below.

t (s)	T_s (°C)	ΔT_e (°C)	Nu_D	h (W/m²·K)
0	600	500	228	1800
1	562	462	162	1424
2	533	434	166	1138
5	467	367	169	978
20	325	225	168	502
49	221	121	160	352

DISCUSSION

Radiation is negligible compared to convection, because the emissivity of the nickel rod is low. For cases when $h_{rad} << h_{conv}$, a simpler relation for the combined heat transfer coefficient may be used,

$$h = h_{conv} + \tfrac{3}{4} h_{rad}$$

which yields the same results as the transcendental relation.

❒ ❒ ❒

PROBLEM 10.9 Forced Convection Boiling for a Stainless Steel Cylinder

A long stainless steel cylinder of diameter 4.0 cm is exposed to a 6-m/s cross flow of saturated water at atmospheric pressure. The surface of the cylinder, which is mechanically polished, is maintained at 125°C. Find the heat transfer from the cylinder.

DIAGRAM

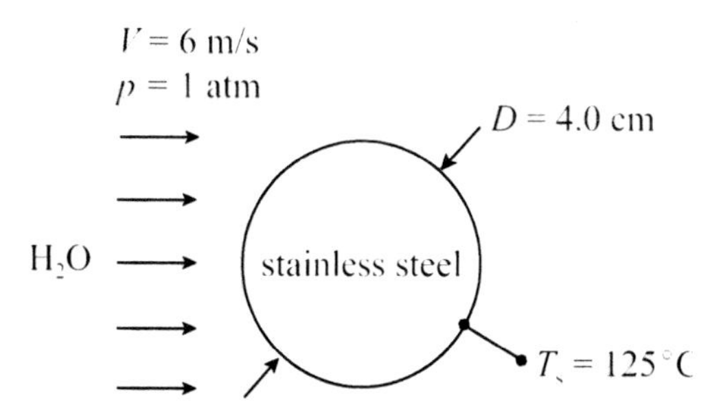

ASSUMPTIONS

1. Steady conditions.
2. Surface of cylinder is isothermal.
3. Water is isothermal at saturation temperature.
4. Radiation is neglected.
5. Constant properties.

PROPERTIES

saturated water, liquid, 100°C: $\rho_l = 957.9$ kg/m^3, $h_{fg} = 2257$ kJ/kg, $\mu_l = 279 \times 10^{-6}$ Pa·s, $Pr_l = 1.76$, $h_{fg} = 2257$ kJ/kg, $\sigma = 58.9 \times 10^{-3}$ N/m

$T_f = (T_s + T_{sat})/2 = (125°C + 100°C)/2 = 112.5°C = 386$ K

saturated water, liquid, 112.5°C: $\nu = 2.549 \times 10^{-7}$ m^2/s, $k = 0.686$ W/m·K, $Pr = 1.49$

saturated water, vapor, 100°C: $\rho_v = 0.5956$ kg/m^3

ANALYSIS

One approach to a forced convection boiling problem is to assume that the heat transfer due to forced convection can simply be added to the heat transfer due to boiling, effectively splitting the analysis into two separate parts. The total heat flux is the sum of the heat fluxes for forced convection and boiling,

$$q'' = q''_c + q''_b$$

We first consider forced convection, for which properties at the film temperature are used. The Reynolds number is

$$Re_D = \frac{VD}{\nu}$$

$$Re_D = \frac{(6 \text{ m/s})(0.040 \text{ m})}{2.549 \times 10^{-7} \text{ m}^2\text{/s}}$$

$$= 9.415 \times 10^5$$

The average Nusselt number may be found using the empirical correlation

$$Nu_D = 0.3 + \frac{0.62 Re_D^{1/2} Pr^{1/3}}{\left[1 + (0.4/Pr)^{2/3}\right]^{1/4}} \left[1 + \left(\frac{Re_D}{282{,}000}\right)^{5/8}\right]^{4/5}$$

$$Nu_D = 0.3 + \frac{0.62(9.415 \times 10^5)^{1/2}(1.49)^{1/3}}{[1 + (0.4/1.49)^{2/3}]^{1/4}} [1 + (9.415 \times 10^5/282{,}000)^{5/8}]^{4/5}$$

$$= 1567$$

so average heat transfer coefficient is

$$h = \frac{Nu_D k}{D}$$

$$h = \frac{(1567)(0.686 \text{ W/m·K})}{0.040 \text{ m}}$$

$$= 2.687 \times 10^4 \text{ W/m}^2\text{·K}$$

and the heat flux is

$$q''_c = h(T_s - T_{sat})$$

$q''_c = (2.687 \times 10^4 \text{ W/m}^2\cdot\text{K})(125 - 100)°\text{C}$

$= 6.718 \times 10^5 \text{ W/m}^2$

Next, we consider boiling. The excess temperature is

$$\Delta T_e = T_s - T_{sat}$$

$\Delta T_e = (125 - 100)°\text{C}$

$= 25°\text{C} = 25 \text{ K}$

Nucleate pool boiling occurs in the range $5°\text{C} < \Delta T_e < 30°\text{C}$. The recommended correlation for calculating the surface heat flux in this range is

$$q''_b = \mu_l h_{fg} \left[\frac{g(\rho_l - \rho_v)}{\sigma} \right]^{1/2} \left(\frac{c_{p,l} \Delta T_e}{C_{s,f} h_{fg} Pr_l^n} \right)^3$$

where the values of the quantities $C_{s,f}$ and n depend on the surface-fluid combination. For a mechanically polished stainless steel surface and water, $C_{s,f} = 0.0130$ and $n = 1.0$. The surface heat flux is

$$q''_b = (279 \times 10^{-6})(2257 \times 10^3) \left[\frac{(9.81)(957.9 - 0.5956)}{58.9 \times 10^{-3}} \right]^{1/2} \left(\frac{(4217)(25)}{(0.0130)(2257 \times 10^3)(1.76)} \right)^3$$

$= 2.139 \times 10^6 \text{ W/m}^2$

The total heat flux is

$q'' = q''_c + q''_b$

$= (6.718 \times 10^5 + 2.139 \times 10^6) \text{ W/m}^2$

$= 2.811 \times 10^6 \text{ W/m}^2$

Hence, the heat transfer per unit length of cylinder is

$$q' = q'' \pi D$$

$q' = (2.811 \times 10^6 \text{ W/m}^2)\pi(0.040 \text{ m})$

$= \underline{\underline{3.532 \times 10^5 \text{ W/m}}}$

DISCUSSION

The superposition of the forced convection heat flux and the nucleate boiling heat flux illustrated here is simple, perhaps overly so. Experiments have shown that, depending on the boiling regime, the surface heat flux is essentially independent of velocity. More rigorous methods for handling forced convection boiling problems are found in the heat transfer literature.

❐ ❐ ❐

PROBLEM 10.10 Condensation on a Vertical Surface

A vertical surface measuring 20 cm high and 80 cm wide is maintained at 40°C while exposed to saturated water vapor at atmospheric pressure. Find the heat transfer and the condensation rate.

DIAGRAM

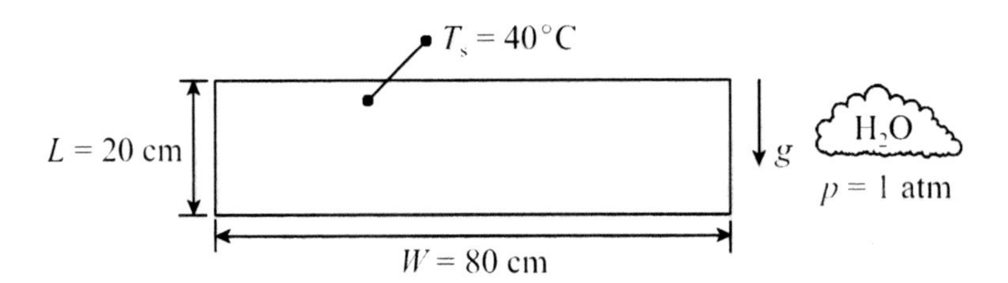

ASSUMPTIONS

1. Steady conditions.
2. Surface is isothermal.
3. Water vapor is isothermal at saturation temperature.
4. Laminar film condensation.
5. Radiation is neglected.
6. Constant properties.

PROPERTIES

$T_f = (T_s + T_{sat})/2 = (40°C + 100°C)/2 = 70°C = 343$ K

saturated water, liquid, 70°C: $\rho_l = 977.7$ kg/m^3, $c_{p,l} = 4190$ J/kg·K, $\mu_l = 401 \times 10^{-6}$ Pa·s, $k_l = 0.665$ W/m·K, $\nu_l = 4.102 \times 10^{-7}$ m^2/s

saturated water, vapor, 100°C: $\rho_v = 0.5956$ kg/m^3, $h_{fg} = 2257$ kJ/kg

ANALYSIS

Consistent with our assumption of laminar film condensation on a vertical surface, we calculate the average Nusselt number using the relation

$$Nu_L = \frac{hL}{k_l} = 0.943\left[\frac{\rho_l g(\rho_l - \rho_v)h'_{fg}L^3}{\mu_l k_l(T_{sat} - T_s)}\right]^{1/4}$$

where,

$$h'_{fg} = h_{fg} + 0.68c_{p,l}(T_{sat} - T_s)$$

$$h'_{fg} = 2257 \times 10^3 \text{ J/kg} + 0.68(4190 \text{ J/kg·K})(100 - 40)°\text{C}$$

$$= 2428 \times 10^3 \text{ J/kg}$$

Thus, the average Nusselt number is

$$Nu_L = 0.943\left[\frac{(977.7)(9.81)(977.7 - 0.5956)(2428 \times 10^3)(0.20)^3}{(401 \times 10^{-6})(0.665)(100 - 40)}\right]^{1/4}$$

$$= 1732$$

so the average heat transfer coefficient is

$$h = \frac{Nu_L k_l}{L}$$

$$h = \frac{(1732)(0.665 \text{ W/m·K})}{0.20 \text{ m}}$$

$$= 5759 \text{ W/m}^2\text{·K}$$

and the heat transfer to the surface is

$$q = hWL(T_{sat} - T_s)$$

$$q = (5759 \text{ W/m}^2\text{·K})(0.80 \text{ m})(0.20 \text{ m})(100 - 40)°\text{C}$$

$$= 5.529 \times 10^4 \text{ W} = \underline{\underline{55.3 \text{ kW}}}$$

The rate of condensation, i.e., the rate at which liquid water forms on the surface, is the heat transfer divided by the modified latent heat of vaporization,

$$\dot{m} = \frac{q}{h'_{fg}}$$

$$\dot{m} = (5.529 \times 10^4 \text{ W})/(2428 \times 10^3 \text{ J/kg})$$

$$= \underline{\underline{0.0228 \text{ kg/s}}}$$

Our assumption of laminar film condensation may be checked by calculating the Reynolds number based on the film thickness δ,

$$Re_\delta = \frac{4\dot{m}}{\mu_l W}$$

$$Re_\delta = \frac{4(0.0228 \text{ kg/s})}{(401 \times 10^{-6} \text{ Pa·s})(0.80 \text{ m})}$$

$$= 284$$

For the wave-free laminar region, $Re_\delta < 30$, and for the wavy-laminar region, $30 < Re_\delta < 1800$. Hence, our assumption is somewhat questionable since for $Re_\delta = 284$, a portion of the condensate film is wavy.

DISCUSSION

A correlation for the average heat transfer coefficient that applies to the wavy-laminar region is

$$\frac{h(\nu_l^2/g)^{1/3}}{k_l} = \frac{Re_\delta}{1.08 Re_\delta^{1.22} - 5.2}$$

Combining the two equations,

$$\dot{m} = \frac{q}{h'_{fg}} = \frac{hWL(T_{sat} - T_s)}{h'_{fg}}$$

$$Re_\delta = \frac{4\dot{m}}{\mu_l W}$$

the correlation may be expressed as

$$Re_\delta = \left[\frac{3.704\, L k_l (T_{sat} - T_s)}{h'_{fg}\mu_l(\nu_l^2/g)^{1/3}} + 4.815\right]^{0.820}$$

from which we obtain $Re_\delta = 331$, $\dot{m} = 0.0265$ kg/s and $q = 64.3$ kW. The condensation rate and heat transfer based on the wavy-laminar correlation exceed those based on the wave-free correlation by 16 percent.

❒ ❒ ❒

PROBLEM 10.11 Condensation on a Horizontal Tube

A long thin-walled copper tube runs horizontally through a region containing saturated water vapor at one atmosphere pressure. A coolant flowing through the tube maintains the tube surface at 65°C. If the diameter of the tube is 1.8 cm, find the heat transfer to and condensation rate on the outside surface of the tube.

DIAGRAM

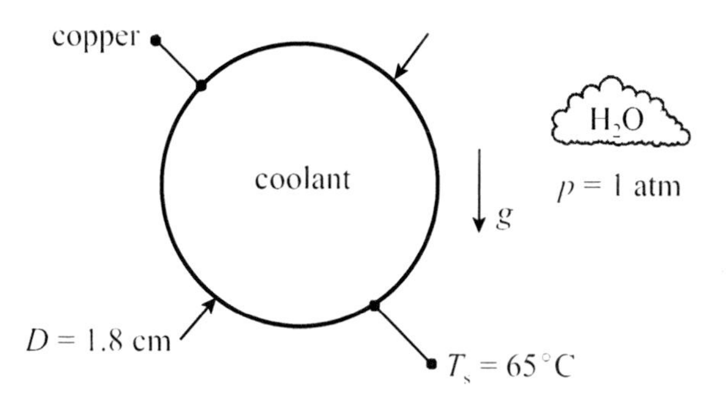

ASSUMPTIONS

1. Steady conditions.
2. Surface is isothermal.
3. Water vapor is isothermal at saturation temperature.
4. Laminar film condensation.
5. Radiation is neglected.
6. Constant properties.

PROPERTIES

$T_f = (T_s + T_{sat})/2 = (65°C + 100°C)/2 = 82.5°C = 356$ K

saturated water, liquid, 70°C: $\rho_l = 970.1$ kg/m^3, $c_{p,l} = 4200$ J/kg·K, $\mu_l = 339 \times 10^{-6}$ Pa·s, $k_l = 0.672$ W/m·K, $\nu_l = 3.495 \times 10^{-7}$ m^2/s

saturated water, vapor, 100°C: $\rho_v = 0.5956$ kg/m^3, $h_{fg} = 2257$ kJ/kg

ANALYSIS

A well known correlation for the average heat transfer coefficient for laminar film condensation on spheres and horizontal tubes is

$$h = C\left[\frac{g\rho_l(\rho_l - \rho_v)k_l^3 h'_{fg}}{\mu_l(T_{sat} - T_s)D}\right]^{1/4}$$

where $C = 0.826$ and 0.729 for spheres and horizontal tubes, respectively, and h'_{fg} is a corrected heat of vaporization,

$$h'_{fg} = h_{fg} + 0.68c_{p,l}(T_{sat} - T_s)$$

$h'_{fg} = 2257 \times 10^3$ J/kg + 0.68(4200 J/kg·K)(100 − 65)°C

$= 2357 \times 10^3$ J/kg

The average heat transfer coefficient is

$$h = 0.729\left[\frac{(9.81)(970.1)(970.1 - 0.5956)(0.672)^3(2357 \times 10^3)}{(339 \times 10^{-6})(100 - 65)(0.018)}\right]^{1/4}$$

$= 9665$ W/m²·K

so the heat transfer per unit length of tube is

$$q' = h\pi D(T_{sat} - T_s)$$

$q' = (9665$ W/m²·K$)\pi(0.018$ m$)(100 - 65)$°C

$= 1.913 \times 10^4$ W/m = 19.1 kW/m

The rate of condensation per unit length of tube is the heat transfer divided by the modified latent heat of vaporization,

$$\dot{m} = \frac{q'}{h'_{fg}}$$

$\dot{m} = (1.913 \times 10^4$ W/m$)/(2357 \times 10^3$ J/kg$)$

$= 8.12 \times 10^{-3}$ kg/s·m

DISCUSSION

It is instructive to mention an important dimensionless heat transfer quantity called the Jakob number, defined as

$$Ja \equiv \frac{c_p(T_s - T_{sat})}{h_{fg}}$$

Physically, the Jakob number is the ratio of sensible energy to latent energy involved in a liquid-vapor phase change. This definition is valid for boiling and condensation, but for condensation the surface and saturation temperature terms are interchanged so as to give a positive value. Studies have shown that the errors associated with the correlation used here are less than 3 percent if $Ja < 0.1$. In this analysis, $Ja = 0.0651$, so our assumption of laminar film condensation is reasonable.

❒ ❒ ❒

PROBLEM 10.12 A Steam Condenser

A steam condenser consisting of a square array of 64 tubes is exposed to saturated water vapor at 51 kPa pressure. The outside diameter of the tubes is 6.35 mm. A coolant flowing through the tubes maintains the surface of the tubes at 30 °C. If the tubes are 40 cm long, what is the total heat transfer and condensation rate for the condenser?

DIAGRAM

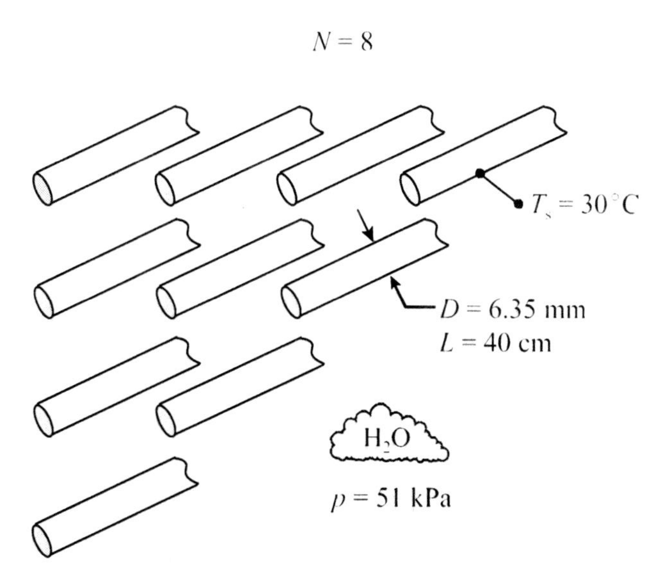

ASSUMPTIONS

1. Steady conditions.
2. Surface of tubes is isothermal.
3. Water vapor is isothermal at saturation temperature.
4. Laminar film condensation.
5. Radiation is neglected.
6. Constant properties.

PROPERTIES

$p = p_{sat} = 51$ kPa: $T = T_{sat} \approx 82°C$

$T_f = (T_s + T_{sat})/2 = (30°C + 82°C)/2 = 56°C = 329$ K

saturated water, liquid, 56°C: $\rho_l = 986.6$ kg/m^3, $c_{p,l} = 4184$ J/kg·K, $\mu_l = 520 \times 10^{-6}$ Pa·s, $k_l = 0.646$ W/m·K, $\nu_l = 5.271 \times 10^{-7}$ m^2/s

saturated water, vapor, 82°C: $\rho_v = 0.3145$ kg/m^3, $h_{fg} = 2304$ kJ/kg

ANALYSIS

A correlation for the average heat transfer coefficient for laminar film condensation on a single horizontal tube is

$$h_1 = C\left[\frac{g\rho_l(\rho_l - \rho_v)k_l^3 h'_{fg}}{\mu_l(T_{sat} - T_s)D}\right]^{1/4}$$

where $C = 0.729$, and h'_{fg} is a corrected heat of vaporization,

$$h'_{fg} = h_{fg} + 0.68c_{p,l}(T_{sat} - T_s)$$

$$h'_{fg} = 2304 \times 10^3 \text{ J/kg} + 0.68(4184 \text{ J/kg·K})(82 - 56)°\text{C}$$

$$= 2378 \times 10^3 \text{ J/kg}$$

For a vertical tier of N horizontal tubes, the average heat transfer coefficient is

$$h_N = h_1 N^{-1/4}$$

The average heat transfer coefficient for a single tube is

$$h_l = 0.729\left[\frac{(9.81)(986.6)(986.6 - 0.3145)(0.646)^3(2378 \times 10^3)}{(520 \times 10^{-6})(82 - 56)(6.35 \times 10^{-3})}\right]^{1/4}$$

$= 1.191 \times 10^4$ W/m²·K

so the average heat transfer coefficient for a vertical tier of 8 tubes is

$h_N = (1.191 \times 10^4 \text{ W/m}^2\cdot\text{K})(8)^{-1/4}$

$= 7.083 \times 10^3 \text{ W/m}^2\cdot\text{K}$

The reduction in h with increasing N is due to an increase in the film thickness for successive tubes. The total heat transfer for the condenser is

$$q = N^2 h_N \pi D L (T_{sat} - T_s)$$

$q = (8)^2 (7.083 \times 10^3 \text{ W/m}^2\cdot\text{K})\pi(6.35 \times 10^{-3} \text{ m})(0.40 \text{ m})(82 - 56)°\text{C}$

$= 9.404 \times 10^4 \text{ W} = \underline{\underline{94.0 \text{ kW}}}$

and the condensation rate is

$$\dot{m} = \frac{q}{h'_{fg}}$$

$\dot{m} = (9.404 \times 10^4 \text{ W})/(2378 \times 10^3 \text{ J/kg})$

$= \underline{\underline{0.0395 \text{ kg/s}}}$

DISCUSSION

The Jakob number is

$$Ja = \frac{c_{c,l}(T_{sat} - T_s)}{h_{fg}}$$

$$Ja = \frac{(4184 \text{ J/kg}\cdot\text{K})(82 - 56)°\text{C}}{2304 \times 10^3 \text{ J/kg}}$$

$= 0.0472$

Since $Ja < 0.1$, our assumption of laminar film condensation is valid.

❒ ❒ ❒

PROBLEM 10.13 Condensation Inside a Horizontal Tube

A horizontal, 2.5-cm diameter thin-walled tube carries saturated water at a mean velocity of 1.4 m/s. The temperature of the tube wall is maintained at 55 °C. If the pressure of the water vapor is 75 kPa, find the heat transfer and condensation rate.

DIAGRAM

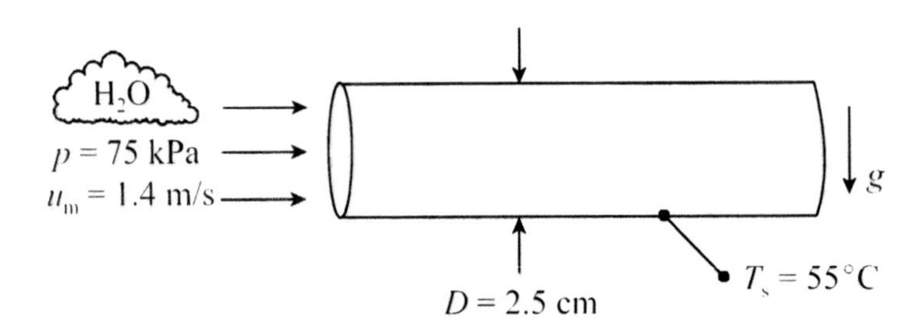

ASSUMPTIONS

1. Steady conditions.
2. Surface of tube is isothermal.
3. Water is isothermal at saturation temperature.
4. Film condensation.
5. Constant properties.

PROPERTIES

$p = p_{sat} = 75$ kPa: $\quad T = T_{sat} \approx 92$ °C

$T_f = (T_s + T_{sat})/2 = (55\,°C + 92\,°C)/2 = 73.5\,°C = 347$ K

saturated water, liquid, 73.5 °C: $\rho_l = 975.4$ kg/m^3, $c_{p,l} = 4193$ J/kg·K, $\mu_l = 379 \times 10^{-6}$ Pa·s, $k_l = 0.668$ W/m·K, $\nu_l = 3.886 \times 10^{-7}$ m^2/s

saturated water, vapor, 92 °C: $\rho_v = 0.4521$ kg/m^3, $\mu_v = 11.69 \times 10^{-6}$ Pa·s, $h_{fg} = 2278$ kJ/kg

ANALYSIS

First, we calculate the Reynolds number for the vapor flow,

$$Re_D = \frac{\rho_v u_m D}{\mu_v}$$

$$Re_D = \frac{(0.4521\ \text{kg/m}^3)(1.4\ \text{m/s})(0.025\ \text{m})}{11.69 \times 10^{-6}\ \text{Pa·s}}$$

$$= 1354$$

For low vapor velocities such that $Re_D < 35{,}000$, a recommended correlation for the average heat transfer coefficient for film condensation inside horizontal tubes is

$$h = 0.555\left[\frac{g\rho_l(\rho_l - \rho_v)k_l^3 h'_{fg}}{\mu_l(T_{sat} - T_s)D}\right]^{1/4}$$

where h'_{fg} is a modified heat of vaporization,

$$h'_{fg} = h_{fg} + \tfrac{3}{8}c_{p,l}(T_{sat} - T_s)$$

$$h'_{fg} = 2278 \times 10^3\ \text{J/kg} + (3/8)(4193\ \text{J/kg·K})(92 - 55)°\text{C}$$

$$= 2336 \times 10^3\ \text{J/kg}$$

The average heat transfer coefficient is

$$h = 0.555\left[\frac{(9.81)(975.4)(975.4 - 0.4521)(0.668)^3(2336 \times 10^3)}{(379 \times 10^{-6})(92 - 55)(0.025)}\right]^{1/4}$$

$$= 6.475 \times 10^3\ \text{W/m}^2\text{·K}$$

so the heat transfer per unit length of tube is

$$q' = h\pi D(T_{sat} - T_s)$$

$$q' = (6.475 \times 10^3\ \text{W/m}^2\text{·K})\pi(0.025\ \text{m})(92 - 55)°\text{C}$$

$$= 1.882 \times 10^4\ \text{W/m} = \underline{\underline{18.8\ \text{kW/m}}}$$

and rate of condensation per unit length of tube is

$$\dot{m} = \frac{q'}{h'_{fg}}$$

$$\dot{m} = (1.882 \times 10^4\ \text{W/m})/(2336 \times 10^3\ \text{J/kg})$$

$$= \underline{\underline{8.057 \times 10^{-3}\ \text{kg/s·m}}}$$

DISCUSSION

Conditions within horizontal tubes depend primarily on the mean velocity of the vapor. For low vapor velocities, the vapor occupies the upper portion of the tube, and the condensate collects in the lower portion, as depicted in the tube cross section below.

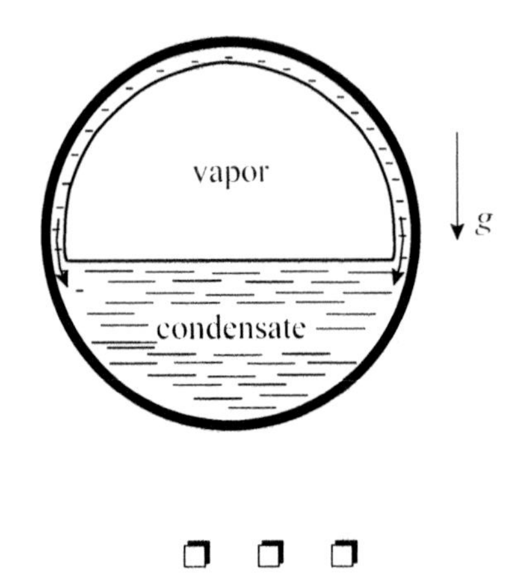

❐ ❐ ❐

PROBLEM 10.14 Condensation on the Outside Surface of a Fluid-Carrying Horizontal Tube

A long, horizontal stainless steel (k = 13.4 W/m·K) tube with an inside and outside diameter of 10.21 mm and 12.70 mm, respectively, carries ethylene glycol at a mean velocity of 8 m/s. The tube is surrounded by saturated water at atmospheric pressure. If the mean temperature of the ethylene glycol is 40°C, find the heat transfer and the condensation rate of water on the outside surface of the tube.

DIAGRAM

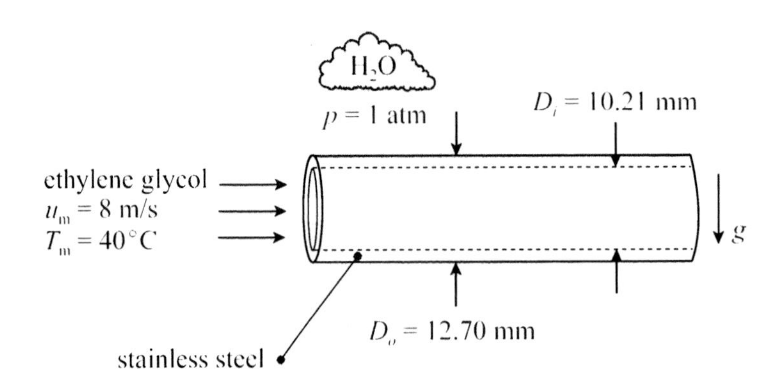

ASSUMPTIONS

1. Steady conditions.
2. Water is isothermal at saturation temperature.
3. Laminar film condensation.
4. Fully developed flow.
5. One-dimensional steady conduction in tube wall.
6. Constant properties.

PROPERTIES

ethylene glycol

ethylene glycol, liquid, 40°C: $\rho_l = 1101$ kg/m^3, $c_{p,l} = 2474$ J/kg·K, $\mu_l = 0.976 \times 10^{-2}$ Pa·s, $k_l = 0.256$ W/m·K, $Pr = 94.2$

Water

$p = p_{sat} = 1$ atm: $T = T_{sat} = 100$°C

Estimate: $T_{s,o} = 80$°C

$T_f = (T_{s,o} + T_{sat})/2 = (80°C + 100°C)/2 = 90°C = 363$ K

saturated water, liquid, 90°C: $\rho_l = 964.9$ kg/m^3, $c_{p,l} = 4207$ J/kg·K, $\mu_l = 313 \times 10^{-6}$ Pa·s, $k_l = 0.676$ W/m·K, $\nu_l = 3.244 \times 10^{-7}$ m^2/s

saturated water, vapor, 100°C: $\rho_v = 0.5956$ kg/m^3, $h_{fg} = 2257$ kJ/kg

stainless steel: $k = 13.4$ W/m·K

ANALYSIS

First, we perform an analysis for the outside surface of the tube that is in contact with the saturated water. The ethylene glycol is colder than the saturated water surrounding the tube, so condensation occurs on the outside surface. A correlation for the average heat transfer coefficient for laminar film condensation on a horizontal tube is

$$h_o = C\left[\frac{g\rho_l(\rho_l - \rho_v)k_l^3 h'_{fg}}{\mu_l(T_{sat} - T_{s,o})D_o}\right]^{1/4}$$

where $C = 0.729$, and h'_{fg} is a corrected heat of vaporization,

$$h'_{fg} = 2257 \times 10^3 \text{ J/kg} + h'_{fg} = h_{fg} + 0.68c_{p,l}(T_{sat} - T_{s,o})\ 0.68(4207 \text{ J/kg·K})(100 - 80)°\text{C}$$

$$= 2314 \times 10^3 \text{ J/kg}$$

The average heat transfer coefficient is

$$h_o = 0.729\left[\frac{(9.81)(964.9)(964.9 - 0.5956)(0.676)^3(2314 \times 10^3)}{(313 \times 10^{-6})(100 - 80)(12.70 \times 10^{-3})}\right]^{1/4}$$

$$= 1.234 \times 10^4 \text{ W/m}^2\text{·K}$$

Second, we analyze the thermal conditions for the inside surface of the tube. The Reynolds number for the flow of ethylene glycol is

$$Re_D = \frac{\rho_l u_m D_i}{\mu_l}$$

$$Re_D = \frac{(1101 \text{ kg/m}^3)(8 \text{ m/s})(10.21 \times 10^{-3} \text{ m})}{0.976 \times 10^{-2} \text{ Pa·s}}$$

$$= 9214$$

The flow may be considered turbulent, so we find the average Nusselt number using the correlation,

$$Nu_D = \frac{h_i D_i}{k_l} = 0.023 Re_D^{4/5} Pr^{0.4}$$

$$Nu_D = (0.023)(9214)^{4/5}(94.2)^{0.4}$$

$$= 210$$

Thus, the average heat transfer coefficient is

$$h_i = \frac{Nu_D k_l}{D_i}$$

$$h_i = \frac{(210)(0.256 \text{ W/m·K})}{10.21 \times 10^{-3} \text{ m}}$$

$$= 5265 \text{ W/m}^2\text{·K}$$

Now that the heat transfer coefficients have been found, we can calculate the heat transfer per unit length of tube,

$$q' = \frac{T_{sat} - T_m}{\dfrac{1}{h_i \pi D_i} + \dfrac{\ln(D_o/D_i)}{2\pi k} + \dfrac{1}{h_o \pi D_o}}$$

$$q' = \frac{(100 - 40)}{\dfrac{1}{(5265)\pi(10.21 \times 10^{-3})} + \dfrac{\ln[(12.70)/(10.21)]}{2\pi(13.4)} + \dfrac{1}{(1.234 \times 10^{4})\pi(12.70 \times 10^{-3})}}$$

$= 5690$ W/m

Our initial estimate of the outside surface temperature must be checked. Rearranging Newton's law of cooling, the temperature of the outside surface of the tube is

$$T_{s,o} = T_{sat} - \frac{q'}{h_o \pi D_o}$$

$$T_{s,o} = 100°\text{C} - \frac{5690 \text{ W/m}}{(1.234 \times 10^{4} \text{ W/m}^2\cdot\text{K})\pi(12.70 \times 10^{-3} \text{ m})}$$

$= 88.4°$C

which is only 8.4°C higher than the initial estimate, so a property correction would yield only a minor improvement in accuracy. The condensation rate of water is

$$\dot{m} = \frac{q'}{h'_{fg}}$$

$\dot{m} = (5690 \text{ W/m})/(2314 \times 10^{3} \text{ J/kg})$

$= \underline{\underline{2.459 \times 10^{-3} \text{ kg/s}\cdot\text{m}}}$

DISCUSSION

Due to condensing water vapor, the surface temperature is constant, but the temperature of the ethylene glycol is not constant throughout the tube. Its mean temperature, T_m, will increase as a function of axial position, x, according to the relation

$$\frac{T_{s,i} - T_m(x)}{T_{s,i} - T_{m,i}} = \exp\left(-\frac{\pi D_i h_i}{\dot{m} c_p} x\right)$$

where $T_{s,i}$ is the inside surface temperature, $T_{m,i}$ is the mean temperature at an inlet location defined by $x = 0$, and $\dot{m}$ is mass flow rate.

Chapter 11

Heat Exchangers

PROBLEM 11.1 Counterflow Concentric Tube Heat Exchanger

A counterflow, concentric tube heat exchanger consists of a 3-cm diameter thin-walled copper tube within a 4.5-cm diameter tube. The inner tube carries water at a mass flow rate of 0.3 kg/s, while the annulus carries engine oil at a mass flow rate of 0.7 kg/s. The inlet and outlet temperatures of the water are 10°C and 45°C, respectively, and the inlet temperature of the engine oil is 80°C. Find the heat transfer and the required heat exchanger length.

DIAGRAM

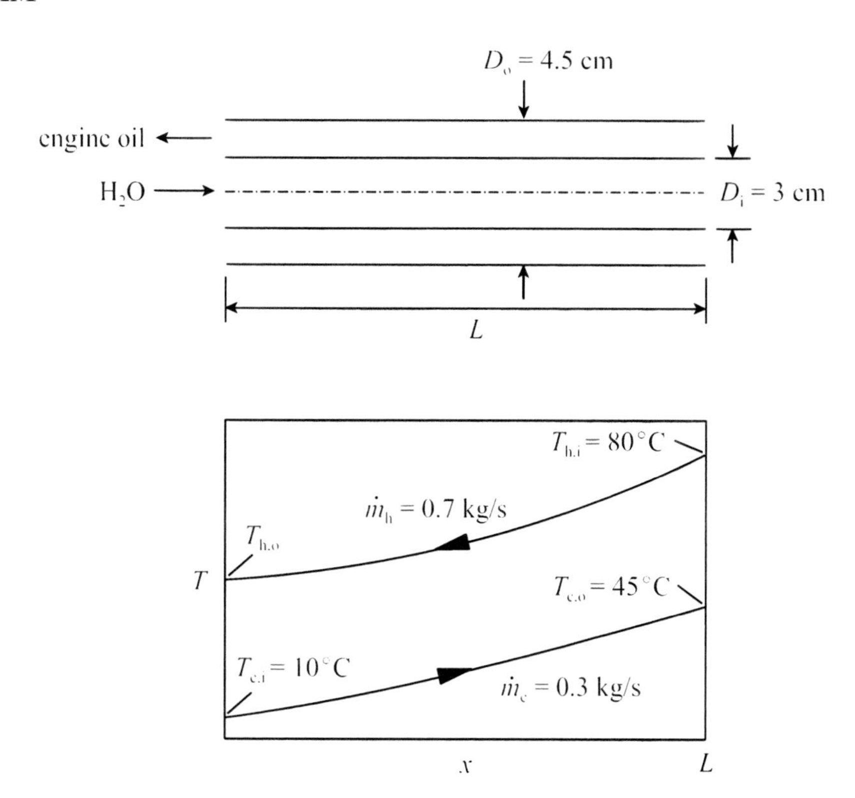

ASSUMPTIONS

1. Steady conditions.
2. Outer tube is insulated from surroundings.
3. Fully developed flow.
4. Fouling is neglected.
5. Surfaces are smooth.
6. Thermal resistance of tube wall is neglected.
7. Constant properties.

PROPERTIES

$T_{av} = (T_{c,i} + T_{c,o})/2 = (10°C + 45°C)/2 = 27.5°C = 301$ K

water: $c_{p,c} = 4179$ J/kg·K, $k = 0.614$ W/m·K, $\nu = 8.409 \times 10^{-7}$ m²/s, $Pr = 5.70$, $\mu = 838 \times 10^{-6}$ Pa·s

Assume $T_{h,o} = 55°C$. $T_{av} = (T_{h,i} + T_{h,o})/2 = (80°C + 55°C)/2 = 67.5°C = 341$ K

engine oil: $\rho = 859.3$ kg/m³, $c_{p,h} = 2080$ J/kg·K, $k = 0.139$ W/m·K, $Pr = 768$ $\mu = 5.135 \times 10^{-2}$ Pa·s, $\nu = 59.76 \times 10^{-6}$ m²/s

ANALYSIS

Knowing the inlet and outlet temperatures of the water, we can calculate the heat transfer using the energy balance,

$$q = \dot{m}_c c_{p,c}(T_{c,o} - T_{c,i})$$

$$q = (0.3 \text{ kg/s})(4179 \text{ J/kg·K})(45 - 10)°\text{C}$$

$$= 4.388 \times 10^4 \text{ W} = \underline{\underline{43.9 \text{ kW}}}$$

Applying an energy balance to the engine oil,

$$q = \dot{m}_h c_{p,h}(T_{h,i} - T_{h,o})$$

the outlet temperature of the engine oil is

$$T_{h,o} = T_{h,i} - \frac{q}{\dot{m}_h c_{p,h}}$$

$$T_{h,o} = 80°\text{C} - \frac{4.388 \times 10^4 \text{ W}}{(0.7 \text{ kg/s})(2080 \text{ J/kg·K})}$$

$$= 49.9°\text{C}$$

This value is close enough to our initial estimate that a correction of properties is not required. The required heat exchanger length is calculated from the relation

$$q = UA\Delta T_m$$

where U is the overall heat transfer coefficient, $A = \pi D_i L$, and ΔT_m is the log mean temperature difference,

$$\Delta T_m = \frac{(T_{h,i} - T_{c,o}) - (T_{h,o} - T_{c,i})}{\ln[(T_{h,i} - T_{c,o})/(T_{h,o} - T_{c,i})]}$$

$$\Delta T_m = \frac{(80 - 45)°C - (49.9 - 10)°C}{\ln[(80 - 45)°C/(49.9 - 10)°C}$$

$$= 37.4°C$$

Neglecting the thermal resistance of the tube wall, the overall heat transfer coefficient is given by the relation

$$U = \frac{1}{1/h_i + 1/h_o}$$

where h_i and h_o are the heat transfer coefficients for the inside and outside surfaces, respectively, of the inner tube. The Reynolds number for the water is

$$Re_D = \frac{4\dot{m}_c}{\pi D_i \mu}$$

$$Re_D = \frac{4(0.3 \text{ kg/s})}{\pi(0.03 \text{ m})(838 \times 10^{-6} \text{ Pa·s})}$$

$$= 1.519 \times 10^4$$

The flow is turbulent, so we use the following relation for the Nusselt number,

$$Nu_D = \frac{h_i D_i}{k} = 0.023 Re_D^{4/5} Pr^{0.4}$$

$$Nu_D = 0.023(1.519 \times 10^4)^{4/5}(5.70)^{0.4}$$

$$= 102.2$$

so the heat transfer coefficient is

$$h_i = \frac{Nu_D k}{D_i}$$

$$h_i = \frac{(102.2)(0.614 \text{ W/m·K})}{0.03 \text{ m}}$$

$$= 2092 \text{ W/m}^2\text{·K}$$

The hydraulic diameter of the annulus is

$$D_h = \frac{4A}{P} = \frac{4\pi(D_o^2 - D_i^2)/4}{\pi(D_o + D_i)} = D_o - D_i$$

$$D_h = (0.045 - 0.03) \text{ m}$$

$$= 0.015 \text{ m}$$

so the Reynolds number for the oil is

$$Re_{D_h} = \frac{u_m D_h}{\nu}$$

where u_m is the mean velocity

$$u_m = \frac{\dot{m}_h}{\rho A}$$

$$u_m = \frac{0.7 \text{ kg/s}}{(859.3 \text{ kg/m}^3)(8.836 \times 10^{-4} \text{ m}^2)}$$

$$= 0.9219 \text{ m/s}$$

The Reynolds number is

$$Re_{D_h} = \frac{u_m D_h}{\nu}$$

$$Re_{Dh} = \frac{(0.9219 \text{ m/s})(0.015 \text{ m})}{59.76 \times 10^{-6} \text{ m}^2\text{/s}}$$

$$= 231.4$$

Thus, the flow of engine oil in the annulus is laminar. The Nusselt number for fully developed laminar flow in an annulus with one surface insulated is a constant that depends only on the ratio of diameters. For $D_i/D_o = (3 \text{ cm})/(4.5 \text{ cm}) = 0.667$ and the outer surface insulated, the Nusselt number is $Nu_{Dh} \approx 5.45$. Hence, the heat transfer coefficient for the annulus is

$$h_o = \frac{Nu_{D_h} k}{D_h}$$

$$h_o = \frac{(5.45)(0.139 \text{ W/m}\cdot\text{K})}{0.015 \text{ m}}$$

$$= 50.5 \text{ W/m}^2\cdot\text{K}$$

The overall heat transfer coefficient is

$$U = \frac{1}{1/(2092\ \text{W/m}^2\cdot\text{K}) + 1/(50.5\ \text{W/m}^2\cdot\text{K})}$$

$$= 49.3\ \text{W/m}^2\cdot\text{K}$$

Finally, the required heat exchanger length is

$$L = \frac{q}{U\pi D_i \Delta T_m}$$

$$L = \frac{4.388 \times 10^4\ \text{W}}{(49.3\ \text{W/m}^2\cdot\text{K})\pi(0.030\ \text{m})(37.4°\text{C})}$$

$$= \underline{\underline{253\ \text{m}}}$$

DISCUSSION

This is an extremely long concentric tube heat exchanger. There are other types of heat exchangers with a higher surface area per unit volume that would handle the same heat transfer for the same two fluids. Nevertheless, a concentric tube heat exchanger of this length could perhaps be accommodated if the tube was formed into a circular coil. For example, a coil with $N = 50$ wraps yields a coil diameter of

$$D_{coil} = \frac{L}{\pi N}$$

$$D_{\text{coil}} = \frac{253\ \text{m}}{50\pi}$$

$$= 1.61\ \text{m}$$

Assuming that adjacent wraps are in contact with each other, the height of this coil is

$$H = ND_o$$

$$H = (50)(0.045\ \text{m})$$

$$= 2.25\ \text{m}$$

❒ ❒ ❒

PROBLEM 11.2 Surface Area Comparison for Four Different Heat Exchangers

Consider the four heat exchanger configurations: (a) concentric tube, parallel flow, (b) concentric tube, counterflow, (c) shell-and-tube, one shell pass, two tube passes, and (d) shell-and-tube, two shell passes, four tube passes. For each heat exchanger configuration, the overall heat transfer coefficient is $U = 200$ W/m^2·K, and the heat transfer is 75 kW. The inlet and outlet temperatures of the fluids are $T_{h,i} = 185\,°C$, $T_{h,o} = 120\,°C$, $T_{c,i} = 40\,°C$ and $T_{c,o} = 95\,°C$. Find the required surface area for each heat exchanger configuration.

DIAGRAM

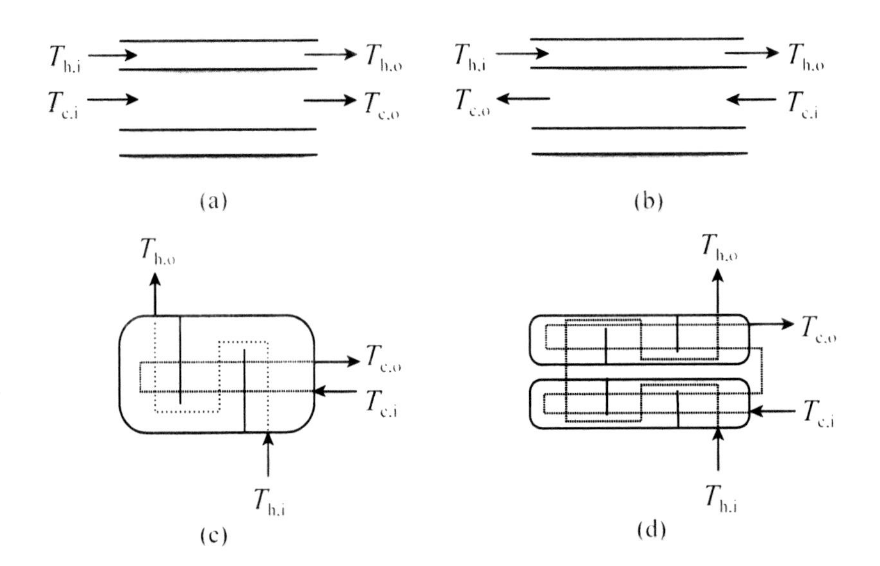

ASSUMPTIONS

1. Steady conditions.
2. Overall heat transfer coefficient is equal for all configurations.
3. Heat transfer is equal for all configurations.
4. Inlet and outlet fluid temperatures are equal for all configurations.

PROPERTIES

not applicable

ANALYSIS

For all four configurations, we will use the log mean temperature difference method of analysis. The log mean temperature difference is given by the relation

$$\Delta T_m = \frac{\Delta T_1 - \Delta T_2}{\ln(\Delta T_1/\Delta T_2)}$$

where, for parallel flow,

$$\Delta T_1 = T_{h,i} - T_{c,i} \quad , \quad \Delta T_2 = T_{h,o} - T_{c,o}$$

and for counterflow,

$$\Delta T_1 = T_{h,i} - T_{c,o} \quad , \quad \Delta T_2 = T_{h,o} - T_{c,i}$$

The second set of relations must be used to calculate the log mean temperature difference for all heat exchanger configurations except parallel flow. The heat transfer relation for a heat exchanger in terms of the log mean temperature difference is

$$q = FUA\Delta T_m$$

where, $U = 200$ W/m²·K, A is surface area and F is a correction factor for multipass and cross-flow heat exchangers. For heat exchanger (a),

$$\Delta T_1 = (185 - 40)°\text{C}$$

$$= 145°\text{C}$$

$$\Delta T_2 = (120 - 95)°\text{C}$$

$$= 25°\text{C}$$

Thus,

$$\Delta T_m = \frac{(145 - 25)°\text{C}}{\ln(145°\text{C}/25°\text{C})}$$

$$= 68.3°\text{C}$$

Noting that $F = 1$, the required surface area of heat exchanger (a) is

$$A = \frac{75 \times 10^3 \text{ W}}{(200 \text{ W/m}^2\cdot\text{K})(68.3°\text{C})}$$

$$= \underline{\underline{5.49 \text{ m}^2}}$$

For heat exchangers (b), (c) and (d),

$$\Delta T_1 = (185 - 95)°\text{C}$$

$$= 90°\text{C}$$

$$\Delta T_2 = (120 - 40)°\text{C}$$

$$= 80°\text{C}$$

Thus,

$$\Delta T_m = \frac{(90 - 80)°\text{C}}{\ln(90°\text{C}/80°\text{C})}$$

$$= 84.9°\text{C}$$

Noting again that $F = 1$, the required surface area of heat exchanger (b) is

$$A = \frac{75 \times 10^3 \text{ W}}{(200 \text{ W/m}^2\cdot\text{K})(84.9°\text{C})}$$

$$= \underline{\underline{4.42 \text{ m}^2}}$$

For heat exchanger (c), the correction factor is $F \approx 0.91$, which is typically found using a graph constructed from a somewhat long algebraic relation. Such graphs are found in any elementary heat transfer text. The required surface area of heat exchanger (c) is

$$A = \frac{75 \times 10^3 \text{ W}}{(0.91)(200 \text{ W/m}^2\cdot\text{K})(84.9°\text{C})}$$

$$= \underline{\underline{4.85 \text{ m}^2}}$$

For heat exchanger (d), the correction factor is $F \approx 0.98$, which is found as before but using a different graph. The required surface area of heat exchanger (d) is

$$A = \frac{75 \times 10^3 \text{ W}}{(0.98)(200 \text{ W/m}^2\cdot\text{K})(84.9°\text{C})}$$

$$= \underline{\underline{4.51 \text{ m}^2}}$$

To summarize, the required surface areas are:

(a) 5.49 m^2 (b) 4.42 m^2 (c) 4.85 m^2 (d) 4.51 m^2

DISCUSSION

Parallel flow is the least effective flow configuration and therefore has the highest surface area of all the heat exchangers considered. Counterflow is the most effective flow configuration, so this heat exchanger requires the least amount of surface area. The surface areas for the other two configurations lie between the parallel flow and counterflow configurations because the flow in shell-and-tube heat exchangers is neither purely parallel flow nor purely counterflow, but a complex combination of the two.

❒ ❒ ❒

PROBLEM 11.3 **Effectiveness-NTU Method for a Counterflow Concentric Tube Heat Exchanger**

A concentric tube heat exchanger operating in the counterflow mode carries glycerin in the inner tube and water in the annulus. The glycerin enters the heat exchanger at 80 °C with a mass flow rate of 4 kg/s, whereas the water enters the heat exchanger at 20 °C with a mass flow rate of 1.5 kg/s. The outside diameter of the inner tube is 8 cm, the overall heat transfer coefficient based on the outside surface area of the inner tube is 350 W/m²·K, and the heat exchanger is 50 m long. Find the heat transfer and the outlet temperatures of the glycerin and water.

DIAGRAM

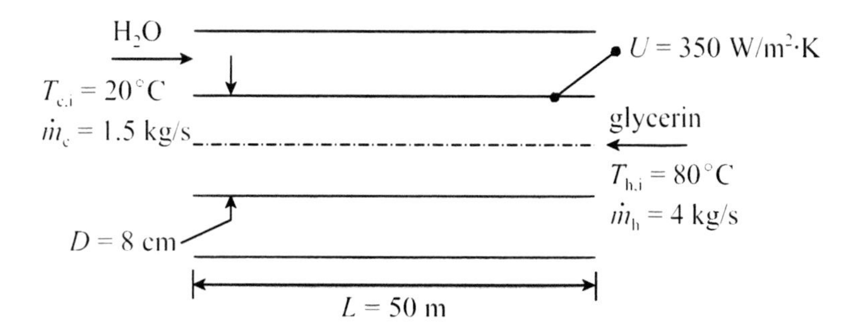

ASSUMPTIONS

1. Steady conditions.
2. Outer tube is insulated from surroundings.
3. Constant properties.

PROPERTIES

Assume $T_{h,o} = 50°C$. $T_{av} = (T_{h,i} + T_{h,o})/2 = (80°C + 50°C)/2 = 65°C = 338$ K

glycerin: $c_{p,h} = 2588$ J/kg·K

Assume $T_{c,o} = 40°C$. $T_{av} = (T_{c,i} + T_{c,o})/2 = (20°C + 40°C)/2 = 30°C = 303$ K

water: $c_{p,c} = 4178$ J/kg·K

ANALYSIS

Because only the inlet fluid temperatures are known, we use the effectiveness-*NTU* method of analysis. Heat capacity rate, C, is defined as

$$C \equiv \dot{m} c_p$$

so the heat capacity rate for the glycerin (hot fluid) is

$$C_h = \dot{m}_h c_{p,h}$$

$$= (4 \text{ kg/s})(2588 \text{ J/kg·K})$$

$$= 1.035 \times 10^4 \text{ W/K}$$

and the heat capacity rate for the water (cold fluid) is

$$C_c = \dot{m}_c c_{p,c}$$

$$= (1.5 \text{ kg/s})(4178 \text{ J/kg·K})$$

$$= 6267 \text{ W/K}$$

Thus, $C_{max} = C_h$ and $C_{min} = C_c$. The heat capacity rate ratio is defined as

$$C_r \equiv C_{min}/C_{max}$$

so we have

$$C_r = (6267 \text{ W/K})/(1.035 \times 10^4 \text{ W/K})$$

$$= 0.6055$$

The number of transfer units is defined as

$$NTU \equiv \frac{UA}{C_{min}}$$

where $A = \pi DL$ for a concentric tube heat exchanger. The value of NTU is

$$NTU = \frac{(350 \text{ W/m}^2\text{·K})\pi(0.08 \text{ m})(50 \text{ m})}{6267 \text{ W/K}}$$

$$= 0.7018$$

The effectiveness of our counterflow concentric tube heat exchanger is

$$\varepsilon = \frac{1-\exp[-NTU(1-C_r)]}{1-C_r\exp[-NTU(1-C_r)]}$$

$$\varepsilon = \frac{1 - \exp[-0.7018(1 - 0.6055)]}{1 - 0.6055\exp[-0.7018(1 - 0.6055)]}$$

$= 0.447$

The heat transfer is

$$q = \varepsilon C_{\min}(T_{h,i} - T_{c,i})$$

$q = (0.447)(6267 \text{ W/K})(80 - 20)°\text{C}$

$= 1.681 \times 10^5$ W

Now that the heat transfer is known, we can find the outlet temperatures. An energy balance on each fluid yields

$$q = C_h(T_{h,i} - T_{h,o}) = C_c(T_{c,o} - T_{c,i})$$

Hence, the outlet temperature of the glycerin is

$$T_{h,o} = T_{h,i} - q/C_h$$

$T_{h,o} = 80°\text{C} - (1.681 \times 10^5 \text{ W})/(1.035 \times 10^4 \text{ W/K})$

$= 63.8°\text{C}$

and the outlet temperature of the water is

$$T_{c,o} = T_{c,i} + q/C_c$$

$T_{c,o} = 20°\text{C} + (1.681 \times 10^5 \text{ W})/(6267 \text{ W/K})$

$= 46.8°\text{C}$

The calculated outlet temperatures differ somewhat from our initial estimates, so we correct the properties. The new average temperatures for the glycerin and water are 71.9°C and 33.4°C, respectively, which yields new specific heats, $c_{p,h} = 2626$ J/kg·K and $c_{p,c} = 4178$ J/kg·K. Because the specific heat of water is unchanged, C_{min} and NTU have the same values as before. However, $C_{max} = 1.050 \times 10^4$ W/K, so

$C_r = (6267 \text{ W/K})/(1.050 \times 10^4 \text{ W/K})$

$= 0.5966$

Thus, the heat exchanger effectiveness is

$$\varepsilon = \frac{1 - \exp[-0.7018(1 - 0.5966)]}{1 - 0.5966\exp[-0.7018(1 - 0.5966)]}$$

$= 0.448$

which is essentially identical to our first result. Thus, a correction of properties did not produce a significant increase in accuracy, so our final answers are

$$q = \underline{\underline{168 \text{ kW}}} \quad , \quad T_{h,o} = \underline{\underline{63.8°C}} \quad , \quad T_{c,o} = \underline{\underline{46.8°C}}$$

DISCUSSION

The effectiveness of our concentric tube heat exchanger, if operating in the parallel flow mode, would be

$$\varepsilon = \frac{1 - \exp[-NTU(1 + C_r)]}{1 + C_r}$$

$$\varepsilon = \frac{1 - \exp[-0.7018(1 + 0.5966)]}{1 + 0.5966}$$

$= 0.422$

and the heat transfer would be

$$q = \varepsilon C_{\min}(T_{h,i} - T_{c,i})$$

$$q = (0.422)(6267 \text{ W/K})(80 - 20)°C$$

$= 1.587 \times 10^5 \text{ W} = 159 \text{ kW}$

This heat transfer is 5.4 percent lower than that for the counterflow mode.

❒ ❒ ❒

PROBLEM 11.4 Automobile Radiator

An automobile radiator is a cross-flow heat exchanger with both fluids unmixed. The coolant, a 50 percent by volume mixture of ethylene glycol and water, flows through the radiator tubes at a mass flow rate of 0.25 kg/s. Atmospheric air flows across the finned tubes at a mass flow rate of 1.0 kg/s. The inlet temperatures of the coolant and air are 90°C and 27°C, respectively. If the overall heat transfer coefficient is 225 W/m^2·K and the air exits the radiator at 50°C, what is the surface area?

DIAGRAM

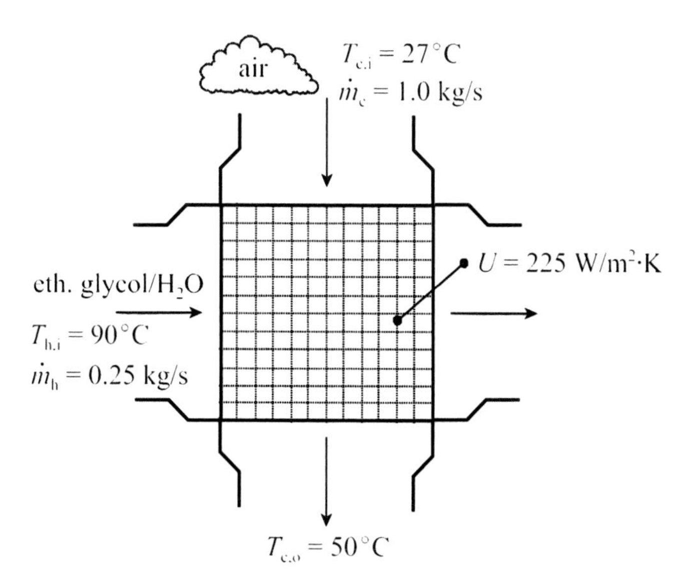

ASSUMPTIONS

1. Steady conditions.
2. Radiator is insulated from surroundings.
3. Constant properties.

PROPERTIES

$T_{av} = (T_{c,i} + T_{c,o})/2 = (27°C + 50°C)/2 = 38.5°C = 312$ K

atmospheric air: $c_{p,c} = 1008$ J/kg·K

Assume $T_{h,o} = 60°C$. $T_{av} = (T_{h,i} + T_{h,o})/2 = (90°C + 60°C)/2 = 75°C = 348$ K

ethylene glycol/H_2O: $c_{p,h} = 3493$ J/kg·K

ANALYSIS

The inlet and outlet temperatures of the air are known, so we can calculate the heat transfer by writing an energy balance for the air,

$$q = \dot{m}_c c_{p,c}(T_{c,o} - T_{c,i})$$

$q = (1.0 \text{ kg/s})(1008 \text{ J/kg·K})(50 - 27)°C$

$= 2.318 \times 10^4$ W

Using the same energy balance for the ethylene glycol and water mixture, the outlet temperature of the coolant is

$$T_{h,o} = T_{h,i} - \frac{q}{\dot{m}_h c_{p,h}}$$

$$T_{h,o} = 90°\text{C} - \frac{2.318 \times 10^4 \text{ W}}{(0.25 \text{ kg/s})(3493 \text{ J/kg·K})}$$

$= 63.5°$C

which is very close to our initial estimate. The relation for log mean temperature difference is

$$\Delta T_m = \frac{\Delta T_1 - \Delta T_2}{\ln(\Delta T_1/\Delta T_2)}$$

where, for counterflow conditions,

$$\Delta T_1 = T_{h,i} - T_{c,o} \quad , \quad \Delta T_2 = T_{h,o} - T_{c,i}$$

Thus,

$\Delta T_1 = (90 - 50)°$C

$= 40°$C

$\Delta T_2 = (63.5 - 27)°$C

$= 36.5°$C

The log mean temperature difference is

$$\Delta T_m = \frac{(40 - 36.5)°\text{C}}{\ln(40°\text{C}/36.5°\text{C})}$$
$= 38.2°$C

The heat transfer relation for a heat exchanger in terms of the log mean temperature difference is

$$q = FUA\Delta T_m$$

where, $U = 225$ W/m^2·K, A is surface area and F is a correction factor for multipass and cross-flow heat exchangers. Using a graph, $F \approx 0.96$. Hence, the surface area is

$$A = \frac{q}{FU\Delta T_m}$$

$$A = \frac{2.318 \times 10^4 \text{ W}}{(0.96)(225 \text{ W/m}^2\cdot\text{K})(38.2\,°\text{C})}$$

$$= \underline{\underline{2.81 \text{ m}^2}}$$

DISCUSSION

This problem could have been worked using the effectiveness-*NTU* method of analysis.

❒ ❒ ❒

PROBLEM 11.5 Shell-and-Tube Heat Exchanger for Cooling Lubricating Oil

A shell-and-tube heat exchanger is utilized for cooling lubricating oil in a marine diesel engine. The oil temperature must be reduced from 115°C to 70°C using 18°C seawater. The heat transfer is 210 kW, and the overall heat transfer coefficient is 180 W/m²·K. If the outlet temperature of the seawater is 52°C, what is the required surface area if the heat exchanger has one shell pass and four tube passes?

DIAGRAM

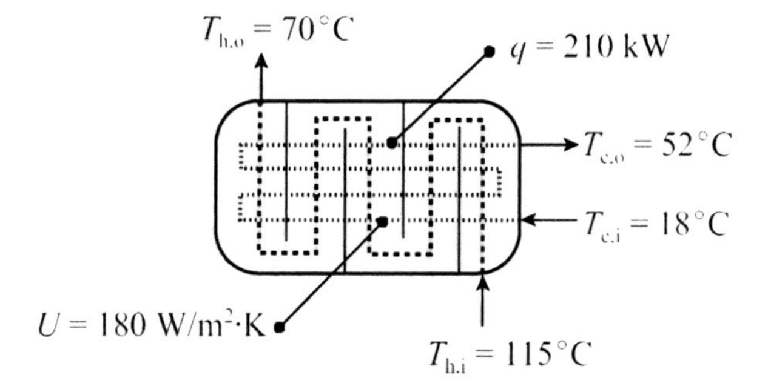

ASSUMPTIONS

1. Steady conditions.
2. Heat exchanger is insulated from surroundings.

PROPERTIES

not applicable

ANALYSIS

Because the inlet and outlet fluid temperatures are known, we use the log mean temperature difference method of analysis. The log mean temperature difference is given by the relation

$$\Delta T_m = \frac{\Delta T_1 - \Delta T_2}{\ln(\Delta T_1/\Delta T_2)}$$

where, for counterflow,

$$\Delta T_1 = T_{h,i} - T_{c,o} \quad , \quad \Delta T_2 = T_{h,o} - T_{c,i}$$

The heat transfer relation for a heat exchanger in terms of the log mean temperature difference is

$$q = FUA\Delta T_m$$

where, U = 180 W/m²·K, A is surface area and F is a correction factor for multipass and cross-flow heat exchangers. The temperature differences are

$$\Delta T_1 = (115 - 52)°\text{C}$$

$$= 63°\text{C}$$

$$\Delta T_2 = (70 - 18)°\text{C}$$

$$= 52°\text{C}$$

Thus,

$$\Delta T_m = \frac{(63 - 52)°\text{C}}{\ln(63°\text{C}/52°\text{C})}$$

$$= 57.3°\text{C}$$

Using the given inlet and outlet fluid temperatures and a graph of the correction factor for one shell pass and any multiple of two tube passes, we find that $F \approx 0.92$. Thus, the required surface area is

$$A = \frac{q}{FU\Delta T_m}$$

$$A = \frac{(210 \times 10^3 \text{ W})}{(0.92)(180 \text{ W/m}^2\cdot\text{K})(57.3°\text{C})}$$

$= \underline{\underline{22.1 \text{ m}^2}}$

DISCUSSION

As with Problem 11.4, this problem could have been worked using the effectiveness-*NTU* method of analysis. Note that neither the mass flow rates nor the thermal properties of the oil and seawater are required to solve this problem because the inlet and outlet temperatures of both fluids are given, and the log mean temperature difference method of analysis does not require these quantities.

❐ ❐ ❐

PROBLEM 11.6 Miniature Shell-and-Tube Heat Exchanger for Cooling Blood

In a medical research application, a miniature shell-and-tube heat exchanger cools blood from 36.2°C to 28.0°C using 15°C deionized water. The mass flow rates of the blood and water are 0.025 kg/s and 0.13 kg/s, respectively. The heat exchanger consists of 24 thin-walled stainless steel tubes with a diameter of 2.8 mm. If the blood makes two passes through the tubes and the water makes a single pass through the shell, what length of heat exchanger is required if the overall heat transfer coefficient is 775 W/m²·K? Assume that blood has the same thermal properties as water.

DIAGRAM

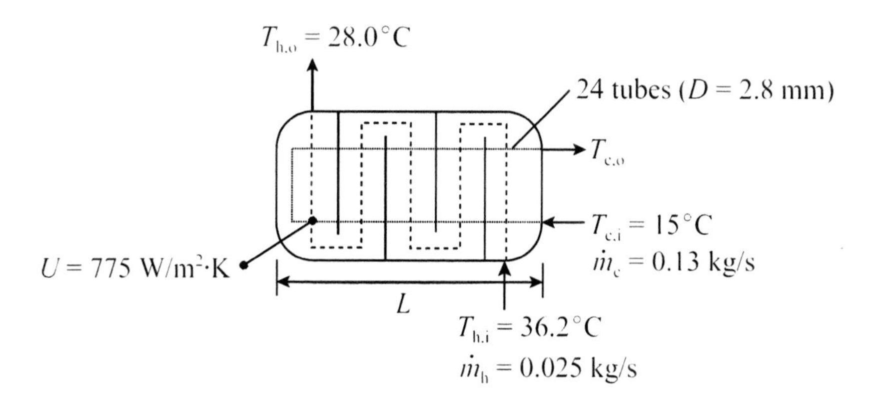

ASSUMPTIONS

1. Steady conditions.
2. Heat exchanger is insulated from surroundings.
3. Blood has the same thermal properties as water.
4. Lengths of tube bends are neglected.
5. Constant properties.

PROPERTIES

$T_{av} = (T_{h,i} + T_{h,o})/2 = (36.2°C + 28.0°C)/2 = 32.1°C = 305$ K

blood (assume H_2O): $c_{p,h} = 4178$ J/kg·K

Assume $T_{c,o} = 17°C$. $T_{av} = (T_{c,i} + T_{c,o})/2 = (15°C + 17°C)/2 = 16.0°C = 289$ K

water: $c_{p,c} = 4185$ J/kg·K

ANALYSIS

The inlet and outlet temperatures of the blood are known, so we can calculate the heat transfer by writing an energy balance for the air,

$$q = \dot{m}_h c_{p,h}(T_{h,i} - T_{h,o})$$

$$q = (0.025 \text{ kg/s})(4178 \text{ J/kg·K})(36.2 - 28.0)°C$$

$$= 856.5 \text{ W}$$

Using the same energy balance for the deionized water, the outlet temperature of the water is

$$T_{c,o} = T_{c,i} + \frac{q}{\dot{m}_c c_{p,c}}$$

$$T_{c,o} = 15°C + \frac{856.5 \text{ W}}{(0.13 \text{ kg/s})(4185 \text{ J/kg·K})}$$

$$= 16.6°C$$

which is very close to our initial estimate. The relation for log mean temperature difference is

$$\Delta T_m = \frac{\Delta T_1 - \Delta T_2}{\ln(\Delta T_1/\Delta T_2)}$$

where, for counterflow conditions,

$$\Delta T_1 = T_{h,i} - T_{c,o} \quad , \quad \Delta T_2 = T_{h,o} - T_{c,i}$$

Thus,

$$\Delta T_1 = (36.2 - 16.6)°C$$

$$= 19.6°C$$

$$\Delta T_2 = (28.0 - 15)°C$$

$$= 13°C$$

The log mean temperature difference is

$$\Delta T_m = \frac{(19.6 - 13)°C}{\ln(19.6°C/13°C)}$$

$$= 16.1°C$$

The heat transfer relation for a heat exchanger in terms of the log mean temperature difference is

$$q = FUA\Delta T_m$$

where, $U = 775$ W/m^2·K, A is surface area and F is a correction factor for multipass and cross-flow heat exchangers. Using a graph for a one-shell pass, two-tube pass heat exchanger, we find that $F \approx 0.99$. Hence, the surface area is

$$A = \frac{q}{FU\Delta T_m}$$

$$A = \frac{856.5 \text{ W}}{(0.99)(775 \text{ W/m}^2\cdot\text{K})(16.1°C)}$$

$$= 0.0693 \text{ m}^2$$

Neglecting the lengths of the tube bends, the total surface area of the tube is

$$A = 2N\pi DL$$

where the multiplier 2 is for the number of tube passes, and N is the number of tubes. Thus, the heat exchanger length is

$$L = \frac{0.0693 \text{ m}^2}{2(24)\pi(2.8 \times 10^{-3} \text{ m})}$$

$$= 0.164 \text{ m} = \underline{\underline{16.4 \text{ cm}}}$$

DISCUSSION

Our result affirms that the heat exchanger is miniature. Most shell-and-tube heat exchangers, particularly off-the-shelf units, are substantially larger than 16 cm in length. This miniature heat exchanger would most likely be a custom made unit or a product manufactured by a company that specializes in miniature heat exchangers.

PROBLEM 11.7 Water-Methane Cross-Flow Heat Exchanger

Methane flows across the tubes of a cross-flow heat exchanger at a mass flow rate of 0.05 kg/s, and water flows in the tubes at a mass flow rate of 1.7 kg/s. The tubes are baffled such that the methane cannot mix in the transverse direction. For a heat transfer of 12 kW and an overall heat transfer coefficient of 85 W/m²·K, find the required surface area if the inlet temperatures of the methane and water are 150°C and 30°C, respectively.

DIAGRAM

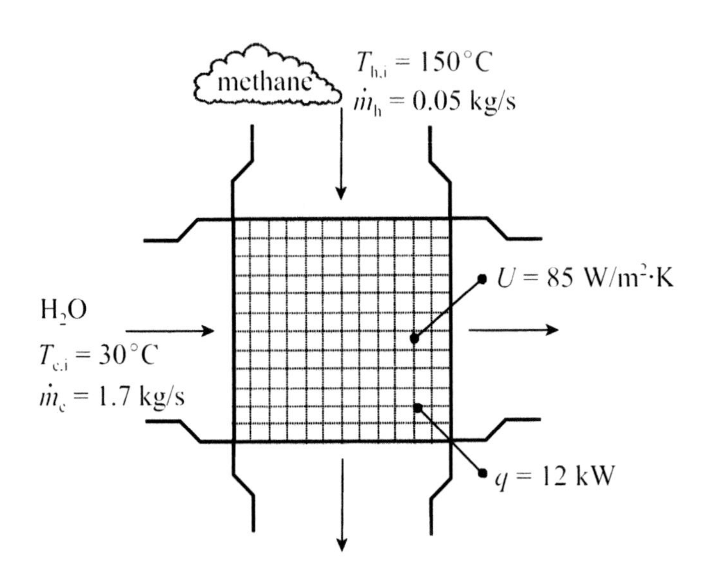

ASSUMPTIONS

1. Steady conditions.
2. Heat exchanger is insulated from surroundings.
3. Methane is at atmospheric pressure.
4. Constant properties.

PROPERTIES

Assume $T_{h,o} = 60°C$. $T_{av} = (T_{h,i} + T_{h,o})/2 = (150°C + 60°C)/2 = 105°C = 378$ K

atmospheric methane: $c_{p,h} = 2467$ J/kg·K

Assume $T_{c,o} = 35°C$. $T_{av} = (T_{c,i} + T_{c,o})/2 = (30°C + 35°C)/2 = 32.5°C = 306$ K

water: $c_{p,c} = 4178$ J/kg·K

ANALYSIS

For this problem we use the effectiveness-*NTU* method of analysis. Heat capacity rate, C, is defined as

$$C \equiv \dot{m} c_p$$

so the heat capacity rate for the methane (hot fluid) is

$$C_\mathrm{h} = \dot{m}_\mathrm{h} c_\mathrm{p,h}$$

$$= (0.05 \text{ kg/s})(2467 \text{ J/kg·K})$$

$$= 123.4 \text{ W/K}$$

and the heat capacity rate for the water (cold fluid) is

$$C_\mathrm{c} = \dot{m}_\mathrm{c} c_\mathrm{p,c}$$

$$= (1.7 \text{ kg/s})(4178 \text{ J/kg·K})$$

$$= 7103 \text{ W/K}$$

Energy balances on the methane and water can be used to calculate the outlet temperatures and, if necessary, correct thermal properties.

$$q = C_h(T_{h,i} - T_{h,o}) = C_c(T_{c,o} - T_{c,i})$$

The outlet temperature of the methane is

$$T_\mathrm{h,o} = T_\mathrm{h,i} - q/C_\mathrm{h}$$

$$= 150°\mathrm{C} - (12 \times 10^3 \text{ W})/(123.4 \text{ W/K})$$

$$= 52.8°\mathrm{C}$$

and the outlet temperature of the water is

$$T_\mathrm{c,o} = T_\mathrm{c,i} + q/C_\mathrm{c}$$

$$= 30°\mathrm{C} + (12 \times 10^3 \text{ W})/(7103 \text{ W/K})$$

$$= 31.7°\mathrm{C}$$

The calculated temperatures are sufficiently close to our initial estimates so that no thermal property corrections are required. Based on the relative sizes of the heat capacity rates, we have $C_\mathrm{max} = C_\mathrm{c}$ and $C_\mathrm{min} = C_\mathrm{h}$. The heat capacity rate ratio is defined as

$$C_r \equiv C_{min}/C_{max}$$

so

C_r = (123.4 W/K)/(7103 W/K)

= 0.0174

The heat transfer may be expressed by the relation

$$q = \varepsilon C_{min}(T_{h,i} - T_{c,i})$$

from which the heat exchanger effectiveness, ε, may be found.

$$\varepsilon = \frac{12 \times 10^3 \text{ W}}{(123.4 \text{ W/K})(150 - 30)°\text{C}}$$

= 0.8104

The relationship between ε and *NTU* for a cross-flow heat exchanger with both fluids unmixed is

$$\varepsilon = 1 - \exp\left[\left(\frac{1}{C_r}\right)(NTU)^{0.22}\left\{\exp\left[-C_r(NTU)^{0.78}\right] - 1\right\}\right]$$

Numerically solving this relation for *NTU*,

NTU = 1.685

The number of transfer units is defined as

$$NTU \equiv \frac{UA}{C_{min}}$$

so the surface area is

$$A = \frac{C_{min} NTU}{U}$$

$$= \frac{(123.4 \text{ W/K})(1.685)}{85 \text{ W/m}^2\cdot\text{K}}$$

$$= \underline{\underline{2.45 \text{ m}^2}}$$

DISCUSSION

The log mean temperature difference method of analysis can also be used to find the surface area.

$$\Delta T_1 = T_{h,i} - T_{c,o} \quad , \quad \Delta T_2 = T_{h,o} - T_{c,i}$$

Thus,

$$\Delta T_1 = (150 - 31.7)°\text{C}$$

$$= 118.3°\text{C}$$

$$\Delta T_2 = (52.8 - 30\,)°\text{C}$$

$$= 22.8°\text{C}$$

The log mean temperature difference is

$$\Delta T_\text{m} = \frac{(118.3 - 22.8)°\text{C}}{\ln(118.3°\text{C}/22.8°\text{C})}$$

$$= 58.0°\text{C}$$

The heat transfer relation is

$$q = FUA\Delta T_m$$

where $F \approx 1.0$. Hence, the surface area is

$$A = \frac{12 \times 10^3\ \text{W}}{(1.0)(85\ \text{W/m}^2\cdot\text{K})(58.0°\text{C})}$$

$$= 2.43\ \text{m}^2$$

which, within numerical roundoff, is in excellent agreement with our first result.

❒ ❒ ❒

PROBLEM 11.8 Shell-and-Tube Condenser

Saturated water vapor at 90 kPa is condensed in a shell-and-tube heat exchanger with one shell pass and two tube passes. There are 140 thin-walled stainless steel tubes with a diameter of 12 mm and a length of 1.75 m per pass. Cooling water, which enters at 20°C, flows through the tubes at a total mass flow rate of 30 kg/s. If the heat transfer coefficient for condensation on the outside surfaces of the tubes is 14 kW/m^2·K, find the heat transfer, the outlet temperature of the cooling water and the rate of condensation.

DIAGRAM

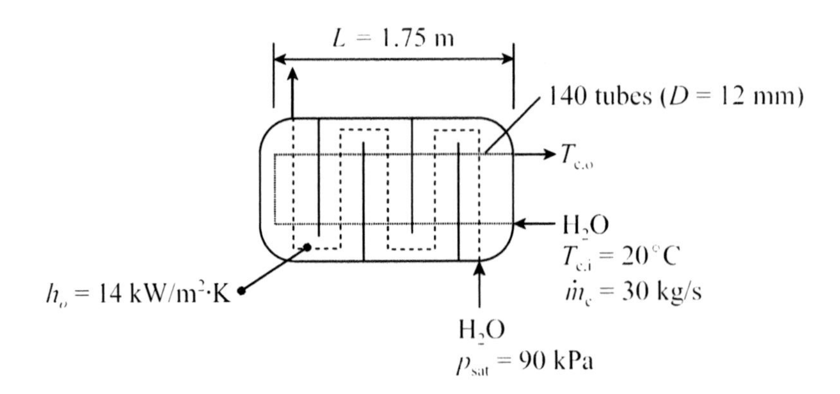

ASSUMPTIONS

1. Steady conditions.
2. Heat exchanger is insulated from surroundings.
3. Fully developed flow in tubes.
4. One-dimensional steady conduction in tube wall.
5. Thermal resistance of tube wall is neglected.
6. Lengths of tube bends are neglected.
7. Fouling is neglected.
8. Constant properties.

PROPERTIES

saturated water vapor: $p_{sat} = 90$ kPa, $T_{sat} = T_{h,i} = T_{h,o} = 97°C = 370$ K

$h_{fg} = 2265$ kJ/kg

Assume $T_{c,o} = 60°C$. $T_{av} = (T_{c,i} + T_{c,o})/2 = (20°C + 60°C)/2 = 40°C = 313$ K

water: $\rho = 991.9$ kg/m^3, $c_{p,c} = 4179$ J/kg·K, $k = 0.632$ W/m·K, $Pr = 4.34$, $\mu = 657 \times 10^{-6}$ Pa·s, $\nu = 6.624 \times 10^{-7}$ m^2/s

ANALYSIS

In order to find the overall heat transfer coefficient, we must find the heat transfer coefficient for the inside surface of the tubes. Neglecting the thermal resistance of the tube wall, the overall heat transfer coefficient is given by the relation

$$U = \frac{1}{1/h_i + 1/h_o}$$

where h_i and h_o are the heat transfer coefficients for the inside and outside surfaces, respectively. The mass flow rate per tube is

$$\dot{m} = \dot{m}_c/140$$

$$= (30 \text{ kg/s})/140$$

$$= 0.2143 \text{ kg/s}$$

so the Reynolds number for the flow of water in the tubes is

$$Re_D = \frac{4\dot{m}}{\pi D \mu}$$

$$Re_D = \frac{4(0.2143 \text{ kg/s})}{\pi(0.012 \text{ m})(657 \times 10^{-6} \text{ Pa·s})}$$

$$= 3.461 \times 10^4$$

The flow is turbulent, so we use the following relation for the Nusselt number,

$$Nu_D = \frac{h_i D}{k} = 0.023 Re_D^{4/5} Pr^{0.4}$$

$$Nu_D = 0.023(3.461 \times 10^4)^{4/5}(4.34)^{0.4}$$

$$= 177.0$$

so the heat transfer coefficient is

$$h_i = \frac{Nu_D k}{D}$$

$$h_i = \frac{(177.0)(0.632 \text{ W/m·K})}{0.012 \text{ m}}$$

$$= 9322 \text{ W/m}^2\text{·K}$$

Hence, overall heat transfer coefficient is

$$U = \frac{1}{1/(9322 \text{ W/m}^2\text{·K}) + 1/(14 \times 10^3 \text{ W/m}^2\text{·K})}$$

$$= 5596 \text{ W/m}^2\text{·K}$$

Neglecting the tube bends, the total surface area of the tubes is

$$A = 2N\pi DL$$

$A = 2(140)\pi(0.012 \text{ m})(1.75 \text{ m})$

$= 18.47 \text{ m}^2$

For a heat exchanger in which a phase change occurs, i.e., for a condenser or evaporator, the fluid that is not changing phase is the minimum heat capacity fluid. Thus,

$$C_{\text{min}} = \dot{m}_c c_{p,c}$$

$C_{\text{min}} = (30 \text{ kg/s})(4179 \text{ J/kg·K})$

$= 1.254 \times 10^5 \text{ W/K}$

The number of transfer units is defined as

$$NTU \equiv \frac{UA}{C_{\text{min}}}$$

$$NTU = \frac{(5596 \text{ W/m}^2\text{·K})(18.47 \text{ m}^2)}{1.254 \times 10^5 \text{ W/K}}$$

$= 0.8244$

Because $C_{\text{max}} \to \infty$ for a condenser, $C_r = C_{\text{min}}/C_{\text{max}} \to 0$. The relation for heat exchanger effectiveness for this special case is

$$\varepsilon = 1 - \exp(-NTU)$$

$\varepsilon = 1 - \exp(-0.8244)$

$= 0.5615$

The heat transfer is

$$q = \varepsilon C_{\text{min}}(T_{h,i} - T_{c,i})$$

$q = (0.5615)(1.254 \times 10^5 \text{ W/K})(97 - 20)°\text{C}$

$= 5.422 \times 10^6 \text{ W} = \underline{\underline{5.42 \text{ MW}}}$

Now that the heat transfer has been calculated, we find the outlet temperature of the cooling water.

$T_{c,o} = T_{c,i} + q/C_{\text{min}}$

$= 20°C + (5.422 \times 10^6 \text{ W})/(1.253 \times 10^5 \text{ W/K})$

$= \underline{\underline{63.3°C}}$

The calculated outlet temperature is sufficiently close to our initial estimate that no property corrections are needed. Finally, the condensation rate of steam is

$$\dot{m}_h = \frac{q}{h_{fg}}$$

$\dot{m}_h = (5.422 \times 10^6 \text{ W})/(2265 \times 10^3 \text{ J/kg})$

$= \underline{\underline{2.39 \text{ kg/s}}}$

DISCUSSION

As shown in the graph, steam condensation rate increases with cooling water flow rate.

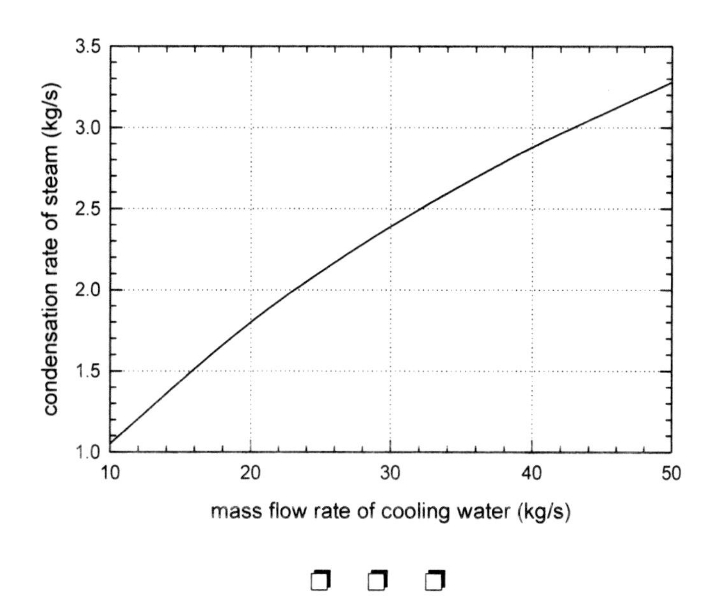

❒ ❒ ❒

PROBLEM 11.9 Exhaust Stack

A tall stack carries exhaust gas from a copper smelting operation. Exhaust gas enters the stack at a temperature of 440°C, while a steady 8 m/s wind blows across it. The tapered, 76 m high stack is constructed of concrete (k_c = 1.9 W/m·K) and has a mean inside and outside diameter of 5.0 m and 6.8 m, respectively. If the temperature of the outdoor air is 10°C, and the mean velocity of the exhaust gas in the stack is 2.5 m/s, what is the temperature of the exhaust gas as it exits the stack?

DIAGRAM

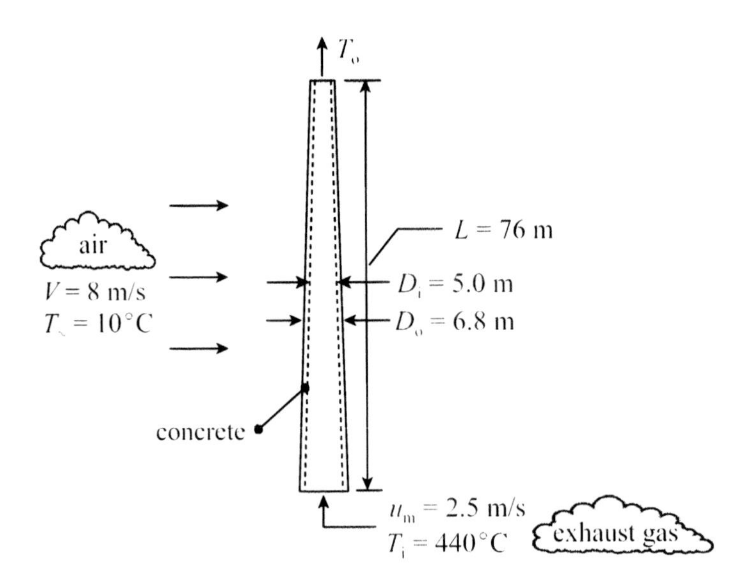

ASSUMPTIONS

1. Steady conditions.
2. Approximate tapered stack as a hollow cylinder with mean diameters.
3. Fully developed flow in stack.
4. One-dimensional steady conduction in stack wall.
5. Outside surface of stack is smooth.
6. Approximate exhaust gas properties with those of atmospheric air.
7. Surfaces of stack are isothermal.
8. Radiation from the stack is neglected.
9. Constant properties.

PROPERTIES

Assume $T_s = 30°C$. $T_f = (T_s + T_\infty)/2 = (30 + 10)°C/2 = 20°C = 293$ K

atmospheric air: $k = 0.0257$ W/m·K, $\nu = 15.27 \times 10^{-6}$ m^2/s, $Pr = 0.709$

Assume $T_o = 400°C$. $T_{av} = (T_i + T_o)/2 = (440 + 400)/2 = 420°C = 693$ K

exhaust gas (air): $\rho = 0.5028$ kg/m^3, $c_p = 1073$ J/kg·K, $k = 0.0520$ W/m·K, $Pr = 0.694$, $\mu = 336.5 \times 10^{-7}$ Pa·s, $\nu = 67.00 \times 10^{-6}$ m^2/s

concrete: $k_c = 1.9$ W/m·K

ANALYSIS

The stack is a cross-flow heat exchanger of sorts. As cold outdoor air (wind) flows over the stack, hot exhaust gas flows through it. These two fluids exchange heat across the concrete stack wall. But the standard heat exchanger relations will not be used here. Instead, the relations for external and internal forced convection will be used to find the outlet temperature of the exhaust gas.

We need to calculate the overall heat transfer coefficient, which means that we must find the heat transfer coefficient for the inside and outside surfaces of the stack. First, we analyze the outside surface conditions. The Reynolds number is

$$Re_D = \frac{VD_o}{\nu}$$

$$Re_D = \frac{(8 \text{ m/s})(6.8 \text{ m})}{15.27 \times 10^{-6} \text{ m}^2/\text{s}}$$

$$= 3.563 \times 10^6$$

For the Nusselt number, we use the following relation, which is recommend for all $Re_D Pr > 0.2$,

$$Nu_D = \frac{h_o D_o}{k} = 0.3 + \frac{0.62 Re_D^{1/2} Pr^{1/3}}{\left[1 + (0.4/Pr)^{2/3}\right]^{1/4}} \left[1 + \left(\frac{Re_D}{282{,}000}\right)^{5/8}\right]^{4/5}$$

$$Nu_D = 0.3 + \frac{0.62(3.563 \times 10^6)^{1/2}(0.709)^{1/3}}{[1 + (0.4/0.709)^{2/3}]^{1/4}} \left[1 + \left(\frac{3.563 \times 10^6}{282{,}000}\right)^{5/8}\right]^{4/5}$$

$$= 3781$$

Thus, the heat transfer coefficient is

$$h_o = \frac{Nu_D k}{D_o}$$

$$h_o = \frac{(3781)(0.0257 \text{ W/m·K})}{6.8 \text{ m}}$$

$$= 14.3 \text{ W/m}^2\text{·K}$$

Now we analyze conditions for the inside surface of the stack. The Reynolds number is

$$Re_D = \frac{u_m D_i \rho}{\mu}$$

$$Re_D = \frac{(2.5 \text{ m/s})(5.0 \text{ m})(0.5028 \text{ kg/m}^3)}{336.5 \times 10^{-7} \text{ Pa·s}}$$

$$= 1.868 \times 10^5$$

A typical value for the roughness of concrete is $\varepsilon = 1.5$ mm. The friction factor is

$$f = \frac{0.25}{\left[\log\left(\frac{\varepsilon}{3.7D_i} + \frac{5.74}{Re_D^{0.9}}\right)\right]^2}$$

$$f = \frac{0.25}{\left[\log\left(\frac{1.5 \times 10^{-3} \text{ m}}{3.7(5.0 \text{ m})} + \frac{5.74}{(1.868 \times 10^5)^{0.9}}\right)\right]^2}$$

$$= 0.0179$$

and the average Nusselt number is

$$Nu_D = \frac{h_i D_i}{k} = \frac{(f/8)(Re_D - 1000)Pr}{1 + 12.7(f/8)^{1/2}(Pr^{2/3} - 1)}$$

$$Nu_D = \frac{(0.0179/8)(1.868 \times 10^5 - 1000)(0.694)}{1 + 12.7(0.0179/8)^{1/2}(0.694^{2/3} - 1)}$$

$$= 331.6$$

The heat transfer coefficient is

$$h_i = \frac{Nu_D k}{D_i}$$

$$h_i = \frac{(331.6)(0.0520 \text{ W/m·K})}{5.0 \text{ m}}$$

$$= 3.45 \text{ W/m}^2\text{·K}$$

Now that both heat transfer coefficients have been calculated, we calculate the overall heat transfer coefficient. Arbitrarily, we base the calculation on the inside surface area of the stack.

$$U_i = \frac{1}{\frac{1}{h_i} + D_i \frac{\ln(D_o/D_i)}{k_c} + \frac{D_i}{D_o}\frac{1}{h_o}}$$

$$U_i = \cfrac{1}{\cfrac{1}{3.45\ \text{W/m}^2\cdot\text{K}} + (5.0\ \text{m})\cfrac{\ln(6.8\ \text{m}/5.0\ \text{m})}{1.9\ \text{W/m}\cdot\text{K}} + \cfrac{5.0\ \text{m}}{6.8\ \text{m}}\cfrac{1}{14.3\ \text{W/m}^2\cdot\text{K}}}$$

$$= 1.99\ \text{W/m}^2\cdot\text{K}$$

An energy balance for the exhaust gas is

$$\frac{T_\infty - T_o}{T_\infty - T_i} = \exp\left(-\frac{U_i A_i}{\dot{m} c_p}\right)$$

where $\dot{m}$ is mass flow rate of the exhaust gas,

$$\dot{m} = \rho u_m \pi D_i^2/4$$

$$\dot{m} = (0.5028\ \text{kg/m}^3)(2.5\ \text{m/s})\pi(5.0\ \text{m})^2/4$$

$$= 24.68\ \text{kg/s}$$

Solving the energy balance for the outlet temperature,

$$T_o = T_\infty - (T_\infty - T_i)\exp\left(-\frac{U_i 2\pi D_i L}{\dot{m} c_p}\right)$$

$$T_o = 10°\text{C} - (10°\text{C} - 440°\text{C})\exp\left(-\frac{(1.99\ \text{W/m}^2\cdot\text{K})2\pi(5.0\ \text{m})(76\ \text{m})}{(24.68\ \text{kg/s})(1073\ \text{J/kg}\cdot\text{K})}\right)$$

$$= \underline{\underline{369°\text{C}}}$$

DISCUSSION

Our calculated outlet temperature of the exhaust gas is 31 °C lower than the initial estimate, but a correction of properties would yield only a modest improvement in accuracy because the properties of air do not change significantly over this temperature range.

The heat transfer through the stack wall is

$$q = \dot{m} c_p (T_i - T_o)$$

$$q = (24.68\ \text{kg/s})(1073\ \text{J/kg}\cdot\text{K})(440 - 369)°\text{C}$$

$$= 1.880 \times 10^6\ \text{W} = 1.88\ \text{MW}$$

Of course, heat is also lost to the surroundings in the exhaust gas as it exits the stack. It is left as an exercise for the student to calculate the mean surface temperatures of the stack and to determine if a correction of properties for the outdoor air is warranted.

❒ ❒ ❒

PROBLEM 11.10 Shell-and-Tube Heat Exchanger for Cooling Engine Oil

A shell-and-tube heat exchanger with one shell pass and two tube passes is used to cool engine oil from 110°C to 30°C. The heat exchanger has 40 thin-walled tubes with a diameter of 1.6 cm and a length of 4.5 m per pass. Water enters the tubes at 15°C, and the overall heat transfer coefficient is 310 W/m²·K. If the mass flow rate of the engine oil is 0.6 kg/s, find the required mass flow rate of cooling water.

DIAGRAM

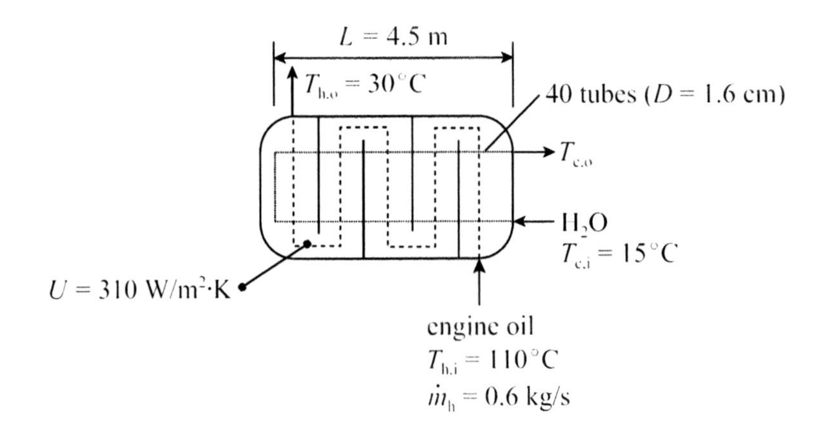

ASSUMPTIONS

1. Steady conditions.
2. Heat exchanger is insulated from surroundings.
3. Lengths of tube bends are neglected.
4. Constant properties.

PROPERTIES

$T_{av} = (T_{h,i} + T_{h,o})/2 = (110 + 30)°C/2 = 70°C = 343$ K

engine oil: $c_{p,h} = 2089$ J/kg·K

Assume $T_{c,o} = 40°C$. $T_{av} = (T_{c,i} + T_{c,o})/2 = (15 + 40)°C/2 = 27.5°C = 301$ K

water: $c_{p,c} = 4179$ J/kg·K

ANALYSIS

The heat transfer can be calculated by applying an energy balance to the engine oil,

$$q = \dot{m}_h c_{c,h}(T_{h,i} - T_{h,o})$$

q = (0.6 kg/s)(2089 J/kg·K)(110 – 30)°C

= 1.003 × 10^5 W = 100 kW

Neglecting the lengths of tube bends, the total surface area is

$$A = 2N\pi DL$$

A = 2(40)π(0.016 m)(4.5 m)

= 18.10 m^2

Because neither the mass flow rate nor the outlet temperature of the water is known, we use the effectiveness-*NTU* method of analysis. The heat transfer is given by the relation

$$q = \varepsilon C_{\min}(T_{h,i} - T_{c,i})$$

where, for a shell-and-tube heat exchanger with one shell pass and two tube passes, the effectiveness is given by the relation

$$\varepsilon = 2\left\{1 + C_r + (1 + C_r^2)^{1/2} \cdot \frac{1 + \exp\left[-NTU(1 + C_r^2)^{1/2}\right]}{1 - \exp\left[-NTU(1 + C_r^2)^{1/2}\right]}\right\}^{-1}$$

where the heat capacity ratio is

$$C_r = \frac{C_{\min}}{C_{\max}}$$

and the number of transfer units is

$$NTU = \frac{UA}{C_{\min}}$$

The heat capacity for the engine oil (hot fluid) and water (cold fluid), respectively, are given by the relations

$$C_h = \dot{m}_h c_{p,h} \quad , \quad C_c = \dot{m}_c c_{p,c}$$

where one of the fluids will have the minimum heat capacity ($C_{\min}$), and one will have the maximum heat capacity ($C_{\max}$). In order to find the required mass flow rate of cooling water, the foregoing relations must be solved simultaneously. An iterative approach may be used in which a mass flow rate for the cooling water is estimated, yielding values for C_c, C_{min}, C_r, NTU and ε. The calculated value of ε is substituted into the relation for heat transfer, and the calculated value of q is compared with the known value. The cooling water mass flow rate is adjusted accordingly, and the other quantities are calculated again. This process is continued until convergence is achieved. The following results are obtained:

$$C_{min} = 1253.4 \text{ W/K (engine oil)}, \; C_r = 0.3089, \; NTU = 4.476, \; \varepsilon = 0.842, \; T_{c,o} = 39.7°\text{C}$$

The required mass flow rate of cooling water is

$$\dot{m}_c = \underline{\underline{0.971 \text{ kg/s}}}$$

DISCUSSION

It is instructive to see how tube length affects the required mass flow rate of cooling water. Flow rate decreases significantly over the range 2.25 m < L < 5.5 m, as shown in the graph below.

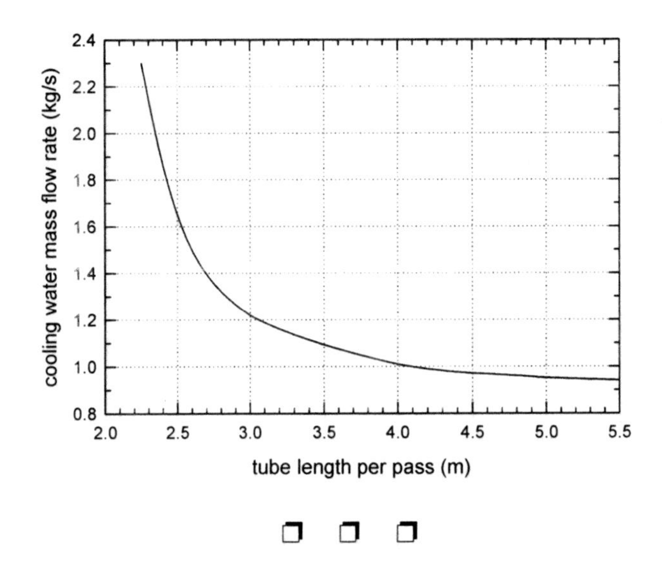

❒ ❒ ❒

PROBLEM 11.11 Plate-and-Frame Heat Exchanger for Pasteurizing Milk

A plate-and-frame heat exchanger is used in a pasteurization process to heat milk from 25°C to 70°C. The heat exchanger frame has the capacity for up to 75 plates measuring 1.4 m × 2.3 m. Water, which enters the heat exchanger at 90°C and 18 kg/s, is used as the heating medium. The mass flow rate of the milk is 8 kg/s, and the overall heat transfer coefficient is 185 W/m^2·K. If the heat exchanger operates in the counterflow mode, how many plates are required?

DIAGRAM

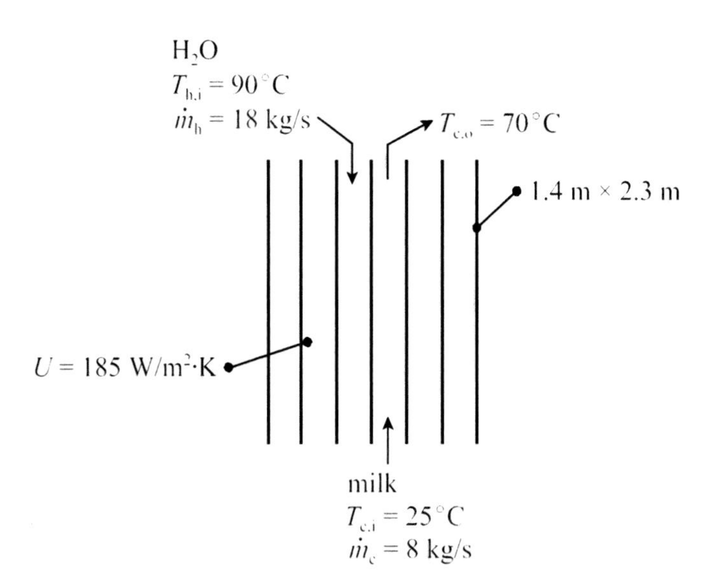

ASSUMPTIONS

1. Steady conditions.
2. Heat exchanger is insulated from surroundings.
3. Approximate properties of milk with those of water.
4. Constant properties.

PROPERTIES

$T_{av} = (T_{c,i} + T_{c,o})/2 = (25 + 70)°C/2 = 47.5°C = 321$ K

milk (water): $c_{p,c} = 4180$ J/kg·K

Assume $T_{h,o} = 65°C$. $T_{av} = (T_{h,i} + T_{h,o})/2 = (90 + 65)°C/2 = 77.5°C = 351$ K

water: $c_{p,h} = 4196$ J/kg·K

ANALYSIS

The heat transfer may be calculated by an energy balance on the milk,

$$q = \dot{m}_c c_{p,c}(T_{c,o} - T_{c,i})$$

$q = (8 \text{ kg/s})(4180 \text{ J/kg·K})(70 - 25)°C$

$= 1.505 \times 10^6 \text{ W} = 1.51 \text{ MW}$

Applying a similar energy balance on the water, the outlet temperature of the water is

$$T_{h,o} = T_{h,i} - \frac{q}{\dot{m}_h c_{p,h}}$$

$$T_{h,o} = 90°C - \frac{1.505 \times 10^6 \text{ W}}{(18 \text{ kg/s})(4196 \text{ J/kg·K})}$$

$$= 70.1°C$$

Now that all inlet and outlet temperatures are known, we can use the log mean temperature difference method of analysis. We have

$$\Delta T_m = \frac{\Delta T_1 - \Delta T_2}{\ln(\Delta T_1/\Delta T_2)}$$

where, for counterflow conditions,

$$\Delta T_1 = T_{h,i} - T_{c,o} \quad , \quad \Delta T_2 = T_{h,o} - T_{c,i}$$

Thus,

$$\Delta T_1 = (90 - 70)°C$$

$$= 20.0°C$$

$$\Delta T_2 = (70.1 - 25)°C$$

$$= 45.1°C$$

The log mean temperature difference is

$$\Delta T_m = \frac{(20.0 - 45.1)°C}{\ln(20.0°C/45.1°C)}$$

$$= 30.9°C$$

The heat transfer may be expressed by the relation

$$q = UA\Delta T_m$$

where the surface area is $A = 2NLW$, the multiplier of 2 denotes two heat transfer surfaces per plate (excluding the plates on the ends), and N is the total number of plates in the frame. Hence, the number of plates required is

$$N = \frac{q}{2LWU\Delta T_m}$$

$$N = \frac{1.505 \times 10^6 \text{ W}}{2(1.4 \text{ m})(2.3 \text{ m})(185 \text{ W/m}^2\text{·K})(30.9°C)}$$

$= 40.9 \approx \underline{\underline{41}}$

DISCUSSION

The frame has a capacity for 75 plates, so it is suitable for this application.

❒ ❒ ❒

PROBLEM 11.12 Cross-Flow Heat Exchanger for Waste Heat Recovery

In a metal casting facility, waste heat is recovered from the flue gas of a furnace to augment the primary heating system of an adjacent machine shop. Using a large blower, 130°C flue gas is drawn across a baffled bank of 60 thin-walled stainless steel tubes at a mass flow rate of 28 kg/s. The length and diameter of the tubes are 2.75 m and 20 cm, respectively. Return air from the machine shop at 25°C enters the tubes at a total mass flow rate of 20 kg/s. The warm air exiting the tubes is directed into the machine shop. If the overall heat transfer coefficient is 18 W/m²·K, what is the heat input to the machine shop?

DIAGRAM

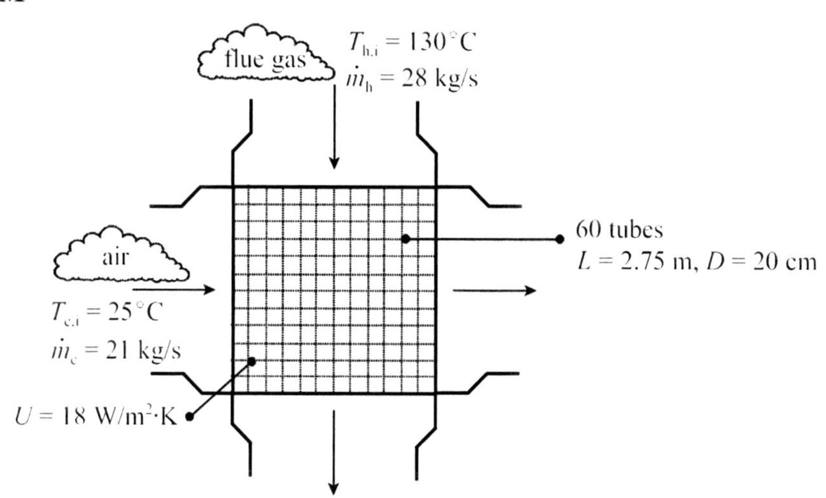

ASSUMPTIONS

1. Steady conditions.
2. Heat exchanger is insulated from surroundings.
3. Approximate properties of flue gas with those of atmospheric air.
4. Constant properties.

PROPERTIES

Assume $T_{h,o}$ = 120°C. $T_{av} = (T_{h,i} + T_{h,o})/2 = (130 + 120)°C/2 = 125°C = 398$ K

flue gas (air): $c_{p,h}$ = 1014 J/kg·K

Assume $T_{c,o} = 30°C$. $T_{av} = (T_{c,i} + T_{c,o})/2 = (25 + 30)°C/2 = 27.5°C = 301$ K

air: $c_{p,c} = 1007$ J/kg·K

ANALYSIS

Because only the inlet fluid temperatures are known, we use the effectiveness-*NTU* method of analysis. Heat capacity rate, *C*, is defined as

$$C \equiv \dot{m} c_p$$

so the heat capacity rate for the flue gas (hot fluid) is

$$\begin{aligned} C_h &= \dot{m}_h c_{p,h} \\ &= (28 \text{ kg/s})(1014 \text{ J/kg·K}) \\ &= 2.839 \times 10^4 \text{ W/K} \end{aligned}$$

and the heat capacity rate for the air (cold fluid) is

$$\begin{aligned} C_c &= \dot{m}_c c_{p,c} \\ &= (21 \text{ kg/s})(1007 \text{ J/kg·K}) \\ &= 2.115 \times 10^4 \text{ W/K} \end{aligned}$$

Thus, $C_{max} = C_h$ and $C_{min} = C_c$. The heat capacity rate ratio is defined as

$$C_r \equiv C_{min}/C_{max}$$

so we have

$$\begin{aligned} C_r &= (2.115 \times 10^4 \text{ W/K})/(2.839 \times 10^4 \text{ W/K}) \\ &= 0.7449 \end{aligned}$$

The number of transfer units is defined as

$$NTU \equiv \frac{UA}{C_{min}}$$

where $A = N\pi DL$ and N is the number of tubes. The value of *NTU* is

$$NTU = \frac{(18 \text{ W/m}^2\text{·K})(60)\pi(0.20 \text{ m})(2.75 \text{ m})}{2.115 \times 10^4 \text{ W/K}}$$

$= 0.0882$

There are no baffles on the tubes, so the flue gas is mixed, and the air is unmixed. The effectiveness of our cross-flow heat exchanger is

$$\varepsilon = \left(\frac{1}{C_r}\right)\left(1 - \exp\left\{-C_r\left[1 - \exp(-NTU)\right]\right\}\right)$$

$\varepsilon = (1/0.7449)(1 - \exp\{-0.7449[1 - \exp(-0.0882)]\})$

$= 0.0818$

The heat transfer is

$$q = \varepsilon C_{\text{min}}(T_{h,i} - T_{c,i})$$

$q = (0.0818)(2.115 \times 10^4 \text{ W/K})(130 - 25)°\text{C}$

$= 1.817 \times 10^5 \text{ W} = 182 \text{ kW}$

Now that the heat transfer is known, we can find the outlet temperatures. An energy balance on each fluid yields

$$q = C_h(T_{h,i} - T_{h,o}) = C_c(T_{c,o} - T_{c,i})$$

Hence, the outlet temperature of the flue gas is

$$T_{h,o} = T_{h,i} - q/C_h$$

$T_{h,o} = 130°\text{C} - (1.817 \times 10^5 \text{ W})/(2.839 \times 10^4 \text{ W/K})$

$= 123.6°\text{C}$

and the outlet temperature of the air is

$$T_{c,o} = T_{c,i} + q/C_c$$

$T_{c,o} = 25°\text{C} + (1.817 \times 10^5 \text{ W})/(2.115 \times 10^4 \text{ W/K})$

$= 33.6°\text{C}$

The calculated outlet temperatures are sufficiently close to our initial estimates, so no property corrections are needed. The heat input to the machine shop is the heat transferred from the flue gas to the air. Thus,

$q = \underline{\underline{182 \text{ kW}}}$

DISCUSSION

If the flow rate of air is varied, the heat input to the machine shop and the temperature of the air entering the machine shop can be controlled to some extent. The relationship between air mass flow rate and heat input is illustrated in the graph below. The corresponding outlet air temperatures for mass flow rates of 10 kg/s and 28 kg/s, respectively, are 42.2°C and 31.5°C.

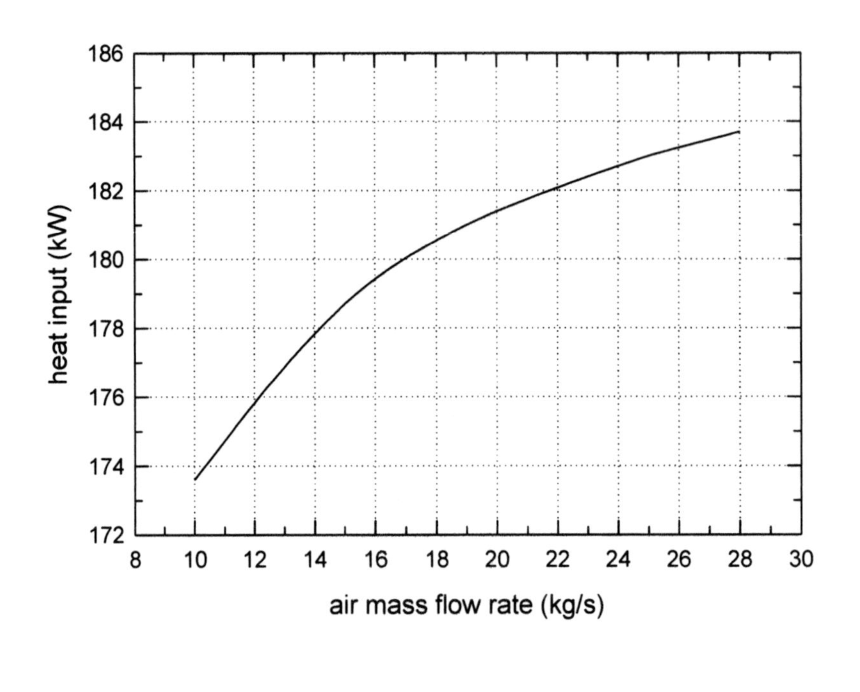

❒ ❒ ❒

PROBLEM 11.13 Design of a Recuperator

A recuperator is a heat exchanger used to preheat combustion air for an incinerator. Combustion gas from the incinerator enters the tubes of the recuperator at 750°C at a mass flow rate of 5 kg/s, and atmospheric air at 20°C enters the shell of the unit at a mass flow rate of 1.8 kg/s. When the heat transfer surfaces of the unit are clean, the overall heat transfer coefficient is 30 $W/m^2 \cdot K$. After a few years of operation, the surfaces experience a fouling resistance of 0.002 $K \cdot m^2/W$. A 1 percent fuel savings is achieved for every 16°C the combustion air to the incinerator exceeds 20°C. If a fuel savings of 30 percent is desired, what is the total heat transfer surface area required? Present the results in terms of a square array of tubes with diameters of 1.0, 2.0, 4.0, and 5.0 cm.

DIAGRAM

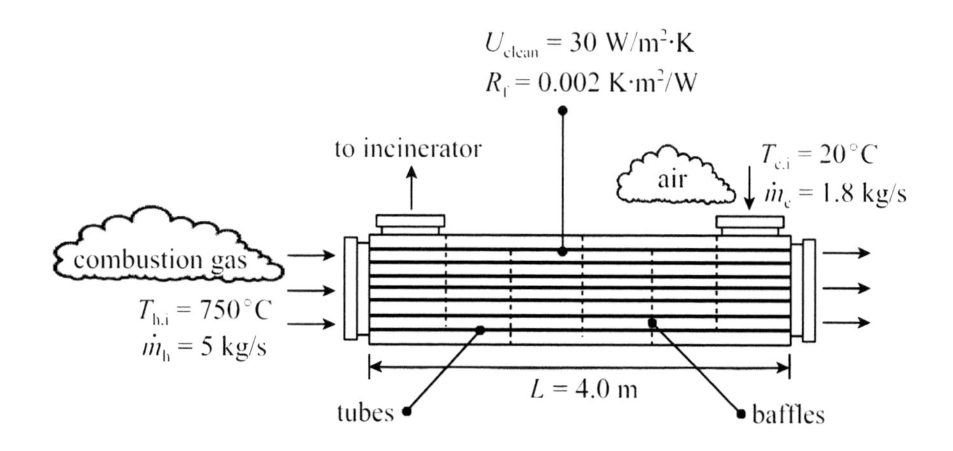

ASSUMPTIONS

1. Steady conditions.
2. Heat exchanger is insulated from surroundings.
3. Approximate properties of combustion gas with those of atmospheric air.
4. Constant properties.

PROPERTIES

Assume $T_{h,o}$ = 575°C. T_{av} = $(T_{h,i} + T_{h,o})/2$ = (750 + 575)°C/2 = 662.5°C = 936 K

combustion gas (air): $c_{p,h}$ = 1128 J/kg·K

$T_{c,o}$ = (0.30/0.01)(16°C) + 20°C = 500°C. T_{av} = $(T_{c,i} + T_{c,o})/2$ = (20 + 500)°C/2 = 260°C = 533 K

air: $c_{p,c}$ = 1037 J/kg·K

ANALYSIS

Knowing the inlet and outlet temperatures of the air, the heat transfer is

$$q = \dot{m}_c c_{p,c}(T_{c,o} - T_{c,i})$$

$$q = (1.8\ \text{kg/s})(1037\ \text{J/kg·K})(500 - 20)°\text{C}$$

$$= 8.960 \times 10^5\ \text{W}$$

so the outlet temperature of the combustion gas is

$$T_{h,o} = T_{h,i} - \frac{q}{\dot{m}_h c_{p,h}}$$

$$T_{h,o} = 750\,°C - \frac{8.960 \times 10^5 \text{ W}}{(5 \text{ kg/s})(1128 \text{ J/kg·K})}$$

$$= 591\,°C$$

which is close enough to our initial estimate that property corrections are not required. The log mean temperature difference is given by the relation

$$\Delta T_m = \frac{\Delta T_1 - \Delta T_2}{\ln(\Delta T_1/\Delta T_2)}$$

where, for counterflow conditions,

$$\Delta T_1 = T_{h,i} - T_{c,o} \quad , \quad \Delta T_2 = T_{h,o} - T_{c,i}$$

The heat transfer relation for a heat exchanger in terms of the log mean temperature difference is

$$q = UA\Delta T_m$$

where the overall heat transfer coefficient for the fouled heat exchanger is

$$U = \frac{1}{R_f + 1/U_{clean}}$$

$$U = \frac{1}{0.002 \text{ K·m}^2\text{/W} + 1/(30 \text{ W/m}^2\text{·K})}$$

$$= 28.3 \text{ W/m}^2\text{·K}$$

A correction factor F is not required for the log mean temperature difference because the recuperator is essentially a baffled, concentric tube heat exchanger operating in the counterflow mode, albeit the unit has multiple tubes. The temperature differences are

$$\Delta T_1 = (750 - 500)°C$$

$$= 250°C$$

$$\Delta T_2 = (591 - 20)°C$$

$$= 571\,°C$$

Thus,

$$\Delta T_{\mathrm{m}} = \frac{(250 - 571)°\mathrm{C}}{\ln(250°\mathrm{C}/571°\mathrm{C})}$$

$$= 389°\mathrm{C}$$

Hence, the required surface area is

$$A = \frac{q}{U\Delta T_m}$$

$$A = \frac{8.960 \times 10^5 \mathrm{\ W}}{(28.3 \mathrm{\ W/m^2\cdot K})(389°\mathrm{C})}$$

$$= \underline{\underline{81.4 \mathrm{\ m}^2}}$$

The number of tubes, N, is the surface area divided by the surface area of a single tube,

$$N = \frac{A}{\pi DL}$$

The results are summarized in the following table, where the number of rows and columns in the square array for each tube diameter has been rounded up to the next integer.

D (cm)	N	Rows/Columns
1.0	648	26
2.0	324	18
4.0	162	13
5.0	130	12

DISCUSSION

It is important to understand that although a constant overall heat transfer coefficient was assumed, this quantity is a function of the number of tubes, tube diameter, and other factors. Furthermore, the mass flow rate may be a function of these factors, which also affects the overall heat transfer coefficient.

PROBLEM 11.14 **Shell-and-Tube Heat Exchanger with Two Shell Passes and Eight Tube Passes**

A shell-and-tube heat exchanger with two shell passes and eight tube passes is used to cool gasoline in a refinery. Gasoline enters the tubes at a temperature and mass flow rate of 110°C and 12 kg/s, respectively, while water enters the shell at a temperature and mass flow rate of 15°C and 20 kg/s, respectively. There are 25 thin-walled tubes with a diameter of 6.5 cm and a length of 7.4 m per pass. If the overall heat transfer coefficient is 280 W/m²·K, what is the heat transfer and the outlet temperature of the gasoline and water?

DIAGRAM

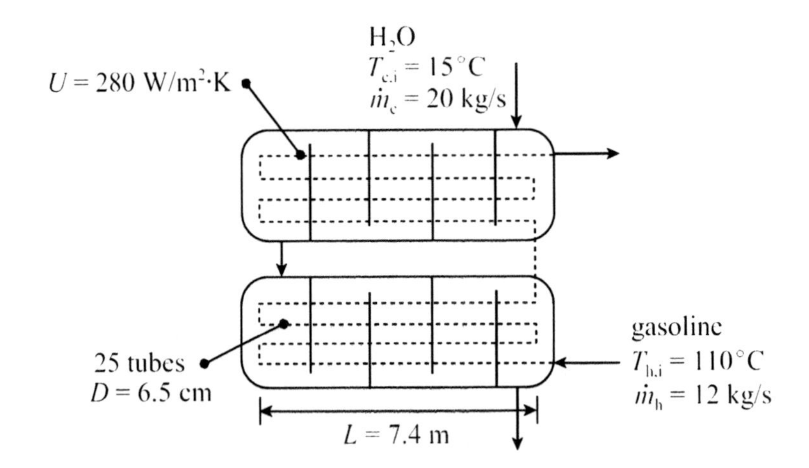

ASSUMPTIONS

1. Steady conditions.
2. Heat exchanger is insulated from surroundings.
3. Lengths of tube bends are neglected.
4. Constant properties.

PROPERTIES

Assume $T_{h,o} = 30°C$. $T_{av} = (T_{h,i} + T_{h,o})/2 = (110 + 30)°C/2 = 70°C = 343$ K

gasoline: $c_{p,h} = 2295$ J/kg·K

Assume $T_{c,o} = 40°C$. $T_{av} = (T_{c,i} + T_{c,o})/2 = (15 + 40)°C/2 = 27.5°C = 301$ K

water: $c_{p,c} = 4179$ J/kg·K

ANALYSIS

Because only the inlet fluid temperatures are known, we use the effectiveness-*NTU* method of analysis. Heat capacity rate, C, is defined as

$$C \equiv \dot{m} c_p$$

so the heat capacity rate for the gasoline (hot fluid) is

$$\begin{aligned} C_{\mathrm{h}} &= \dot{m}_{\mathrm{h}} c_{\mathrm{p,h}} \\ &= (12\ \mathrm{kg/s})(2295\ \mathrm{J/kg \cdot K}) \\ &= 2.754 \times 10^4\ \mathrm{W/K} \end{aligned}$$

and the heat capacity rate for the water (cold fluid) is

$$\begin{aligned} C_{\mathrm{c}} &= \dot{m}_{\mathrm{c}} c_{\mathrm{p,c}} \\ &= (20\ \mathrm{kg/s})(4179\ \mathrm{J/kg \cdot K}) \\ &= 8.358 \times 10^4\ \mathrm{W/K} \end{aligned}$$

Thus, $C_{max} = C_c$ and $C_{min} = C_h$. The heat capacity rate ratio is defined as

$$C_r \equiv C_{\min}/C_{\max}$$

so we have

$$\begin{aligned} C_{\mathrm{r}} &= (2.754 \times 10^4\ \mathrm{W/K})/(8.358 \times 10^4\ \mathrm{W/K}) \\ &= 0.3295 \end{aligned}$$

The number of transfer units is defined as

$$NTU \equiv \frac{UA}{C_{\min}}$$

where $A = 4N\pi DL$, the heat transfer surface area for one shell. The factor 4 is the number of tube passes per shell, and N is the number of tubes. The value of NTU is

$$\begin{aligned} NTU &= \frac{(4)(25)(280\ \mathrm{W/m^2 \cdot K})\pi(0.065\ \mathrm{m})(7.4\ \mathrm{m})}{2.754 \times 10^4\ \mathrm{W/K}} \\ &= 1.536 \end{aligned}$$

The heat exchanger effectiveness is given by the relation

$$\varepsilon = \left[\left(\frac{1-\varepsilon_1 C_r}{1-\varepsilon_1}\right)^n - 1\right]\left[\left(\frac{1-\varepsilon_1 C_r}{1-\varepsilon_1}\right)^n - C_r\right]^{-1}$$

where ε_1 is the effectiveness of a shell-and-tube heat exchanger with one shell pass and any multiple of two tube passes, given by the relation

$$\varepsilon_1 = 2\left\{1 + C_r + (1+C_r^2)^{1/2} \cdot \frac{1+\exp\left[-NTU(1+C_r^2)^{1/2}\right]}{1-\exp\left[-NTU(1+C_r^2)^{1/2}\right]}\right\}^{-1}$$

Thus,

$$\varepsilon_1 = 2\left\{1 + 0.3295 + (1 + 0.3295^2)^{½} \cdot \frac{1 + \exp[-1.536(1 + 0.3295^2)^{½}]}{1 - \exp[-1.536(1 + 0.3295^2)^{½}]}\right\}^{-1}$$

$$= 0.7014$$

For a heat exchanger with two shell passes and any multiple of four tube passes, $n = 2$. Thus, the effectiveness, ε, is

$$\varepsilon = \left[\left(\frac{1 - (0.7014)(0.3295)}{1 - 0.7114}\right)^2 - 1\right]\left[\left(\frac{1 - (0.7014)(0.3295)}{1 - 0.7014}\right)^2 - 0.3295\right]^{-1}$$

$$= 0.8936$$

The heat transfer is

$$q = \varepsilon C_{\min}(T_{h,i} - T_{c,i})$$

$$q = (0.8936)(2.754 \times 10^4 \text{ W/K})(110 - 15)°\text{C}$$

$$= 2.338 \times 10^6 \text{ W} = \underline{\underline{2.34 \text{ MW}}}$$

Now that the heat transfer is known, we can find the outlet temperatures. An energy balance on each fluid yields

$$q = C_h(T_{h,i} - T_{h,o}) = C_c(T_{c,o} - T_{c,i})$$

Hence, the outlet temperature of the gasoline is

$$T_{h,o} = T_{h,i} - q/C_h$$

$$T_{\text{h,o}} = 110°\text{C} - (2.338 \times 10^6 \text{ W})/(2.754 \times 10^4 \text{ W/K})$$

$= 25.1°\text{C}$

and the outlet temperature of the water is

$$T_{c,o} = T_{c,i} + q/C_c$$

$$T_{c,o} = 15°\text{C} + (2.338 \times 10^6\ \text{W})/(8.358 \times 10^4\ \text{W/K})$$

$$= 43.0°\text{C}$$

A property correction is not required. Hence,

$$T_{h,o} = \underline{\underline{25.1°\text{C}}} \ , \quad T_{c,o} = \underline{\underline{43.0°\text{C}}}$$

DISCUSSION

In this system, the temperature of the gasoline is reduced from a value that would cause severe burns to a safe room temperature value.

❒ ❒ ❒

PROBLEM 11.15 Heat Exchanger for Preheating Water in a Truck Washing Facility

In an effort to conserve energy in a large truck washing facility, the warm dirty water that flows into the floor drain is recirculated through a shell-and-tube heat exchanger to preheat the water used for cleaning. The mass flow rates of the clean and dirty water are 35 kg/s and 30 kg/s, respectively, and the overall heat transfer coefficient is 1900 W/m^2·K. The inlet temperatures of the clean and dirty water are 20°C and 65°C, respectively. If the heat exchanger has one shell pass and two tube passes, what heat transfer surface area is required to achieve a 10°C increase in the clean water as it enters the electric boiler? What is the hourly cost savings if the price of electricity is \$0.085/kW·h?

DIAGRAM

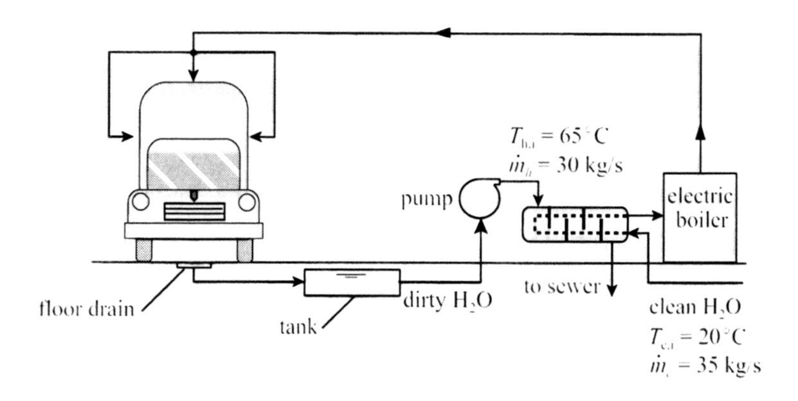

ASSUMPTIONS

1. Steady conditions.
2. Heat exchanger is insulated from surroundings.
3. At a given temperature, properties of dirty water are equivalent to those of clean water.
4. Constant properties.

PROPERTIES

$T_{av} = (T_{c,i} + T_{c,o})/2 = (20 + 30)°C/2 = 25°C = 298$ K

clean water: $c_{p,h} = 4179$ J/kg·K

Assume $T_{h,o} = 55°C$. $T_{av} = (T_{h,i} + T_{h,o})/2 = (65 + 55)°C/2 = 60°C = 333$ K

dirty water: $c_{p,c} = 4186$ J/kg·K

ANALYSIS

The heat transfer can be calculated by an energy balance on the clean water,

$$q = \dot{m}_c c_{p,c}(T_{c,o} - T_{c,i})$$

$q = (35 \text{ kg/s})(4179 \text{ J/kg·K})(30 - 20)°C$

$= 1.463 \times 10^6$ W

By a similar energy balance, the outlet temperature of the dirty water is

$$T_{h,o} = T_{h,i} - \frac{q}{\dot{m}_h c_{p,h}}$$

$$T_{h,o} = 65°C - \frac{1.463 \times 10^6 \text{ W}}{(30 \text{ kg/s})(4186 \text{ J/kg·K})}$$

$= 53.4°C$

No property correction for the dirty water is required. The log mean temperature difference is given by the relation

$$\Delta T_m = \frac{\Delta T_1 - \Delta T_2}{\ln(\Delta T_1/\Delta T_2)}$$

where, for counterflow conditions,

$$\Delta T_1 = T_{h,i} - T_{c,o} \quad , \quad \Delta T_2 = T_{h,o} - T_{c,i}$$

Thus,

$$\Delta T_1 = (65 - 30)°C$$

$$= 35°C$$

$$\Delta T_2 = (53.4 - 20)°C$$

$$= 33.4°C$$

The log mean temperature difference is

$$\Delta T_m = \frac{(35 - 33.4)°C}{\ln(35°C/33.4°C)}$$

$$= 34.2°C$$

The heat transfer relation for a heat exchanger in terms of the log mean temperature difference is

$$q = FUA\Delta T_m$$

where F is a correction factor for multipass and cross-flow heat exchangers, $U = 1900$ W/m²·K, and A is heat transfer surface area. Using a graph for a one-shell pass, two-tube pass heat exchanger, we find that $F \approx 0.98$. Thus, the required surface area is

$$A = \frac{q}{FU\Delta T_m}$$

$$A = \frac{1.463 \times 10^6 \text{ W}}{(0.98)(1900 \text{ W/m}^2\cdot\text{K})(34.2°C)}$$

$$= \underline{\underline{23.0 \text{ m}^2}}$$

and the hourly cost savings is

$$S = \$0.085/\text{kW}\cdot\text{h} \times 1.463 \times 10^6 \text{ W} \times 1 \text{ kW}/10^3 \text{ W}$$

$$= \underline{\underline{\$124.36/\text{h}}}$$

DISCUSSION

Recirculating the warm dirty water through a heat exchanger, preheating it before it enters the electric boiler, saves the facility \$124.36 for each hour of operation. Suppose that the initial cost of the system is \$60,000 and that the system operates an average of four hours every weekday.

The simple payback period is

$$\text{payback} = \$60{,}000 \times 1\ \text{h}/\$124.36 \times \text{day}/4\ \text{h} \times 1\ \text{week}/5\ \text{day}$$

$$\approx 24\ \text{weeks}$$

Thus, the system pays for itself in approximately six months.

❒ ❒ ❒

Chapter 12

Principles of Radiation

PROBLEM 12.1 Diffuse Emission from a Small Surface

A small surface of area $A_1 = 2 \times 10^{-4}$ m^2 emits diffusely with a total, hemispherical emissive power of 10^5 W/m^2. A second small surface of area $A_2 = 8 \times 10^{-4}$ m^2 is oriented with respect to A_1 as shown in the diagram. Find the rate at which A_2 intercepts the emission from A_1 and the irradiation on A_2.

DIAGRAM

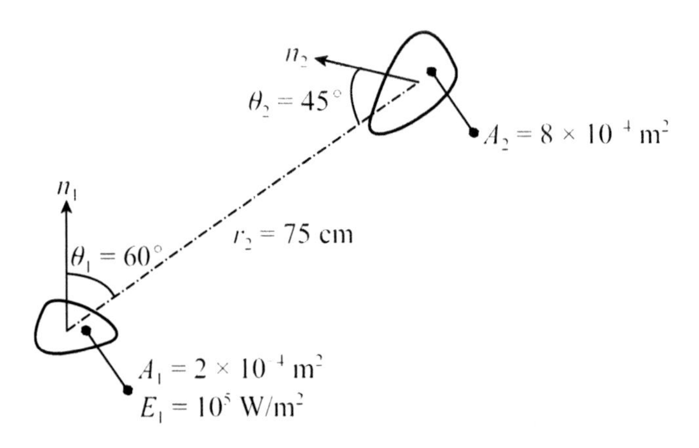

ASSUMPTIONS

1. Surface A_1 emits diffusely.
2. Approximate A_1 as a differential area.

PROPERTIES

not applicable

ANALYSIS

The rate at which surface A_2 intercepts emission from surface A_1 is expressed by the relation

$$q_{1-2} = I_{e,1}(\theta,\phi) A_1 \cos\theta_1 d\omega_{2-1}$$

where $I_{e,1}(\theta,\varphi)$ is the total intensity and $d\omega_{2-1}$ is the differential solid angle subtended by A_2 with respect to A_1. The angles θ and φ define the hypothetical hemisphere centered at a point on surface A_1. Surface A_1 is diffuse, so the total intensity is

$$I_{e,1}(\theta,\phi) = \frac{E_1}{\pi}$$

where $E_1 = 10^5$ W/m^2. The quantity π in this relation has units of steradians (sr). The solid angle, which also has units of steradians, subtended by A_2 with respect to A_1 may be approximated by the relation,

$$d\omega_{2-1} \approx \frac{A_2 \cos\theta_2}{r_2^2}$$

Substituting the last two relations into the first relation,

$$q_{1-2} = \frac{E_1 A_1 \cos\theta_1 A_2 \cos\theta_2}{\pi r_2^2}$$

$$q_{2-1} = \frac{(10^5 \text{ W/m}^2)(2 \times 10^{-4} \text{ m}^2)\cos(60°)(8 \times 10^{-4} \text{ m}^2)\cos(45°)\text{sr}}{(\pi \text{ sr})(0.75 \text{ m})^2}$$

$$= \underline{\underline{3.201 \times 10^{-3} \text{ W}}}$$

The irradiation on surface A_2 is the rate at which radiation is incident upon the surface per unit area,

$$G_2 = \frac{q_{1-2}}{A_2}$$

$$G_2 = \frac{3.201 \times 10^{-3} \text{ W}}{8 \times 10^{-4} \text{ m}^2}$$

$$= \underline{\underline{4.00 \text{ W/m}^2}}$$

DISCUSSION

The variation of G_2 with r_2 and θ_2 is shown in the graph below.

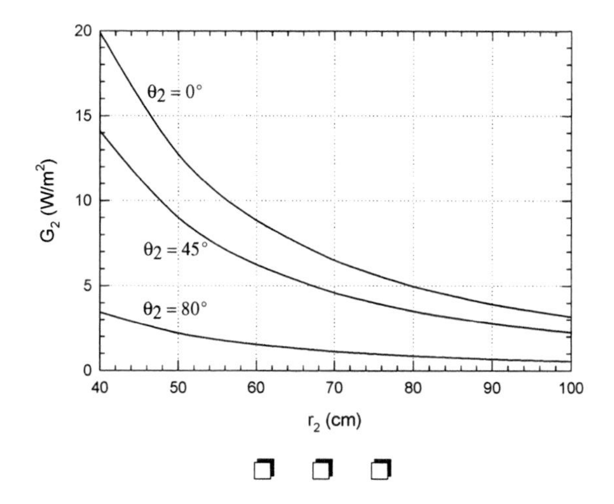

❒ ❒ ❒

PROBLEM 12.2 **Total Emissive Power from an Approximate Spectral Distribution**

The spectral distribution of the radiation emitted by a diffuse surface is approximated in the diagram below. Find the total, hemispherical emissive power and the total intensity emitted in the normal direction and at an angle of 45° from the normal.

DIAGRAM

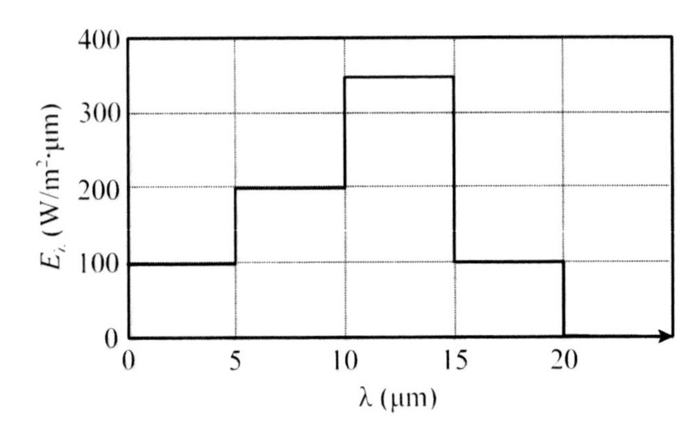

ASSUMPTIONS

1. Surface emits diffusely.

PROPERTIES

not applicable

ANALYSIS

The total, hemispherical emissive power is the rate at which radiation is emitted per unit area at all wavelengths in all directions. Thus, we integrate the spectral, hemispherical emissive power over all wavelengths,

$$E = \int_0^\infty E_\lambda(\lambda)d\lambda$$

$$E = \int_0^5 (100)d\lambda + \int_5^{10} (200)d\lambda + \int_{10}^{15} (350)d\lambda + \int_{15}^{20} (100)d\lambda + \int_{20}^\infty (0)d\lambda$$

$$E = (100)(5-0) + (200)(10-5) + (350)(15-10) + (100)(20-15) + 0$$

$= \underline{\underline{3750 \text{ W/m}^2}}$

For a diffuse surface, the emitted radiation is independent of direction, so the total intensity emitted in the normal direction and at an angle of 45° from the normal is

$$I_e = \frac{E}{\pi}$$

$$I_e = \frac{3750 \text{ W/m}^2}{\pi \text{ sr}}$$

$$= \underline{\underline{1194 \text{ W/m}^2\cdot\text{sr}}}$$

DISCUSSION

For simplicity, the spectral emissive power was approximated using a step function that was broken into parts and integrated. Any spectral emissive power distribution can be handled in a similar manner by using step functions or any functions that can be readily integrated.

❒ ❒ ❒

PROBLEM 12.3 Irradiation on a Surface Within an Enclosure

The entire inside surface of an evacuated, cubical enclosure measuring 3 m on a side is coated with carbon black. If this surface is maintained at 400°C, what is the irradiation on a surface placed within the enclosure? If the inside surface of the enclosure is not coated, what is the irradiation on the enclosed surface?

DIAGRAM

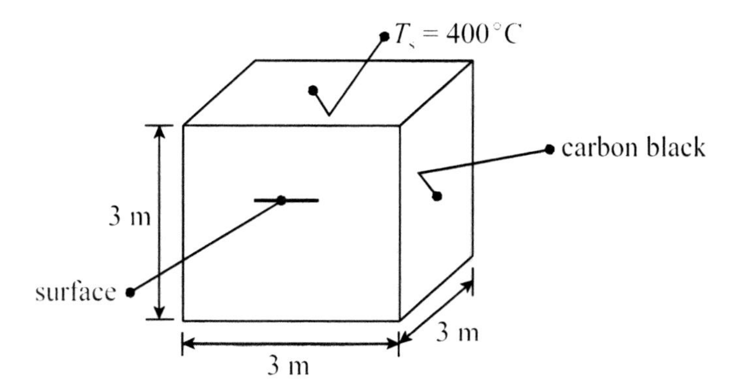

ASSUMPTIONS

1. Inside surface of enclosure is isothermal.
2. Enclosed surface area is much smaller than surface area of enclosure.
3. Enclosure is evacuated.

PROPERTIES

not applicable

ANALYSIS

The isothermal cubical enclosure behaves as a blackbody. The irradiation on a small surface within the enclosure is equivalent to the blackbody emissive power at the surface temperature of the enclosure.

$$G = E_b = \sigma T_s^4$$

$$G = (5.669 \times 10^{-8}\ \text{W/m}^2\cdot\text{K}^4)(400 + 273)^4\ \text{K}^4$$

$$= \underline{\underline{1.163 \times 10^4\ \text{W/m}^2}}$$

The irradiation is independent of the surface properties of the enclosure.

DISCUSSION

It is instructive to note that the irradiation is a function of enclosure surface temperature only. The second assumption allows for reflections to occur within the enclosure so that the thermal radiation field is diffuse. If the enclosure contained a non-participating gas such as air, the irradiation on the enclosed surface would be the same.

❐ ❐ ❐

PROBLEM 12.4 Spectral Blackbody Emissive Power of an Orange-Red Source

Find the spectral blackbody emissive power of a 1200-K source emitting orange-red light at a wavelength of 0.63 μm.

DIAGRAM

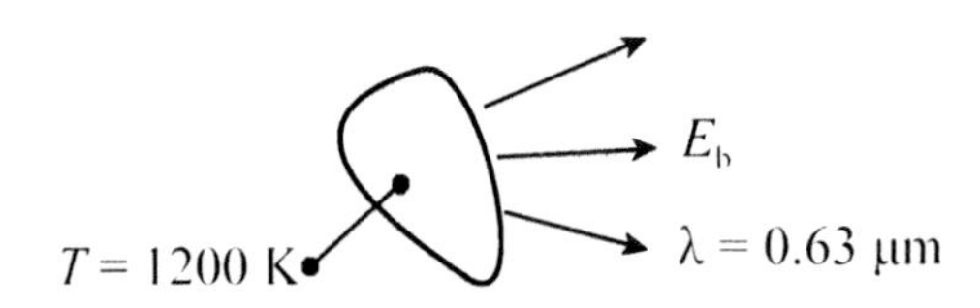

ASSUMPTIONS

1. Source is a blackbody.
2. Source is isothermal.

PROPERTIES

not applicable

ANALYSIS

The spectral blackbody emissive power, which is a function of wavelength and temperature, is given by Planck's distribution,

$$E_b(\lambda, T) = \frac{C_1}{\lambda^5\left[\exp(C_2/\lambda T) - 1\right]}$$

where $C_1 = 3.742 \times 10^8$ W·μm^4/m^2 and $C_2 = 1.439 \times 10^4$ μm·K. Hence,

$$E_b(\lambda,T) = \frac{3.742 \times 10^8 \text{ W·μm}^4/\text{m}^2}{(0.63 \text{ μm})^5\left[\exp(1.439 \times 10^4 \text{ μm·K}/(0.63 \text{ μm})(1200 \text{ K})) - 1\right]}$$

$$= \underline{\underline{20.4 \text{ W/m}^2\text{·μm}}}$$

DISCUSSION

The graph below shows the variation of spectral blackbody emissive power with temperature.

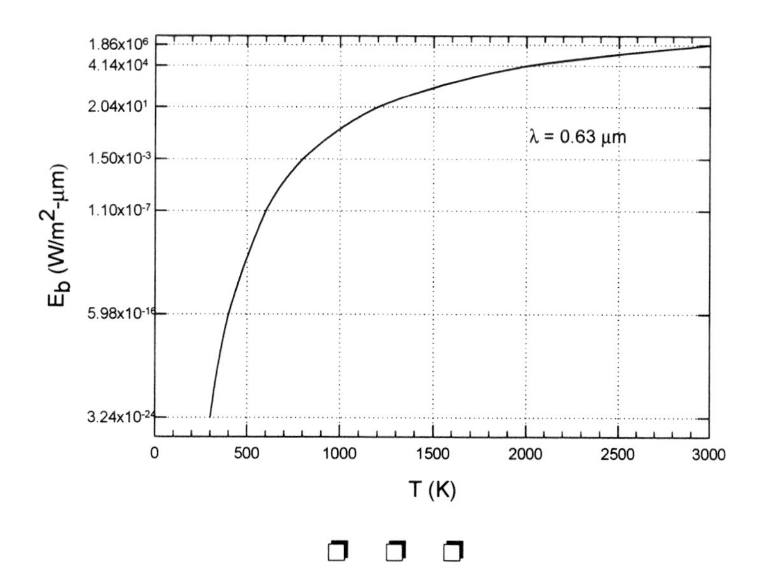

❒ ❒ ❒

PROBLEM 12.5 Radiation Emitted by a Current-Carrying Wire

A current-carrying wire of diameter 1.8 mm has a surface temperature of 375°C. If the wire radiates as a blackbody, find the radiation emitted per meter of wire length.

DIAGRAM

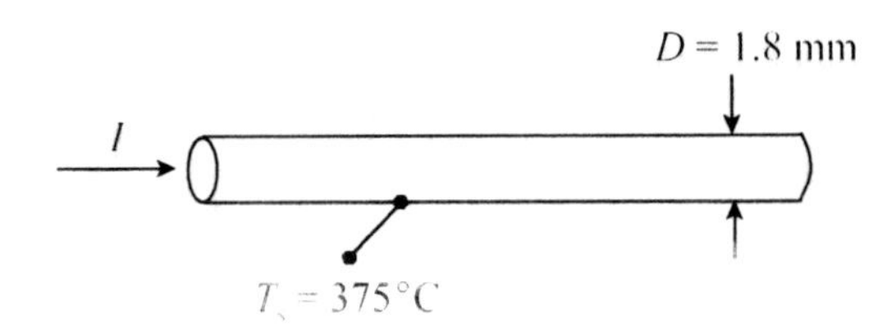

ASSUMPTIONS

1. Wire behaves as a blackbody.
2. Wire surface is isothermal.

PROPERTIES

not applicable

ANALYSIS

The radiation emitted by the wire per meter of length is

$$q' = \pi D \sigma T_s^4$$

$$q' = \pi(1.8 \times 10^{-3}\ \text{m})(5.669 \times 10^{-8}\ \text{W/m}^2\cdot\text{K}^4)(375 + 273)^4\ \text{K}^4$$

$$= \underline{\underline{56.5\ \text{W/m}}}$$

DISCUSSION

Emitted radiation from a blackbody is independent of surface properties. If the wire did not behave as a blackbody, the relation above would be modified to include the emissivity of the wire surface,

$$q' = \varepsilon \pi D \sigma T_s^4$$

For most metals, $0.01 < \varepsilon < 0.2$, unless the surface is highly oxidized, for which $0.7 < \varepsilon < 0.95$.

❐ ❐ ❐

PROBLEM 12.6 Fraction of Radiation Emitted as Visible Light from a Filament

The filament of an incandescent lamp has a temperature of 3000 K. Calculate the fraction of radiation emitted in the visible light band if the filament is approximated as a blackbody.

DIAGRAM

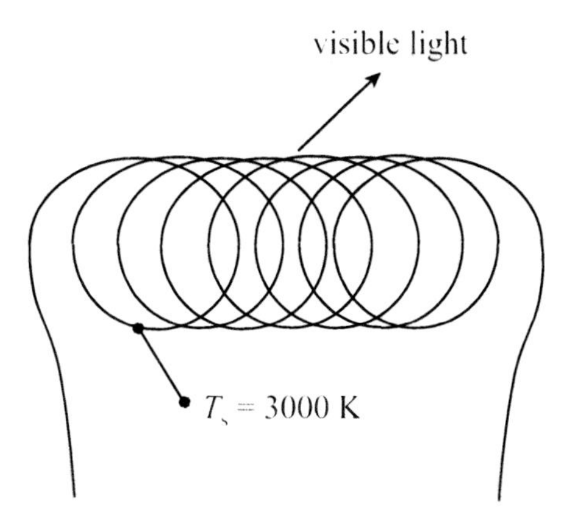

ASSUMPTIONS

1. Filament behaves as a blackbody.

PROPERTIES

not applicable

ANALYSIS

The visible light band has a range of wavelengths from $\lambda_1 = 0.4$ μm to $\lambda_2 = 0.76$ μm. Hence, the wavelength-temperature products are

$$\lambda_1 T = (0.4\ \mu\text{m})(3000\ \text{K})$$

$$= 1200\ \mu\text{m·K}$$

$$\lambda_2 T = (0.76\ \mu\text{m})(3000\ \text{K})$$

$$= 2280\ \mu\text{m·K}$$

The blackbody radiation functions corresponding to these quantities are

$$F_{0-1} = 0.002134 \quad , \quad F_{0-2} = 0.116635$$

A table of blackbody radiation functions is typically given in a heat transfer text. The fraction of radiation emitted by the filament in the visible light band is

$$F_{1-2} = F_{0-2} - F_{0-1}$$

$F_{1-2} = 0.116635 - 0.002134$

$= \underline{\underline{0.114501}}$

DISCUSSION

Assuming that the filament behaves as a blackbody, about 11.5 percent of the radiation emitted by the filament is in the visible light band. This fraction increases with filament temperature, as illustrated in the graph below. As shown in the graph, the filament must have a temperature of at least 1600 K to emit a non-significant amount of radiation in the visible light band of wavelengths. An inflection point occurs at about 4200 K, where the fraction of radiation emitted as visible light is about 28 percent.

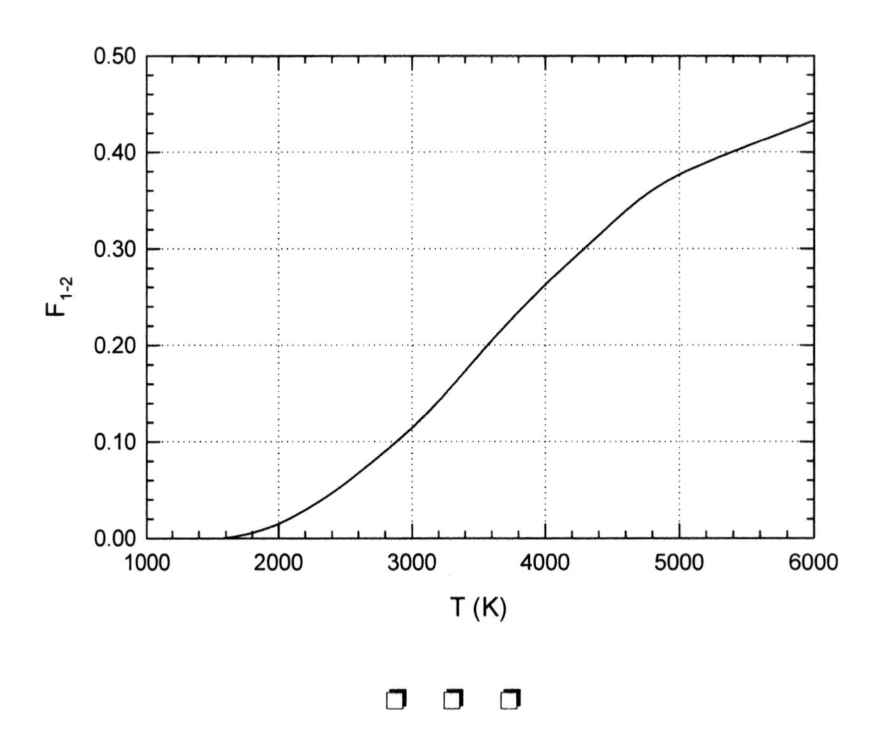

❒ ❒ ❒

PROBLEM 12.7 Radiation Emerging from the Aperture of a Furnace

Consider a furnace that operates at 1600 K. Radiation emerges from a small aperture of the furnace, behaving as a blackbody. Find the blackbody emissive power of the aperture and the spectral emissive power of the aperture at wavelengths of 0.6, 4 and 20 μm.

DIAGRAM

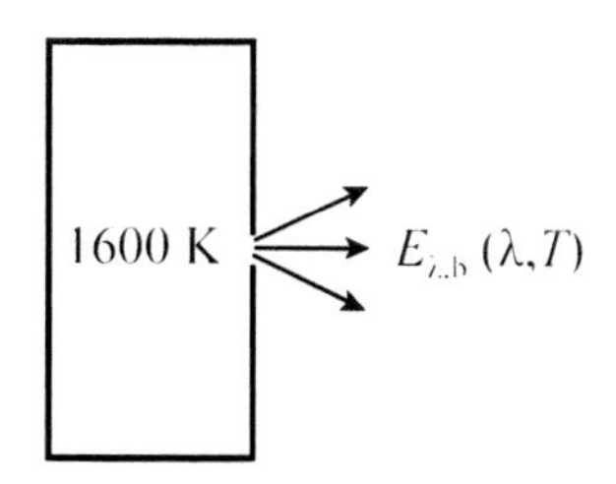

ASSUMPTIONS

1. Aperture behaves as a blackbody at 1600 K.

PROPERTIES

not applicable

ANALYSIS

Blackbody emissive power is given by the Stefan-Boltzmann law,

$$E_b = \sigma T^4$$

$E_b = (5.669 \times 10^{-8}\ \text{W/m}^2\cdot\text{K}^4)(1600\ \text{K})^4$

$= \underline{\underline{3.72 \times 10^5\ \text{W/m}^2}}$

To calculate the spectral emissive power at a given wavelength, we use the Planck distribution,

$$E_{\lambda,b}(\lambda,T) = \frac{C_1}{\lambda^5\left[\exp(C_2/\lambda T)-1\right]}$$

where $C_1 = 3.742 \times 10^8$ W·μm^4/m^2 and $C_2 = 1.439 \times 10^4$ μm·K. The spectral emissive power at $\lambda = 0.6$ μm is

$$E_{\lambda,b}(\lambda,T) = \frac{3.742 \times 10^8\ \text{W}\cdot\mu\text{m}^4/\text{m}^2}{(0.6\ \mu\text{m})^5\ [\exp(1.439 \times 10^4\ \mu\text{m}\cdot\text{K}/(0.6\ \mu\text{m})(1600\ \text{K})) - 1]}$$

$= \underline{\underline{1487\ \text{W/m}^2}}$

The spectral emissive power at $\lambda = 4\ \mu\text{m}$ is

$$E_{\lambda,b}(\lambda,T) = \frac{3.742 \times 10^8\ \text{W}\cdot\mu\text{m}^4/\text{m}^2}{(4\ \mu\text{m})^5\,[\exp(1.439 \times 10^4\ \mu\text{m}\cdot\text{K}/(4\ \mu\text{m})(1600\ \text{K})) - 1]}$$

$$= \underline{\underline{4.31 \times 10^4\ \text{W/m}^2}}$$

The spectral emissive power at $\lambda = 20\ \mu\text{m}$ is

$$E_{\lambda,b}(\lambda,T) = \frac{3.742 \times 10^8\ \text{W}\cdot\mu\text{m}^4/\text{m}^2}{(20\ \mu\text{m})^5\,[\exp(1.439 \times 10^4\ \mu\text{m}\cdot\text{K}/(20\ \mu\text{m})(1600\ \text{K})) - 1]}$$

$$= \underline{\underline{206\ \text{W/m}^2}}$$

DISCUSSION

The Planck distribution for a 1600-K blackbody is shown in the graph below.

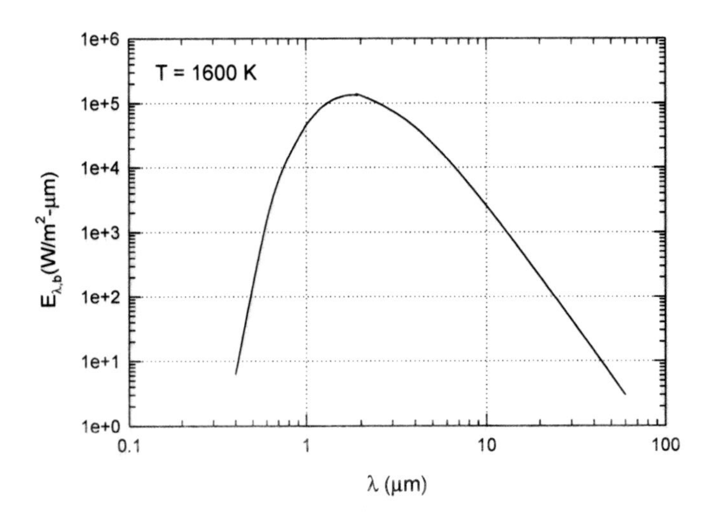

According to Wein's displacement law,

$$\lambda_{max} T = 2897.8$$

the wavelength at which the aperture emits the maximum power is

$$\lambda_{max} = (2897.8\ \mu\text{m}\cdot\text{K})/(1600\ \text{K})$$

$$= 1.81\ \mu\text{m}$$

❒ ❒ ❒

PROBLEM 12.8 **Finding Total Emissivity from a Graph of Spectral Emissivity**

A surface has a spectral, hemispherical emissivity that is approximated by the distribution shown in the diagram below, and the temperature of the surface is 2200 K. Find the total hemispherical emissivity and the total emissive power of the surface.

DIAGRAM

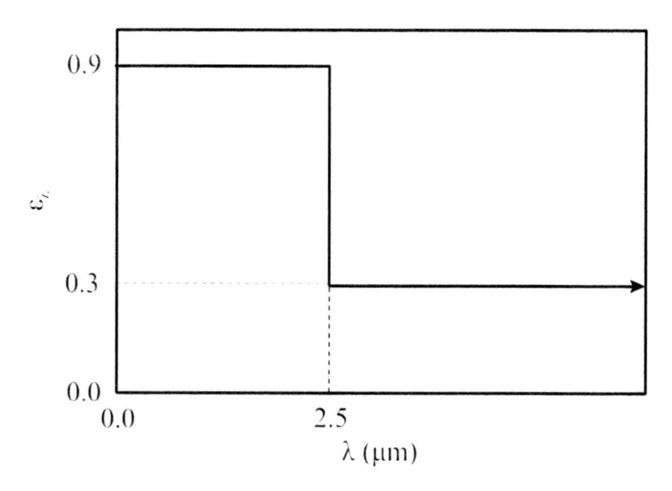

ASSUMPTIONS

1. Spectral, hemispherical emissivity follows a step function, as shown in the diagram.

PROPERTIES

$0.0 < \lambda < 2.5$ μm: $\varepsilon_1 = 0.9$
$\lambda > 2.5$ μm: $\varepsilon_2 = 0.3$

ANALYSIS

The total hemispherical emissivity at temperature T is given by the relation

$$\varepsilon(T) = \frac{\int_0^{\infty} \varepsilon_\lambda(\lambda,T) E_{\lambda,b}(\lambda,T) d\lambda}{E_b(T)}$$

This integral can be broken into two parts, writing the total hemispherical emissivity as

$$\varepsilon(T) = \frac{\varepsilon_1 \int_0^{2.5} E_{\lambda,b}(\lambda,T) d\lambda}{E_b(T)} + \frac{\varepsilon_2 \int_{2.5}^{\infty} E_{b,\lambda}(\lambda,T) d\lambda}{E_b(T)}$$

which can be written in terms of blackbody radiation functions,

$$\varepsilon(T) = \varepsilon_1 F_{0-2.5} + \varepsilon_2 (1 - F_{0-2.5})$$

We have,

$\lambda T = (2.5\ \mu\text{m})(2200\ \text{K})$

$= 5500\ \mu\text{m}\cdot\text{K}$

so,

$\varepsilon(T) = (0.9)(0.690703) + (0.3)(1 - 0.690703)$

$= \underline{\underline{0.714}}$

Thus, the emissive power is

$$E(T) = \varepsilon(T)E_b = \varepsilon(T)\sigma T^4$$

$E(T) = (0.714)(5.669 \times 10^{-8}\ \text{W/m}^2\cdot\text{K})(2200\ \text{K})^4$

$= \underline{\underline{9.48 \times 10^5\ \text{W/m}^2}}$

DISCUSSION

Spectral, hemispherical emissivity is often a complex function of wavelength. The single step function used here is a simple approximation. For higher accuracy, a multi-step function may be used, as illustrated in the example below.

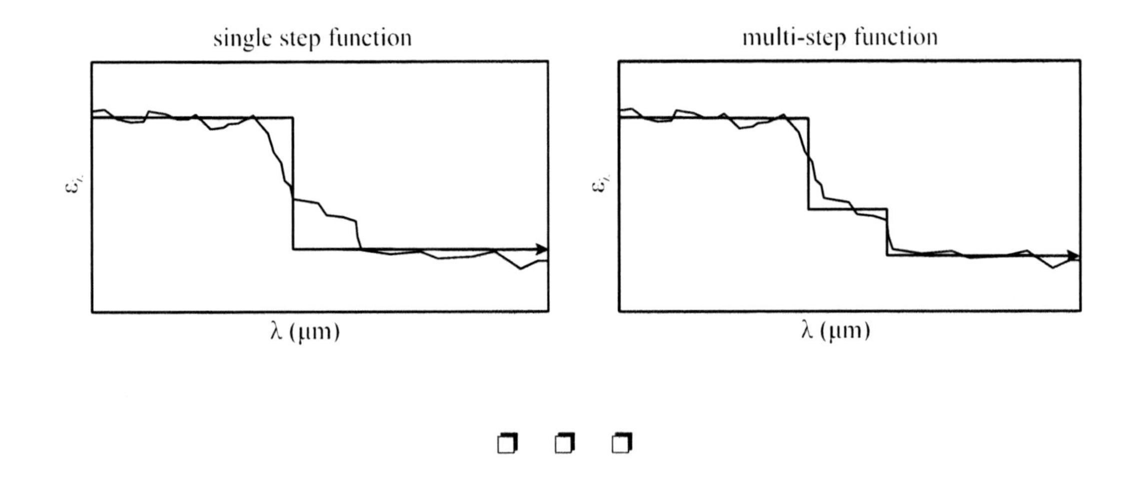

❒ ❒ ❒

PROBLEM 12.9 Surface Temperature of a Mars Space Probe

A spherical space probe on its way to Mars is subjected to a solar heat flux of 850 W/m^2. The solar absorptivity and emissivity of the surface of the probe are 0.40 and 0.90, respectively, and onboard electronics generates 175 W of power. If the diameter of the probe is 1.25 m, find the surface temperature of the probe.

DIAGRAM

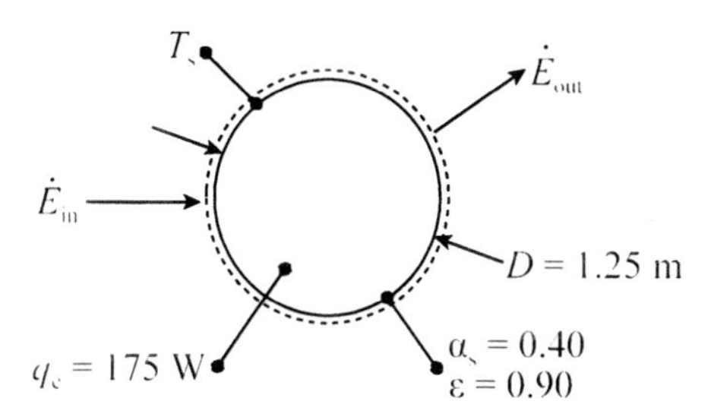

ASSUMPTIONS

1. Steady conditions.
2. Solar heat flux acts uniformly over probe surface.
3. Probe surface is isothermal.
4. Effective temperature of space is 0 K.

PROPERTIES

$\alpha_s = 0.40$ $\varepsilon = 0.90$

ANALYSIS

A surface energy balance on the space probe yields

$$\dot{E}_{in} + q_e = \dot{E}_{out}$$

where

$$\dot{E}_{in} = q_s'' A \alpha_s$$

$$\dot{E}_{out} = \varepsilon \sigma A (T_s^4 - T_{sur}^4)$$

The solar heat flux is $q''_s = 850$ W/m^2 at the location of the probe between Earth and Mars, $A = \pi D^2$ is the surface area of the probe, and consistent with the fourth assumption, $T_{sur} = 0$ K. The quantity q_e is the power generated by the onboard electronics. Substituting these expressions into the energy balance and solving for T_s,

$$T_s = \left(\frac{q_s''\alpha_s + q_e / \pi D^2}{\varepsilon\sigma} \right)^{1/4}$$

$$T_s = \left(\frac{(850 \text{ W/m}^2)(0.40) + 175 \text{ W}/[\pi(1.25 \text{ m})^2]}{(0.90)(5.669 \times 10^{-8} \text{ W/m}^2\cdot\text{K}^4)} \right)^{1/4}$$

$$= 292.9 \text{ K} = \underline{\underline{19.7°\text{C}}}$$

DISCUSSION

The effects of solar absorptivity α_s and emissivity ε are readily seen in the solution for T_s. In the first exceptional case of a perfectly reflecting surface, $\alpha_s = 0$, and the only heat source is the onboard electronics. For this case, the surface temperature is

$$T_s = \left(\frac{q_e}{\varepsilon\sigma\pi D^2} \right)^{1/4}$$

$$T_s = \left(\frac{175 \text{ W}}{(0.90)(5.669 \times 10^{-8} \text{ W/m}^2\cdot\text{K}^4)\pi(1.25 \text{ m})^2} \right)^{1/4}$$

$$= 162.6 \text{ K} = -111°\text{C}$$

In the second exceptional case of a highly absorptive surface, $\alpha_s = 0.95$, and a poorly emitting surface, $\varepsilon = 0.05$, the surface temperature is

$$T_s = \left(\frac{(850 \text{ W/m}^2)(0.95) + 175 \text{ W}/[\pi(1.25 \text{ m})^2]}{(0.05)(5.669 \times 10^{-8} \text{ W/m}^2\cdot\text{K}^4)} \right)^{1/4}$$

$$= 738.5 \text{ K} = 465°\text{C}$$

This temperature suggests widespread thermal-related failure of electronic and optical components and large thermal-induced structural deformations. In the design of spacecraft, careful attention must be paid to radiation surface properties in order to maintain suitable operating temperatures. Generally, a small value for the ratio α_s/ε is desirable, which can be achieved by proper material selection and the use of special surface coatings.

❒ ❒ ❒

PROBLEM 12.10 Wavelength of Maximum Emission for Surfaces at Various Temperatures

Estimate the wavelength corresponding to the maximum emission from the following surfaces: the sun at 5800 K, a fireclay brick at 2750 K, a nichrome filament at 1800 K, a steel forging at 1300 K, and a slab of concrete at 310 K.

DIAGRAM

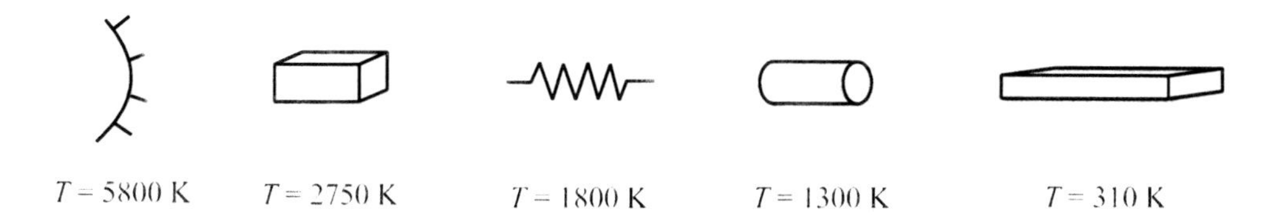

ASSUMPTIONS

1. Each surface emits as a blackbody.

ANALYSIS

The spectral blackbody emissive power is given by the Planck distribution,

$$E_{\lambda,b}(\lambda,T) = \frac{C_1}{\lambda^5\left[\exp(C_2/\lambda T) - 1\right]}$$

where $C_1 = 3.742 \times 10^8$ W·μm^4/m^2 and $C_2 = 1.439 \times 10^4$ μm·K. A graph of the Planck distribution is shown below for several values of temperature. We can see that the blackbody spectral distrubution shows a maximum wavelength λ_{max} that depends on temperature T. The relationship between λ_{max} and T may be found by differentiating the Planck distribution with respect to λ and setting the result equal to zero. This operation leads to Wien's displacement law,

$$\lambda_{max}\, T = 2897.8 \text{ μm·K}$$

which is the locus of points shown by the dashed line in the graph. The following table shows the maximum wavelength for each surface at the temperature indicated.

Surface	Sun (5800 K)	Brick (2750 K)	Filament (1800 K)	Forging (1300 K)	Concrete (310 K)
λ_{max} (μm)	0.500	1.054	1.610	2.229	9.348

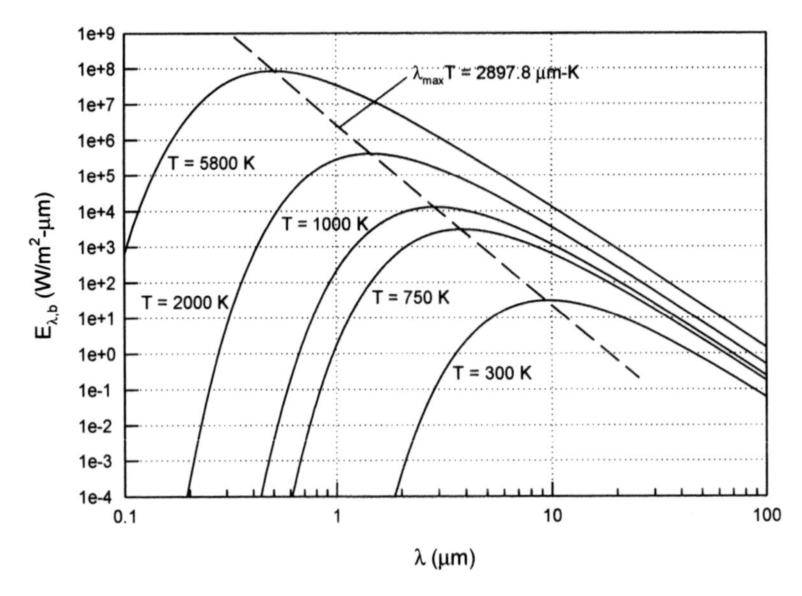

DISCUSSION

The visible light region of the electromagnetic spectrum ranges from about 0.4 to 0.7 µm. The maximum wavelength for the sun falls into this wavelength band, whereas the wavelengths for the other materials fall into the infrared region.

❒ ❒ ❒

PROBLEM 12.11 Total Irradiation from a Spectral Irradiation Distribution

A spectral distribution of surface irradiation is given in the diagram below. What is the total irradiation?

DIAGRAM

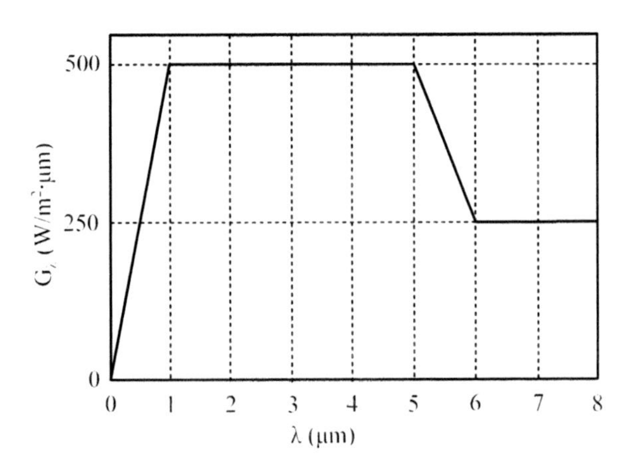

ASSUMPTIONS

1. Spectral surface irradiation is approximated by the distribution shown in the diagram.

PROPERTIES

not applicable

ANALYSIS

Spectral irradiation G_λ is the radiation of a given wavelength λ incident upon a surface, and total irradiation G is the radiation from all directions and at all wavelengths incident upon a surface. Thus,

$$G = \int_0^\infty G_\lambda(\lambda)d\lambda$$

The integral can be evaluated by breaking it into parts.

$$G = \int_0^1 G_\lambda d\lambda + \int_1^5 G_\lambda d\lambda + \int_5^6 G_\lambda d\lambda + \int_6^8 G_\lambda d\lambda$$

$$\begin{aligned} G = {} & \tfrac{1}{2}(1-0)\mu\text{m}(500\ \text{W/m}^2\cdot\mu\text{m}) + (5-1)\mu\text{m}(500-0)\text{W/m}^2\cdot\mu\text{m} \\ & + \tfrac{1}{2}(6-5)\mu\text{m}(500-250)\text{W/m}^2\cdot\mu\text{m} + (6-5)\mu\text{m}(250-0)\text{W/m}^2\cdot\mu\text{m} \\ & + (8-6)\mu\text{m}(250-0)\text{W/m}^2\cdot\mu\text{m} \end{aligned}$$

$$= \underline{\underline{3125\ \text{W/m}^2}}$$

DISCUSSION

The irradiation spectral distribution given here should be considered an approximation of a more complex spectral distribution, an example of which is shown below. Straight-line relationships were used to simplify the integration.

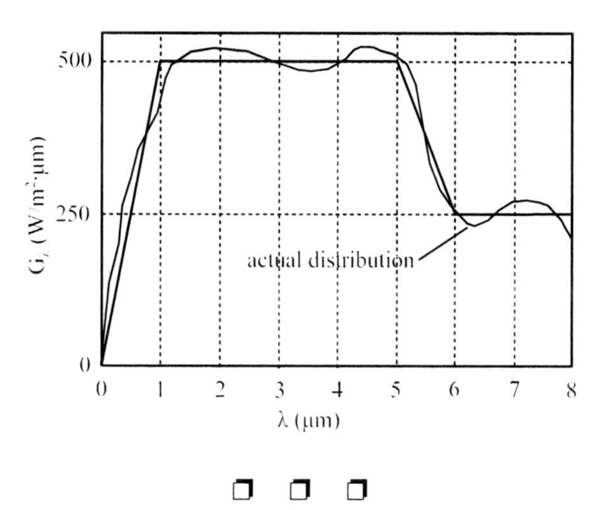

❒ ❒ ❒

PROBLEM 12.12 **Absorptivity and Radiative Heat Flux from Spectral Reflectivity and Irradiation Distributions**

An opaque surface has the spectral, hemispherical reflectivity shown in the diagram. This surface is subjected to the spectral irradiation shown. Determine the spectral, hemispherical absorptivity distribution, the total irradiation, the radiative heat flux absorbed by the surface, and the total, hemispherical absorptivity of the surface.

DIAGRAM

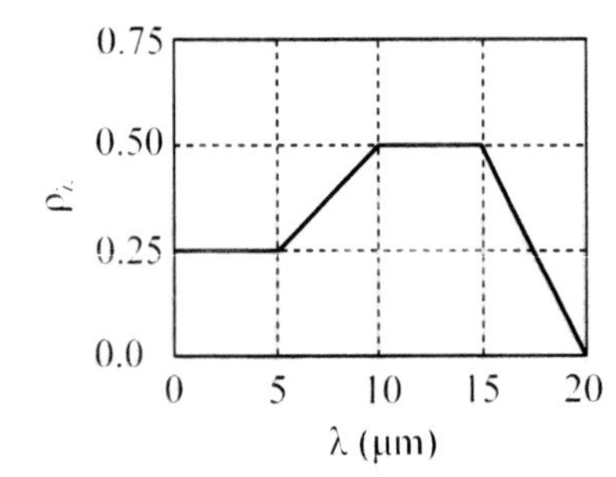

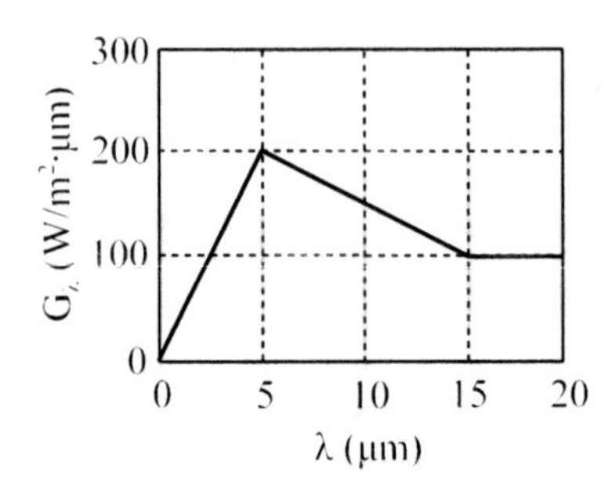

ASSUMPTIONS

1. Spectral, hemispherical reflectivity is approximated by the distribution shown in the diagram.
2. Spectral surface irradiation is approximated by the distribution shown in the diagram.
3. Surface is opaque.

PROPERTIES

ρ_λ varies as shown in the diagram

ANALYSIS

Because the surface is opaque, the spectral absorptivity α_λ and reflectivity ρ_λ are related by

$$\alpha_\lambda + \rho_\lambda = 1$$

Hence, the spectral, hemispherical absorptivity distribution looks like the following:

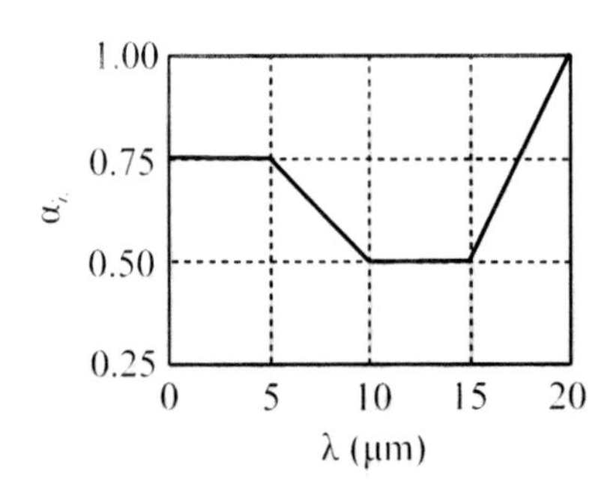

Total irradiation G is the radiation from all directions and at all wavelengths incident upon a surface. Thus,

$$G = \int_0^{\infty} G_\lambda d\lambda$$

The integral can be evaluated by breaking it into parts.

$$G = \int_0^5 G_\lambda d\lambda + \int_5^{15} G_\lambda d\lambda + \int_{15}^{20} G_\lambda d\lambda$$

$$G = \tfrac{1}{2}(5 - 0)\mu\text{m}(200 - 0)\text{W/m}^2\cdot\mu\text{m} + \tfrac{1}{2}(15 - 5)\mu\text{m}(200 - 100)\text{W/m}^2\cdot\mu\text{m}$$
$$+ (15 - 5)\mu\text{m}(100 - 0)\text{W/m}^2\cdot\mu\text{m} + (20 - 15)\mu\text{m}(100 - 0)\text{W/m}^2\cdot\mu\text{m}$$

$$= \underline{\underline{2500\ \text{W/m}^2}}$$

The irradiation absorbed by the surface is

$$G_{abs} = \int_0^{\infty} \alpha_\lambda G_\lambda d\lambda$$

This integral can be evaluated by breaking it into four parts,

$$G_{abs} = \alpha_1 \int_0^5 G_\lambda d\lambda + \int_5^{10} \alpha_\lambda G_\lambda d\lambda + \alpha_3 \int_{10}^{15} G_\lambda d\lambda + G_4 \int_{15}^{20} \alpha_\lambda d\lambda$$

where $\alpha_1 = 0.75$, $\alpha_3 = 0.50$ and $G_4 = 100$ W/m^2·μm. The first integral in the above expression, denoted I_1, is

$$I_1 = (0.75)\tfrac{1}{2}(5 - 0)\mu\text{m}(200 - 0)\text{W/m}^2\cdot\mu\text{m}$$

$$= 375.0\ \text{W/m}^2$$

For the second integral, we must first obtain functions for α_λ and G_λ. This is readily accomplished by examining the distributions for the spectral irradiation and spectral, hemispherical absorptivity. Hence, the second integral, denoted I_2, is

$$I_2 = \int_5^{10} (-0.05\lambda + 1.0)(-10\lambda + 250)d\lambda = \int_5^{10} \left(0.5\lambda^2 - 22.5\lambda + 250\right)d\lambda$$

$$= \left(\frac{0.5\lambda^3}{3} - \frac{22.5\lambda^2}{2} + 250\lambda\right)\Bigg|_5^{10}$$

$$= (1541.7 - 989.6)\text{W/m}^2$$

$$= 552.1\ \text{W/m}^2$$

The third integral, denoted I_3, is

$$I_3 = (0.50)[\tfrac{1}{2}(15 - 10)\mu\text{m}(150 - 100)\text{W/m}^2\cdot\mu\text{m} + (15 - 10)\mu\text{m}(100 - 0)\text{W/m}^2\cdot\mu\text{m}]$$

$= 312.5\ \mathrm{W/m^2}$

The fourth integral, denoted I_4, is

$$I_4 = (100\ \mathrm{W/m^2 \cdot \mu m})[\tfrac{1}{2}(20 - 15)\mu\mathrm{m}(1.0 - 0.5) + (20 - 15)\mu\mathrm{m}(0.5 - 0)]$$

$$= 375.0\ \mathrm{W/m^2}$$

Thus, the irradiation absorbed by the surface, i.e., the radiative heat flux absorbed by the surface, is

$$G_{\mathrm{abs}} = (375.0 + 552.1 + 312.5 + 375.0)\mathrm{W/m^2}$$

$$= \underline{\underline{1614.6\ \mathrm{W/m^2}}}$$

The total hemispherical absorptivity is defined as the fraction of the total irradiation absorbed by the surface.

$$\alpha \equiv \frac{G_{abs}}{G}$$

$$\alpha = \frac{1614.6\ \mathrm{W/m^2}}{2500\ \mathrm{W/m^2}}$$

$$= \underline{\underline{0.646}}$$

DISCUSSION

In general, irradiation interacts with a semitransparent medium, such as glass or a clear liquid. Hence,

$$\alpha_\lambda + \rho_\lambda + \tau_\lambda = 1$$

Because the material under consideration is opaque, $\tau_\lambda = 0$. Upon inspection of the spectral, hemispherical absorptivity distribution, our answer for the total hemispherical absorptivity looks correct.

❒ ❒ ❒

PROBLEM 12.13 Radiosity from Spectral Reflectivity and Irradiation Distributions

An opaque surface has the spectral, hemispherical reflectivity shown in the diagram. This surface is subjected to the spectral irradiation shown. If the surface emission is 750 W/m^2, what is the radiosity?

DIAGRAM

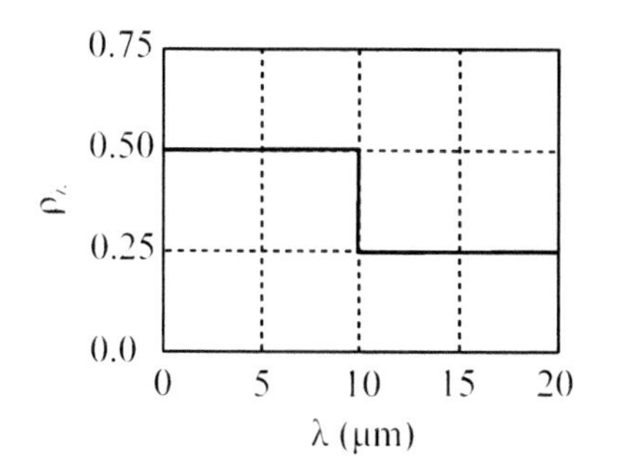

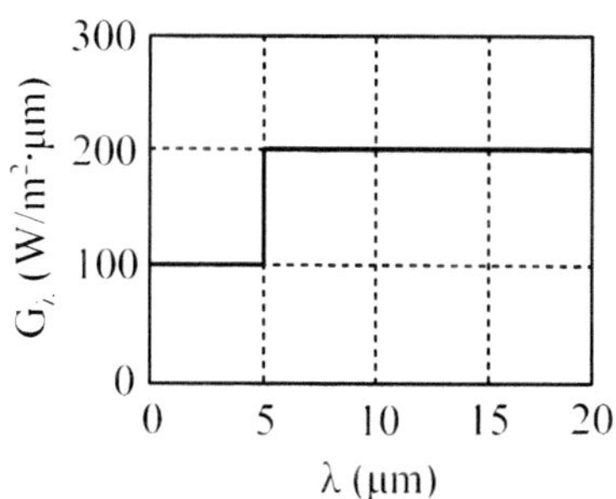

ASSUMPTIONS

1. Spectral, hemispherical reflectivity is approximated by the distribution shown in the diagram.
2. Spectral surface irradiation is approximated by the distribution shown in the diagram.
3. Surface is opaque.

PROPERTIES

ρ_λ varies as shown in the diagram

ANALYSIS

Total irradiation G is the radiation from all directions and at all wavelengths incident upon a surface. Thus,

$$G = \int_0^\infty G_\lambda d\lambda$$

The integral can be evaluated by breaking it into parts.

$$G = \int_0^5 G_\lambda d\lambda + \int_5^{20} G_\lambda d\lambda$$

$$G = (5 - 0)\mu\text{m}(100 - 0)\text{W/m}^2\cdot\mu\text{m} + (20 - 5)\mu\text{m}(200 - 0)\text{W/m}^2\cdot\mu\text{m}$$

$$= 3500 \text{ W/m}^2$$

The irradiation reflected by the surface is

$$G_{ref} = \int_0^\infty \rho_\lambda G_\lambda d\lambda$$

$$G_{ref} = (0.5)[(5 - 0)\mu\text{m}(100 - 0)\text{W/m}^2\cdot\mu\text{m}] + (0.5)[(10 - 5)\mu\text{m}(200 - 0)\text{W/m}^2\cdot\mu\text{m}]$$

$$+ (0.25)[(20 - 10)\mu\text{m}(200 - 0)\text{W/m}^2\cdot\mu\text{m}]$$

$$= 1250 \text{ W/m}^2$$

Radiosity J is defined as the total radiation leaving a surface, which includes the emitted radiation plus the reflected portion of the irradiation. Thus,

$$J \equiv E + G_{ref}$$

$J = (750 + 1250)\text{W/m}^2$

$= \underline{\underline{2000\ \text{W/m}^2}}$

DISCUSSION

Total, hemispherical reflectivity ρ is defined as the fraction of the total irradiation reflected by the surface.

$$\rho \equiv \frac{G_{ref}}{G}$$

$\rho = (1250\ \text{W/m}^2)/(3500\ \text{W/m}^2)$

$= 0.357$

❒ ❒ ❒

PROBLEM 12.14 Transmitted Radiation Through a Thin Film of Aluminum

A thin film of aluminum is subjected to a total, hemispherical irradiation of 4000 W/m^2. It has been demonstrated experimentally that the film transmits radiation over a very narrow range of wavelengths, as shown below in a graph of the spectral, hemispherical transmissivity. Find the radiation that is transmitted through the aluminum film.

DIAGRAM

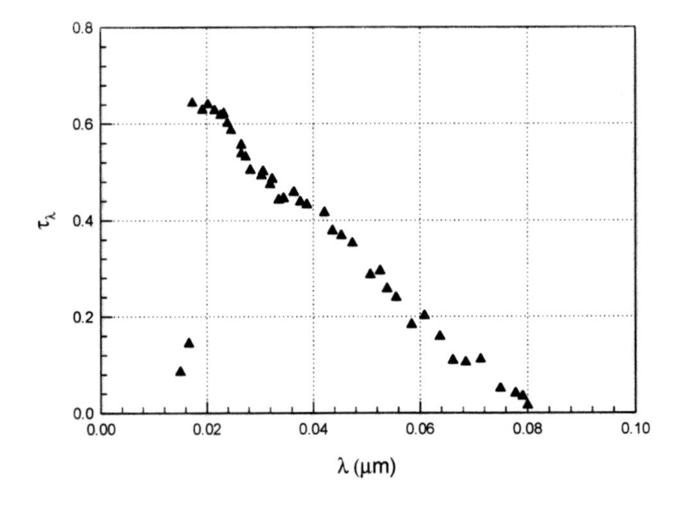

ASSUMPTIONS

1. Spectral, hemispherical transmissivity is given by the data shown in the diagram.
2. Irradiation is not a function of wavelength.

PROPERTIES

τ_λ varies as shown in the diagram

ANALYSIS

The transmitted irradiation is given by the relation

$$G_{tr} = \int_0^\infty \tau_\lambda G_\lambda d\lambda$$

which, because the irradiation is not a function of wavelength, can be expressed as

$$G_{tr} = G\int_0^\infty \tau_\lambda d\lambda$$

In order to integrate the transmissivity, a functional approximation of the data points is made. The data points suggest a step function followed by a linear function, as shown below.

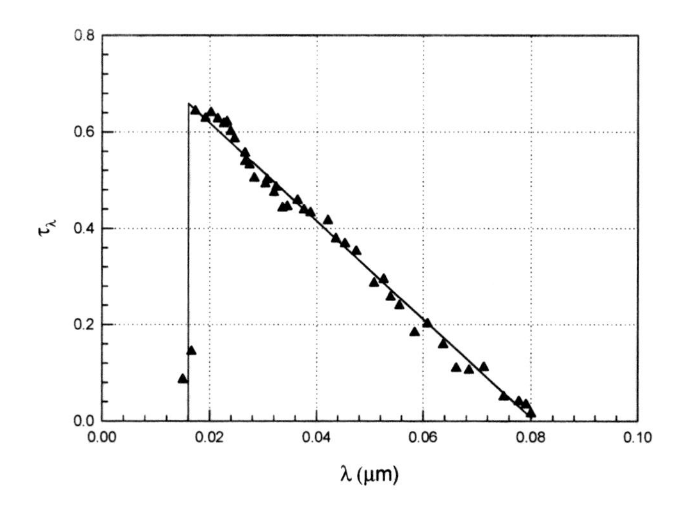

After finding the slope and the intercept of the linear function, the transmitted irradiation may be expressed as

$$G_{tr} = G\int_{0.016}^{0.08}(-10.31\lambda + 0.825)d\lambda$$

Hence,

$$G_{tr} = (4000)\left(-\frac{10.31\lambda^2}{2} + 0.825\lambda\right)\Bigg|_{0.016}^{0.08}$$

$$= (4000)(0.0330 - 0.0119)\ \text{W/m}^2$$

$$= \underline{\underline{84.5\ \text{W/m}^2}}$$

DISCUSSION

Total, hemispherical transmissivity τ is defined as the fraction of irradiation transmitted

$$\tau \equiv \frac{G_{tr}}{G}$$

$$\tau = (84.5\ \text{W/m}^2)/(4000\ \text{W/m}^2)$$

$$= 0.021$$

The transmissivity of the aluminum film would be much larger if the range of wavelengths over which the thin film transmits radiation was larger. The film reflects and absorbs 97.9 percent of the irradiation.

❒ ❒ ❒

PROBLEM 12.15 Radiation Properties for an Opaque Surface Subjected to Solar Irradiation

An opaque surface that measures 1 m² is maintained at 500 K while subjected to a solar irradiation of 875 W/m². The surface is diffuse and it's spectral absorptivity follows the distribution shown in the diagram. Find the absorbed irradiation, emissive power, radiosity and net radiation heat transfer from the surface.

DIAGRAM

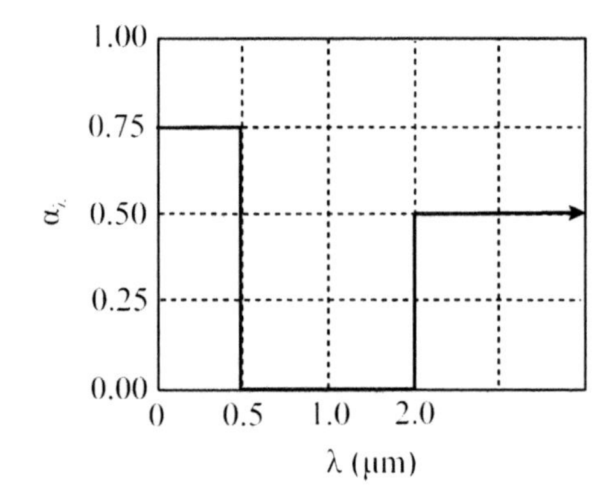

ASSUMPTIONS

1. Surface is opaque and diffuse.
2. Spectral distribution of solar irradiation corresponds to emission from a blackbody at 5800 K.

PROPERTIES

α_λ varies as shown in the diagram

ANALYSIS

The absorptivity of the surface for solar irradiation is

$$\alpha_s = \frac{\int_0^\infty \alpha_\lambda G_\lambda d\lambda}{G}$$

which, by the second assumption, may be expressed as

$$\alpha_s = \frac{\int_0^\infty \alpha_\lambda E_{\lambda,b}(5800\ \text{K}) d\lambda}{E_b}$$

Using blackbody radiation functions, the solar absorptivity may be written as

$$\alpha_s = \alpha_1 F_{0-\lambda_1} + \alpha_2 F_{\lambda_2-\infty}$$

where $\alpha_1 = 0.75$, $\alpha_2 = 0.50$, $\lambda_1 = 0.5$ μm and $\lambda_2 = 2.0$ μm.

$\lambda_1 T = (0.5\ \mu\text{m})(5800\ \text{K})$

$= 2900\ \mu\text{m}\cdot\text{K}$

$\lambda_2 T = (2.0\ \mu\text{m})(5800\ \text{K})$

$= 11{,}600\ \mu\text{m}\cdot\text{K}$

From a table of blackbody radiation functions,

$F_{0-\lambda 1} = 0.250565$, $F_{0-\lambda 2} = 0.940987$

Hence,

$\alpha_s = (0.75)(0.250565) + (0.50)(1 - 0.940987)$

$= 0.217$

The absorbed irradiation is

$$G_{abs} = \alpha_s G_s$$

$G_{abs} = (0.217)(875\ \text{W/m}^2)$

$= \underline{\underline{190\ \text{W/m}^2}}$

The emissivity of the surface is

$$\varepsilon = \frac{\int_0^\infty \varepsilon_\lambda E_{\lambda,b}(500\,\text{K})d\lambda}{E_b}$$

Using blackbody radiation functions, the emissivity may be written as

$$\varepsilon = \varepsilon_1 F_{0-\lambda_1} + \varepsilon_2 F_{\lambda_2-\infty}$$

where $\varepsilon_1 = \alpha_1 = 0.75$ and $\varepsilon_2 = \alpha_2 = 0.50$.

$\lambda_1 T = (0.5\ \mu\text{m})(500\ \text{K})$

$= 250\ \mu\text{m·K}$

$\lambda_2 T = (2.0\ \mu\text{m})(500\ \text{K})$

$= 1000\ \mu\text{m·K}$

From a table of blackbody radiation functions,

$F_{0-\lambda 1} = 0.000000$, $F_{0-\lambda 2} = 0.000321$

Hence,

$\varepsilon = (0.75)(0.000000) + (0.50)(1 - 0.0000321)$

$= 0.500$

The emissive power is

$$E = \varepsilon\sigma T_s^4$$

$E = (0.500)(5.669 \times 10^{-8}\ \text{W/m}^2\text{·K}^4)(500\ \text{K})^4$

$= \underline{\underline{1772\ \text{W/m}^2}}$

Radiosity J is the total radiation leaving a surface, which includes the emitted radiation plus the reflected portion of the irradiation. Thus,

$$J = E + (1 - \alpha_s)G_s$$

$J = 1772 \text{ W/m}^2 + (1 - 0.217)(875 \text{ W/m}^2)$

$= \underline{\underline{2457 \text{ W/m}^2}}$

The net radiation heat transfer from the surface is the difference of the radiation emitted from the surface and the radiation absorbed by the surface.

$$q_{net} = (E - G_{abs})A$$

$q_{net} = [(1772 - 190)\text{W/m}^2](1 \text{ m}^2)$

$= \underline{\underline{1582 \text{ W}}}$

DISCUSSION

Because the net radiation heat transfer from the surface is a positive quantity, the surface temperature will decrease with time unless the surface is supplied with 1582 W of power, from a fluid, for example.

❒ ❒ ❒

PROBLEM 12.16 Cooling Time of a Heat-Treated Steel Rod with a Known Spectral, Hemispherical Emissivity Distribution

A 2-cm diameter steel rod emerges from a heat treating oven at 1300 K into a room where the temperature of the surroundings is 300 K. The density and specific heat of the steel are 7850 kg/m^3 and 450 J/kg·K, respectively. As the rod cools, it's surface oxidizes, giving rise to a total, hemispherical emissivity of the form $\varepsilon(T) = \varepsilon(T_i)(T_i/T)$, where $\varepsilon(T_i)$ is the emissivity of the surface at $T_i = 1300$ K, the initial temperature of the rod. Neglecting convection, how long will it take for the rod to reach a temperature of 500 K if the spectral, hemispherical emissivity at the initial temperature follows the distribution shown in the diagram?

DIAGRAM

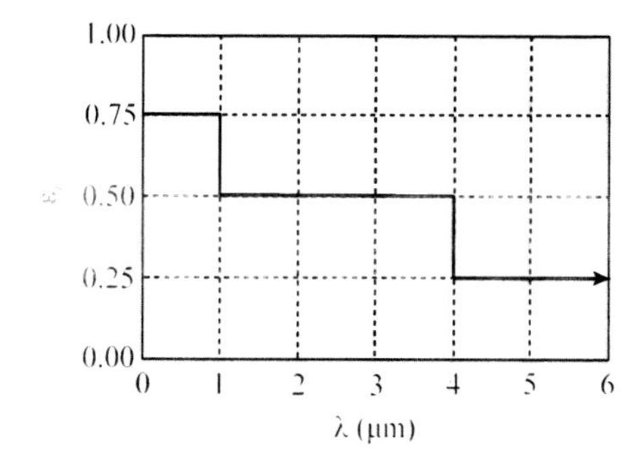

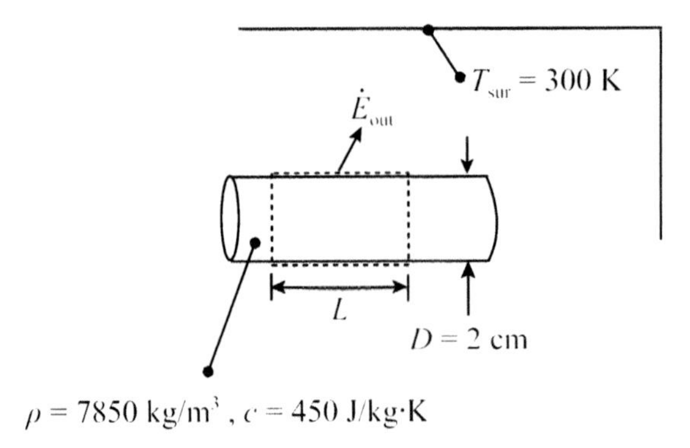

ASSUMPTIONS

1. The rod is isothermal.
2. Convection is neglected.
3. Radiation from ends of rod is neglected.
4. The surroundings act as a large enclosure.

PROPERTIES

steel: ρ = 7850 kg/m^3, c_p = 450 J/kg·K
ε_λ varies as shown in the diagram

ANALYSIS

The total hemispherical emissivity of the rod at temperature T_i is given by the relation

$$\varepsilon(T_i) = \frac{\int_0^\infty \varepsilon_\lambda(\lambda, T_i) E_{\lambda,b}(\lambda, T_i) d\lambda}{E_b(T_i)}$$

This integral can be broken into three parts, writing the total hemispherical emissivity as

$$\varepsilon(T_i) = \frac{\varepsilon_1 \int_0^1 E_{\lambda,b}(\lambda, T_i) d\lambda}{E_b(T_i)} + \frac{\varepsilon_2 \int_1^4 E_{\lambda,b}(\lambda, T_i) d\lambda}{E_b(T_i)} + \frac{\varepsilon_3 \int_4^\infty E_{\lambda,b}(\lambda, T_i) d\lambda}{E_b(T_i)}$$

which can be written in terms of blackbody radiation functions,

$$\varepsilon(T_i) = \varepsilon_1 F_{0-1} + \varepsilon_2 (F_{0-4} - F_{0-1}) + \varepsilon_3 (1 - F_{0-4})$$

We have

$$\lambda_1 T_i = (1\ \mu\text{m})(1300\ \text{K})$$

$$= 1300\ \mu\text{m}\cdot\text{K}$$

$$\lambda_2 T_i = (4\ \mu\text{m})(1300\ \text{K})$$

$$= 5200\ \mu\text{m}\cdot\text{K}$$

Using a table of blackbody radiation functions,

$$\varepsilon(T_i) = (0.75)(0.004962) + (0.50)(0.658970 - 0.004962) + (0.25)(1 - 0.658970)$$

$$= 0.416$$

An energy balance on a control volume of the rod yields

$$\dot{E}_{in} - \dot{E}_{out} = \dot{E}_{st}$$

The rate of energy transfer into the control volume is zero, and the rate of energy transfer out of the control volume is

$$\dot{E}_{out} = \varepsilon\sigma A\left(T^4 - T_{sur}^4\right)$$

where $A = \pi DL$, the surface area of the rod, and $T_{sur} = 300$ K. The rate of energy storage within the rod is

$$\dot{E}_{st} = \rho c V \frac{dT}{dt}$$

where $V = \pi D^2 L$, the volume of the control volume. Substituting these quantities into the energy balance and simplifying,

$$\frac{dT}{dt} = -C\varepsilon(T_i)T_i\left(T^3 - T_{sur}^4/T\right)$$

where,

$$C = \frac{4\sigma}{\rho c D}$$

$$C = \frac{4(5.669 \times 10^{-8}\ \text{W/m}^2\cdot\text{K}^4)}{(7850\ \text{kg/m}^3)(450\ \text{J/kg}\cdot\text{K})(0.02\ \text{m})}$$

$$= 3.2096 \times 10^{-12}\ \text{K}^3/\text{s}$$

Separating variables and integrating,

$$\int_{T_i}^{T} \frac{dT}{T^3 - T_{sur}^4/T} = -C\varepsilon(T_i)T_i \int_0^t dt$$

Evaluating the integrals, the cooling time is

$$t = \frac{1}{2C\varepsilon(T_i)T_i T_{sur}^2}\left[\tanh^{-1}\left(\frac{T_{sur}^2}{T^2}\right) - \tanh^{-1}\left(\frac{T_{sur}^2}{T_i^2}\right)\right]$$

$$t = \frac{1}{2(3.2096 \times 10^{-12}\ \mathrm{K^3/s})(0.416)(1300\ \mathrm{K})(300\ \mathrm{K})^2}\left[\tanh^{-1}\left(\frac{300\ \mathrm{K}}{500\ \mathrm{K}}\right)^2 - \tanh^{-1}\left(\frac{300\ \mathrm{K}}{1300\ \mathrm{K}}\right)^2\right]$$

$$= 1036\ \mathrm{s} = \underline{\underline{17.3\ \mathrm{min}}}$$

DISCUSSION

Upon inspection of the spectral, hemispherical emissivity distribution, the total, hemispherical emissivity appears to be correct. The cooling time for the steel rod would be lower if the second assumption was removed. If convection was included in the analysis, the rate of energy transfer out of the control volume would be

$$\dot{E}_{out} = hA(T - T_\infty) + \varepsilon\sigma A(T^4 - T_{sur}^4)$$

where T_∞ is the free stream fluid temperature. The inclusion of convection would render the integration more difficult, probably requiring a numerical approach.

❒ ❒ ❒

PROBLEM 12.17 Net Heat Flux from a Radiating Surface with Convection

A flat plate that behaves as an opaque, gray surface at 30°C is subjected to an irradiation of 1200 W/m². Atmospheric air at 15°C flows across the surface, giving rise to a heat transfer coefficient of 22 W/m²·K. If the surface of the plate reflects 750 W/m², what is the net heat flux from the surface?

DIAGRAM

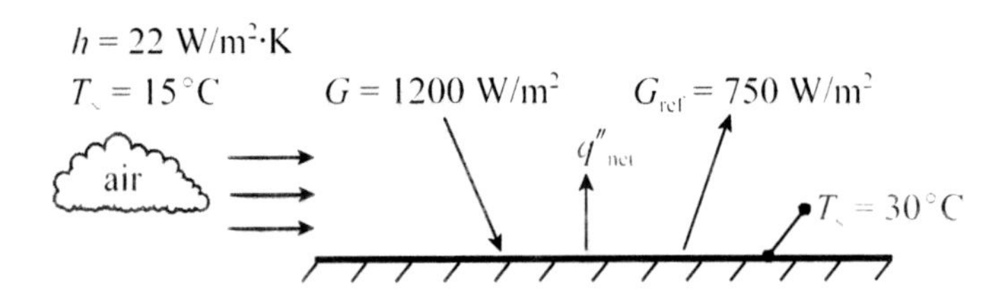

ASSUMPTIONS

1. The surface is isothermal and diffuse.
2. The surface is opaque and gray.
3. Irradiation includes all thermal effects of surroundings.

PROPERTIES

not applicable

ANALYSIS

From an energy balance on the surface of the plate, the net heat flux from the surface is

$$q''_{net} = \frac{\left(\dot{E}_{out} - \dot{E}_{in}\right)}{A}$$

The rate of energy transfer into the surface is the irradiation

$$\dot{E}_{in} = GA$$

and the rate of energy transfer out of the surface is

$$\dot{E}_{out} = hA\left(T_s - T_\infty\right) + \varepsilon\sigma A T_s^4 + G_{ref} A$$

Substituting these expressions into the energy balance,

$$q''_{net} = h\left(T_s - T_\infty\right) + \varepsilon\sigma T_s^4 + G_{ref} - G$$

Because the surface is opaque and gray,

$$\varepsilon = \alpha = 1 - \rho = 1 - G_{ref}/G$$

$\varepsilon = 1 - (750\ \mathrm{W/m^2})/(1200\ \mathrm{W/m^2})$

$= 0.375$

Hence, the net heat flux out of the surface is

$q''_{net} = (22\ \mathrm{W/m^2{\cdot}K})(30 - 15)\mathrm{K} + (0.375)(5.669 \times 10^{-8}\ \mathrm{W/m^2{\cdot}K})[(30 + 273)\mathrm{K}]^4$

$+ 750\ \mathrm{W/m^2} - 1200\ \mathrm{W/m^2}$

$= \underline{\underline{59.2\ \mathrm{W/m^2}}}$

DISCUSSION

The radiosity is the sum of the reflected radiation and emitted radiation from the surface

$$J = G_{ref} + \varepsilon\sigma T_s^4$$

$$J = 750\ \text{W/m}^2 + (0.375)(5.669 \times 10^{-8}\ \text{W/m}^2\text{·K})[(30 + 273)\text{K}]^4$$

$$= 929\ \text{W/m}^2$$

❒ ❒ ❒

PROBLEM 12.18 Combined Heat Transfer for a Nylon Plate

The top surface of a 1.2-cm thick plate of nylon ($k = 0.25$ W/m·K) is subjected to an irradiation of 1800 W/m^2 and exposed to stagnant air at 20°C, giving rise to a heat transfer coefficient of 15 W/m^2·K. The bottom surface of the plate is subjected to the flow of 10°C air, giving rise to a heat transfer coefficient of 40 W/m^2·K. If the spectral, hemispherical emissivity of the top surface has the distribution shown in the diagram, find the surface temperatures of the nylon plate. The nylon is opaque, and radiation from the bottom surface can be neglected.

DIAGRAM

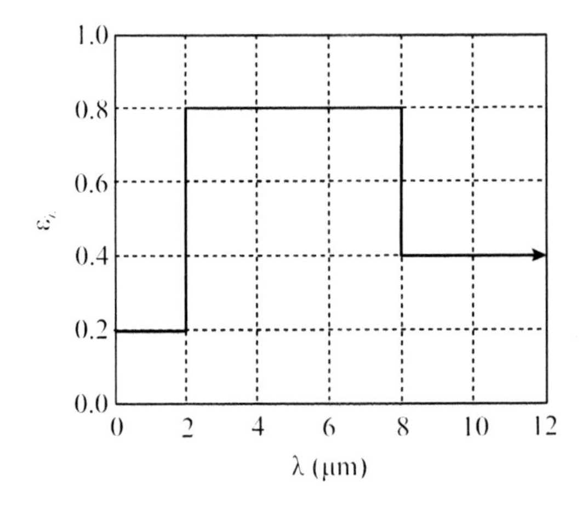

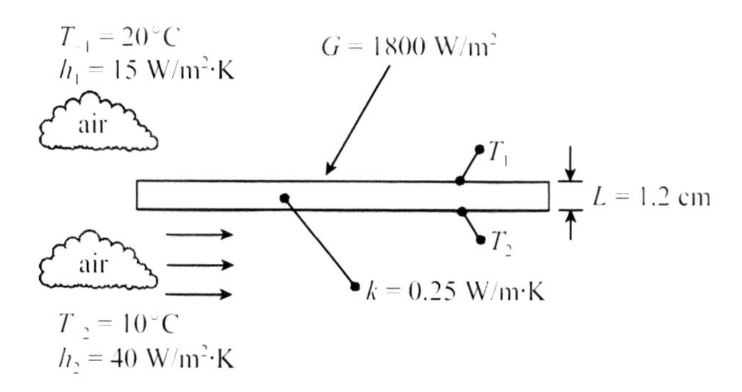

ASSUMPTIONS

1. Steady conditions.
2. The nylon is opaque.
3. The air near the top surface constitutes the surroundings for radiation.
4. Spectral, hemispherical emissivity has the distribution shown in the diagram.
5. Radiation from the bottom surface is neglected.

PROPERTIES

nylon: $k = 0.25$ W/m·K
ε_λ varies as shown in the diagram

ANALYSIS

The total hemispherical emissivity of the top surface at temperature T_1 is given by the relation

$$\varepsilon(T_1) = \frac{\int_0^\infty \varepsilon_\lambda(\lambda, T_1) E_{\lambda,b}(\lambda, T_1) d\lambda}{E_b(T_1)}$$

This integral can be broken into three parts, writing the total hemispherical emissivity as

$$\varepsilon(T_1) = \frac{\varepsilon_1 \int_0^2 E_{\lambda,b}(\lambda, T_1) d\lambda}{E_b(T_1)} + \frac{\varepsilon_2 \int_2^8 E_{\lambda,b}(\lambda, T_1) d\lambda}{E_b(T_1)} + \frac{\varepsilon_3 \int_8^\infty E_{\lambda,b}(\lambda, T_1) d\lambda}{E_b(T_1)}$$

which can be written in terms of blackbody radiation functions,

$$\varepsilon(T_1) = \varepsilon_1 F_{0-2} + \varepsilon_2 (F_{0-8} - F_{0-2}) + \varepsilon_3 (1 - F_{0-8})$$

We do not yet know the surface temperature T_1, so we estimate its value in order to calculate the total, hemispherical emissivity. We perform a surface energy balance and then make a correction to T_1 if required. Beginning with a value of $T_1 = 350$ K, we have

$$\lambda_1 T_1 = (2\ \mu\text{m})(350\ \text{K})$$

$$= 700\ \mu\text{m}\cdot\text{K}$$

$$\lambda_2 T_1 = (8\ \mu\text{m})(350\ \text{K})$$

$$= 2800\ \mu\text{m}\cdot\text{K}$$

Using a table of blackbody radiation functions,

$$\varepsilon(T_1) = (0.20)(0.000008) + (0.80)(0.227897 - 0.000008) + (0.40)(1 - 0.227897)$$

$$= 0.491$$

An energy balance on the top surface of the plate is written as

$$\dot{E}_{in} - \dot{E}_{out} = 0$$

where the rate of energy transfer into the surface is

$$\dot{E}_{in} = GA$$

and the rate of energy transfer out of the surface is

$$\dot{E}_{out} = \varepsilon\sigma A\left(T_1^4 - T_{sur}^4\right) + h_1 A\left(T_1 - T_{\infty 1}\right)$$

Setting $T_{\text{sur}} = T_{\infty 1}$, substituting these expressions into the energy balance, and numerically solving for T_1,

$$T_1 = 386\ \text{K}$$

Our estimate was $T_1 = 350$ K, so we recalculate the total hemispherical emissivity. We have

$$\lambda_1 T_1 = (2\ \mu\text{m})(386\ \text{K})$$

$$= 772\ \mu\text{m}\cdot\text{K}$$

$$\lambda_2 T_1 = (8\ \mu\text{m})(386\ \text{K})$$

$$= 3088\ \mu\text{m}\cdot\text{K}$$

Using a table of blackbody radiation functions,

$$\varepsilon(T_1) = (0.20)(0.000014) + (0.80)(0.292975 - 0.000014) + (0.40)(1 - 0.292975)$$

$= 0.517$

Based on this corrected emissivity value, the new surface temperature is

$T_1 = 385 \text{ K} = \underline{\underline{112°\text{C}}}$

No further corrections are required. Now that T_1 is known, we can find T_2 by setting the heat conducted through the plate equal to the heat convected from the bottom surface,

$$k\frac{(T_1 - T_2)}{L} = h_2(T_2 - T_{\infty 2})$$

Solving for T_2,

$$T_2 = \frac{kT_1/L + h_2 T_{\infty 2}}{h_2 + k/L}$$

$$T_2 = \frac{(0.25 \text{ W/m·K})(385 \text{ K})/(0.012 \text{ m}) + (40 \text{ W/m}^2\text{·K})(283 \text{ K})}{40 \text{ W/m}^2\text{·K} + (0.25 \text{ W/m·K})/(0.012 \text{ m})}$$

$= 318 \text{ K} = \underline{\underline{44.9°\text{C}}}$

DISCUSSION

Had radiation from the top surface been neglected ($\varepsilon = 0$), the surface temperatures of the plate would have been

$T_1 = 140°\text{C}$, $T_2 = 54.4°\text{C}$

❒ ❒ ❒

PROBLEM 12.19 Distance between a Blackbody Source and a Thermopile Detector

A blackbody source consisting of a laboratory furnace is used to calibrate a thermopile with a detector area of 30 mm^2. The furnace has an operating temperature of 2000 K and an aperture area of 40 mm^2. A filter with the spectral transmissivity shown in the diagram is placed between the furnace and the detector. At what distance from the furnace should the detector be placed to provide an irradiation at the detector of 75 W/m^2?

DIAGRAM

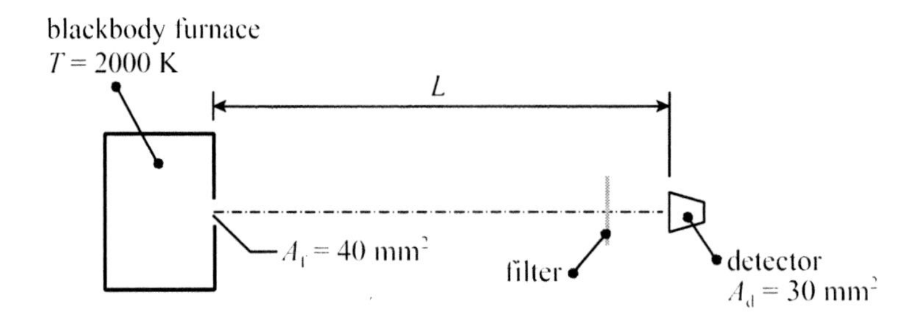

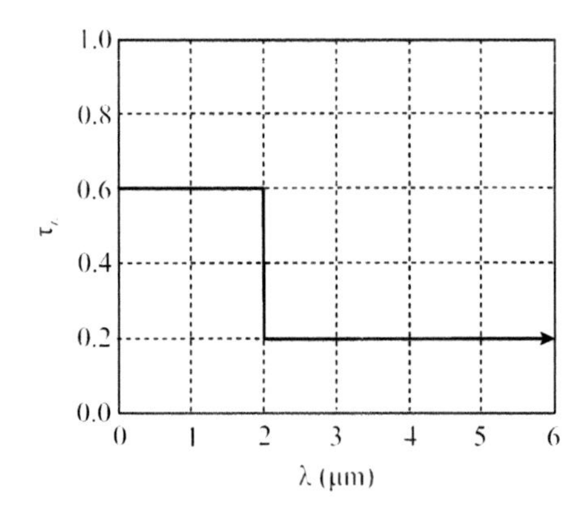

ASSUMPTIONS

1. Blackbody furnace aperture emits diffusely.
2. Spectral transmissivity of the filter has the distribution shown in the diagram.
3. $A_d^2 << L^2$.

PROPERTIES

τ_λ varies as shown in the diagram

ANALYSIS

The total transmissivity at temperature T is given by the relation

$$\tau(T) = \frac{\int_0^\infty \tau_\lambda E_{\lambda,b}(\lambda, T)d\lambda}{E_b(T)}$$

This integral can be broken into two parts, writing the total transmissivity as

$$\tau(T) = \frac{\tau_1 \int_0^2 E_{\lambda,b}(\lambda,T)d\lambda}{E_b(T)} + \frac{\tau_2 \int_2^\infty E_{\lambda,b}(\lambda,T)d\lambda}{E_b(T)}$$

which can be written in terms of blackbody radiation functions,

$$\tau(T) = \tau_1 F_{0-2} + \tau_2 (1 - F_{0-\infty})$$

We have

$$\lambda_1 T = (2\ \mu\text{m})(1250\ \text{K})$$

$$= 2500\ \mu\text{m}\cdot\text{K}$$

Using a table of blackbody radiation functions,

$$\tau(T) = (0.60)(0.161688) + (0.20)(1 - 0.161688)$$

$$= 0.265$$

The irradiation on the detector is the heat flux received by the detector divided by its surface area,

$$G_d = \frac{q}{A_d}$$

The heat flux received by the detector is expressed as

$$q = I_f A_f \cos\theta \omega \tau(T)$$

where the intensity of the radiation leaving the furnace is

$$I_f = \frac{\sigma T^4}{\pi}$$

and the solid angle ω is

$$\omega = \frac{A_d}{L^2}$$

A_f is the area of the furnace aperture, and $\theta \approx 0°$. Substituting these expressions into the relation for irradiation and solving for L,

$$L = \left(\frac{\sigma T^4 A_f \tau(T)}{\pi G_d} \right)^{1/2}$$

$$L = \left(\frac{(5.669 \times 10^{-8}\ \text{W/m}^2\cdot\text{K}^4)(2000\ \text{K})^4(40 \times 10^{-6}\ \text{m}^2)(0.265)}{\pi(75\ \text{W/m}^2)} \right)^{1/2}$$

$= 0.2020 \text{ m} = \underline{\underline{20.2 \text{ cm}}}$

DISCUSSION

Note that the solution for L does not contain the detector area A_d. However, we should check that the third assumption, $A_d << L^2$, is satisfied.

$$30 \times 10^{-6} \text{ m}^2 \stackrel{?}{<<} (0.244 \text{ m})^2 << 0.0595 \text{ m}^2$$

It is instructive to examine the relationship between detector irradiation G_d and spacing L. The graph below shows this relationship for three values of the blackbody source temperature.

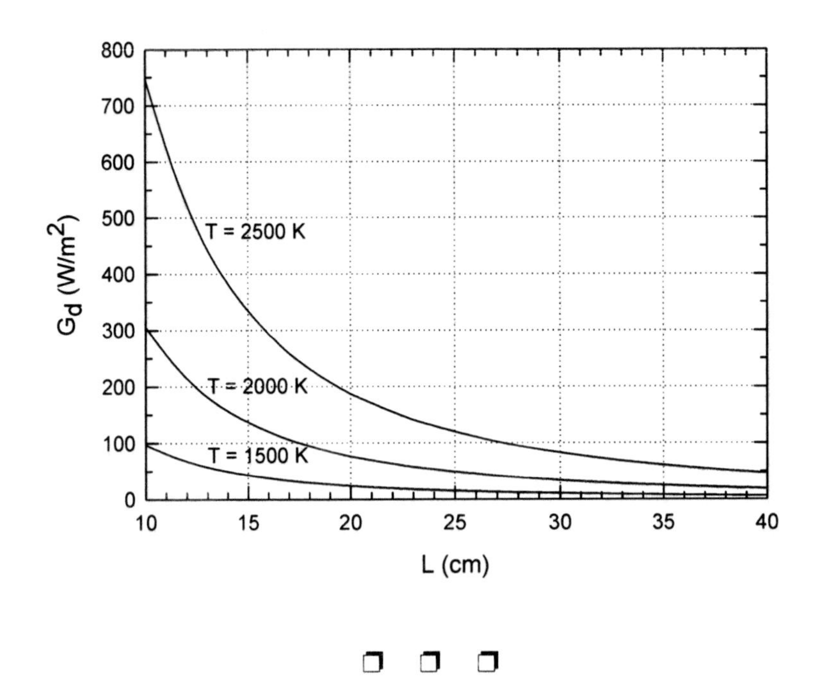

❒ ❒ ❒

PROBLEM 12.20 Radiation Through a Quartz Viewing Window on a Boiler

A 8 cm × 12 cm fused quartz window installed in the wall of gas-fired boiler permits the observation of the flame and conditions of the interior. The interior surfaces of the boiler are maintained at a temperature of 1600 K. If the spectral transmissivity of the quartz window has the distribution shown in the diagram, what is the radiative heat loss through the window?

DIAGRAM

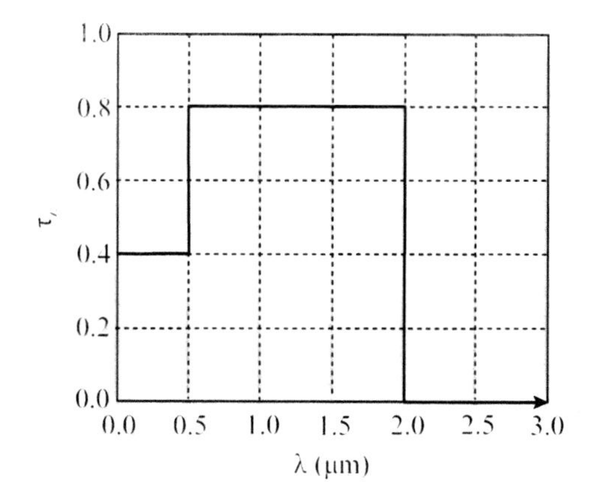

ASSUMPTIONS

1. Surface area of viewing window is much smaller than surface area of interior of boiler.
2. Interior of boiler is isothermal.
3. Boiler emits as a blackbody (large isothermal enclosure).

PROPERTIES

τ_λ varies as shown in the diagram

ANALYSIS

The total transmissivity of the quartz window at temperature T is given by the relation

$$\tau(T) = \frac{\int_0^\infty \tau_\lambda(\lambda, T) E_{\lambda,b}(\lambda, T) d\lambda}{E_b(T)}$$

This integral can be broken into three parts, writing the total hemispherical transmissivity as

$$\tau(T) = \frac{\tau_1 \int_0^{0.5} E_{\lambda,b}(\lambda, T) d\lambda}{E_b(T)} + \frac{\tau_2 \int_{0.5}^{2} E_{\lambda,b}(\lambda, T) d\lambda}{E_b(T)} + \frac{\tau_3 \int_2^\infty E_{\lambda,b}(\lambda, T) d\lambda}{E_b(T)}$$

which can be written in terms of blackbody radiation functions,

$$\tau(T) = \tau_1 F_{0-0.5} + \tau_2 (F_{0-2} - F_{0-0.5}) + \tau_3 (1 - F_{0-2})$$

We have

$\lambda_1 T = (0.5\ \mu\text{m})(1600\ \text{K})$

$= 800\ \mu m \cdot K$

$\lambda_2 T = (2\ \mu m)(1600\ K)$

$= 3200\ \mu m \cdot K$

Using a table of blackbody radiation functions,

$\tau(T) = (0.40)(0.000016) + (0.80)(0.318102 - 0.000016) + (0.0)(1 - 0.318102)$

$= 0.255$

The radiative heat loss through the viewing window is

$$q = \tau A \sigma T^4$$

where A is the surface area of the window, and T is the temperature of the boiler interior. Thus,

$q = (0.255)(0.08\ m)(0.12\ m)(5.669 \times 10^{-8}\ W/m^2 \cdot K^4)(1600\ K)^4$

$= \underline{\underline{909\ W}}$

DISCUSSION

Without the quartz viewing window the transmissivity of the opening in the boiler wall is unity, so the heat loss would be

$$q = A \sigma T^4$$

$q = (0.08\ m)(0.12\ m)(5.669 \times 10^{-8}\ W/m^2 \cdot K^4)(1600\ K)^4$

$= 3567\ W$

❒ ❒ ❒

PROBLEM 12.21 Temperature of a Metal Roof Subjected to Solar Irradiation

A metal roof is subjected to a solar irradiation of 1050 W/m^2 on a day when the wind blowing across the roof produces a heat transfer coefficient of 40 W/m^2·K. The solar absorptivity and emissivity of the metal roof are 0.85 and 0.40, respectively. If the temperature of the ambient air is 30°C, what is the temperature of the roof? Assume that the underside of the roof is well insulated.

DIAGRAM

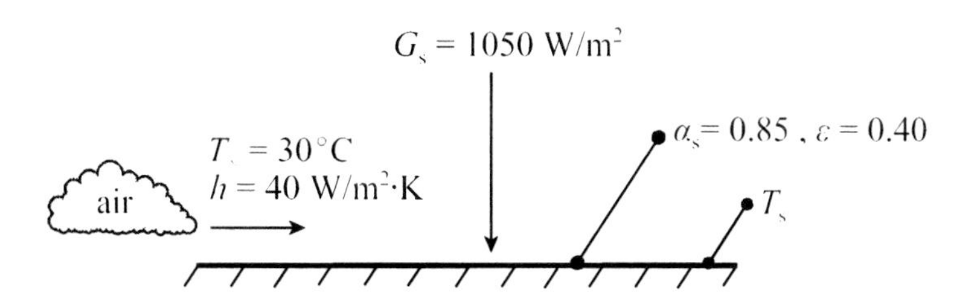

ASSUMPTIONS

1. Steady conditions.
2. Roof is isothermal.
3. Underside of roof is insulated.
4. Sky emission to roof is neglected.

PROPERTIES

metal roof: $\alpha_s = 0.85$, $\varepsilon = 0.40$

ANALYSIS

A surface energy balance on the roof yields

$$\dot{E}_{in} - \dot{E}_{out} = 0$$

where,

$$\dot{E}_{in} = A\alpha_s G_s$$

$$\dot{E}_{out} = A(q''_{conv} + \varepsilon E_b)$$

Substituting these expressions into the energy balance, and noting that

$$q''_{conv} = h(T_s - T_\infty)$$

$$E_b = \sigma T_s^4$$

we have

$$\alpha_s G_s - h(T_s - T_\infty) - \varepsilon\sigma T_s^4 = 0$$

Note that the surface area A divides out. Numerically solving this equation for T_s,

$$T_s = 319.6 \text{ K} = \underline{\underline{46.5°\text{C}}}$$

DISCUSSION

The roof temperature, which can be minimized by minimizing the ratio α_s/ε, is a function of both α_s and ε. In the trivial case where $\alpha_s = \varepsilon = 0$, $T_s = T_\infty$. If radiation from the roof is neglected, which is equivalent to saying that $\varepsilon = 0$, the surface temperature reduces to the expression

$$T_s = \frac{\alpha_s G_s}{h} + T_\infty$$

The graph below shows how h, α_s and ε affect surface temperature T_s. As the heat transfer coefficient increases, the surface temperature approaches the ambient air temperature for all values of α_s and ε. Surface temperatures below ambient air temperature are possible for low values of the ratio α_s/ε.

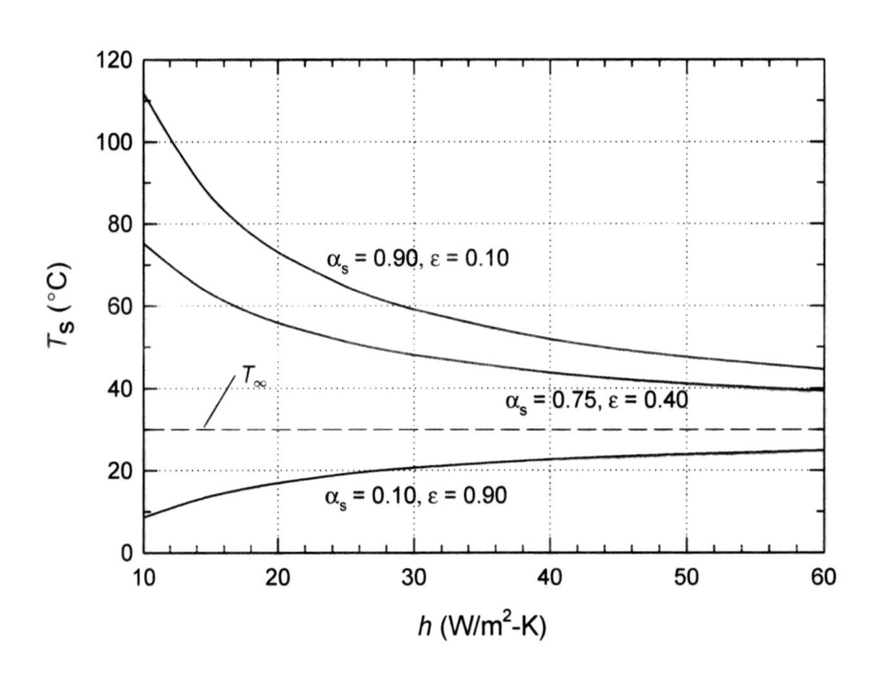

❐ ❐ ❐

PROBLEM 12.22 **Surface Temperature of a Shallow Puddle of Water in the Desert**

After a late afternoon rain shower in a desert region, a shallow puddle of water forms in a dry river bed. As night falls, the ambient air temperature drops to 12°C, and the still air creates a heat transfer coefficient at the puddle surface of 8 W/m^2·K. If the effective sky temperature is −35°C, what is the surface temperature of the water? For the emissivity of water, use $\varepsilon = 0.96$.

DIAGRAM

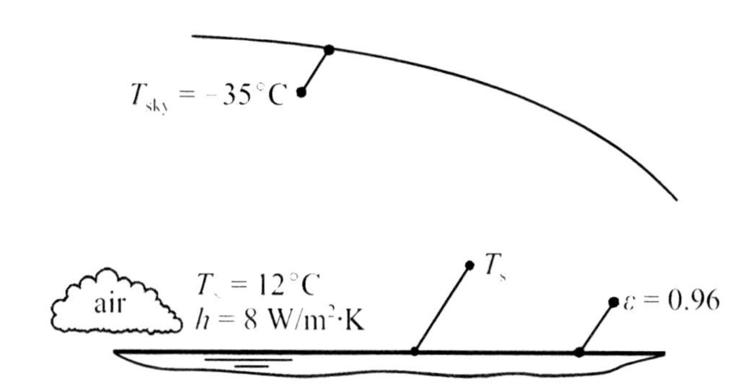

ASSUMPTIONS

1. Steady conditions.
2. Bottom of puddle is insulated.
3. Water surface is gray and diffuse.
4. Sky radiates as a blackbody.

PROPERTIES

water: $\varepsilon = 0.96$

ANALYSIS

A surface energy balance on the water yields

$$\dot{E}_{in} - \dot{E}_{out} = 0$$

where,

$$\dot{E}_{in} = A\alpha G_{sky}$$

$$\dot{E}_{out} = A\varepsilon E_b + hA(T_s - T_\infty)$$

The surface is gray and diffuse, so $\alpha = \varepsilon$. Substituting these expressions into the energy balance and noting that

$$E_b = \sigma T_s^4$$

$$G_{sky} = \sigma T_{sky}^4$$

we have

$$\varepsilon\sigma(T_{sky}^4 - T_s^4) - h(T_s - T_\infty) = 0$$

The surface area A divides out. Numerically solving this equation for T_s,

$T_s = 270.6\text{ K} = \underline{\underline{-2.6°\text{C}}}$

DISCUSSION

The surface temperature is lower than 0°C, so the water will freeze under the conditions given. If the heat transfer coefficient were to increase due to wind, convective heat transfer to the puddle surface would increase such that freezing might not occur. The graph below shows the variation of surface temperature with heat transfer coefficient with all other quantities fixed at their given values.

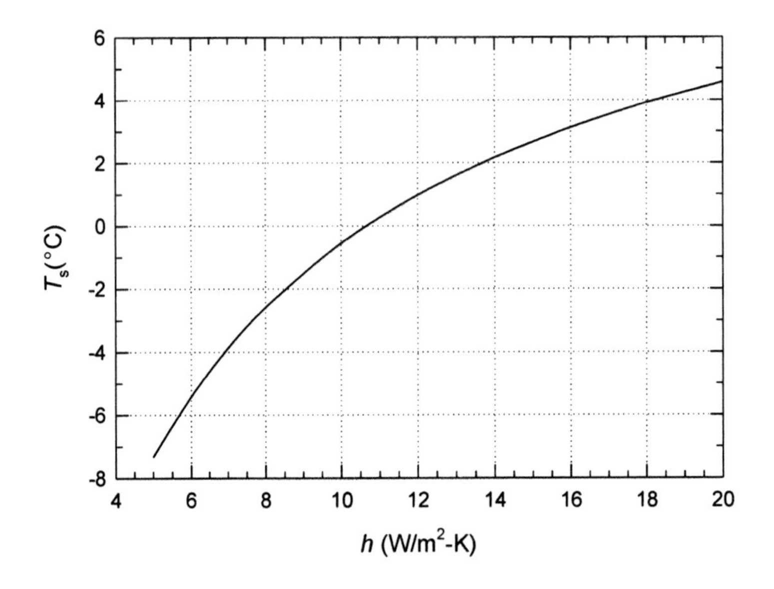

❒ ❒ ❒

PROBLEM 12.23 Temperatures of an Orbiting Satellite in Two Locations

A spherical satellite of diameter D in an equatorial orbit around the earth has a surface coating that provides a solar absorptivity of 0.30 and an absorptivity for long wavelength radiation of 0.70. When the earth is between the satellite and the sun, the satellite is exposed to an irradiation of 360 W/m^2 from the earth only, but when the satellite is between the earth and the sun, the satellite is exposed to a solar irradiation of 1353 W/m^2 plus earth irradiation. What are the steady temperatures of the satellite in these two locations? Assume that the sun and earth radiate as blackbodies.

DIAGRAM

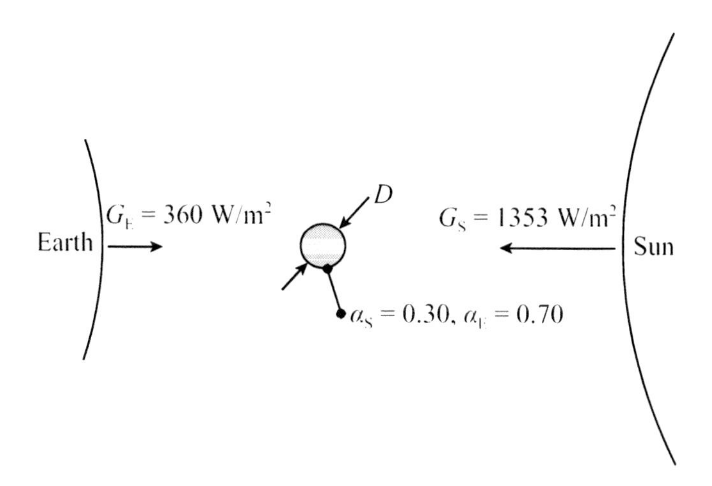

ASSUMPTIONS

1. Steady conditions.
2. Satellite surface is diffuse and gray.
3. Sun and earth radiate as blackbodies.
4. Temperatures of the earth and satellite are below 400 K.
5. Effective temperature of outer space is 0 K.

PROPERTIES

satellite surface: $\alpha_S = 0.30$, $\alpha_E = 0.70$

ANALYSIS

A surface energy balance on the satellite yields

$$\dot{E}_{in} - \dot{E}_{out} = 0$$

where,

$$\dot{E}_{in} = (\pi D^2/4)(\alpha_S G_S + \alpha_E G_E)$$

$$\dot{E}_{out} = \pi D^2 \varepsilon E_b$$

The first expression is the rate of energy transfer to the satellite when the satellite is exposed to both sun and earth irradiations. Note that the area in the first expression is the projected surface area of the satellite because this is the area that "sees" the sun and earth irradiations. But the area in the second expression is the total surface area of a sphere because the entire satellite surface radiates to space. Because the satellite surface is diffuse and gray, $\alpha_E = \varepsilon$. Substituting these expressions into the energy balance and noting that

$$E_b = \sigma T_s^4$$

we have

$$\alpha_S G_S + \alpha_E G_E - 4\varepsilon\sigma T_s^4 = 0$$

Solving for surface temperature,

$$T_s = \left(\frac{\alpha_S G_S + \alpha_E G_E}{4\sigma\varepsilon}\right)^{1/4}$$

When the earth is between the satellite and the sun, we set $G_S = 0$ and obtain

$$T_s = \left(\frac{(0.70)(360\ \text{W/m}^2)}{4(5.669 \times 10^{-8}\ \text{W/m}^2\cdot\text{K}^4)(0.30)}\right)^{1/4}$$

$$= \underline{\underline{246.7\ \text{K}}}$$

When the satellite receives earth and sun irradiations,

$$T_s = \left(\frac{(0.30)(1353\ \text{W/m}^2) + (0.70)(360\ \text{W/m}^2)}{4(5.669 \times 10^{-8}\ \text{W/m}^2\cdot\text{K}^4)(0.30)}\right)^{1/4}$$

$$= \underline{\underline{313.6\ \text{K}}}$$

DISCUSSION

To minimize satellite temperature, the ratio α_S/ε should be minimized. The temperatures of the satellite are independent of diameter.

❒ ❒ ❒

Chapter 13

Radiation Between Surfaces

PROBLEM 13.1 View Factors for a Long Semicircular Duct

Consider a long semicircular duct of diameter D. Find the view factors F_{12}, F_{21} and F_{22}.

DIAGRAM

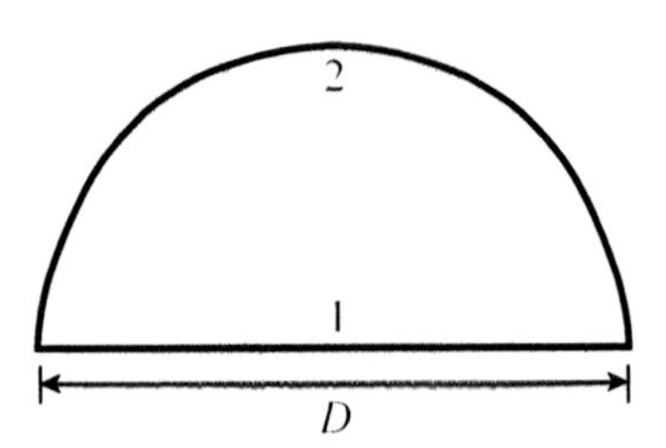

ASSUMPTIONS

1. Duct is long: $L >> D$.
2. Surfaces are diffuse.

PROPERTIES

not applicable

ANALYSIS

Surface 1 sees surface 2 only, so

$$F_{12} = \underline{\underline{1}}$$

View factor F_{21} can be found using the reciprocity relation

$$A_1 F_{12} = A_2 F_{21}$$

Solving for F_{21},

$$F_{21} = \left(\frac{A_1}{A_2}\right) F_{12} = \left(\frac{DL}{\pi DL/2}\right) F_{12}$$

$$F_{21} = (2/\pi)(1)$$

$$= \underline{\underline{0.637}}$$

Invoking the summation rule for surface 2,

$$F_{21} + F_{22} = 1$$

$$F_{22} = 1 - F_{21}$$
$$= 1 - 0.637$$
$$= \underline{\underline{0.363}}$$

DISCUSSION

The reciprocity relation and the summation rule were sufficient to find the view factors for this enclosure. Note that the view factors are independent of duct diameter.

❐ ❐ ❐

PROBLEM 13.2 View Factor for Two Aligned, Long Parallel Plates of Unequal Width

Two long parallel plates of widths 4 cm and 20 cm have their centers aligned, as shown in the diagram. If the plates are spaced 7 cm apart, what is the view factor F_{12}?

DIAGRAM

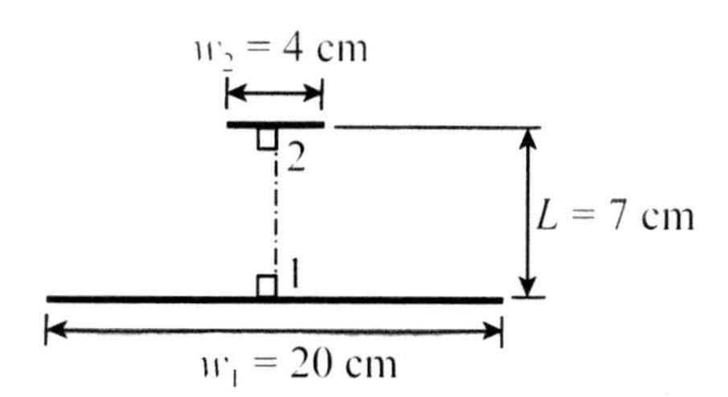

ASSUMPTIONS

1. Plates are long, parallel and aligned at their centers.
2. Surfaces are diffuse.

PROPERTIES

not applicable

ANALYSIS

The view factor relation for this configuration is

$$F_{12} = \frac{[(W_1 + W_2)^2 + 4]^{1/2} - [(W_2 - W_1)^2 + 4]^{1/2}}{2W_1}$$

where,

$$W_1 = w_1/L \, , W_2 = w_2/L$$

We have

$W_1 = (20 \text{ cm})/(7 \text{ cm})$

$= 2.8571$

$W_2 = (4 \text{ cm})/(7 \text{ cm})$

$= 0.5714$

Thus,

$$F_{12} = \frac{[(2.8571 + 0.5714)^2 + 4]^{1/2} - [(0.5714 - 2.8571)^2 + 4]^{1/2}}{2(2.8571)}$$

$= \underline{\underline{0.163}}$

DISCUSSION

A second approach for finding view factors for long parallel surfaces is the *crossed string* method. Consider the generic long parallel surfaces below with imaginary strings stretched between the points A, B, C and D. The crossed string method applies to surfaces that are flat, concave, convex or irregular. In terms of string lengths, the relation for F_{12} is

$$F_{12} = \frac{\Sigma\,(\text{crossed strings}) - \Sigma\,(\text{uncrossed strings})}{2(\text{string on surface 1})}$$

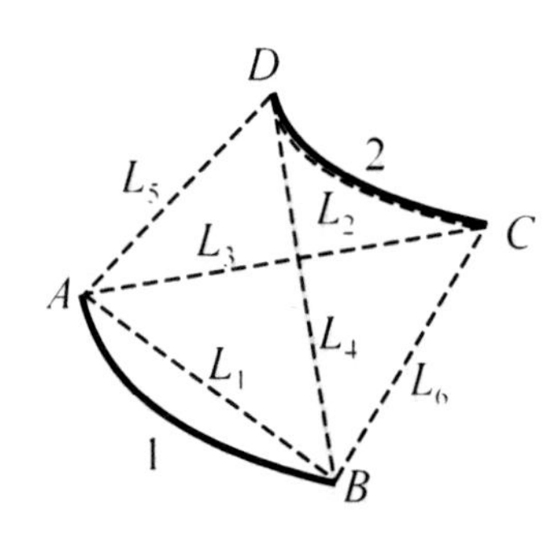

which can be written more compactly as

$$F_{12} = \frac{(L_3 + L_4) - (L_5 + L_6)}{2L_1}$$

From geometry, these lengths are

$$L_1 = 20 \text{ cm} , L_3 = L_4 = 13.89 \text{ cm} , L_5 = L_6 = 10.63 \text{ cm}$$

Thus,

$$F_{12} = \frac{(13.89 \text{ cm} + 13.89 \text{ cm}) - (10.63 \text{ cm} + 10.63 \text{ cm})}{2(20 \text{ cm})}$$

$$= 0.163$$

which is in agreement with our previous result.

❒ ❒ ❒

PROBLEM 13.3 View Factors for a Long Cylinder Parallel to a Long Plate

A long cylinder is aligned with and parallel to a long rectangular plate, as shown in the diagram. Find the view factors F_{12} and F_{21}.

DIAGRAM

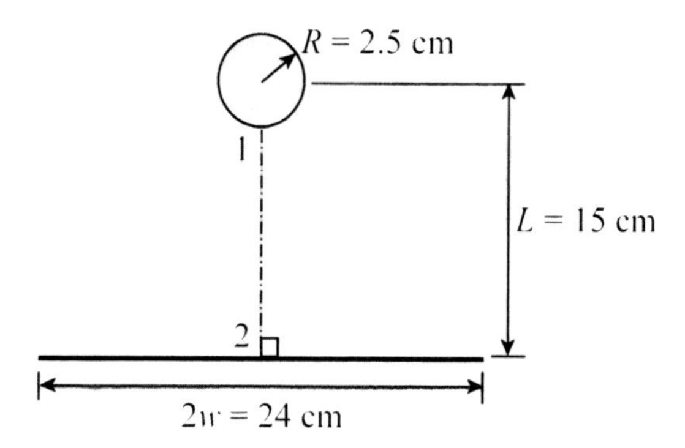

ASSUMPTIONS

1. Cylinder and plate are long, parallel and aligned at their centers.
2. Surfaces are diffuse.

PROPERTIES

not applicable

ANALYSIS

The view factor relation for this configuration is

$$F_{12} = \frac{1}{\pi}\tan^{-1}(w/L)$$

$F_{12} = (1/\pi)\tan^{-1}(12\ \text{cm}/15\ \text{cm})$

$= \underline{\underline{0.215}}$

Using reciprocity,

$$A_1 F_{12} = A_2 F_{21}$$

the view factor F_{21} is

$$F_{21} = \left(\frac{A_1}{A_2}\right) F_{12} = \left(\frac{2\pi RL}{2wL}\right) F_{12}$$

$$F_{21} = \frac{\pi(2.5\ \text{cm})}{12\ \text{cm}}(0.215)$$

$= \underline{\underline{0.141}}$

DISCUSSION

We see that F_{12} is independent of cylinder radius, whereas F_{21} is not. It is instructive to note that for the limiting case of $(w/L) \rightarrow \infty$, we obtain $F_{12} \rightarrow 0.5$.

❐ ❐ ❐

PROBLEM 13.4 View Factor for Two Long, Perpendicular Rectangular Plates with a Common Edge

Consider two long, perpendicular rectangular plates with a common edge, as shown in the diagram. Find the view factor F_{12}.

DIAGRAM

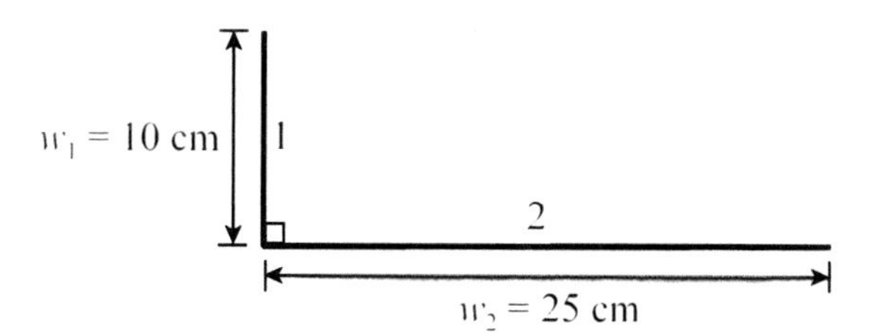

ASSUMPTIONS

1. Plates are long, perpendicular and have a common edge.
2. Surfaces are diffuse.

PROPERTIES

not applicable

ANALYSIS

The view factor relation for this configuration is

$$F_{12} = \frac{1 + (w_2/w_1) - [1 + (w_2/w_1)^2]^{1/2}}{2}$$

Hence,

$$F_{12} = \frac{1 + (25\text{ cm}/10\text{ cm}) - [1 + (25\text{ cm}/10\text{ cm})^2]^{1/2}}{2}$$

$$= \underline{\underline{0.404}}$$

DISCUSSION

The crossed string method could also be used to find F_{12}. By reciprocity, the view factor F_{21} is

$$F_{21} = (w_1/w_2)F_{12}$$

$$= (10\text{ cm}/25\text{ cm})(0.404)$$

$$= 0.162$$

❒ ❒ ❒

PROBLEM 13.5 **View Factors for Perpendicular, Rectangular Plates with a Common Edge**

For the configuration shown in the diagram, find the view factors F_{23} and F_{13}.

DIAGRAM

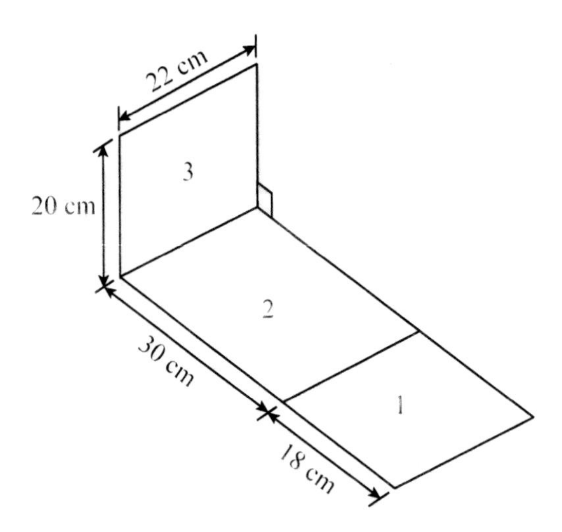

ASSUMPTIONS

1. Plates are perpendicular and have a common edge.
2. Surfaces are diffuse.

PROPERTIES

not applicable

ANALYSIS

The view factor relation for F_{23} is

$$F_{23} = \frac{1}{\pi W}\left(\begin{array}{l} W\tan^{-1}\dfrac{1}{W} + H\tan^{-1}\dfrac{1}{H} - (H^2+W^2)^{1/2}\tan^{-1}\dfrac{1}{(H^2+W^2)^{1/2}} \\ + \dfrac{1}{4}\ln\left\{ \dfrac{(1+W^2)(1+H^2)}{1+W^2+H^2}\left[\dfrac{W^2(1+W^2+H^2)}{(1+W^2)(W^2+H^2)}\right]^{W^2}\left[\dfrac{H^2(1+H^2+W^2)}{(1+H^2)(H^2+W^2)}\right]^{H^2}\right\} \end{array} \right)$$

where,

$H = (20 \text{ cm}/22 \text{ cm})$

$= 0.9091$

$W = (30 \text{ cm}/22 \text{ cm})$

$= 1.3636$

Substituting these values into the relation for F_{23},

$F_{23} = \underline{\underline{0.154}}$

Using reciprocity,

$$A_2 F_{23} = A_3 F_{32}$$

the view factor F_{32} is

$F_{32} = (30 \text{ cm}/20 \text{ cm})(0.154)$

$= 0.231$

Using the same relation, we find the view factor $F_{3(1+2)}$, where

$H = (48 \text{ cm}/22 \text{ cm})$

$= 2.1818$

$W = (20 \text{ cm}/22 \text{ cm})$

$= 0.9091$

Thus,

$F_{3(1+2)} = 0.247$

To introduce F_{31} into the calculation, we use the superposition rule

$$F_{3(1+2)} = F_{31} + F_{32}$$

and the reciprocity relation

$$A_1 F_{13} = A_3 F_{31}$$

Solving for F_{13},

$$F_{13} = (A_3/A_1)(F_{3(1+2)} - F_{32})$$

$F_{13} = (20 \text{ cm}/18 \text{ cm})(0.247 - 0.231)$

$= \underline{\underline{0.0178}}$

DISCUSSION

Because of the length and complexity of view factor relations for some configurations, such as the one here, graphs are provided in most heat transfer texts.

❒ ❒ ❒

PROBLEM 13.6 View Factor for a Cylinder on the Center of a Disk

Consider a cylinder of radius R_1 and length L centered on and perpendicular to a disk of radius R_2, as shown in the diagram. The end of the cylinder is on the disk. Find the view factor F_{12}.

DIAGRAM

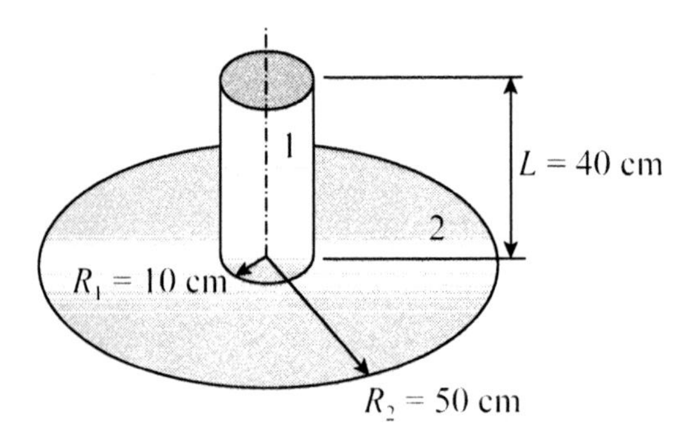

ASSUMPTIONS

1. Cylinder is concentric with and perpendicular to disk.
2. End of cylinder is on disk.
3. Surfaces are diffuse.

PROPERTIES

not applicable

ANALYSIS

The view factor relation for this configuration is

$$F_{12} = \frac{D}{8AB} + \frac{1}{2\pi}\left\{\cos^{-1}\left(\frac{C}{D}\right) - \frac{1}{2B}\left[\frac{(C+2)^2}{A^2} - 4\right]^{1/2}\cos^{-1}\left(\frac{AC}{D}\right) - \frac{C}{2AB}\sin^{-1}A\right\}$$

where,

$$A = R_1/R_2\ ,\ B = L/R_2\ ,\ C = B^2 + A^2 - 1\ ,\ D = B^2 - A^2 + 1$$

We have

A = (10 cm/50 cm)

= 0.2

B = (40 cm/50 cm)

= 0.8

$C = 0.8^2 + 0.2^2 - 1$

$= -0.32$

$D = 0.8^2 - 0.2^2 + 1$

= 1.6

Substituting these values into the relation for F_{12},

F_{12} = 0.257

DISCUSSION

Using reciprocity,

$$A_1F_{12} = A_2F_{21}$$

the view factor F_{21} is

$$F_{21} = \left(\frac{A_1}{A_2}\right)F_{12} = \left(\frac{2\pi R_1 L}{\pi(R_2^2 - R_1^2)}\right)F_{12}$$

$$F_{21} = \left(\frac{2\pi(10\text{ cm})(40\text{ cm})}{\pi[(50\text{ cm})^2 - (10\text{ cm})^2]} \right)(0.257)$$

$$= 0.0857$$

❐ ❐ ❐

PROBLEM 13.7 **View Factor for a Sphere Aligned with the Corner of a Rectangular Plate**

Consider a sphere of radius R whose center is aligned with and a distance L away from the corner of a rectangular plate, as shown in the diagram. A line from the center of the sphere to the corner of the plate is perpendicular to the plate. Find the view factor F_{12}.

DIAGRAM

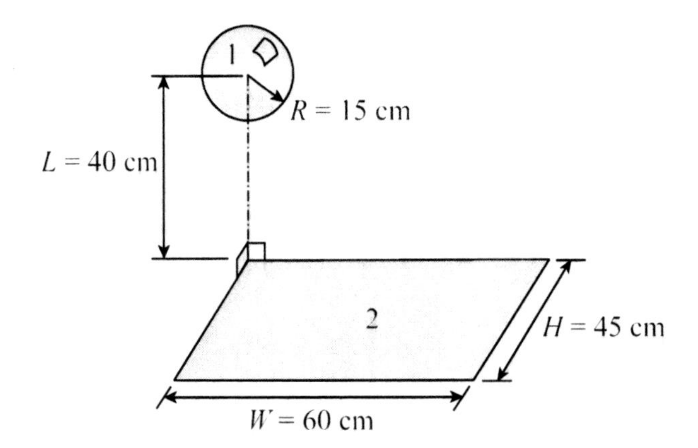

ASSUMPTIONS

1. Center of sphere is aligned with corner of plate.
2. A line from center of sphere is perpendicular to plate.
3. $R < L$.
4. Surfaces are diffuse.

PROPERTIES

not applicable

ANALYSIS

The view factor relation for this configuration, subject to assumption 3, is

$$F_{12} = \frac{1}{4\pi}\tan^{-1}\left(\frac{1}{A^2 + B^2 + A^2B^2}\right)^{1/2}$$

where,

$$A = L/W \,, \; B = L/H$$

We have

$A = (40 \text{ cm}/60 \text{ cm})$

$= 0.6667$

$B = (40 \text{ cm}/45 \text{ cm})$

$= 0.8889$

Hence,

$$F_{12} = \frac{1}{4\pi}\tan^{-1}\left(\frac{1}{0.6667^2 + 0.8889^2 + (0.6667)^2(0.8889)^2}\right)^{1/2}$$

$= \underline{\underline{0.0534}}$

DISCUSSION

It is instructive to examine the view factor relation in the limiting case of a semi-infinite plate, i.e., a plate that has two edges aligned with the center line through the sphere but extends to infinity in the other directions. In this case, $A \to 0$ and $B \to 0$, giving $F_{12} \to 1/8$. This result is expected because half the radiation emitted by the sphere is intercepted by the surroundings above the sphere. The other half of the emitted radiation is split by the fractions 3/4, the surroundings below the sphere, and 1/4, the semi-infinite plate, thereby yielding a view factor of $F_{12} = 1/8$.

The other limiting case is $L \to \infty$, for which $F_{12} \to 0$.

❐ ❐ ❐

PROBLEM 13.8 View Factor for Identical, Parallel, Directly Opposed Rectangles

Find the view factor F_{12} for identical, parallel, directly opposed rectangles.

DIAGRAM

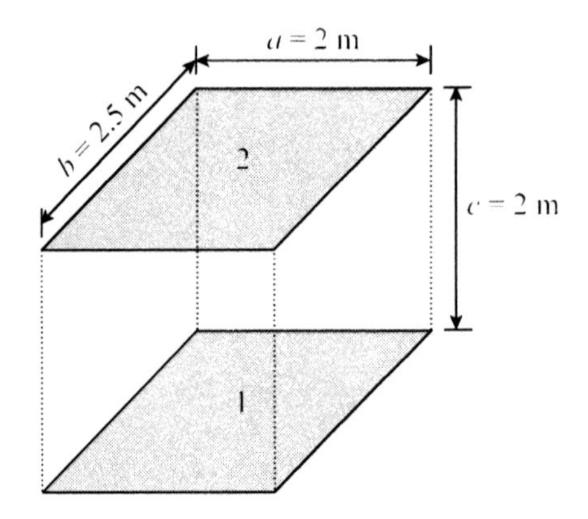

ASSUMPTIONS

1. Rectangles are identical, parallel and directly opposed.
2. Surfaces are diffuse.

PROPERTIES

not applicable

ANALYSIS

The view factor relation for this configuration is

$$F_{12} = \frac{2}{\pi XY}\left\{ \begin{array}{l} \ln\left[\dfrac{(1+X^2)(1+Y^2)}{1+X^2+Y^2}\right]^{1/2} + X(1+Y^2)^{1/2}\tan^{-1}\left(\dfrac{X}{(1+Y^2)^{1/2}}\right) \\ + Y(1+X^2)^{1/2}\tan^{-1}\left(\dfrac{Y}{(1+X^2)^{1/2}}\right) - X\tan^{-1}X - Y\tan^{-1}Y \end{array} \right\}$$

where,

$$X = a/c\,,\, Y = b/c$$

We have

$X = (2\text{ m}/2\text{ m})$

$= 1$

$Y = (2.5\text{ m}/2\text{ m})$

$= 1.25$

Substituting these values into the relation for F_{12},

$F_{12} = \underline{\underline{0.229}}$

DISCUSSION

In the limiting case of $X \to \infty$ and $Y \to \infty$, $F_{12} \to 1$. We also note that if $a = b = c$, $F_{12} = 1/5$. Furthermore, if $a = b = c$, the view factor between any two surfaces formed by the cubical enclosure is 1/5.

❐ ❐ ❐

PROBLEM 13.9 View Factor for an Infinite Plane and Row of Cylinders

Consider an infinite plane and a row of cylinders of diameter $D = 5$ cm with a center-to-center spacing of $L = 7.5$ cm, as shown in the diagram. Find the view factor F_{12}.

DIAGRAM

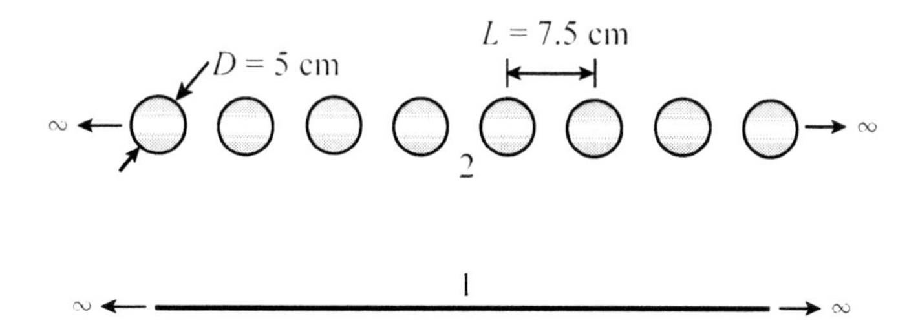

ASSUMPTIONS

1. Plane is infinite, and cylinders are infinitely long and infinite in number.
2. Cylinders are equally spaced and equidistant from plane.
3. Surfaces are diffuse.

PROPERTIES

not applicable

ANALYSIS

The view factor relation for this configuration is

$$F_{12} = 1 - (1 - X^2)^{1/2} + X \tan^{-1}\left(\frac{1 - X^2}{X^2}\right)^{1/2}$$

where,

$$X = D/L$$

We have

$X = (5 \text{ cm}/7.5 \text{ cm})$

$= 0.6667$

Hence,

$$F_{12} = 1 - (1 - 0.6667^2)^{1/2} + 0.6667 \tan^{-1}\left(\frac{1 - 0.6667^2}{0.6667^2}\right)^{1/2}$$

$= \underline{\underline{0.904}}$

DISCUSSION

Upon examination of the view factor relation, two limiting cases may be considered. First, for very large spacing between cylinders, $X \to 0$, so $F_{12} \to 0$, because all radiation emitted by the plane would pass through the spaces between the cylinders. In the second case of $X \to 1$, $F_{12} \to 1$, all radiation emitted by the plane would be intercepted by the cylinders because there are no spaces between them.

❒ ❒ ❒

PROBLEM 13.10 Radiation in a Long Semicircular Duct

Consider a long semicircular duct of diameter $D = 50$ cm. The plane surface is maintained at 600 K and has an emissivity of 0.90. The other surface is maintained at 300 K and has an emissivity of 0.30. Find the net radiation per unit length of duct. Assume that the surfaces are diffuse and gray.

DIAGRAM

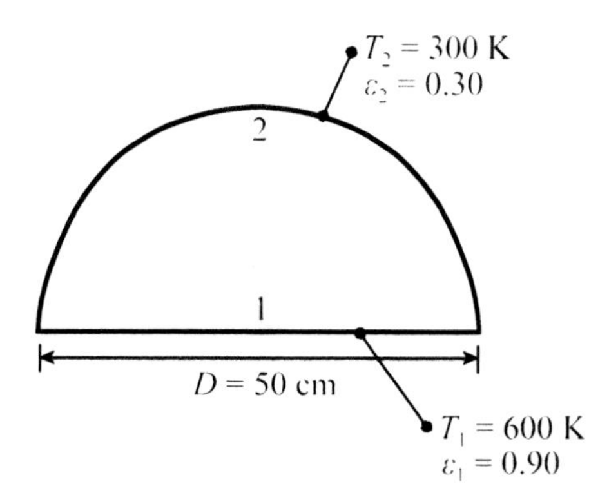

ASSUMPTIONS

1. Duct is infinitely long.
2. Steady conditions.
3. Surfaces are diffuse and gray.

PROPERTIES

$\varepsilon_1 = 0.90$, $\varepsilon_2 = 0.30$

ANALYSIS

The relation for radiation between diffuse, gray surfaces that form an enclosure is

$$q_{12} = \frac{\sigma(T_1^4 - T_2^4)}{\dfrac{1-\varepsilon_1}{\varepsilon_1 A_1} + \dfrac{1}{A_1 F_{12}} + \dfrac{1-\varepsilon_2}{\varepsilon_2 A_2}}$$

where,

$$A_1 = DL\ ,\ A_2 = \pi DL/2$$

We have

$$A_1 = (0.50 \text{ m})(1 \text{ m})$$

$$= 0.50 \text{ m}^2$$

$$A_2 = \pi(0.50 \text{ m})(1 \text{ m})/2$$

$$= 0.7854 \text{ m}^2$$

Surface 1 sees surface 2 only, so $F_{12} = 1$. Thus, the net radiative heat transfer is

$$q_{12} = \frac{(5.669 \times 10^{-8}\ \text{W/m}^2\cdot\text{K}^4)[(600\ \text{K})^4 - (300\ \text{K})^4]}{\dfrac{1 - 0.90}{(0.90)(0.50\ \text{m}^2)} + \dfrac{1}{(0.50\ \text{m}^2)(1)} + \dfrac{1 - 0.30}{(0.30)(0.7854\ \text{m}^2)}}$$

$$= \underline{\underline{1326\ \text{W}}}$$

DISCUSSION

The relation for radiation between diffuse, gray surfaces that form an enclosure can be used for some special configurations, one of which is illustrated in Problem 13.11.

❐ ❐ ❐

PROBLEM 13.11 Radiation Between Concentric Spheres

Consider concentric spheres with radii $R_1 = 1$ m and $R_2 = 2$ m, as shown in the diagram. The temperature and emissivity of the inner sphere are 750 K and 0.80, respectively, whereas the temperature and emissivity of the outer sphere are 400 K and 0.25, respectively. Assuming that the surfaces are diffuse and gray, find the net radiation.

DIAGRAM

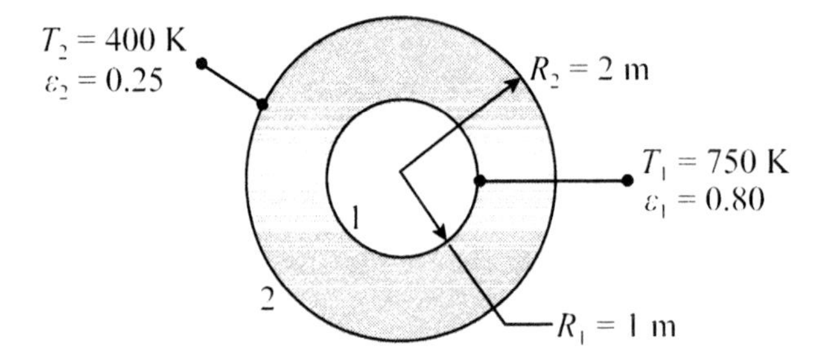

ASSUMPTIONS

1. Spheres are concentric.
2. Steady conditions.
3. Surfaces are diffuse and gray.

PROPERTIES

$\varepsilon_1 = 0.80$, $\varepsilon_2 = 0.25$

ANALYSIS

The relation for net radiation between diffuse, gray surfaces that form an enclosure is

$$q_{12} = \frac{\sigma(T_1^4 - T_2^4)}{\dfrac{1-\varepsilon_1}{\varepsilon_1 A_1} + \dfrac{1}{A_1 F_{12}} + \dfrac{1-\varepsilon_2}{\varepsilon_2 A_2}}$$

where, for concentric spheres,

$$A_1 = 4\pi R_1^2 \ , \ A_2 = 4\pi R_2^2$$

The inner sphere sees the outer sphere only, so $F_{12} = 1$. Substituting these expressions and view factor into the heat transfer relation and simplifying,

$$q_{12} = \frac{4\pi R_1^2 \sigma(T_1^4 - T_2^4)}{\dfrac{1}{\varepsilon_1} + \dfrac{1-\varepsilon_2}{\varepsilon_2}\left(\dfrac{R_1}{R_2}\right)^2}$$

$$q_{12} = \frac{4\pi(1\ \text{m})^2(5.669 \times 10^{-8}\ \text{W/m}^2\cdot\text{K}^4)[(750\ \text{K})^4 - (400\ \text{K})^4]}{\dfrac{1}{0.80} + \dfrac{1-0.25}{0.25}\left(\dfrac{1\ \text{m}}{2\ \text{m}}\right)^2}$$

$$= \underline{\underline{1.04 \times 10^5\ \text{W}}}$$

DISCUSSION

Let's examine the limiting case of $R_2 \to \infty$. The relation for heat transfer between two concentric spheres reduces to

$$q_{12} = 4\pi R_1^2 \varepsilon_1 \sigma(T_1^4 - R_2^4)$$

The equivalent relation for any object within a large enclosure is thus

$$q_{12} = A_1 \varepsilon_1 \sigma(T_1^4 - T_2^4)$$

where A_1 is the surface area of the object within the enclosure. Note that the net radiation for this limiting case is independent of the emissivity of the enclosure and the specific shape of the object enclosed.

PROBLEM 13.12 Radiation in a Long Triangular Enclosure with Black Surfaces

Consider a long three-sided enclosure in the shape of an isosceles triangle, as shown in the diagram. If the surfaces are black and have temperatures of 1200 K, 500 K and 1400 K, find the net radiation for each surface of the enclosure.

DIAGRAM

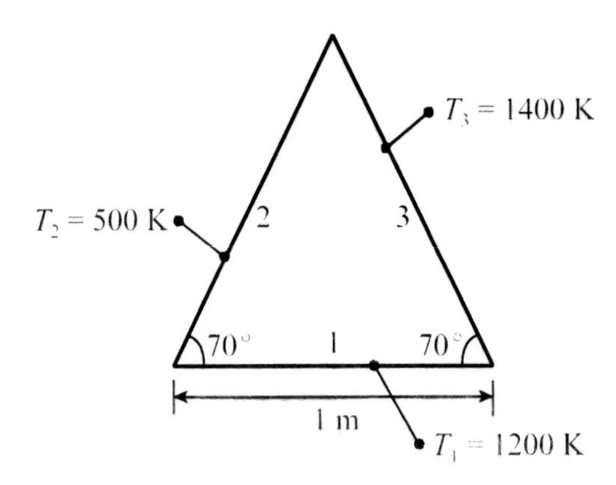

ASSUMPTIONS

1. Steady conditions.
2. Enclosure is long.
3. Surfaces are black.

PROPERTIES

not applicable

ANALYSIS

Using the law of sines for the isosceles triangle,

$$\frac{\sin 70^\circ}{a} = \frac{\sin 40^\circ}{1}$$

the length of the large sides is

$$a = (1\text{ m})(\sin 70^\circ/\sin 40^\circ)$$

$$= 1.462\text{ m}$$

The general relation for net radiation from black surface i of an enclosure to other black surfaces j of the enclosure is

$$q_i = \sum_{j=1}^{N} q_{ij} = \sum_{j=1}^{N} A_i F_{ij} \sigma (T_i^4 - T_j^4)$$

where F_{ij} are the view factors between surfaces i and j and N is the number of surfaces. By symmetry, $F_{12} = F_{13} = 0.5$. Using reciprocity, the view factors F_{21} and F_{31} are

$$F_{21} = F_{31} = \left(\frac{A_1}{A_2} \right) F_{12}$$

$F_{21} = F_{31} = (1 \text{ m}/1.462 \text{ m})(0.5)$

$= 0.342$

and by the summation rule,

$$F_{23} = F_{32} = 1 - F_{21}$$

$F_{23} = F_{32} = 1 - 0.342$

$= 0.658$

The net radiation per unit length of enclosure for surface 1 is

$$q_1 = A_1 \sigma [F_{12}(T_1^4 - T_2^4) + F_{13}(T_1^4 - T_3^4)]$$

$q_1 = (1 \text{ m}^2)(5.669 \times 10^{-8} \text{ W/m}^2{\cdot}\text{K}^4)\{(0.5)[(1200 \text{ K})^4 - (500 \text{ K})^4]$

$+ (0.5)[(1200 \text{ K})^4 - (1400 \text{ K})^4]\}$

$= 5.700 \times 10^4 \text{ W} - 5.011 \times 10^4 \text{ W}$

$= \underline{\underline{6891 \text{ W}}}$

The net radiation per unit length of enclosure for surface 2 is

$$q_2 = A_2 \sigma [F_{21}(T_2^4 - T_1^4) + F_{23}(T_2^4 - T_3^4)]$$

$q_2 = (1.462 \text{ m}^2)(5.669 \times 10^{-8} \text{ W/m}^2{\cdot}\text{K}^4)\{(0.342)[(500 \text{ K})^4 - (1200 \text{ K})^4]$

$+ (0.658)[(500 \text{ K})^4 - (1400 \text{ K})^4]\}$

$= -5.70 \times 10^4 \text{ W} - 2.06 \times 10^5 \text{ W}$

$= \underline{\underline{-\ 7.76 \times 10^5 \text{ W}}}$

The net radiation per unit length of enclosure for surface 3 is

$$q_3 = A_3\sigma[F_{31}(T_3^4 - T_1^4) + F_{32}(T_3^4 - T_2^4)]$$

$$q_3 = (1.462 \text{ m}^2)(5.669 \times 10^{-8} \text{ W/m}^2\cdot\text{K}^4)\{(0.342)[(1400 \text{ K})^4 - (1200 \text{ K})^4]$$

$$+ (0.658)[(1400 \text{ K})^4 - (500 \text{ K})^4]\}$$

$$= 5.01 \times 10^4 \text{ W} + 2.06 \times 10^5 \text{ W}$$

$$= \underline{\underline{2.56 \times 10^5 \text{ W}}}$$

DISCUSSION

Note that the radiation for surface 2 is negative, which means that the net radiative heat flow is *to* this surface and not *from* it, as is the case for surfaces 1 and 3. Clearly, $F_{11} = F_{22} = F_{33} = 0$, and each surface is isothermal, so no surface radiates to itself.

❐ ❐ ❐

PROBLEM 13.13 Power Requirement for a Plastic Curing System

A system for curing plastic disks consists of a circular radiation source positioned 10 cm away from, parallel to and concentric with the plastic disk, as shown in the diagram. The radii of the radiation source and plastic disk are 6 cm and 8 cm, respectively, and the temperature of the source is 650 K. The temperature of the surroundings is 300 K. If the radiation source and plastic disk approximate blackbody behavior, find the electrical power requirement for the source if the surface temperature of the plastic disk is to be 475 K. Assume that the source and disk are insulated on their back sides, and neglect convection.

DIAGRAM

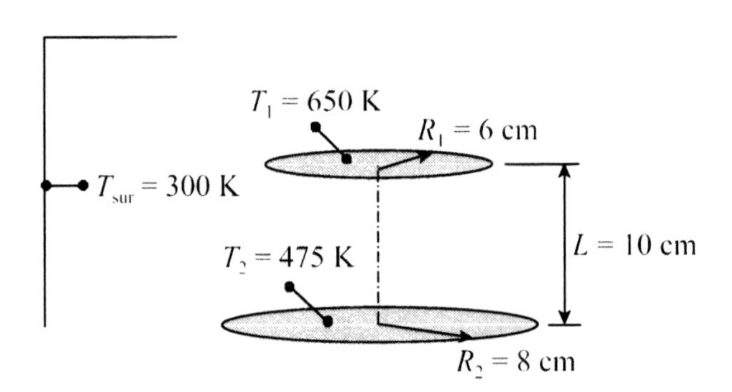

ASSUMPTIONS

1. Steady conditions.
2. Source and disk are coaxial and parallel.
3. Source and disk are insulated on their back sides.
4. Convection is neglected.
5. Surfaces are black.

PROPERTIES

not applicable

ANALYSIS

The view factor relation for two coaxial parallel disks of unequal radius is

$$F_{12} = \frac{1}{2}\left\{C - \left[C^2 - 4(B/A)^2\right]^{1/2}\right\}$$

where,

$$A = R_1/L\,,\ B = R_2/L\,,\ C = 1 + \frac{1 + B^2}{A^2}$$

We have

$$A = (6\text{ cm}/10\text{ cm})$$

$$= 0.6$$

$$B = (8\text{ cm}/10\text{ cm})$$

$$= 0.8$$

$$C = 1 + \frac{1 + (0.8)^2}{(0.6)^2}$$

$$= 5.556$$

Thus, the view factor is

$$F_{12} = \tfrac{1}{2}\,\{5.556 - [5.556^2 - 4(0.8/0.6)^2]^{1/2}\}$$

$$= 0.341$$

The radiation from the source, and thus the electrical power requirement, is

$$q_1 = A_1\sigma[F_{12}(T_1^4 - T_2^4) + (1 - F_{12})(T_1^4 - T_{sur}^4)]$$

$$q_1 = \pi(0.06 \text{ m})^2(5.669 \times 10^{-8} \text{ W/m}^2\cdot\text{K}^4)\{0.341[(650 \text{ K})^4 - (475 \text{ K})^4]$$

$$+ (1 - 0.341)[(650 \text{ K})^4 - (300 \text{ K})^4]\}$$

$$= \underline{\underline{99.9 \text{ W}}}$$

DISCUSSION

For design purposes, it is helpful to examine the effect of spacing L on power requirement. The graph below shows the variation of q_1 with L, holding all other quantities constant.

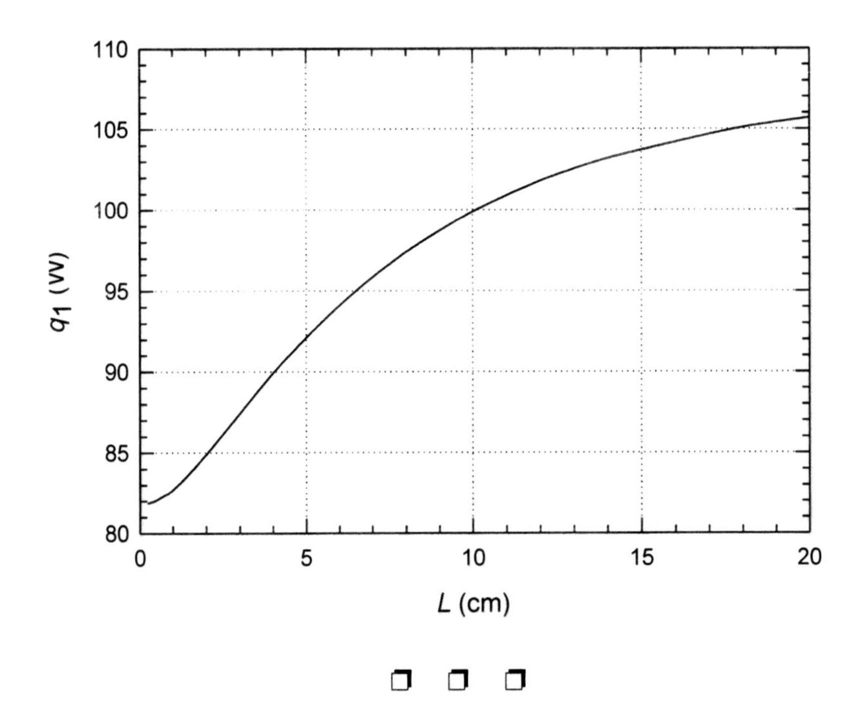

❒ ❒ ❒

PROBLEM 13.14 Radiation in a Long Rectangular Channel

Consider the long rectangular channel shown in the diagram. The bottom surface is maintained at 1200 K, and the sides are maintained at 500 K. If the temperature of the surroundings is 300 K, find the radiative heat loss to the surroundings. All surfaces, including the surroundings, can be approximated as blackbodies, and the back sides of the channel are insulated. Neglect convection.

DIAGRAM

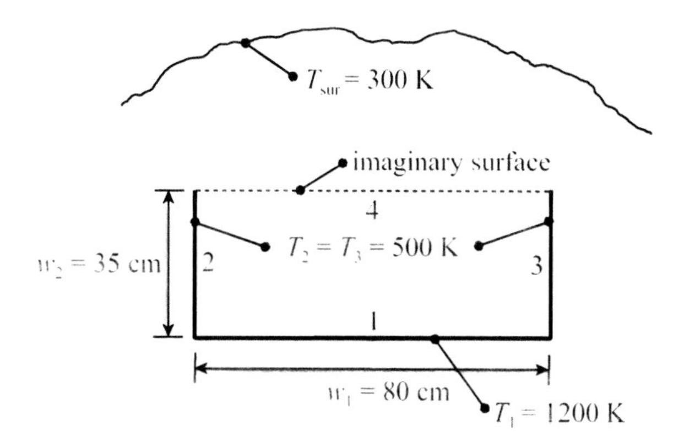

ASSUMPTIONS

1. Steady conditions.
2. Channel is long.
3. Channel surfaces are insulated on their back sides.
4. Convection is neglected.
5. Surfaces are black.

PROPERTIES

not applicable

ANALYSIS

An imaginary surface is shown in the diagram to represent the surroundings, thereby rendering the channel a four-surface enclosure. Two view factors are required, F_{14} and F_{24}. By symmetry, $F_{24} = F_{34}$ and $F_{12} = F_{13} = F_{42} = F_{43}$. The view factor relation for F_{14} is

$$F_{14} = (1 + H^2)^{1/2} - H$$

where,

$$H = w_2/w_1$$

We have

$$H = (35\text{ cm}/80\text{ cm})$$

$$= 0.4375$$

Hence,

$F_{14} = (1 + 0.4375^2)^{½} - 0.4375$

$= 0.6540$

Using the summation rule,

$$F_{12} + F_{13} + F_{14} = 2F_{12} + F_{14} = 1$$

Thus,

$$F_{12} = F_{13} = \frac{1 - F_{14}}{2}$$

$F_{12} = (1 - 0.6540)/2$

$= 0.1730$

By reciprocity,

$$A_2 F_{24} = A_4 F_{42}$$

Hence,

$F_{24} = (80 \text{ cm}/35 \text{ cm})(0.1730)$

$= 0.3954$

The heat loss from the channel is the radiation from surfaces 1, 2 and 3 to surface 4. For a unit length of channel,

$$q = \sigma[F_{14} w_1 (T_1^4 - T_4^4) + 2F_{24} w_2 (T_2^4 - T_4^4)]$$

$q = (5.669 \times 10^{-8} \text{ W/m}^2\cdot\text{K}^4)\{(0.6540)(0.80 \text{ m}^2)[(1200 \text{ K})^4 - (300 \text{ K})^4]$

$+ 2(0.3954)(0.35 \text{ m})^2[(500 \text{ K})^4 - (300 \text{ K})^4]\}$

$= \underline{\underline{6.04 \times 10^4 \text{ W}}}$

DISCUSSION

Heat loss from the channel would be lower if the surfaces were real, i.e., non-black.

❐ ❐ ❐

PROBLEM 13.15 Radiation in a Long Triangular Furnace

Consider a long furnace in the shape of an equilateral triangle, as shown in the diagram. The surfaces are maintained at temperatures 1400 K, 800 K and 600 K and have the emissivities indicated. Find the net radiative heat transfer for each surface of the furnace. The back sides of the furnace are insulated, and convection can be neglected.

DIAGRAM

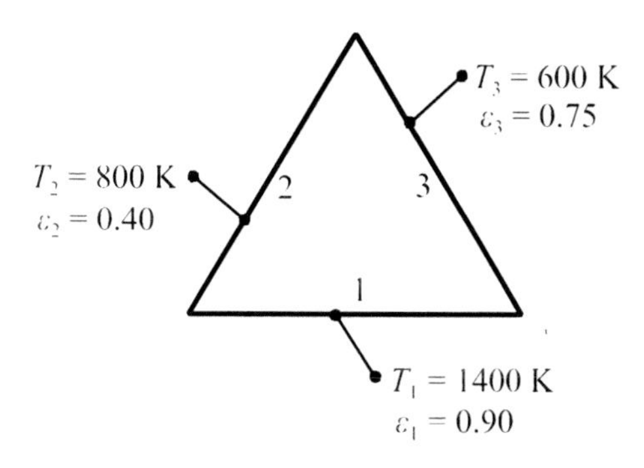

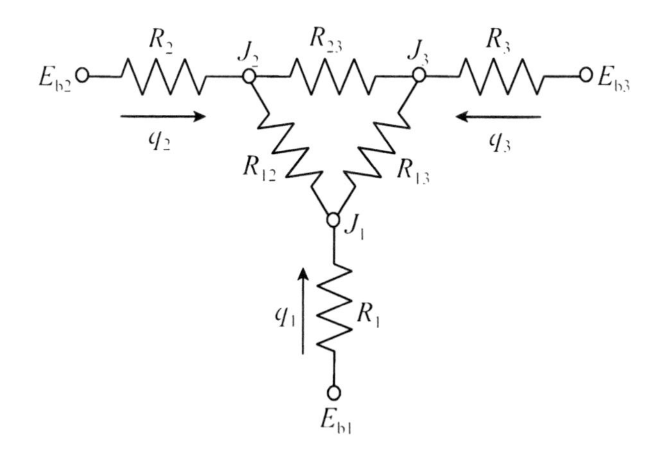

ASSUMPTIONS

1. Steady conditions.
2. Furnace is long.
3. Surfaces are diffuse and gray.
4. Furnace surfaces are insulated on their back sides.
5. Convection is neglected.

PROPERTIES

$\varepsilon_1 = 0.90$, $\varepsilon_2 = 0.40$, $\varepsilon_3 = 0.75$

ANALYSIS

An electrical network analogy is used to model radiation in the enclosure, as illustrated in the diagram, where J denotes radiosity and E_b denotes blackbody emissive power. The surface resistances are given by the relations

$$R_1 = \frac{1-\varepsilon_1}{A_1\varepsilon_1}, \; R_2 = \frac{1-\varepsilon_2}{A_2\varepsilon_2}, \; R_3 = \frac{1-\varepsilon_3}{A_3\varepsilon_3}$$

and the space resistances are given by the relations

$$R_{12} = \frac{1}{A_1 F_{12}}, \; R_{13} = \frac{1}{A_1 F_{13}}, \; R_{23} = \frac{1}{A_2 F_{23}}$$

By symmetry, all view factors have a value of 0.5, and all sides of the furnace have equal surface area. Based on a unit surface area, the values of the resistances are

$$R_1 = \frac{(1 - 0.90)}{(1 \text{ m}^2)(0.90)}$$

$$= 0.1111 \text{ m}^{-2}$$

$$R_2 = \frac{(1 - 0.40)}{(1 \text{ m}^2)(0.40)}$$

$$= 1.500 \text{ m}^{-2}$$

$$R_3 = \frac{(1 - 0.75)}{(1 \text{ m}^2)(0.75)}$$

$$= 0.3333 \text{ m}^{-2}$$

$$R_{12} = R_{13} = R_{23} = \frac{1}{(1 \text{ m}^2)(0.5)}$$

$$= 2.000 \text{ m}^{-2}$$

The algebraic sum of the currents (net radiation heat transfer) at each node in the network equals zero. Thus, we have the following system of linear equations in terms of radiosities,

$$\frac{E_{b1} - J_1}{R_1} + \frac{J_2 - J_1}{R_{12}} + \frac{J_3 - J_1}{R_{13}} = 0$$

$$\frac{J_1 - J_2}{R_{12}} + \frac{E_{b2} - J_2}{R_2} + \frac{J_3 - J_2}{R_{23}} = 0$$

$$\frac{J_1 - J_3}{R_{13}} + \frac{J_2 - J_3}{R_{23}} + \frac{E_{b3} - J_3}{R_3} = 0$$

where the blackbody emissive power for surface i is given by the relation

$$E_{bi} = \sigma T_i^4$$

Substituting values of known quantities into the system of equations, we obtain the radiosities

$$J_1 = 2.022 \times 10^5 \text{ W/m}^2$$

$$J_2 = 8.226 \times 10^4 \text{ W/m}^2$$

$$J_3 = 4.106 \times 10^4 \text{ W/m}^2$$

The radiative heat transfer from surface i is expressed as

$$q_i = \frac{E_{bi} - J_i}{R_i} = \frac{\sigma T_i^4 - J_i}{R_i}$$

Hence,

$$q_1 = \frac{(5.669 \times 10^{-8} \text{ W/m}^2\cdot\text{K}^4)(1400 \text{ K})^4 - 2.022 \times 10^5 \text{ W/m}^2}{0.1111 \text{ m}^{-2}}$$

$$= \underline{\underline{1.40 \times 10^5 \text{ W}}}$$

$$q_2 = \frac{(5.669 \times 10^{-8} \text{ W/m}^2\cdot\text{K}^4)(800 \text{ K})^4 - 8.226 \times 10^4 \text{ W/m}^2}{1.500 \text{ m}^{-2}}$$

$$= \underline{\underline{-3.94 \times 10^4 \text{ W}}}$$

$$q_3 = \frac{(5.669 \times 10^{-8} \text{ W/m}^2\cdot\text{K}^4)(600 \text{ K})^4 - 4.106 \times 10^4 \text{ W/m}^2}{0.3333 \text{ m}^{-2}}$$

$$= \underline{\underline{-1.01 \times 10^5 \text{ W}}}$$

DISCUSSION

The radiation for surface 1 is positive, indicating that the net heat transfer is *from* the surface, whereas the radiation for surfaces 2 and 3 is negative, indicating that the net heat transfer is *to* the surface.

❒ ❒ ❒

PROBLEM 13.16 Radiation Between a Current-Carrying Nichrome Wire and a Copper Plate

A long 10-gage nichrome wire ($R = 1.29$ mm) is parallel to and centered with a long copper plate of width 4 cm, as shown in the diagram. The wire and plate are oxidized, giving emissivities of 0.3 and 0.7, respectively, and the distance between the wire and plate is 1.5 cm. The temperatures of the copper plate and surroundings are 350 K and 300 K, respectively, and the surroundings may be approximated as a blackbody. If the wire carries a current of 40 A, what is the temperature of the wire? A 10-gage nichrome wire has a resistance of 0.2128 Ω/m. Neglect convection.

DIAGRAM

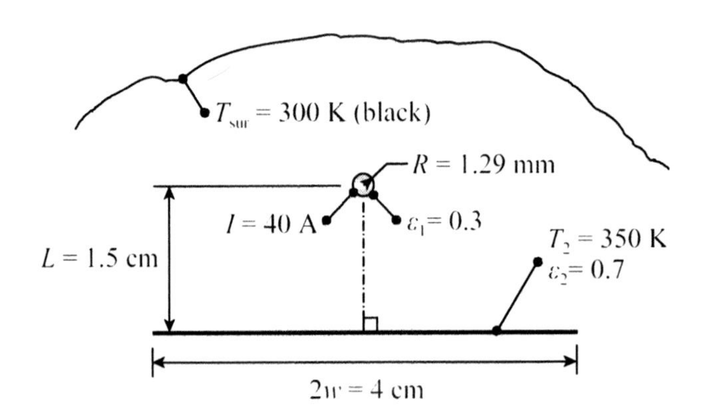

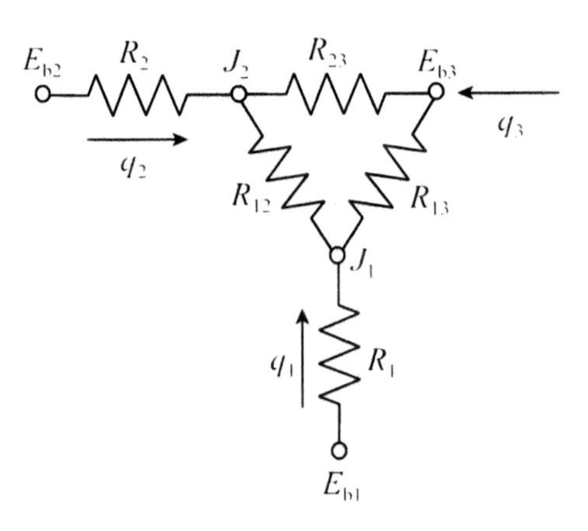

ASSUMPTIONS

1. Steady conditions.
2. Wire and plate are long, parallel and centered.
3. Wire and plate surfaces are diffuse and gray.
4. Surroundings are black.
5. Plate is insulated on the back side.
6. Convection is neglected.

PROPERTIES

nichrome wire: $\varepsilon_1 = 0.3$
copper plate: $\varepsilon_2 = 0.7$

ANALYSIS

In the analysis, subscripts 1, 2 and 3 will be used for the wire, plate and surroundings, respectively. The view factor for the wire and plate is

$$F_{12} = \frac{1}{\pi}\tan^{-1}(w/L)$$

$F_{12} = (1/\pi)\tan^{-1}(2\text{ cm}/1.5\text{ cm})$

$= 0.2952$

An electrical network analogy is used to model radiation between the wire, plate and surroundings, as illustrated in the diagram, where J denotes radiosity and E_b denotes blackbody emissive power. The surface resistances are given by the relations

$$R_1 = \frac{1-\varepsilon_1}{A_1\varepsilon_1},\ R_2 = \frac{1-\varepsilon_2}{A_2\varepsilon_2}$$

Note that because surface 3 is a blackbody, $E_{b3} = J_3$, so $R_3 = 0$. The space resistances are given by the relations

$$R_{12} = \frac{1}{A_1F_{12}},\ R_{13} = \frac{1}{A_1F_{13}},\ R_{23} = \frac{1}{A_2F_{23}}$$

By the summation rule,

$$F_{13} = 1 - F_{12}$$

$F_{13} = 1 - 0.2952$

$= 0.7048$

Using reciprocity,

$$F_{21} = \left(\frac{A_1}{A_2}\right) F_{12}$$

$F_{21} = \{2\pi(1.29 \times 10^{-3}\text{ m})(1\text{ m})/[(0.04\text{ m})(1\text{ m})]\}(0.2952)$

$= 0.0598$

Using the summation rule,

$$F_{23} = 1 - F_{21}$$

$F_{23} = 1 - 0.0598$

$= 0.9402$

Based on a unit length, the values of the resistances are

$$R_1 = \frac{(1 - 0.3)}{2\pi(1.29 \times 10^{-3}\text{ m})(1\text{ m})(0.3)}$$

$= 287.88\text{ m}^{-2}$

$$R_2 = \frac{(1 - 0.7)}{(0.04\text{ m})(1\text{ m})(0.7)}$$

$= 10.714\text{ m}^{-2}$

$$R_{12} = \frac{1}{2\pi(1.29 \times 10^{-3}\text{ m})(1\text{ m})(0.2952)}$$

$= 417.94\text{ m}^{-2}$

$$R_{13} = \frac{1}{2\pi(1.29 \times 10^{-3}\text{ m})(1\text{ m})(0.7048)}$$

$= 175.05\text{ m}^{-2}$

$$R_{23} = \frac{1}{(0.04\text{ m})(1\text{ m})(0.9402)}$$

$= 26.590\text{ m}^{-2}$

The algebraic sum of the currents (net radiation heat transfer) at nodes 1 and 2 in the network equals zero. Thus, we have two linear equations in terms of unknown radiosities J_1 and J_2.

$$\frac{E_{b1} - J_1}{R_1} + \frac{J_2 - J_1}{R_{12}} + \frac{E_{b3} - J_1}{R_{13}} = 0$$

$$\frac{J_1 - J_2}{R_{12}} + \frac{E_{b2} - J_2}{R_2} + \frac{E_{b3} - J_2}{R_{23}} = 0$$

where the blackbody emissive power for surface i is given by the relation

$$E_{bi} = \sigma T_i^4$$

We also need the equation for heat transfer at the wire surface,

$$q_1 = \frac{E_{b1} - J_1}{R_1} = \frac{\sigma T_1^4 - J_1}{R_1}$$

The heat transfer (per unit length) from the wire can be found using the relation

$$q_1 = I^2 R'$$

$q_1 = (40\text{ A})^2(0.2128\ \Omega)$

$= 340.5\text{ W}$

Hence, there are three unknown quantities, J_1, J_2 and T_1. Substituting values of known quantities into the equations above, $J_1 = 4.277 \times 10^4\text{ W/m}^2$, $J_2 = 1493\text{ W/m}^2$ and

$\underline{\underline{T_1 = 1255\text{ K}}}$

DISCUSSION

The graph below shows the variation of wire temperature with electrical current.

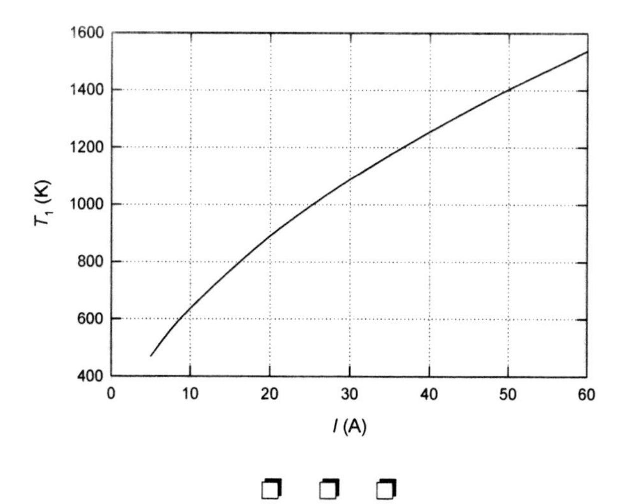

❒ ❒ ❒

PROBLEM 13.17 Radiation Between the Exhaust Pipe and Floor of a Truck

The exhaust pipe of a truck runs parallel to the floor, as shown in the diagram. Under normal operating conditions, the temperature of the exhaust pipe is 460 K, and the emissivity of its surface is 0.8. Considering only a 30-cm wide strip of floor centered on the exhaust pipe, find the temperature of the floor if its upper surface is insulated. The temperature of the surroundings, which may be considered black, is 300 K. Neglect convection.

DIAGRAM

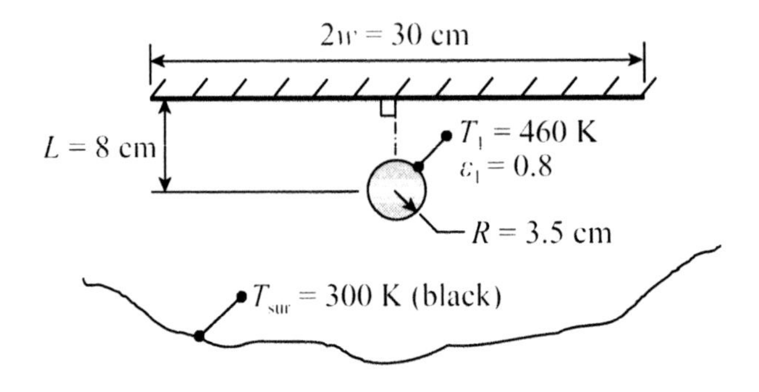

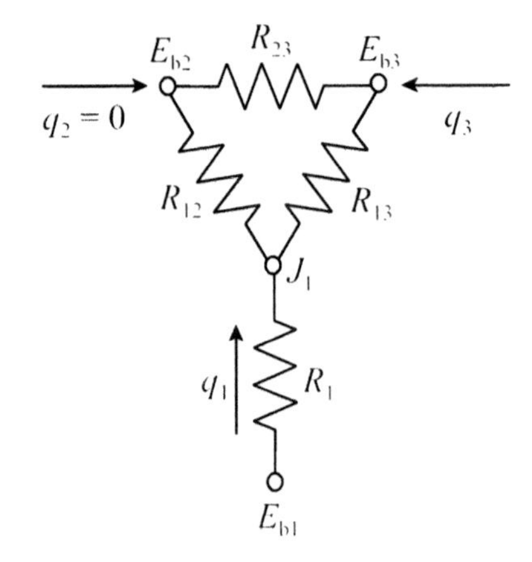

ASSUMPTIONS

1. Steady conditions.
2. Exhaust pipe and floor are long, parallel and centered.
3. Exhaust pipe and floor surfaces are diffuse and gray.
4. Surroundings are black.
5. Upper surface of floor is insulated.
6. Convection is neglected.

PROPERTIES

exhaust pipe: $\varepsilon_1 = 0.8$

ANALYSIS

In the analysis, subscripts 1, 2 and 3 will be used for the exhaust pipe, floor and surroundings, respectively. The view factor for the exhaust pipe and floor is

$$F_{12} = \frac{1}{\pi}\tan^{-1}(w/L)$$

$F_{12} = (1/\pi)\tan^{-1}(15\text{ cm}/8\text{ cm})$

$= 0.3440$

An electrical network analogy is used to model radiation between the exhaust pipe, floor and surroundings, as illustrated in the diagram. Because the floor is insulated and convection is neglected, this surface is a re-radiating surface. Hence, $q_2 = 0$. The surface resistance for the exhaust pipe is given by the relation

$$R_1 = \frac{1-\varepsilon_1}{A_1\varepsilon_1}$$

Note that because surface 3 is a blackbody, $E_{b3} = J_3$, so $R_3 = 0$. The space resistances are given by the relations

$$R_{12} = \frac{1}{A_1F_{12}},\ R_{13} = \frac{1}{A_1F_{13}},\ R_{23} = \frac{1}{A_2F_{23}}$$

By the summation rule,

$$F_{13} = 1 - F_{12}$$

$F_{13} = 1 - 0.3440$

$= 0.6560$

Using reciprocity,

$$F_{21} = \left(\frac{A_1}{A_2}\right)F_{12}$$

$F_{21} = \{2\pi(0.035\text{ m})(1\text{ m})/[(0.30\text{ m})(1\text{ m})]\}(0.3440)$

$= 0.2522$

Using the summation rule,

$$F_{23} = 1 - F_{21}$$

$F_{23} = 1 - 0.2522$

$= 0.7478$

Based on a unit length, the values of the resistances are

$$R_1 = \frac{(1 - 0.8)}{2\pi(0.035\text{ m})(1\text{ m})(0.8)}$$

$= 1.1368\text{ m}^{-2}$

$$R_{12} = \frac{1}{2\pi(0.035\text{ m})(1\text{ m})(0.3440)}$$

$= 13.219\text{ m}^{-2}$

$$R_{13} = \frac{1}{2\pi(0.035\text{ m})(1\text{ m})(0.6560)}$$

$= 6.9318\text{ m}^{-2}$

$$R_{23} = \frac{1}{(0.30\text{ m})(1\text{ m})(0.7478)}$$

$= 4.4575\text{ m}^{-2}$

The algebraic sum of the currents (net radiation heat transfer) at nodes 1 and 2 in the network equals zero. Thus, we have two equations in terms of the unknown quantities J_1 and E_{b2}.

$$\frac{E_{b1} - J_1}{R_1} + \frac{E_{b2} - J_1}{R_{12}} + \frac{E_{b3} - J_1}{R_{13}} = 0$$

$$\frac{J_1 - E_{b2}}{R_{13}} + \frac{E_{b3} - E_{b2}}{R_{23}} = 0$$

where the blackbody emissive power for surface i is given by the relation

$$E_{bi} = \sigma T_i^4$$

Substituting values of known quantities into the equations above and solving for J_1 and E_{b2},

$J_1 = 2151.8\text{ W/m}^2$, $E_{b2} = 886.0\text{ W/m}^2$

Hence,

$$T_2 = \left(\frac{E_{b2}}{\sigma} \right)^{1/4}$$

$$T_2 = \left(\frac{886.0 \text{ W/m}^2}{5.669 \times 10^{-8} \text{ W/m}^2\text{·K}^4} \right)^{1/4}$$

$$= \underline{\underline{353.6 \text{ K}}}$$

DISCUSSION

The floor temperature is 80.5°C (177°F), which could pose a problem for interior carpeting or other vehicle components. A radiation shield would reduce the radiative heat transfer to the truck floor.

Because the floor is a re-radiating surface,

$$q_1 + q_3 = 0$$

Using the network shown in the diagram, we see that the heat transfer for the exhaust pipe is

$$q_1 = \frac{E_{b1} - J_1}{R_1}$$

$$q_1 = \frac{(5.669 \times 10^{-8} \text{ W/m}^2\text{·K}^4)(460 \text{ K})^4 - 2151.8 \text{ W/m}^2}{1.1368 \text{ m}^{-2}}$$

$$= 339.9 \text{ W}$$

Thus, the heat transfer for the surroundings is

$$q_3 = -339.9 \text{ W}$$

Note that our results are independent of the emissivity of the floor surface.

❒ ❒ ❒

PROBLEM 13.18 Radiation Between Parallel Circuit Boards in a Spacecraft

Two square circuit boards measuring 15 cm on a side are mounted parallel and directly opposed to each other in a spacecraft. As shown in the diagram, the boards have an average spacing of 4 cm, as defined by the tops of the devices. One circuit board dissipates 8 W, and the other dissipates 12 W. The emissivity of both boards is 0.6, and the temperature and emissivity of the surrounding surfaces of the spacecraft are 280 K and 0.9, respectively. If the dissipated power of each circuit board is uniformly distributed and the back sides of the boards are insulated, find the surface temperatures of the boards.

DIAGRAM

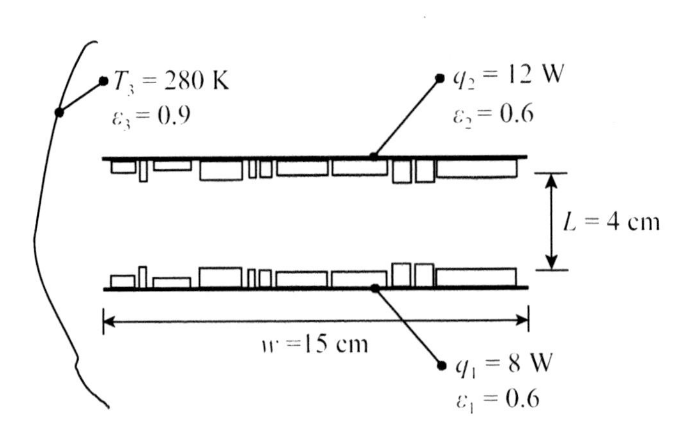

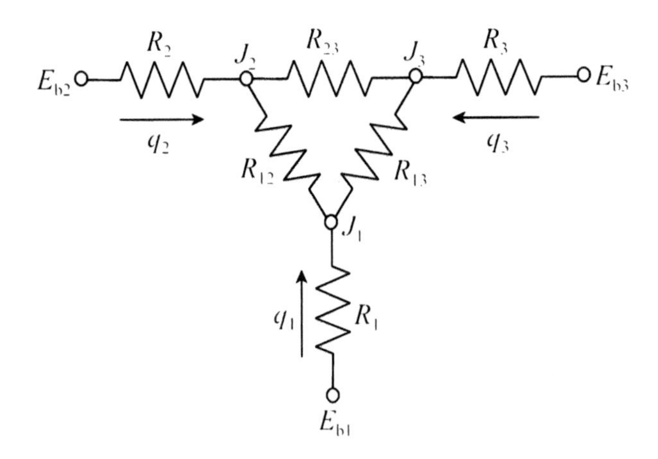

ASSUMPTIONS

1. Steady conditions.
2. Circuit boards are parallel and directly opposed.
3. All surfaces are diffuse and gray.
4. Power dissipations are uniformly distributed.
5. Surfaces of circuit boards are flat.
6. Conduction is neglected.
7. Convection is absent.

PROPERTIES

circuit boards: $\varepsilon_1 = \varepsilon_2 = 0.6$
surroundings: $\varepsilon_3 = 0.9$

ANALYSIS

In the analysis, subscripts 1 and 2 will be used for the circuit boards, and subscript 3 will be used for the surroundings. The view factor relation for the square, parallel and directly opposed circuit boards is

$$F_{12} = \frac{2}{\pi A^2}\left\{ \ln\left[\frac{(1+A^2)^2}{1+2A^2}\right]^{1/2} + 2A(1+A^2)^{1/2}\tan^{-1}\left(\frac{A}{(1+A^2)^{1/2}}\right) - 2A\tan^{-1}A\right\}$$

where,

$$A = w/L$$

We have

$A = (15 \text{ cm}/4 \text{ cm})$

$= 3.75$

Substituting the value of A into the relation above, the view factor for the circuit boards is

$F_{12} = 0.6323$

Recognizing the configuration as a three-surface enclosure and using the summation rule, the view factor for the circuit boards and surroundings is

$$F_{13} = F_{23} = 1 - F_{12}$$

$F_{13} = F_{23} = 1 - 0.6323$

$= 0.3677$

An electrical network analogy is used to model radiation in the enclosure, as illustrated in the diagram, where J denotes radiosity and E_b denotes blackbody emissive power. The surface resistances are given by the relations

$$R_1 = \frac{1-\varepsilon_1}{A_1\varepsilon_1},\ R_2 = \frac{1-\varepsilon_2}{A_2\varepsilon_2},\ R_3 = \frac{1-\varepsilon_3}{A_3\varepsilon_3}$$

and the space resistances are given by the relations

$$R_{12} = \frac{1}{A_1F_{12}}, \; R_{13} = \frac{1}{A_1F_{13}}, \; R_{23} = \frac{1}{A_2F_{23}}$$

The effective surface area of the surroundings is the total surface area of the spaces around the circuit board cavity,

$$A_3 = 4wL$$

$$A_3 = 4(0.15 \text{ m})(0.04 \text{ m})$$

$$= 0.0240 \text{ m}^2$$

The values of the resistances are

$$R_1 = R_2 = \frac{(1 - 0.6)}{(0.15 \text{ m})(0.15 \text{ m})(0.6)}$$

$$= 29.630 \text{ m}^{-2}$$

$$R_3 = \frac{(1 - 0.9)}{(0.0240 \text{ m}^2)(0.9)}$$

$$= 4.6296 \text{ m}^{-2}$$

$$R_{12} = \frac{1}{(0.15 \text{ m})(0.15 \text{ m})(0.6323)}$$

$$= 70.290 \text{ m}^{-2}$$

$$R_{13} = R_{23} = \frac{1}{(0.15 \text{ m})(0.15 \text{ m})(0.3677)}$$

$$= 120.87 \text{ m}^{-2}$$

The algebraic sum of the currents (net radiation heat transfer) at each node in the network equals zero. Thus, we have the following system of linear equations in terms of radiosities,

$$\frac{E_{b1} - J_1}{R_1} + \frac{J_2 - J_1}{R_{12}} + \frac{J_3 - J_1}{R_{13}} = 0$$

$$\frac{J_1 - J_2}{R_{12}} + \frac{E_{b2} - J_2}{R_2} + \frac{J_3 - J_2}{R_{23}} = 0$$

$$\frac{J_1 - J_3}{R_{13}} + \frac{J_2 - J_3}{R_{23}} + \frac{E_{b3} - J_3}{R_3} = 0$$

where the blackbody emissive power for surface i is given by the relation

$$E_{bi} = \sigma T_i^4$$

where $T_3 = 280$ K. The heat transfers at the circuit board surfaces are given by the relations

$$q_1 = \frac{E_{b1} - J_1}{R_1}, \quad q_2 = \frac{E_{b2} - J_2}{R_2}$$

where $q_1 = 8$ W and $q_2 = 12$ W. Substituting values of known quantities into the equations above, we obtain the radiosities and circuit board temperatures,

$J_1 = 1595.3$ W/m^2 , $J_2 = 1704.2$ W/m^2 , $J_3 = 441.0$ W/m^2

$T_1 = \underline{\underline{424.0\text{ K}}}$, $T_2 = \underline{\underline{436.6\text{ K}}}$

DISCUSSION

The graph below shows the effect of circuit board spacing on temperatures, holding all other quantities fixed.

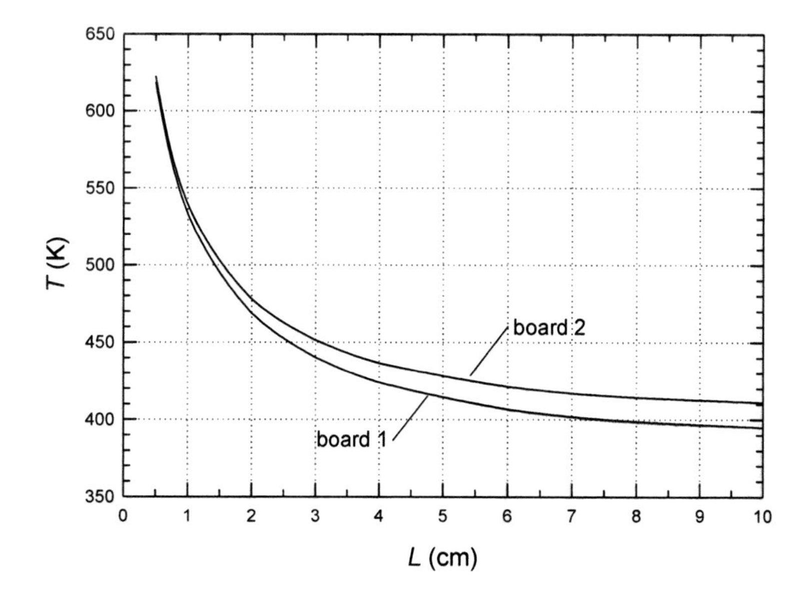

Most electronic devices cannot survive the temperatures calculated here. The circuit board temperatures would be much lower if conduction into connecting hardware had been included. If the spacecraft was pressurized with a gas, convective heat transfer would lower the temperatures further.

❒ ❒ ❒

PROBLEM 13.19   Reflector Temperature of a Radiant Space Heater

A radiant space heater consists of a long cylindrical heating element of radius 3 mm placed parallel to a sector of a long, circular metallic reflector of radius 10 cm, as shown in the diagram. The emissivities of the heating element and reflector are 0.8 and 0.1, respectively, and the back side of the reflector is insulated. The surroundings have a temperature and emissivity of 290 K and 0.95, respectively. The desired radiant heat output of the unit is 575 W per meter of heater length. If the surface temperature of the heating element is 925 K, what is the surface temperature of the reflector? Assume that all surfaces are diffuse and gray, and neglect convection.

DIAGRAM

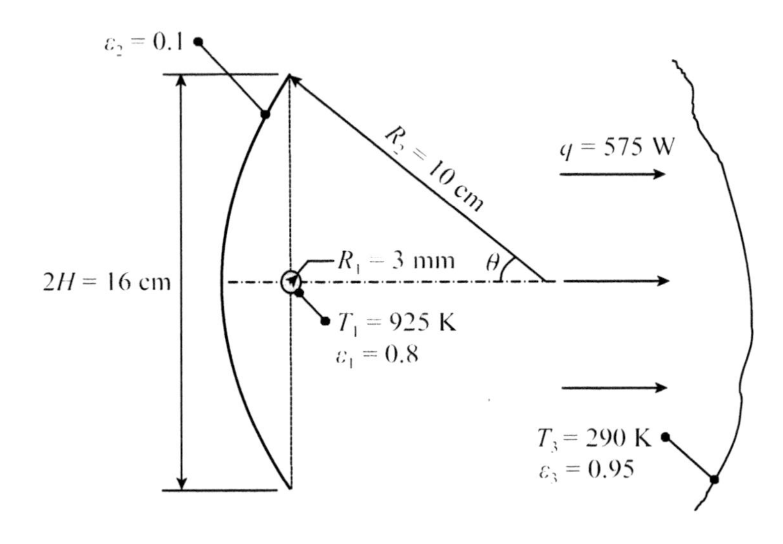

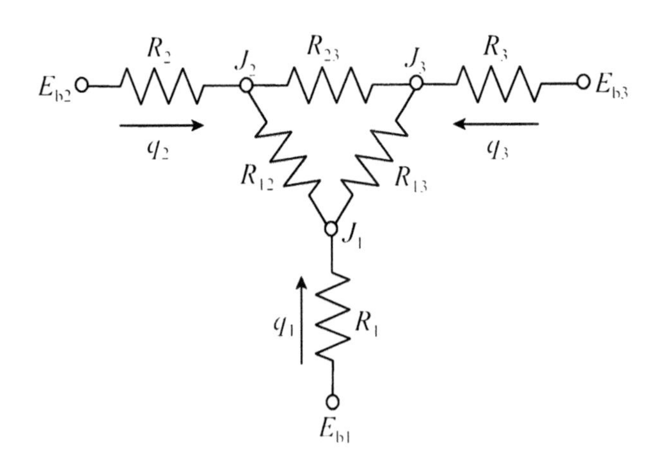

ASSUMPTIONS

1. Steady conditions.
2. Heating element and reflector are long and parallel.
3. All surfaces are diffuse and gray.
4. Back side of reflector is insulated.
5. Convection is neglected.

PROPERTIES

heating element: $\varepsilon_1 = 0.8$
reflector: $\varepsilon_2 = 0.1$
surroundings: $\varepsilon_3 = 0.95$

ANALYSIS

By inspection, the view factor for the heating element and reflector is $F_{12} = 0.5$, and hence the view factor for the heating element and surroundings is $F_{13} = 0.5$. The reciprocity relation for surfaces 1 and 2 is

$$F_{21} = \left(\frac{A_1}{A_2}\right) F_{12}$$

where, for a unit length of heating element,

$$A_1 = 2\pi R_1 L$$

$$A_1 = 2\pi(3.0 \times 10^{-3}\ \text{m})(1\ \text{m})$$

$$= 0.0188\ \text{m}^2$$

For the reflector,

$$A_2 = 2R_2\theta L$$

where,

$$\theta = \sin^{-1}(H/R_2)$$

$$\theta = \sin^{-1}(8\ \text{cm}/10\ \text{cm})$$

$$= 0.9273\ \text{rad}$$

Thus,

$$A_2 = 2(0.1\ \text{m})(0.9273)(1\ \text{m})$$

$= 0.1855 \text{ m}^2$

so we have

$$F_{21} = (0.0188 \text{ m}^2/0.1855 \text{ m}^2)(0.5)$$

$$= 0.0507$$

By the summation rule,

$$F_{23} = 1 - F_{21}$$

$$F_{23} = 1 - 0.0507$$

$$= 0.9493$$

An electrical network analogy is used to model radiation between the surfaces, as illustrated in the diagram, where J denotes radiosity and E_b denotes blackbody emissive power. The surface resistances are given by the relations

$$R_1 = \frac{1-\varepsilon_1}{A_1\varepsilon_1},\ R_2 = \frac{1-\varepsilon_2}{A_2\varepsilon_2},\ R_3 = \frac{1-\varepsilon_3}{A_3\varepsilon_3}$$

and the space resistances are given by the relations

$$R_{12} = \frac{1}{A_1F_{12}},\ R_{13} = \frac{1}{A_1F_{13}},\ R_{23} = \frac{1}{A_2F_{23}}$$

The values of the resistances are

$$R_1 = \frac{(1 - 0.8)}{(0.0188 \text{ m}^2)(0.8)}$$

$$= 13.298 \text{ m}^{-2}$$

$$R_2 = \frac{(1 - 0.1)}{(0.1855 \text{ m}^2)(0.1)}$$

$$= 48.518 \text{ m}^{-2}$$

Treating the three-surface system as an enclosure, the effective surface area of the surroundings is

$$A_3 = 2(H - R_1)L$$

$$A_3 = 2(0.08 \text{ m} - 3.0 \times 10^{-3} \text{ m})(1 \text{ m})$$

$$= 0.1540 \text{ m}^2$$

$$R_3 = \frac{(1 - 0.95)}{(0.1540 \text{ m}^2)(0.95)}$$

$$= 0.3418 \text{ m}^{-2}$$

$$R_{12} = R_{13} = \frac{1}{(0.0188 \text{ m}^2)(0.5)}$$

$$= 106.38 \text{ m}^{-2}$$

$$R_{23} = \frac{1}{(0.1855 \text{ m}^2)(0.9493)}$$

$$= 5.6787 \text{ m}^{-2}$$

The algebraic sum of the currents (net radiation heat transfer) at each node in the network equals zero. Thus, we have the following system of linear equations in terms of radiosities,

$$\frac{E_{b1} - J_1}{R_1} + \frac{J_2 - J_1}{R_{12}} + \frac{J_3 - J_1}{R_{13}} = 0$$

$$\frac{J_1 - J_2}{R_{12}} + \frac{E_{b2} - J_2}{R_2} + \frac{J_3 - J_2}{R_{23}} = 0$$

$$\frac{J_1 - J_3}{R_{13}} + \frac{J_2 - J_3}{R_{23}} + \frac{E_{b3} - J_3}{R_3} = 0$$

where the blackbody emissive power for surface i is given by the relation

$$E_{bi} = \sigma T_i^4$$

where $T_1 = 975$ K and $T_3 = 290$ K. The heat transfers at the surface of the heating element and reflector are given by the relations

$$q_1 = \frac{E_{b1} - J_1}{R_1}, q_2 = \frac{E_{b2} - J_2}{R_2}$$

and the total radiant heat output of the heater is the sum of these two quantities,

$$q = q_1 + q_2$$

where $q = 750$ W. Substituting values of known quantities into the equations above, we obtain the radiosities and surface temperature of the reflector,

$$J_1 = 41{,}502 \text{ W/m}^2, J_2 = 2107.8 \text{ W/m}^2, J_3 = 597.5 \text{ W/m}^2$$

$T_2 = \underline{\underline{334.2\text{ K}}}$

DISCUSSION

A reflector temperature of 334.2 K (61.1 °C) might be a safe touch temperature for a space heater, A good design would use a guard to protect the user from both the heating element and reflector.

❒ ❒ ❒

PROBLEM 13.20　　Radiation in a Long Four-Sided Duct

Consider the long four-sided duct shown in the diagram. The four parallel surfaces have the temperatures and emissivities indicated. Find the net radiative heat transfer for each surface. Neglect convection.

DIAGRAM

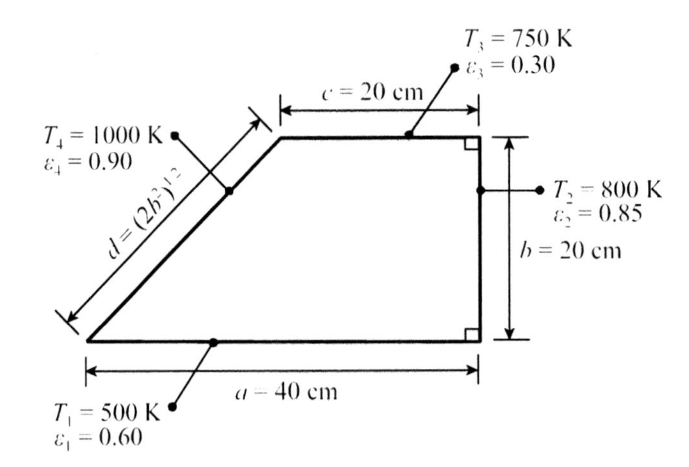

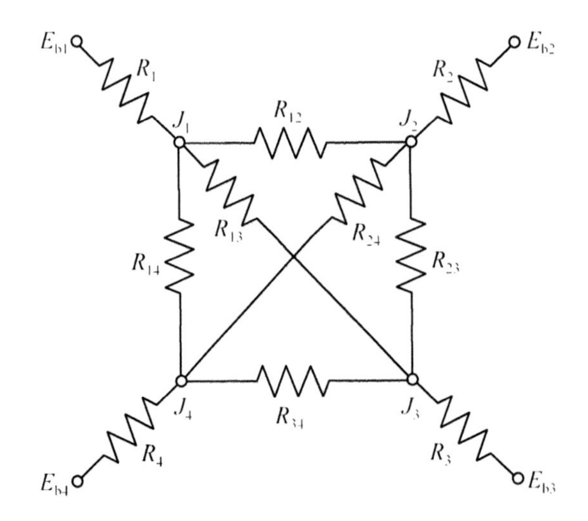

ASSUMPTIONS

1. Steady conditions.
2. Duct is long and surfaces are parallel.
3. All surfaces are diffuse and gray.
4. Back sides of duct are insulated.
5. Convection is neglected.

PROPERTIES

$\varepsilon_1 = 0.60$, $\varepsilon_2 = 0.85$, $\varepsilon_3 = 0.30$, $\varepsilon_4 = 0.90$

ANALYSIS

Before analyzing the equivalent electrical network for this enclosure, we must find the view factors for all surfaces. Because the surfaces are long and parallel, the crossed string method will be used. In terms of string lengths, the crossed string method employs the general relation,

$$F_{ij} = \frac{\sum(\text{crosssed strings}) - \sum(\text{uncrossed strings})}{2(\text{string on surface } i)}$$

Hence, the view factors are

$$F_{12} = \frac{(20 + 40)\text{ cm} - (44.72 + 0)\text{ cm}}{2(40\text{ cm})}$$

$$= 0.1910$$

$$F_{13} = \frac{(28.28 + 44.72)\text{ cm} - (20 + 28.280)\text{ cm}}{2(40\text{ cm})}$$

$$= 0.3090$$

$$F_{14} = 1 - F_{12} - F_{13}$$

$$= 1 - (0.1910 + 0.3090)$$

$$= 0.5000$$

$$F_{23} = \frac{(20 + 20)\text{ cm} - (28.28 + 0)\text{ cm}}{2(20\text{ cm})}$$

$$= 0.2930$$

$$F_{24} = \frac{(44.72 + 28.28)\text{ cm} - (20 + 40)\text{ cm}}{2(20\text{ cm})}$$

$$= 0.3250$$

$$F_{34} = \frac{(28.28 + 20)\text{ cm} - (44.72 + 0)\text{ cm}}{2(20\text{ cm})}$$

$$= 0.0890$$

An electrical network analogy is used to model radiation in the enclosure, as illustrated in the diagram, where J denotes radiosity and E_b denotes blackbody emissive power. The surface resistances are given by the relations

$$R_1 = \frac{1-\varepsilon_1}{A_1\varepsilon_1},\ R_2 = \frac{1-\varepsilon_2}{A_2\varepsilon_2},\ R_3 = \frac{1-\varepsilon_3}{A_3\varepsilon_3},\ R_4 = \frac{1-\varepsilon_4}{A_4\varepsilon_4}$$

and the space resistances are given by the relations

$$R_{12} = \frac{1}{A_1F_{12}},\ R_{13} = \frac{1}{A_1F_{13}},\ R_{14} = \frac{1}{A_1F_{14}},\ R_{23} = \frac{1}{A_2F_{23}},\ R_{24} = \frac{1}{A_2F_{24}},\ R_{34} = \frac{1}{A_3F_{34}}$$

Based on a unit length of duct, the values of the resistances are

$$R_1 = \frac{(1 - 0.60)}{(0.40\text{ m})(1\text{ m})(0.60)}$$

$$= 1.6667\text{ m}^{-2}$$

$$R_2 = \frac{(1 - 0.85)}{(0.20\text{ m})(1\text{ m})(0.85)}$$

$$= 0.8824\text{ m}^{-2}$$

$$R_3 = \frac{(1 - 0.30)}{(0.20\text{ m})(1\text{ m})(0.30)}$$

$$= 11.667\text{ m}^{-2}$$

$$R_4 = \frac{(1 - 0.90)}{(0.2828\text{ m})(1\text{ m})(0.90)}$$

$$= 0.3929\text{ m}^{-2}$$

$$R_{12} = \frac{1}{(0.40\text{ m})(1\text{ m})(0.1910)}$$

$$= 13.089 \text{ m}^{-2}$$

$$R_{13} = \frac{1}{(0.40 \text{ m})(1 \text{ m})(0.3090)}$$

$$= 8.0906 \text{ m}^{-2}$$

$$R_{14} = \frac{1}{(0.40 \text{ m})(1 \text{ m})(0.5000)}$$

$$= 5.0000 \text{ m}^{-2}$$

$$R_{23} = \frac{1}{(0.20 \text{ m})(1 \text{ m})(0.2930)}$$

$$= 17.065 \text{ m}^{-2}$$

$$R_{24} = \frac{1}{(0.20 \text{ m})(1 \text{ m})(0.3250)}$$

$$= 15.385 \text{ m}^{-2}$$

$$R_{34} = \frac{1}{(0.2828 \text{ m})(1 \text{ m})(0.0890)}$$

$$= 39.731 \text{ m}^{-2}$$

The algebraic sum of the currents (net radiation heat transfer) at each node in the network equals zero. Thus, we have the following system of linear equations in terms of radiosities,

$$\frac{E_{b1} - J_1}{R_1} + \frac{J_2 - J_1}{R_{12}} + \frac{J_3 - J_1}{R_{13}} + \frac{J_4 - J_1}{R_{14}} = 0$$

$$\frac{J_1 - J_2}{R_{12}} + \frac{E_{b2} - J_2}{R_2} + \frac{J_3 - J_2}{R_{23}} + \frac{J_4 - J_2}{R_{24}} = 0$$

$$\frac{J_1 - J_3}{R_{13}} + \frac{J_2 - J_3}{R_{23}} + \frac{E_{b3} - J_3}{R_3} + \frac{J_4 - J_3}{R_{34}} = 0$$

$$\frac{J_1 - J_4}{R_{14}} + \frac{J_2 - J_4}{R_{24}} + \frac{J_3 - J_4}{R_{34}} + \frac{E_{b4} - J_4}{R_4} = 0$$

where the blackbody emissive power for surface i is given by the relation

$$E_{bi} = \sigma T_i^4$$

Solving the system of equations for the radiosities,

$$J_1 = 17{,}260 \text{ W/m}^2,\ J_2 = 24{,}266 \text{ W/m}^2,\ J_3 = 21{,}916 \text{ W/m}^2,\ J_4 = 52{,}857 \text{ W/m}^2$$

Finally, the heat transfers at the surfaces are

$$q_1 = \frac{E_{b1} - J_1}{R_1}$$

$$q_1 = [(5.669 \times 10^{-8} \text{ W/m}^2\cdot\text{K}^4)(500 \text{ K})^4 - 17{,}260 \text{ W/m}^2]/(1.6667 \text{ m}^{-2})$$

$$= \underline{\underline{-8230 \text{ W}}}$$

$$q_2 = \frac{E_{b2} - J_2}{R_2}$$

$$q_2 = [(5.669 \times 10^{-8} \text{ W/m}^2\cdot\text{K}^4)(800 \text{ K})^4 - 24{,}266 \text{ W/m}^2]/(0.8824 \text{ m}^{-2})$$

$$= \underline{\underline{-1185 \text{ W}}}$$

$$q_3 = \frac{E_{b3} - J_3}{R_3}$$

$$q_3 = [(5.669 \times 10^{-8} \text{ W/m}^2\cdot\text{K}^4)(750 \text{ K})^4 - 21{,}916 \text{ W/m}^2]/(11.667 \text{ m}^{-2})$$

$$= \underline{\underline{-341 \text{ W}}}$$

$$q_4 = \frac{E_{b4} - J_4}{R_4}$$

$$q_4 = [(5.669 \times 10^{-8} \text{ W/m}^2\cdot\text{K}^4)(1000 \text{ K})^4 - 52{,}857 \text{ W/m}^2]/(0.3929 \text{ m}^{-2})$$

$$= \underline{\underline{9756 \text{ W}}}$$

DISCUSSION

Negative heat transfer values indicate that the net radiative heat flow is *to* the surface.

❒ ❒ ❒

PROBLEM 13.21 **Radiation Between Two Large Parallel Plates with and Without Shields**

Consider two large parallel plates with temperatures of 1600 K and 1100 K and emissivities of 0.8 and 0.5, respectively. How many radiation shields are required to reduce the radiative heat flux between the plates to five percent or less of the heat flux with no shields present if the emissivity of the shields is 0.2? Find the radiative heat flux with no shields and with the number of shields calculated.

DIAGRAM

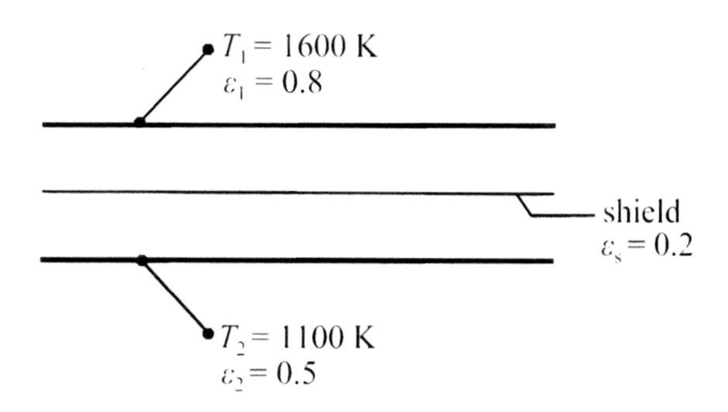

ASSUMPTIONS

1. Steady conditions.
2. Plates and shield are large and parallel.
3. Plates and shield are diffuse and gray.
4. Convection is neglected.

PROPERTIES

plate 1: $\varepsilon_1 = 0.8$
plate 2: $\varepsilon_2 = 0.5$
shield: $\varepsilon_s = 0.2$

ANALYSIS

It can be shown that the net radiative heat flux between two large parallel plates with N shields between them is given by the relation

$$(q''_{12})_N = \frac{\sigma(T_1^4 - T_2^4)}{\left(\frac{1}{\varepsilon_1} + \frac{1}{\varepsilon_2} - 1\right) + N\left(\frac{2}{\varepsilon_s} - 1\right)}$$

If no shields are present ($N = 0$), the radiative heat flux is

$$q''_{12} = \frac{\sigma(T_1^4 - T_2^4)}{\dfrac{1}{\varepsilon_1} + \dfrac{1}{\varepsilon_2} - 1}$$

$$q''_{12} = \frac{(5.669 \times 10^{-8}\ \text{W/m}^2\cdot\text{K}^4)[(1600\ \text{K})^4 - (1100\ \text{K})^4]}{1/0.8 + 1/0.5 - 1}$$

$$= \underline{\underline{1.28 \times 10^5\ \text{W/m}^2}}$$

Thus, the radiative heat flux for N shields, five percent of the value above, is

$$(q''_{12})_N = \underline{\underline{6.41 \times 10^3\ \text{W/m}^2}}$$

Substituting this value into the relation above and solving for N,

$$N = \frac{\dfrac{\sigma(T_1^4 - T_2^4)}{(q''_{12})_N} - \left(\dfrac{1}{\varepsilon_1} + \dfrac{1}{\varepsilon_2} - 1\right)}{\dfrac{2}{\varepsilon_s} - 1}$$

$$N = \frac{(5.669 \times 10^{-8}\ \text{W/m}^2\cdot\text{K}^4)[(1600\ \text{K})^4 - (1100\ \text{K})^4]/(6.41 \times 10^3\ \text{W/m}^2) - (1/0.8 + 1/0.5 - 1)}{2/0.2 - 1}$$

$$N = 4.75$$

Rounding our answer up to an integer,

$$N = \underline{\underline{5\ \text{shields}}}$$

DISCUSSION

The relation for net radiative heat flux between two large parallel plates for N shields between them may be derived by using an equivalent electrical network for the plates and shields. This derivation if left as an exercise for the student.

❒ ❒ ❒

PROBLEM 13.22 **Radiation Error for a Thermocouple in a Cold Duct Carrying a Hot Gas**

A thermocouple inserted into a duct carrying hot CO_2 indicates a temperature of 600 K, and the duct wall is maintained at 350 K. The thermocouple sensor is oxidized, giving it an emissivity of 0.9. If the heat transfer coefficient for the thermocouple sensor is 140 W/m²·K, what is the true temperature of the CO_2? What is the true temperature of the CO_2 if a single radiation shield with an emissivity of 0.1 is placed between the sensor and duct wall?

DIAGRAM

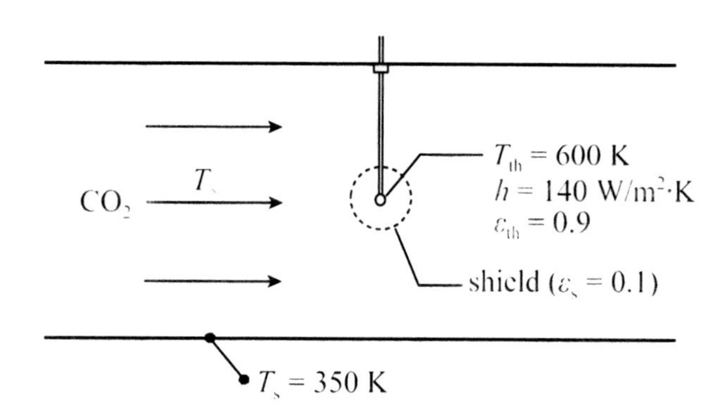

ASSUMPTIONS

1. Steady conditions.
2. Duct surface is large compared to sensor surface.
3. Radiation shield permits CO_2 to flow across sensor; heat transfer coefficient is not affected.

PROPERTIES

thermocouple sensor: $\varepsilon_{th} = 0.9$
radiation shield: $\varepsilon_s = 0.1$

ANALYSIS

Under steady conditions, a surface energy balance for the thermocouple sensor is

$$\dot{E}_{in} - \dot{E}_{out} = 0$$

where the rate of heat transfer into the sensor is

$$\dot{E}_{in} = hA(T_\infty - T_{th})$$

and the rate of heat transfer from the sensor is

$$\dot{E}_{out} = \varepsilon_{th}\sigma A(T_{th}^4 - T_s^4)$$

where A is surface area. Substituting these expressions into the energy balance, and solving for the temperature of the CO_2,

$$T_\infty = T_{th} + \frac{\varepsilon_{th}\sigma(T_{th}^4 - T_s^4)}{h}$$

$$T_\infty = 600\text{ K} + \frac{(0.9)(5.669 \times 10^{-8}\text{ W/m}^2\cdot\text{K}^4)[(600\text{ K})^4 - (350\text{ K})^4]}{140\text{ W/m}^2\cdot\text{K}}$$

$$= \underline{\underline{641.8\text{ K}}}$$

With a shield surrounding the sensor, the rate of heat transfer from the sensor is (see Problem 13.21)

$$\dot{E}_{out} = \frac{\sigma A(T_{th}^4 - T_s^4)}{1/\varepsilon_{th} + 2/\varepsilon_s - 1}$$

which, when substituted into the energy balance, yields

$$T_\infty = T_{th} + \frac{\sigma(T_{th}^4 - T_s^4)}{h(1/\varepsilon_{th} + 2/\varepsilon_s - 1)}$$

$$T_\infty = 600\text{ K} + \frac{(0.9)(5.669 \times 10^{-8}\text{ W/m}^2\cdot\text{K}^4)[(600\text{ K})^4 - (350\text{ K})^4]}{(140\text{ W/m}^2\cdot\text{K})(1/0.9 + 2/0.1 - 1)}$$

$$= \underline{\underline{602.1\text{ K}}}$$

DISCUSSION

Without the shield the radiation error is 41.8 K, whereas with the shield the radiation error is only 2.1 K. Radiation shields are therefore recommended when measuring gas temperatures in cold or hot surroundings.

❐ ❐ ❐

CPSIA information can be obtained at www.ICGtesting.com
Printed in the USA
LVOW11s1218120713

342412LV00004B/599/P

9 781432 730840